Advances in

VIRUS RESEARCH

VOLUME 67

ADVISORY BOARD

Advances in
VIRUS RESEARCH

Edited by

KARL MARAMOROSCH
Department of Entomology
Rutgers University
New Brunswick, New Jersey

AARON J. SHATKIN
Center for Advanced Biotechnology
and Medicine
Piscataway, New Jersey

VOLUME 67
Plant Virus Epidemiology

Edited by

J. M. THRESH

Natural Resources Institute
University of Greenwich
Chatham Maritime
Kent, United Kingdom

ELSEVIER

AMSTERDAM • BOSTON • HEIDELBERG • LONDON
NEW YORK • OXFORD • PARIS • SAN DIEGO
SAN FRANCISCO • SINGAPORE • SYDNEY • TOKYO
Academic Press is an imprint of Elsevier

Academic Press is an imprint of Elsevier
525 B Street, Suite 1900, San Diego, California 92101-4495, USA
84 Theobald's Road, London WC1X 8RR, UK

This book is printed on acid-free paper. 

For information on all Elsevier Academic Press publications
visit our Web site at www.books.elsevier.com

ISBN-13: 978-0-12-039866-9
ISBN-10: 0-12-039866-4

PRINTED IN THE UNITED STATES OF AMERICA
06 07 08 09 9 8 7 6 5 4 3 2 1

CONTENTS

Wild Plants and Viruses: Under-Investigated Ecosystems

Ian Cooper and Roger A. C. Jones

Genetic Diversity of Plant Virus Populations: Towards Hypothesis Testing in Molecular Epidemiology

B. Moury, C. Desbiez, M. Jacquemond, and H. Lecoq

Plant Virus Epidemiology: The Concept of Host Genetic Vulnerability

J. M. Thresh

History and Current Distribution of Begomoviruses in Latin America

Francisco J. Morales

Evolutionary Epidemiology of Plant Virus Disease

M. J. Jeger, S. E. Seal, and F. van den Bosch

Control of Plant Virus Diseases

Roger A. C. Jones

Control of Tropical Plant Virus Diseases

J. M. Thresh

Begomovirus Evolution and Disease Management

S. E. Seal, M. J. Jeger, and F. van den Bosch

Transgenic Papaya: Development, Release, Impact and Challenges

Dennis Gonsalves

Cassava Mosaic Virus Disease in East and Central Africa: Epidemiology and Management of a Regional Pandemic

J. P. Legg, B. Owor, P. Sseruwagi, and J. Ndunguru

Host-Plant Viral Infection Effects on Arthropod-Vector Population Growth, Development and Behaviour: Management and Epidemiological Implications

J. Colvin, C. A. Omongo, M. R. Govindappa, P. C. Stevenson, M. N. Maruthi, G. Gibson, S. E. Seal, and V. Muniyappa

The Migration of Insect Vectors of Plant and Animal Viruses

D. R. Reynolds, J. W. Chapman, and R. Harrington

PREFACE

The Plant Virus Epidemiology Committee of the International Society for Plant Pathology was established in 1978. The intention from the outset was to promote plant virus epidemiology by arranging a series of international conferences. Nine such conferences have been held and the Oxford, United Kingdom, gathering in 1981 and the one in Australia in 1983 led to Proceedings volumes. Special issues of the Elsevier journal *Virus Research* contained papers based on material presented in Almeria, Spain (1999), and Aschersleben, Germany (2002). This led to an invitation to produce a special issue of *Advances in Virus Research* devoted exclusively to plant virus epidemiology and control. These topics have featured only occasionally in previous volumes of the *Advances* series and so this is an important precedent.

Twelve chapters are included here and there are contributors from nine different countries. The chapters cover a wide range of topics and diseases of temperate, subtropical, and tropical crops. However, it is notable that there is considerable emphasis on whiteflies and whitefly-borne viruses. This is understandable and appropriate because of the current importance of these viruses in many regions and the magnitude of the losses caused. Details are provided by F. J. Morales (pp. 127–162) from his experience in many parts of Latin America, and Legg *et al.* (pp. 355–418) from their studies on the current pandemic of cassava mosaic disease (CMD) in East and Central Africa.

Morales describes how many different whitefly-borne viruses have been described in diverse crops in areas where they were previously unknown or unimportant. In many instances, this has been associated with the introduction and establishment of the damaging B biotype of *Bemisia tabaci* that causes direct damage and has a wider host range than the biotype occurring previously. Very severe losses have occurred and these have undermined attempts to boost family incomes and national economies by exports of agricultural and horticultural produce.

Legg *et al.* (pp. 355–418) focus exclusively on the pandemic of the particularly severe form of CMD that has caused devastating losses over very large tracts of sub-Saharan Africa. The pandemic has affected the livelihood and well-being of human populations in areas that were heavily dependent on cassava as the main staple food and source of rural income. Moreover, it continues to spread into new areas and now threatens the important cassava-growing areas of West Africa. The methods used to monitor disease progress are described and also the procedures adopted to introduce and disseminate virus-resistant cultivars as a means of overcoming the problem.

A feature of the cassava mosaic pandemic and many other virus disease outbreaks is that they are associated with unusually high population densities of the insect vectors. In at least some instances, this is due to an effect of the virus responsible in increasing the susceptibility of the host, which becomes particularly suitable for vector survival and reproduction. This topic is discussed by Colvin *et al.* (pp. 419–452) with the emphasis on their studies of the cassava mosaic and tomato leafcurl pathosystems.

Reynolds *et al.* (pp. 453–517) also focus on insect vectors of a diverse range of plant and animal pathogens. They describe how RADAR and other innovative techniques are being used to provide new information on long-standing problems of insect migration. However, they stress the difficulty of obtaining adequate funding for long-term studies of this type. Several of the other contributors also refer to funding difficulties and stress the limited support for applied research at a time of budget restrictions and a diversion of the available resources to molecular studies. This is occurring in developed countries, as noted by R. A. C. Jones in Australia (pp. 205–244) where the health of potato stocks has deteriorated in some areas following a decrease in regulatory activities and personnel. The consequences of cuts in government and donor funding are even more serious in developing countries where research and extension activities are totally inadequate, especially when considered in relation to the magnitude of the problems encountered (J. M. Thresh, pp. 245–295). Similarly, F. J. Morales (pp. 127–162) describes how cuts in research budgets have limited the ability to respond to new problems and develop effective control measures in much of Latin America.

Many different aspects of virus disease control are discussed by several contributors in general conceptual terms (Jeger *et al.*, pp. 163–203) or by reference to particular diseases. R. A. C. Jones (pp. 205–244) emphasises the need for an integrated approach, and J. M. Thresh (pp. 245–295) deals exclusively with tropical virus diseases, four of which are considered in detail as specific ‘case histories’. Both authors and also Legg *et al.* (pp. 355–418) refer to transgenic forms of virus resistance. This approach to control is considered in detail by D. Gonsalves (pp. 317–354) from his experience with papaya ringspot disease in Hawaii and elsewhere. The chapter is notable in that it sets out the many obstacles that must be overcome before transgenics can be adopted commercially. Even now, transgenic papaya from Hawaii cannot be exported to some countries and papaya is one of only two crops in which virus-resistant transgenics are being used commercially in the United States.

J. I. Cooper and R. A. C. Jones (pp. 1–47) consider viruses of wild plants and emphasise the dearth of information on this neglected topic. These and several other authors refer to the difference between crops and natural vegetation and the way in which modern cropping practices facilitate virus spread and lead to damaging epidemics. This has led to considerable debate on the underlying causes of severe epidemics. Some of these are discussed by Seal *et al.* (pp. 297–316), whereas J. M. Thresh (pp. 89–125) deals exclusively with the important role played by host genetic vulnerability.

An important development has been the way in which molecular studies on viral genomes have contributed to epidemiology. This is discussed by Moury *et al.* (pp. 49–87) who show how virus variation has been studied in unprecedented detail in recent years to provide information on virus evolution and behaviour.

Overall, the 12 chapters span a very wide range of topics and show how different approaches are being adopted in addressing epidemiological problems. Progress is being made in developing basic concepts and improved methods of disease control. The continuing losses caused by viruses and their increasing importance in many areas provide powerful incentives for such studies to continue. The need is particularly great in developing countries where populations are heavily dependent on agriculture and it is essential to increase production. Moreover, it is necessary to do so in a sustainable manner and despite declining soil fertility and a decrease in the amount and effectiveness of rural workforces due to urbanisation and the ravages of HIV/AIDS.

WILD PLANTS AND VIRUSES: UNDER-INVESTIGATED ECOSYSTEMS

Ian Cooper* and Roger A. C. Jones†

*Natural Environment Research Council Centre for Ecology and Hydrology
Mansfield Road, Oxford, Oxfordshire OX1 3SR, United Kingdom
†Agricultural Research Western Australia, Locked Bag No. 4, Bentley Delivery Centre
Perth, WA 6983, Australia; and School of Biological Sciences and Biotechnology
Murdoch University, Perth, WA 6150, Australia

This chapter discusses knowledge of viruses in wild fungi, algae and vascular plant communities and highlights some opportunities for further investigations to address, for example, the exploitation of genetic information and the scope for preliminary environmental risk assessment prior to field release of genetically modified plants. Viruses of vascular plants co-evolved with wild plants that were the progenitors of those woody perennials and/or herbaceous species that are now naturally infected and the consequences of past selection are considered. The list of viruses that infect wild plants naturally is long but, in wild plant communities, virus epidemics are usually less obvious and are suspected to be less prevalent than in cultivated crops. Reasons why this impression is probably false are discussed as is the sometimes circumstantial and equivocal evidence for the role of wild plants as reservoirs of viruses that damage cultivated plants. The examples and subjects illustrate opportunities and limitations in the available data and ways in which such knowledge can be

0065-3527/06 $35.00
DOI: 10.1016/S0065-3527(06)67001-2

exploited to test, for example, the significance of vertical transmission of virus in birch or virus impacts on wild plant fitness.

I. Introduction

Since prehistoric times, mankind has exploited wild plants for food, fibre, ornament or medicine. Wild plant communities have also been manipulated, for example, through the use of fire to stimulate new growth of species that feed wild animals hunted for food. Over extended periods, such exploitation has applied potent selection pressure to wild plant communities. Domestication of the wild plant ancestors of modern cultivated plants started 10,000–15,000 years ago and gave rise to land races that persist today in primitive agricultural systems in the centres of origin of cultivated species in different continents (e.g. Harlan, 1965, 1971, 1981). More recently, land races and wild ancestors have been used to breed modern cultivated plants for food for man and domestic animals, fibre, ornament and medicinal purposes. Moreover, cultivated plants have been dispersed away from their centres of origin into other regions thereby creating opportunities for new encounters between hosts and pathogens (*sensu* Buddenhagen, 1977). Wild plant species have been dispersed too, either as seed contaminants or when introduced deliberately by man, and become established, often as nuisance plants (=weeds), in distant locations. Plant viruses co-evolved with wild plants, including the ancestors of cultivated plants, in their centres of origin. Human activities have had a profound impact in facilitating new encounter situations between viruses and plants. Some of these have resulted in epidemics of virus diseases with serious implications for biodiversity and the successful economic exploitation of cultivated plants. Such new encounters occur not only between the endemic wild species and the viruses introduced inadvertently with plants, but also between indigenous viruses emerging from native wild plant communities and spreading to the plant species introduced. Moreover, with the rapidly expanding nature of world trade in plants and plant products, the increasing rapidity of modern transport systems, and the increasing intensity of cultivation to feed the human population, this process is accelerating (e.g. Bos, 1992; Thresh, 1980b).

The study of viruses infecting wild plant communities is much neglected. Priority continues to be given to the study of viruses in food plants and to a lesser extent fibre plants in forestry rather than to those infecting wild species. Of course, one person's wild plant may be another's crop, particularly so in the case of medicinal plants.

These may be a nuisance as agricultural competitors and virus reservoirs but also beneficial to those who know how to use them. The principal aims of this chapter are to highlight some areas of ignorance and to reiterate the opportunities for sustainable virus management through use of host plant resistance, cultural and biological measures (Broadbent, 1964; Jones, 2004; Thresh, 2003).

Viruses are widely distributed in the Plant Kingdom and have been recorded in algae (e.g. Brown, 1972; Muller and Stache, 1992; Muller *et al.*, 1990, 1996; Van Etten and Meints, 1999; Van Etten *et al.*, 1991), fungi (Hollings, 1978), cycads (Kusunoki *et al.*, 1986), mosses and ferns (e.g. Hull, 1968) as well as in flowering plants (e.g. Brunt *et al.*, 1990; Smith, 1972). However, there are very few records of viruses infecting rhodophyte or chlorophyte algae, bryophytes or pteridophytes. Inevitably virus symptoms are recorded only when perceptive and suitably equipped people see something that interests them. The first recorded sub-Antarctic plant virus (Skotnicki *et al.*, 2003) seems to be in that category. It may be unrealistic to expect many resources to be made available to study 'phenomena' such as the tobamovirus-like agent in the aquatic alga (*Chara australis*) that is suspected to be a progenitor of land plants (Krantz *et al.*, 1995; Skotnicki *et al.*, 1976). Nevertheless, there is an increasing awareness of the opportunities for developing and applying evolutionary and ecological theory to plant–virus systems (e.g. Nee, 2000; Thresh, 1982), and there is a crucial need to test hypotheses using real data which can only be obtained using wild species that regenerate naturally. Herbarium specimens allowed Fraile *et al.* (1997) to span a century when following the evolution of *Tobacco mild green mosaic virus* and *Tobacco mosaic virus* (TMV) in *Nicotiana glauca* but shorter term empirical approaches are also appropriate (Naylor *et al.*, 2003).

This chapter discusses knowledge of viruses in wild plant communities. It also briefly considers the unfulfilled drive to exploit genetic information inferred as a result of computer-based analyses in the base sequences of viral genomes, the need for preliminary 'risk' assessment required prior to field release of genetically manipulated plants and the search for agents of environmental improvement. We seek to provide new insights rather than cover information discussed in previous reviews on the topic (Bos, 1981; Thresh, 1981) and have not attempted to be comprehensive. The subjects covered in the principal component of our assessment are a somewhat arbitrary selection from the available literature chosen to illustrate opportunities and limitations in the current data. The secondary component part (Section IX) picks out three ways in which such knowledge can be exploited to test hypotheses.

II. Viruses in Wild Fungi or Algae

Knowledge of the range of viruses in wild fungi is limited, in part because virus-like nucleic acid seems to be commonplace and virus-like particles exceptional (e.g. Chu *et al.*, 2002; Preisig *et al.*, 2000). Nevertheless, viruses detected in some wild fungi justified sustained investigation aimed at amplification and re-release, to create or enhance prevalence of hypovirulence in plant pathogen populations (Choi and Nuss, 1992; Dawe and Nuss, 2001; Hillman and Suzuki, 2004; Hong *et al.*, 1998; Milgroom and Cortesi, 2004; Nuss, 1996; Tavantzis, 2002; Wei *et al.*, 2003). Diverse viruses were detected in other plant pathogenic fungi (e.g. Ghabrial, 1998; Heiniger and Rigling, 1994; Strauss *et al.*, 2000). Ecological context cannot be suggested for most of these, although a theoretical framework is being developed to explain the epidemiology of hypovirulence (e.g. Brasier, 1986; Liu *et al.*, 1999). Other fungi are likely to be more difficult to study and analyse given the diversity and likely complexity of their mating systems (e.g. Brasier, 2000).

The relatively big and stable virions with large dsDNA genomes obtained from marine and freshwater algae across the World are amenable in some instances to maintenance in laboratory cultures and this undoubtedly facilitates their study (Van Etten *et al.*, 2002). Importantly, the genomes of these viruses contain potentially useful genes with diverse affinities. Furthermore, some of these viruses lyse algal cells and, unsurprisingly, specific viruses have been associated with the clearing of harmful algal blooms (Tarutani *et al.*, 2000; Wilson *et al.*, 2002). The viruses concerned presumably play a role in nutriment cycling in aqueous environments but, despite the substantial progress made in characterising a few (possibly an unrepresentative few) of these algal viruses, none can yet be given an ecological context.

III. Viruses in Terrestrial Wild Plants

Cooper (1979) commented that most of the viruses then recorded in woody perennials (=trees and shrubs) were 'opportunists' that infect broad ranges of plants naturally and that trend is also apparent in virus records from herbaceous wild species. However, his prediction that future investigation would reveal greater numbers of 'specialist' viruses that infect few species naturally has not been supported by subsequent events.

A. Herbaceous Plants

In her assessment of the available literature, Mackenzie (1985) found 224 reports of virus infection in wild plants and fewer than half of these described infections in wholly natural situations. Most of the work she surveyed described chance detection following observation of grossly abnormal plant growth. In wild plants virus-induced changes are not common. Indeed, with few exceptions, infections are symptomless or are associated with slight or indistinct abnormalities. During the past 20 years, the virus detection technology of first choice has changed and bioassays have been largely superseded by serological methods augmented by nucleic acid hybridisation. Bioassays are undoubtedly time-consuming but they do potentially provide valuable information not obtained with the faster approaches. The list of viruses that infect wild plants naturally is now very long and the list of those infectible experimentally is even longer (e.g. Brunt *et al.*, 1990, 1996). Some of these records are of the 'grind and find' type that result in brief descriptions that are often hard to put into context except, perhaps, in devising a nation's phytosanitary policy. Many other studies were undertaken in searches for alternative hosts of viruses that are damaging to cultivated plants. Such searches usually record virus occurrence in wild hosts and the degree to which they suffer damage and so may have ecological relevance for the cultivated plant hosts and also for the wild plant communities that act as virus reservoirs. There are numerous examples from different climate zones.

Studies to determine whether viruses of wild herbaceous plants play significant roles as reservoirs of viruses that spread to and damage cultivated plants necessarily range from 'specialist' viruses such as *Celery mosaic virus*, *Carrot virus Y*, *Pea seed-borne mosaic virus*, *Subterranean clover mottle virus* and *Wild cucumber mosaic virus*, to 'generalist' viruses, such as *Arabis mosaic virus*, *Beet western yellows virus* (BWYV), *Cucumber mosaic virus* (CMV), *Turnip mosaic virus* (TuMV) and *Tomato spotted wilt virus* (TSWV). The latter are 'opportunists' that quickly invade and exploit new sites and recover quickly from the drastic decreases in incidence that occur in cycles of rapid epidemic increase and subsequent decline (Harrison, 1981; Jones, 2004; Thresh, 1980a, 1981). Hygiene measures involving destruction of weed and wild plant hosts of virus and vector have long been key components of management plans for both virus types in economically important plants (e.g. Jones, 2004; Thresh, 2003). Studies on viruses thought to be restricted to wild herbaceous plants are fewer and differ in the extent to which ecological issues have been addressed. An example from South

America is *Wild potato mosaic virus* which infects wild potatoes that grow during winter in areas moistened by low cloud in the dry Peruvian coastal desert. The virus persists through the dry summer period in dormant tubers underground (Jones and Fribourg, 1979).

Australia is rich in examples of viruses apparently restricted to wild plants. Probably this is because cultivated plants were only introduced to the continent in the last 200 years and many viruses that are known elsewhere have not yet arrived (e.g. Gibbs and Guy, 1979). Among the indigenous viruses are Cardamine latent virus and *Cardamine chlorotic fleck virus* of *Cardamine* sp. growing in lowland and alpine regions, respectively (Guy and Gibbs, 1985; Scotnicki *et al.*, 1992); *Kennedya yellow mosaic virus* from *Kennedya rubicunda* growing in coastal regions (Gibbs, 1978); *Solanum nodiflorum mottle virus* in *Solanum nodiflorum* growing in northeastern Australia (Greber and Randles, 1986); and *Velvet tobacco mottle virus* and *Nicotiana velutina mosaic virus* both infecting the wild tobacco species *Nicotiana velutina* (Randles *et al.*, 1976, 1981). The deserts of the southwestern United States offer other examples with their xerophytes (e.g. *Saguaro cactus virus*; Nelson *et al.*, 1975), although distribution patterns are not always clear because cacti have been disseminated widely through the international horticultural trade.

Where sufficient searches for virus resistance have been made among the close wild relatives and ancestors of cultivated herbaceous plants, the impact of past encounters with viruses is clearly evident. One of the best known examples is potato. Here, such searches revealed that the abundant wild tuberous potato flora in the centre of origin of the crop in the Andean region of South America (Jones, 1981) contains many single-gene resistances to viruses, some of which are strain specific and others are not. The viruses concerned all presently challenge cultivated potato and most were dispersed to other parts of the world with the crop, including *Potato virus A* (PVA), *Potato virus X* (PVX), *Potato virus Y* (PVY), *Potato virus S*, *Potato virus M* (PVM) and *Potato leafroll virus* (PLRV). For example, *Solanum acaule* contains two genes (Rx_{acl} and Nx_{acl}) that provide extreme and hypersensitive resistance to PVX, respectively. *Solanum stoloniferum* contains Ra_{sto} providing extreme resistance to PVA, Ry_{sto} giving extreme resistance to PVA, PVY and *Potato virus V* (PVV), and Ny_{sto} giving hypersensitive resistance to PVY. *Solanum demissum* contains the gene Ny_{dms} that provides hypersensitive resistance to PVA and PVY, and Nv_{dms} giving hypersensitive resistance to PVV. *Solanum chacoense* has the genes Nx_{cha} and Ny_{cha} that give hypersensitive resistance to PVX and PVY, respectively; *Solanum berthaultii* has Nl_{ber} conferring hypersensitive resistance to PLRV;

and *Solanum megistacrolobum* has Nm_{mga} conferring hypersensitive resistance to PVM. Virus resistance genes are also known to be present in *Solanum hougasii, Solanum gourlayi, Solanum microdontum, Solanum sparsipilum, Solanum spegazzinii* and *Solanum sucrense,* and in the non-tuber bearing wild relatives *Solanum etuberosum* and *Solanum brevidens* (Cockerham, 1970; Solomon-Blackburn and Barker, 2001; Valkonen, 1994; Valkonen *et al.*, 1996). Among many other examples for other cultivated plants is *Lupinus hispanicus*, a close relative of the cultivated species yellow lupin (*Lupinus luteus*) that originated in the Iberian peninsula of western Europe. Here, although mendelian segregation studies are lacking, two distinct strain-specific hypersensitive specificities to CMV were identified in *L. hispanicus*, suggesting the presence of two distinct 'resistance' genes (Jones and Latham, 1996).

B. Woody Perennials

Robinia pseudoacacia infected with *Robinia mosaic virus* in Hungary is 'said' (unpublished observation) to be particularly prone to winter injury that necessitates premature replacement. Infection by the virus is undoubtedly commonplace but 'cause and effect' has not been proven. Similarly, cause and effect remains unproven in the severely damaging mollicute-associated diseases that are often linked to virus infection, such as witches' broom of bamboos in South-East Asia (Lin *et al.*, 1981; Su and Tsai, 1983) and *Acacia* species in Australia. These apart, there have been few studies that provide evidence of a threat to native woodland from any virus. Natural tree/shrub-based ecosystems tend to develop in a disordered fashion in which pattern and change is hard to discern, and so detailed examination is needed to detect subtle alterations in species diversity or abundance. This scenario contrasts with the visually arresting devastation associated with and sometimes directly caused to such ecosystems by some fungal pathogens (e.g. Brasier, 1983; Gibbs, 1978; Gibbs *et al.*, 1984; Hardy *et al.*, 2001; McDougall *et al.*, 2002; Pinon, 1979).

Cooper (1993) catalogued many records of viruses from trees and shrubs and provided context where possible. Consequently, only a few of these issues are raised here. In common with experience concerning non-woody plants, the published data are biased because failure to detect a virus is not usually publishable, notwithstanding the presence of virus-like symptoms.

When dieback and decline syndromes were recognised as affecting coniferous and broadleaved tree species, viruses were targeted specifically in the search for causation, even though the pathology was not wholly consistent with their involvement. A few viruses with 'interesting' properties were recognised in individual trees (e.g. Bertioli *et al.*, 1993; Cooper *et al.*, 1983), but there was no consistent association of any virus with the dieback syndromes. Local site factors likely to lessen water supply and to facilitate invasion by pathogenic fungi were associated with some of these conditions and mollicutes with others (e.g. Sinclair *et al.*, 1990), but pollutants (unknown) were also suspected. There is a proven potential for viruses to exacerbate the harmful effects of pollutants (Heagle, 1973), but it is impossible without very long-term and costly experimentation to show causal associations with diseases of trees.

Many reports are impossible to classify. Thus, the description of tobamovirus-like agents in surface waters draining from forests in New York State (Jacobi and Castello, 1991) and the detection of *Tomato mosaic virus* from *Picea rubens* (Jacobi *et al.*, 1992) in the same region of the Adirondack mountains may reflect infection of trees—but there are other possibilities. Furthermore, some claims, such as the association of PVY with yellowing of needles in *Picea abies* (Ebrahim-Nesbat and Heitefuss, 1989), are more surprising than others (e.g. *Tobacco necrosis virus* in *Pinus sylvestris*; Yarwood, 1959), *Tomato black ring virus* in *Picea sitchensis* (Harrison, 1964), *Arabis mosaic virus* in *Chamaecyparis lawsonsiana* (Harrison, 1964; Thomas, 1970) and *Tomato ringspot virus* in *Cupressus arizonica* (Fulton, 1969)]. All of these records require confirmation and expansion given the extensive distribution and importance of trees in the world's economy and in the landscape.

With the notable exception of the viruses that infect *Theobroma cacao* (cocoa), relatively little is known about viruses that infect trees growing in the tropics. In West Africa, indigenous forest trees, such as *Ceiba pentandra, Sterculia tragacantha* and *Cola chlamydantha*, are naturally infected with *Cacao swollen shoot virus* (Posnette *et al.*, 1950; Tinsley, 1971; Todd, 1951). At a time when symptoms and transmission tests provided the only guide to infection, local epidemics in cocoa trees were observed to occur near infected *C. chlamydantha* and, on this circumstantial basis, this species in particular was suspected to be a reservoir of vectors and inoculum. Now, given the prevalence of infected *T. cacao*, wild hosts probably have little significance as reservoirs of inoculum. Furthermore, the evidence is equivocal that these trees are responsible for frustrating attempts at virus management in the crop; the diversity, polyphagous behaviour and cryptic habits of the virus vectors (pseudococcids), and the size and location of these trees seriously hamper investigation. The epidemiology of *Cacao yellow mosaic virus*, which experimentally infects a similar range of West African tree species and probably has a coleopteran vector, has not been studied in detail (e.g. Brunt, 1970).

IV. Wild Plants as Sources of Virus Diversity

As discussed in Section I, plant viruses co-evolved with wild plants well before they were domesticated (e.g. Lovisolo *et al.*, 2003), so the wild plants are the main source of their diversity and, because of their longevity, trees and shrubs are likely to be particularly important sources in this respect. Evidence of the impact of past encounters with plant viruses and the selection pressures involved is provided by the wealth of virus resistance specificities revealed by searches among the wild ancestors of cultivated plants. However, it is not uncommon, but is frequently forgotten, that changes in the appearance of foliage are inconsistent guides to virus infection. Trees illustrate this phenomenon particularly well (e.g. Cooper and Edwards, 1981) as virus-associated changes in foliage come and go with the seasons making particularly unreliable monitoring programmes based on visual assessments alone.

Virus infections in wild plants often are associated with few or no discernible changes and this fact is sometimes recorded in the virus name, for example, *Dandelion latent virus* (Johns, 1982); Cardamine latent virus (Guy and Gibbs, 1985) and *Spring beauty latent virus* (Valverde, 1985). On other occasions, a specific virus has been found only with others and in circumstances where there were no conspicuous changes in the appearance of infected individuals (e.g. Hammond, 1982; Nelson *et al.*, 1975). However, there are as numerous examples in which disease is implied by the virus name; for example, *Wild cucumber mosaic virus* (van Regenmortel, 1972), *Wild potato mosaic virus* (Jones and Fribourg, 1979), Hypochoeris mosaic virus (Brunt and Stace-Smith, 1978); *Thistle mottle virus* (Donson and Hull, 1983); *Rottboellia yellow mottle virus* (Thottaphilly *et al.*, 1992), *Kennedya yellow mosaic virus* (Gibbs, 1978), *Ononis yellow mosaic* (Gibbs *et al.*, 1966), *Cardamine chlorotic fleck virus* (Skotnicki *et al.*, 1992), *Solanum nodiflorum mottle virus* (Greber and Randles, 1986), *Velvet tobacco mottle virus* and *Nicotiana velutina mosaic virus* (Randles *et al.*, 1976, 1981). Significantly, not all viruses that have been characterised have been categorised by the International Committee for the Taxonomy of Viruses (ICTV). Moreover, latent viruses with novel properties and relationships are still being reported, even in important crops (e.g. Maroon-Lango, 2005).

Against a background of under-reporting and misinterpretation (e.g. Harper, 1988), it is not surprising that there are suspicions about viruses 'emerging' from wild native plants to harm cultivated plants newly introduced into a region, or when previously introduced cultivated plants are grown more intensively than hitherto, or in different ways (e.g. Adlerz, 1981; Mansoor *et al.*, 1999). Inevitably, such concerns are steadily becoming greater with the increasing rapidity of international transport systems, acceleration in world trade in plants and plant products, and increasing use of intensive protected cropping systems (e.g. Bos, 1992; Thresh, 1980b). The concern and the threatening label 'emerging viruses' have also found an appropriate context in human and animal health where there have been notable instances of infections traced to 'silent' infections of wildlife species, as with plant viruses (e.g. Anderson *et al.*, 2004). Although 'new' crop diseases, viruses and 'silent' reservoirs are found regularly, it needs to be emphasised that the phenomenon is not new; it was recognised more than 70 years ago with, for example, yellow fever disease and *Cacao swollen shoot virus* in West Africa (Cooper and Tinsley, 1978; Thresh, 1982, 2003).

V. Virus Prevalence

Determining frequencies of virus infection precisely in wild plants is more difficult than in crops because plant growth stages are rarely uniform within species, and other competing species that are not hosts are usually present. Furthermore, the likelihood of exposure to inoculum depends on location in relation to sources and also to environmental influences that are unstable. Moreover, except during, or following, an epidemic there is usually no routine, structured surveillance of wild plant communities. The prospect of a new pathogen in one or more economic hosts does, however, often trigger surveys based on one tightly defined objective. For example, the damaging prevalence of tospovirus infections has led to many surveys, but one search by Cho *et al.* (1986) for the wild hosts of TSWV is adequately indicative —44 wild species representing 16 families were found to be infected as assessed serologically and 24 new hosts were revealed. A general limitation of their list is that the wild species and individuals in which the virus was not detected may be naturally infectible but had by chance escaped natural inoculation before testing. Another limitation is particularly noteworthy in this context. Although considerable pathotype variation was known (e.g. Finlay, 1952, 1953), no distinct tospovirus serotypes or species were recognised before 1990. Therefore, the possibility that important information may be missing from published lists cannot be excluded, even if, in this instance, most records probably relate to TSWV *sensu stricto*. Similarly, there may be little reassurance to draw from the testing for *Plum pox virus* in Canada (Stobbs *et al.*, 2005). An impressive total of 99,328 wild and cultivated specimens representing 188 species from 53 families were tested and none was found to be infected. However, rare events are difficult to find and this virus was found infecting wild and, in Europe, ornamental plants significantly including both hedgerow shrubs and annual herbs (e.g. Vederevskaja *et al.*, 1985).

In a large structured survey, MacClement and Richards (1956) recorded an average infection of 10% after monitoring a total of 2193 plants in 6 American wild plant communities at fortnightly intervals during 4 years. In a smaller survey of wild grasses (Edwards *et al.*, 1985), 50 different viruses were isolated of which 14 were not readily identified. Hammond (1982) surveyed viruses in *Plantago* species and brought the total number then recorded from the genus worldwide to 26; mixtures of viruses occurred in some individual plants. Similarly, in a survey of viruses infecting wild plants of *Heracleum sphondylium* in Scotland, Bem and Murant (1979a,b) isolated six distinct viruses of which only *Parsnip yellow fleck virus* (PYFV) had been described previously. A second virus, named *Heracleum latent virus*, was found in 6 of 32 symptomlessly infected *H. sphondylium* plants tested and often occurred in mixed infections with one or more of the other five.

'Spill over' of viruses from crops into peri-agricultural communities often occurs and can result in distinct patterns of distribution in wild plant communities. In a series of papers published together in *Annales de Phytopathologie* (Volume 11, pp. 265–475), Quiot and colleagues reported accumulating 953 cucumovirus isolates from 39 of 137 wild species sampled near to vegetable crops in southern France (e.g. Quiot *et al.*, 1979). Furthermore, in central England, CMV was common in a wide range of wild plants. However, this virus was not found 200 km to the north, although *Tobacco rattle virus* was prevalent there, especially in *Stellaria media* which is a common host to both viruses (e.g. Cooper and Harrison, 1973a; Tomlinson *et al.*, 1970). The reason for the different rates of CMV occurrence in Scotland and England are not known but viruses with soil-inhabiting vectors are distributed discontinuously and several factors that explain this were recognised (Cooper, 1971; Cooper and Harrison, 1973b). A relative lack of specific pathogens is often exploited in agriculture and horticulture to propagate healthy stocks of diverse species, as with the production of 'high health' seed potato stocks in regions with small aphid populations in many parts of the world.

Significant differences in virus diversity, in titres as measured serologically and in frequencies of infection were observed in *Brassica* species growing on England's coast. The experimental methods and most results are detailed elsewhere (Maskell *et al.*, 1999; Pallett *et al.*, 2002; Raybould *et al.*, 1999, 2000, 2003; Thurston *et al.*, 2001). Interestingly, CMV was not detected in these surveys, although, 78% of *B. oleracea* and 35% of the *B. nigra* contained at least one virus and multiple infections with different viruses were frequent. Infection with TuMV was rare (only 5 of 597 individual plants tested) in *B. nigra,* yet this virus was common in *B. oleracea* and no evidence of TuMV infection was found in 2644 *B. rapa* plants tested, albeit in a different location. Furthermore, the relative frequencies of seven viruses in the sympatric species, *B. nigra* and *B. oleracea*, were very different. Given such differences between species and among sites, generic interpretation is not possible. Geographical differences in virus frequency might be linked to local vectors, the distribution of virus sources, weather, etc., but, as mentioned in Section I, it is important to emphasise that plant genotypes are much less uniform in wild populations than in their cultivated counterparts.

Most virus taxa are represented in the lists of viruses that have been isolated from wild plants. Ideally, information on virus incidence in wild plants should be obtained by structured testing (e.g. Watson and Gibbs, 1974), as in numerous surveys of wild hosts in south-west Australia, for example, for *Bean yellow mosaic virus* (BYMV) in populations of wild lupins that were introduced into Australia and uncultivated wild legumes that are indigenous (Cheng and Jones, 1999; McKirdy and Jones, 1995; McKirdy *et al.*, 1994); *Alfalfa mosaic virus* and CMV in weeds (McKirdy and Jones, 1994a,b); TSWV in weeds and uncultivated native plants (Latham and Jones, 1997); BWYV in wild radish (Coutts and Jones, 2000); and *Barley yellow dwarf virus* (BYDV) and *Cereal yellow dwarf virus* (CYDV) in over-summering wild grasses, both native and introduced (Hawkes and Jones, 2005; McKirdy and Jones, 1993).

In this context, it is important to reiterate that latent infection with viruses is not uncommon in wild plants and data concerning cryptic viruses are particularly noteworthy, even though tantalizingly limited. Alphacryptoviruses and Betacryptoviruses were so named because they are uniformly inapparent in a wide range of plants (e.g. Luisoni *et al.*, 1987). Furthermore, endornaviruses, as detected in the wild rice *Oryza rufipogon*, are also characteristically inapparent. The cryptic viruses are transmitted through both ova and pollen, but horizontal spread has not been observed and none is known to be mechanically transmissible. This is a serious limitation and, for this reason, the cryptic viruses are as enigmatic as the 'silent' retroviral elements that have been recognised in *Arabidopsis* (Jankowitsch *et al.*, 1999) and other genera (Harper *et al.*, 2002). It is hard to judge the significance of any of these agents other, perhaps, than as factors responsible for unexpected synergies or conceivably at some time in the future, gene capture and transfer.

Past surveys have often tended to focus on non-random elements, and bright leaf colouration in plants undoubtedly attracts attention. For example, the vivid yellow mottle in foliage of *Taraxicum officinale* in United Kingdom and elsewhere in Europe undoubtedly justified sustained investigation that ultimately revealed the sesquivirus *Dandelion yellow mosaic virus* (Bos *et al.*, 1983; Kassanis, 1947). A similar explanation applies to tymoviruses, several of which cause bright yellow mosaic symptoms (e.g. Guy and Gibbs, 1981, 1985). *Dulcamara mottle virus* (Gibbs *et al.*, 1966) is an exception as recognition of the disease this virus causes in nature requires exceptional visual acuity.

Although vivid foliar changes in wild species and surveys for alternative hosts of viruses important in cultivated plants have usually been the attraction to virologists, this has not always been so. For example, the interaction involving PYFV, *Anthriscus yellows virus* (AYV) and aphid vectors including *Cavariella aegopodii* was first noticed in diseased cultivated vegetables and herbs. This recognition was largely based on the detection of naturally infected but asymptomatic *Anthriscus sylvestris* plants. In essence, the fascinating and instructive (yet still incomplete) story is as follows. Numerous isolates of PYFV were detected in wild species in the United Kingdom. The isolates were separated into two main serotypes, one first isolated from *A. sylvestris* and the other from *Pastinaca sativa*. The two serotypes differ in their experimental host ranges (Hemida and Murant, 1989), but seem restricted to natural hosts in the Umbelliferae (notably but not exclusively *A. sylvestris* and *H. sphondylium*). Van Dijk and Bos (1989) in the Netherlands extended the observations. Interestingly, carrot (*Daucus carota)* is immune from infection by AYV and introduction of the virus complex into carrot crops leads to scattered infections because individual aphids retain the transmission-dependent combination for only a few days (Elnagar and Murant, 1976a,b). Bem and Murant (1979a,b) and Murant and Goold (1968) showed that PYFV was transmitted by aphids only when they have first acquired AYV. A variety of analogous helper-dependent complexes are recognised (see review by Pirone and Blanc, 1996), but their significance in the wild plant/crop interface has not been studied in detail. Little is known about mechanisms facilitating virus maintenance within populations or communities and predictive value might result from more knowledge of virus dynamics in wild plants. In this instance, more information about the dynamics in wild umbelifers of PYFV and AYV (if linked to that of their aphid vectors) might enable carrot growers to predict the currently unpredictable epidemics and rationalise their pesticide use when targeting vectors in an environmentally responsible way.

VI. Virus Adaptation to Wild Plants

Viruses co-evolved with wild plants long before any plants were domesticated. Wild plant adapted viruses, for which Harrison (1981) coined the term WILPAD, include those vectored by nematodes, aphids or whiteflies. Tymoviruses may be added to the WILPAD list because they have been isolated almost exclusively from unmanaged, wild dicotyledonous plants. Unfortunately, the classification is not always simple. Some viruses (e.g. species in the *Potyviridae* that are not usually considered to be linked to wild plants particularly) seem sufficiently different from those isolated from cultivated plants to suggest they are WILPAD. *Wild potato mosaic virus* (Jones and Fribourg, 1979) and *Pokeweed mosaic virus* are of this type (Shepherd *et al.*, 1969), but, with some potyviruses, taxonomic issues are complex and unclear. Thus, *Johnsongrass mosaic virus*, which has a number of natural hosts including wild sorghum (*Sorghum verticilliflorum*), was first described as 'Australian' maize dwarf mosaic virus (*sensu lato*) and then as a 'strain' of sugarcane mosaic virus (*sensu lato*), although both *Sugarcane mosaic virus* and *Maize dwarf mosaic virus* are currently considered distinct viruses. The complexities of potyvirus names were explained in Shukla *et al.* (1989, 1994) and in the most recent ICTV report (ICTV; Fauquet *et al.*, 2005). Similarly, with important pathogens such as BYDV and CYDV (regarded as distinct but closely related species of the *Luteoviridae*), 'lumping' is useful in the context of their epidemiology.

VII. Co-occurrence of Interacting Viruses

Multiple virus infection is not uncommon in crops and also in wild plants, but there are only a few records of primary virus isolations from wild plants that indicate transmission-dependent complexes (e.g. Watson and Falk, 1994). However, it is sometimes useful to know how frequently specific viruses occur together, and whether and to what extent the phylogenies of viruses that occur together are correlated. These aspects were studied in the *Carrot mottle virus / Carrot red leaf virus* complex in wild carrot growing in southern England and in synthetic glasshouse populations transmitted between carrots using the aphid vector *Cavariella aegopodii* (Naylor *et al.*, 2003). Those experiences showed that the umbravirus (carrot mottle) decreased the efficiency of transmission of *Carrot red leaf virus* from carrot indicating that the two are not as mutualistic as was first suspected from earlier short-term experiments (Barker, 1989). Elnagar and Murant (1978) reported a few instances of *C. aegopodii*-mediated umbravirus transmission alone and such phenomena should be investigated further in the context of wild plant genotypes because many important questions relating to the key attribute of transmission remain open.

Since the damaging effects of viruses on their wild hosts is a cause of concern because of their potential impact on taxonomic diversity, it is very desirable for sub-genomic agents with potential impacts on pathology to be sought in wild plants. It is notable that five wild solanaceous plant species were shown to harbour viroids (Diener, 1996; Martinez-Soriano *et al.*, 1996) and Diener (2001) argued that viroid reservoirs exist in symptomlessly infected wild plants.

The luteoviruses are characterised by their ability to encapsidate a diverse range of nucleic acids including a viroid (Querci *et al.*, 1997). The process often results in co-transmission of two or more distinct nucleic acids (e.g. Creamer and Falk, 1990; Rochow, 1970; Wen and Lister, 1991). For example, BWYV isolate ST9 associated with an extragenomic RNA first isolated in broccoli was studied in the wild plant species *Capsella bursa pastoris* in which the presence of the extra-genomic RNA was correlated with greatly enhanced disease vis-á-vis that associated with BWYV alone (Sanger *et al.*, 1994). Furthermore, in the presence of the extra-genomic nucleic acid, BWYV concentration in plants increased 10-fold. Ecological/evolutionary significance has been inferred, but appropriate studies have not been done to provide evidence supporting the speculation.

Satellite RNAs (*sensu* Murant and Mayo, 1982) are, by definition, absolutely dependent on another virus for their replication, but their reliance on other viruses for encapsidation and consequent transmission by vectors differs. They have been the subject of many reviews (e.g. Liu and Cooper, 1994; Roosinck *et al.*, 1992). Although distinguishable from viroids, defective interfering replicons and sub-genomic components in multipartite viruses, the limits are sometimes poorly defined and satellite-like molecules that were initially suspected in polerovirus cultures are now known to be sufficiently distinct to justify the special genus-level descriptor 'umbravirus' (Murant, 1990).

Satellite RNAs and satellite viruses (encoding their own coat proteins) may not have been sought systematically in any wild plant species. Nevertheless, they have been observed in wild plants in a few instances, although knowledge of their biological roles is fragmentary (e.g. Cooper, 1993; Francki *et al.*, 1986; Jones and Mayo, 1984; Liu *et al.*, 1991; Randles *et al.*, 1981; Skoric *et al.*, 1997). Satellites sometimes impact dramatically on the severity of disease induced by their transmission-helper viruses. Their effects range from symptom amelioration to exacerbation. Nevertheless, satellites were at times hotly contested intellectual property to exploit for biological control in genetically modified plants or otherwise (e.g. Balcombe *et al.*, 1986; Tien and Wu, 1991), or because of ribozymal activities (e.g. Buzayan *et al.*, 1986; Gerlach *et al.*, 1987).

In agricultural systems, changes in host and pathogen populations are confounded by human influences and cultivated plants are often grown as monocultures far from their centres of origin (e.g. Thresh, 1980). It is often stated that in wild plant communities, virus epidemics are less obvious and prevalent than in cultivated systems (e.g. Bos, 1981; Harper, 1977, 1990). This is attributed to many factors—spatial separation between similar plant communities decreases proximity to external virus inoculum sources, abundance of vectors is lessened by natural predators and parasites, and genotype mixtures and co-occurrence with wild plant species that are non-hosts diminishes spread. Furthermore, rate of virus spread is affected by the presence of natural host resistance and tolerance to viruses and vectors (e.g. Thresh, 1981, 1982). In contrast, viruses tend not to be overtly damaging in communities of wild plants, but they have potential impacts and these should not be underestimated just because they are under-researched. Damaging diseases are most likely to occur as a result of new encounters, as when indigenous viruses contact wild species, for example, *Cardamine chlorotic fleck virus* causing a visually arresting disease in *Cardamine robusta* in alpine areas of southeast Australia (Skotnicki *et al.*, 1992), or obvious disease in the indigenous species *Kennedya prostrata* following infection by the introduced virus BYMV in southwest Australia (Cheng *et al.*, 1999; McKirdy and Jones, 1995; McKirdy *et al.*, 1994). Another example is provided by the visually arresting disease caused by *Turnip yellow mosaic virus* in *Cardamine robusta* (Guy and Gibbs, 1981, 1985). Less obvious and little studied impacts may still be significant (e.g. those due to nematode-transmitted viruses such as *Tomato black ring virus* or *Raspberry ringspot virus*).

In these instances, the wild plants not only provide durable virus reservoirs in the soil's seed bank (Murant and Lister, 1967) but also influence seedling vigour. Thus, *Tomato black ring virus* lessens the rate of germination of *Spergula arvensis* seedlings *vis-a-vis* that of their virus free counterparts (Lister and Murant, 1967). Possible consequences for survival of the infected individuals in competition have not been investigated in this context but were considered in the *Cherry leaf roll virus*–birch pathosystem described later (Section IX).

Except when derived after new encounters, viruses and their principal economic hosts often have common centres of origin (e.g. Lovisolo *et al.*, 2003). However, the pursuit of optimal agricultural productivity has impacted greatly on diverse aspects of virus, vector and host ecology and the complexity is often difficult to unpick. Leppik (1970) was one of the first to recognise the potential utility of gene centres of hereditary variation in plant species. In addition to the wild potato example discussed previously (Section III.A), there are many other instances of wild ancestral species that have been sources of virus tolerance/resistance for plant breeding, either from primary centres where crops and wild relatives co-exist or from secondary centres characterised by extensive sustained cultivation (e.g. Cadman, 1961; Doolittle, 1954; Jones, 1981; Solomon-Blackburn and Barker, 2001; Valkonen, 1994; Valkonen *et al.*, 1996).

VIII. The Role of Wild Plants as Reservoirs

For more than half a century, there have been studies on the reservoir role played by wild plants in the ecology of viruses infecting cultivated plants, and also their vectors (e.g. Anderson, 1959; Bennett, 1952; Bonsquet, 1917; Duffus, 1963; Duffus *et al.*, 1970; Freitag and Severin, 1945; Piemeisel, 1954; Sakimura, 1953; Severin, 1919, 1934; Simons, 1957; Simons *et al.*, 1956; Slykhuis, 1955; Stubbs *et al.*, 1963; Tomlinson *et al.*, 1970). One noteworthy experience in Australia involved the eradication of the herbaceous weed *Sonchus oleraceus* within 150 m of lettuce crops that had the effect of decreasing infection with *Lettuce necrotic yellows virus* from 75% to 6% (Stubbs *et al.*, 1963). There are numerous examples of where removal of wild and weed hosts has contributed greatly to the control of virus epidemics and such hygiene measures are almost invariably key features of control recommendations in many pathosystems (e.g. Jones, 2001, 2004; Thresh, 2003). There are also many instances when the most important sources of virus inoculum are crop species that have escaped from cultivation (e.g. Broadbent *et al.*, 1949; Wallis, 1967; Watson *et al.*, 1951), or neighbouring crops (Duffus, 1963; Hull, 1952, Jones, 2004; Latham and Jones, 2003; Pound, 1946). In addition, volunteer cultivated plants derived from uncollected crop residues that harbour vectors and viruses often initiate and sustain epidemics (e.g. Dunn and Kirkley, 1966; Latham and Jones, 2004; Snyder and Rich, 1942; Stone and Nelson, 1966).

The epidemiological significance of a wild species as host for any particular crop-affecting virus is frequently uncertain because, even when recognised as a natural host of a particular virus, the plant may only play a subsidiary role in virus maintenance or may not be a host to a key vector. Such subtlety has not always been considered. Indeed, some reports are surprisingly superficial. Thus, when suggesting the working hypothesis 'that *Cherry rasp leaf virus* is a virus of balsam root (*Balsamorhiza sagillata*) and possibly other indigenous hosts not yet recognised', Hansen *et al.* (1974) presumed that a wild plant in which they saw abnormalities was more important than its symptomlessly infected counterparts. That judgement was made even when they also had evidence for the occurrence of virus-infected but symptomless *Plantago major* and *T. officinale*. In this instance it seems that they regarded these plants as of less importance because they did not show symptoms! As noted earlier in the context of *Dandelion yellow mosaic virus* and many others, bright leaf colouration attracts attention.

The diverse relationships wild plants may have with the viruses that affect an economically important crop are reflected in the contrasts offered by viruses that infect rice, remembering that there are important epidemiological differences between rice in the tropics (cultivated year-round) and successional/seasonal rice crops in temperate or subtropical regions. In his review of rice viruses, Hibino (1996) identified *Alopecurus aequalis* and other grass weeds as significant overwintering reservoir hosts for *Rice black streaked dwarf virus* in temperate Japan, but wild rice (*Oryza longistaminata*), though naturally infected, was not clearly implicated as important in the ecology of *Rice yellow mottle virus* in tropical Africa. Hibino considered wild grass hosts unimportant reservoirs of rice tungro, *Rice dwarf virus* and *Rice ragged stunt virus* but highlighted *Leersia orizoides* as particularly significant as a reservoir between rice crops for the luteovirus rice galliume virus in northern Italy. The tenuiviruses *Rice grassy stunt virus* and *Hoja blanca virus* seem only to infect rice naturally, but *Rice stripe virus* naturally infects wild graminaceous species that are important over-wintering hosts for the virus and also for the vector planthoppers. The situation concerning the three 'white leaf' (=hoja blanca) viruses is notable and indicates the difficulties that may be encountered. ICTV (Fauquet *et al.*, 2005) listed *Echinochloa hoja blanca virus*, *Urochloa hoja blanca virus* and *Rice hoja blanca virus* as separate species in the *Tenuivirus* genus. On this basis, the genomes of all three viruses have some base sequences in common and are presumed to share in the same gene pool. However, this synonymy is inappropriate since *Echinochloa* hoja blanca and rice hoja blanca (viruses) have distinct ecological niches and vectors (De Miranda *et al.*, 2001; Madriz *et al.*, 1998). Interestingly, the bymovirus isolated once from rice in India differs from *Rice necrosis mosaic virus* from Japan in infecting a number of wild grasses other than rice naturally. Furthermore, it increases the growth of *Ludwigia perennis* (Gosh, 1981, 1982)! Whether this or other wild grasses have a role in the life cycle of this virus is not known.

Winter-grown small-grain cereals and wild graminaceous species in the Mediterranean type climate of south-west Australia provide another example. Over-summering wild native and introduced perennial grasses survive the drought in isolated damp spots, for example, where dew deposited overnight on road surfaces runs off into roadside ditches that remain moist enough to support their growth (Hawkes and Jones, 2005). These grasses harbour BYDV and CYDV and initiate predictable virus epidemics that start in late autumn or early winter (Hawkes and Jones, 2005; Thackray and Jones, 2003).

In the context of potyviruses, *Vicia villosa* is a natural host of *Pea seed-borne mosaic virus* but Stevenson and Hagedorn (1973) showed that weed sources were less important than seed-borne inoculum in pea crops. By contrast, wild relatives of the cucumbers (*Melothria* and *Momordica* spp.) that have economic value in traditional medicine are potent perennial sources for aphid-transmitted viruses that threaten squash production in Florida. Adlerz (1972a,b) recommended that these wild hosts be destroyed to lessen the sources of virus inoculum. Similarly, destruction of the wild cucurbit hosts that harbour damaging potyviruses of cucurbit crops outside the cucurbit growing season was recommended in tropical Northern Australia (Coutts and Jones, 2005), and there are numerous other examples where destruction of a wild host reservoir helps to diminish epidemics (Jones, 2001, 2004; Thresh, 1982, 2003).

The rapid infection of many crops by whitefly transmitted begomoviruses is one line of evidence that strongly indicates that such viruses have 'between crop' hosts among wild plants. For example, *Malva nicaensis* and *Datura stramonium* are both significant over-summering weed hosts of *Tomato yellow leaf curl virus*, although crop residues ('volunteer' tomato and tobacco) also contribute. Furthermore, *Cynanchum acutum* is a natural host of *Tomato yellow leaf curl virus* that is important in the repeated introduction of the virus into tomatoes in the Jordan Valley of Israel (Cohen *et al.*, 1988). Bennett (1952) suggested that most viruses originated in uncultivated plants in which natural selection had tended to encourage the development of tolerance, but begomoviruses do not fit this pattern. *Macroptilium lathyroides* is a common wild plant in Puerto Rico and transmission from individuals with bright yellow mosaic of the leaves revealed a virus that caused symptoms like those associated with *Bean golden mosaic virus*, but *M. lathyroides* was not a host of the latter. The *Macroptilium*-derived isolates seem to be part of the gene pool of, but distinguishable from, *Bean golden mosaic virus* and have been named Macroptilium golden mosaic virus (Idris *et al.*, 1999). Furthermore, *Malva parviflora* is a ubiquitous perennial in regions of southwestern United States where cotton is grown as an annual. Both the crop and the wild plant naturally contain *Cotton leaf crumple virus* with almost all genomic sequences held in common between isolates from the two sources. This genetic information provides support for the view that the *M. parviflora* is an over-wintering host of the virus able to infect the crop. In general, much research effort is required if 'dead end' hosts in wild plants are to be distinguished from important reservoirs, but relations with begomoviruses are particularly problematic. In the Caribbean region, many begomoviruses were first recognised in wild hosts but, with few exceptions, these species seem not to be wild reservoir hosts.

There are particular difficulties interpreting data when insect vectors are involved. Their special needs are not always appreciated fully but it is sometimes possible to provide useful comment. The begomovirus that infects the wild species *Jatropha gossypifolia* naturally in Puerto Rico is transmitted most efficiently by a host-specific 'Jatropha' whitefly biotype (Brown and Bird, 1992). Absolute transmission limits are sometimes hard to confirm with any insect, but are particularly fraught with *Bemisia tabaci* because of its complex host preference behaviour (Frohlich *et al.*, 1999).

When the properties of viruses isolated from wild plants are compared in molecular detail with those isolated from crop hosts, as with *Cotton leaf crumple virus* (Brown, 2002), close similarity or identity may reasonably be assumed to indicate a link between 'source' and 'sink'. Although subtle differences in pathotype may be very important, it is sometimes reasonable to use serological data as a basis for speculation that a wild species is either a reservoir or a 'spill-over' host. However, substantial difficulties sometimes become apparent.

One example due to sub-speciation, has been observed among viruses in the *Luteoviridae* and concerns *Beet mild yellowing virus* and BWYV. The ICTV is finding it necessary to modify the taxonomy to fit better with the natural host range and other peculiarities of these viruses (e.g. Stevens *et al.*, 2005a,b). Complexity in the relationships among luteoviruses that infect grasses and cereals is similarly 'difficult'. As Toko and Bruehl (1959) stated, BYDV (*sensu lato*) 'is a complex of related entities varying in many ways' and this description can hardly be bettered despite much subsequent research and expenditure with the aim of gaining greater understanding. Peters (1991) argued that ecology has failed to become a predictive science because practitioners have concentrated too much on the detailed study of isolated small components, rather than study 'difficult' phenomena holistically. This comment might apply equally to much of plant virology as a lot of description has not resulted in proportionate predictability. This is largely because there has been insufficient research effort. When appropriate research activity has revealed 'the big picture', useful predictions have been possible. As examples, it is useful to draw on experiences in the cool temperate climate of England with the aphid-transmitted 'yellows' viruses in sugar beet (in the United Kingdom sugar beet production is largely confined to England) and the Mediterranean-type climate of south-west Australia with CMV in lupins and BYDV in cereals. In England, survival of aphids during the winter is the key to predicting aphid vector build-up and consequent virus spread, whereas in south-west Australia their survival and build-up over the dry summer and early autumn period is the determining factor. The number of days at cold temperatures during winter is used to forecast virus epidemics in sugar beet (Harrington *et al.*, 1989; Watson *et al.*, 1951, 1975; Werker *et al.*, 1998). By contrast, in south-west Australia, the extent of rainfall in the dry summer and early autumn is used to predict aphid-vector arrival, virus epidemics and yield losses in lupin and cereal crops (Thackray and Jones, 2003; Thackray *et al.*, 2004).

IX. Hypothesis Testing with Wild Plant–Virus Systems

A. *Significance of Vertical Transmission of Virus in Birch*

Seven virus-like agents have been recognised infecting birch trees naturally in Europe or North America (Cooper and Massalski, 1984). Of these, two have been studied in detail; *Apple mosaic virus* in *Betula papyrifera* and *B. alleghaniensis* (Gotlieb, 1975; Hardcastle and Gotlieb, 1980) and *Cherry leaf roll virus* (CLRV) in *Betula verrucosa*, *B. pendula* and *B. pubescens*. Both viruses are characterised by their association with woody perennials and with pollen and seed (Cooper, 1993). The CLRV/*B. pendula* pathosystem was selected for investigation because the virus is transmitted in pollen to the tree pollinated and also to the progeny seedlings, although it was not known whether pollen transmission was essential for the maintenance of this virus in wild tree populations. Birch was chosen because it grows faster than *Juglans regia*, which was a possible alternative candidate for study in this context. Furthermore, with birch, the impact of horticultural practices that facilitate virus maintenance artificially could be avoided.

Fine (1975) proposed an equation that tests the potential for virus maintenance *via* the germ line from parents to progeny. This uses information on the relative proportions of virus-infected individuals in a population, the number of progeny derived from infected as compared to specific virus-free parents and the relative prevalence of virus transmission through the male and the female lines assuming no selective advantage that would help virus-infected plants attain reproductive age. Cooper and co-workers assembled the necessary data concerning CLRV in *B. pendula* in the United Kingdom, where the prevalence of CLRV in unmanaged birch populations is c. 3%, and showed that CLRV would be lost from an inter-breeding birch population within two generations, although this is likely to take many years (Cooper *et al.*, 1984). Stable prevalence rates could only be computed by making assumptions that seemed unrealistic. A large selective advantage for CLRV-infected trees is unlikely since infected birch seedlings grew at a slower rate than their virus-free counterparts. However, the possibility cannot be excluded that CLRV imparts a selective advantage, such as diminished attractiveness to herbivores (*cf.* Gibbs, 1980) that might result from leaf toughness or secondary metabolites initiated by the infection. These observations raised unanswered questions concerning the alternative means of virus transmission indicated. The balance of circumstantial and experimental evidence is against the routine involvement of soil-inhabiting nematodes as vectors of CLRV (e.g. Jones *et al.*, 1981).

B. Virus Impacts on Wild Plant 'Fitness'

In mixed populations of wild plant species, viruses can impact on community structure and dynamics by decreasing the competitive and reproductive ability of infected plants (e.g. Friess and Maillet, 1996;

Harper, 1977, 1990; Malmstrom *et al.*, 2005a,b; Silander, 1985). Indeed, only slight growth impairment due to virus infection may suffice to impact drastically on the competitive ability of the infected plant, as occurred in an annually self-regenerating mixture of capweed (*Arctotheca calendula*) with *Alfalfa mosaic virus*-infected burr medic (*Medicago polymorpha*) in Western Australia (Jones and Nicholas, 1998). A recent study of grassland in California provides a key example. Malmstrom *et al.* (2005a,b) found that BYDV infection markedly impaired native grass 'fitness' by decreasing growth, survivorship and fecundicity, resulting in shifts in community composition in favour of introduced grasses. Similarly, Mackenzie (1985) found that *Primula vulgaris* clones infected with *Arabis mosaic virus* had significantly less growth and survivorship than uninfected clones when each was transplanted into the reciprocal field location. Moreover, Yahara and Oyama (1993) described *Eupatorium chinense* naturally infected with *Tobacco leaf curl virus* as having significantly greater mortality rates and lesser fecundity than uninfected counterparts. The consequences of manual inoculation may differ from naturally occurring infections, but Kelley (1993) reported that clones of *Anthoxanthum odoratum* infected with *Brome mosaic virus* were significantly less fecund and had greater mortality rates than specific virus-free controls. The same pattern was not followed exactly with *Anthoxanthum latent blanching virus* in the same plant species. Here, the virus was associated with lesser growth and greater mortality, but fecundity was somewhat greater than in uninfected plants.

In some pathosystems, wild plants are killed by virus infection, as with BYMV infecting wild *Lupinus angustifolius* (Cheng and Jones, 1999). Alternatively, wild plants may be obviously changed, and examples of this include several mentioned previously (Section II.A) from Australia—*Cardamine chlorotic fleck virus* and *Turnip yellow mosaic virus* in *Cadamine robusta* in the alpine south-east and BYMV in *Kennedya prostrata* and yellow lupin in the dry Mediterranean type environment in the south-west. Also, plants with obvious symptoms may be more attractive to insect vectors than healthy ones, as often reported for infected cultivated plants (e.g. Ajayi and Dewar, 1983; Colvin *et al.*, this volume, pp. 419–452; Magyarosy and Mittler, 1987; Thackray *et al.*, 2000). The many virus resistance genes present in wild plant species provide evidence of host responses to past epidemics. However, although harm to a resident wild plant may be expected, especially when a virus is introduced from elsewhere and a 'new encounter' occurs, there have been insufficient 'in depth' studies like those of Malmstrom *et al.* (2005a,b) to determine the extent to which significant and durable change results. On very rare occasions, virus infection may perhaps actually be beneficial as there are a few (unconfirmed) observations that might be consistent with specific viruses enhancing the survival prospects of their wild hosts through impacts on recently introduced herbivores (Gibbs, 1980). This study is often quoted because it seems to hint at evolutionary significance. However, it would be both interesting and instructive to know if there are similar effects linked to the presence of virus on the feeding behaviour of native (marsupial) herbivores. Duffus (1971) noted incidentally that sowthistle is extremely susceptible to a physiological tip burn injury that severely stunts healthy plants, whereas *Sowthistle yellow vein virus* has a protective effect. It is difficult to judge the real importance of this observation.

Holmes (1950) argued that the original habitat of TMV was likely to be centred on the Andean region of South America where *Nicotiana* species, such as *N. glauca*, native in Peru, and other species (e.g. *N. raimondii*, native in Bolivia) which develop few or no symptoms when infected were candidate long-term reservoirs of the virus. Though plausible, such assessments are not compelling given that *Nicotiana* species native to North America or Australia accumulate greater numbers of virions and thereby may directly contribute to the potency of sources of inoculum for abrasive contacts. Alternatively, subtle elements of leaf phenology and infectibility may be crucial and other properties of infected plants, such as the patterns or intensity of leaf colour may have evolutionary significance (in camouflage) when aerial vectors play key roles in virus dispersal. In the ecological literature, debate has tended to centre on the dilemma of whether plants should 'grow or defend' (Givnich, 1990; Grafen, 1990; Herms and Mattson, 1992). In the virus pathology literature interest has focused almost entirely on the possible role of foliage colour (with physical factors such as hairs or chemical exudates) on virus acquisition or transmission.

Theoreticians have proposed that parasites represent the major selective force favouring the maintenance of genetic variation and sexual reproduction in both plant and animal populations (Hamilton *et al.*, 1990; Lively *et al.*, 1990). For infection to be important in this context, a virus (or perhaps pathogens more generally since they often occur in mixtures) should be prevalent (although not necessarily uniformly distributed) and should lessen the performance of individuals (whether or not symptoms are overt). Kelley (1994) did not provide clear proof but his data support the view that luteovirus infection may generate advantages for sexually reproducing genotypes in communities of sweet vernal grass (*Anthoxanthum odoratum*).

Fluctuating microclimate can induce chaotic flux in parasite populations (Hassel *et al.*, 1991), but conventional wisdom (e.g. Begon and Harper, 1989) anticipates that the evolution of parasitism tends towards commensalism. Other pressures considered by Ewald (1983) may be capable of driving evolution towards severe pathogenicity. On this basis, viruses transmitted 'persistently' by vectors tend to evolve towards benign parasitism or commensalism in vectors and severe pathogenicity in alternative hosts (plant or vertebrate). In parallel, vertically transmitted parasites should evolve towards benign parasitism or mutualism. The transmissibility of a parasite (virus) is a crucial testable component and wild plant species provide appropriate models for study.

Fitness is at the heart of evolutionary biology and connotes differential survival, yet it is a term with many definitions that lack precision (Dawkins, 1978, 1999; Jeger *et al.*, this volume, pp. 163–203). Some ecologists and virologists assume that if plants infected by a specific virus are killed following manual inoculation this indicates a negative fitness impact that 'will' be manifested at the population level (in the field). This is too simplistic and the opposite may be true as when systemic necrosis kills individual plants, thereby preventing them from becoming sources of inoculum that would lead to further virus spread (e.g. Cheng and Jones, 1999; Cheng *et al.*, 2002; Jones, 2005; Jones *et al.*, 2003; Thackray *et al.*, 2002). Such 'sacrificial altruism' is often controlled by hypersensitivity genes operating systemically (Chen *et al.*, 1994; Cockerham, 1970; Jones, 1985, 1990; Jones and Smith, 2005; Ma *et al.*, 1995; Pathipanawat *et al.*, 1996). When developing the useful concept of 'inclusive fitness' Hamilton (1964a,b, 1972) recognised that a gene may have an effect upon the inheritance of copies of that gene possessed by relatives. Dawkins (1999) built on this theme and, with Hamilton's permission, defined inclusive fitness as —'*That property of an organism that appears to be maximised when what is really being maximised is gene survival*' (Dawkins, 1978).

When testing theories regarding plant and viral pathogen interactions there is a need for long-term observations over many generations. This ideally requires organisms with short generation times and species where the genetics are understood. *Arabidopsis thaliana* might seem the ideal candidate, but this extreme ephemeral, although undoubtedly a plant that originates in the wild, is not a relevant model for plants 'in general' except in respect of biochemistry. When ecology, survival, 'fitness' and other whole-plant attributes need to be followed, more 'normal' plants such as *B. rapa* or *B. nigra* or *Medicago truncatula* are more appropriate, although contrasting models including both C3 and C4 species and representing a different habit (e.g. a grass) should certainly be investigated.

C. *Testing an Approach to Environmental Risk Assessment*

Risk assessment is a tool for decision making (Hill and Sendashonga, 2003) and is an approach often used by plant quarantine authorities and others when considering the use of and need for virus control measures. One such option for use in virus management involves genetically modified plants. Since 1986, many DNA sequences derived from virus genomes (and also non-virus derived sequences) have been introduced into crop species to obtain transgenic virus tolerance or resistance (e.g. Cooper and Walsh, 2003). When cultivation of a transgenic crop is envisaged, possible risks to the environment or human health must be assessed. Environmental risk assessment has fitness at its core and is regarded by many to be 'complicated'. It is true that if an increase in 'fitness' of a plant occurs, it will involve diverse environmental variables, but it is not necessary to understand each individual interaction separately and certainly not together. Furthermore, several key objectives in such an assessment are established in law and these are broadly consistent in the major trading nations. For example, the Endangered Species Act in the United States '*prohibits any action that can adversely affect an endangered or threatened species or its habitat*'.

Following a decision to take viruses that infect *Brassica* species as a 'case study in risk assessment' (Cooper and Raybould, 1997), three long-established wild, but probably introduced (Preston *et al.*, 2002), species in the United Kingdom (*B. rapa*, *B. oleracea* and *B. nigra*) were selected as the principal objects of assessment with a fourth, *B. napus*, that grows wild as an escapee from cultivation. *B. napus* as a crop is known colloquially as 'rape' or 'canola' when used for oil, or as 'swede' when grown for food or forage. Confusingly, other *Brassica* species such as *B. rapa* that is a potential recipient of transgenic genes from *B. napus* (e.g. Jorgensen and Anderson, 1994) are also grown as oil crops. Hybridisation between individuals of the same species (crop and wild relatives as with *B. oleracea*) is expected to occur more or less readily.

Risk assessment begins by identifying the potential harmful consequences of a proposed action (e.g. Raybould and Cooper, 2005). Although other outcomes may be envisioned, harm can result from introgression (the ultimate concern in this instance) of virus-derived virus resistance transgenes in crops manifested through ecological release of the wild relative. This is sometimes interpreted in an agricultural context as the generation of 'weeds' (Cooper and Raybould, 1997). The possible harm caused by the ecological release of a wild relative of a transgenic crop is presumed to be analogous to that produced by the spread of an introduced invasive species (Mack *et al.*, 2000). Pimentel *et al.* (2001) identified the main problems caused by invasive plants. In practice, harm through displacement of native species can, in this instance, be defined in terms of population sizes of taxa with legal protection (Raybould, 2005; Raybould and Wilkinson, 2005). However, it is important to appreciate that it is very difficult to measure the population size of any wild species in its natural environment. Furthermore, there is no clear basis for linking the size of a wild plant population with rate or scale of displacement of other species or physical changes in habitat.

Transgenic herbicide-tolerance traits in *Brassica* species are in commercial use and have been notable objects of study. Many of the data obtained in that context were used when testing the utility of a preliminary approach to inform a decision regarding field use of genetically modified *Brassica* containing a sequence from TuMV in England. TuMV damages many crops economically, including *Brassica* (Shattuck, 1992). Transgenic approaches have been used to obtain resistance to TuMV (e.g. Dinant *et al.*, 1993, 1997; Jan *et al.*, 2000). In anticipation of the possible field release of these plants—notably transgenic *B. napus* (Lehmann *et al.*, 1996), it was necessary to assess whether the introgression of transgenic genes for TuMV tolerance could have implications for compatible wild *Brassica* species. A scheme based on studies to determine the cost of conventional major gene resistance to TuMV (Raybould *et al.*, 2003) was developed that seems both useful and cost effective.

1. *Concept of 'Ecological Release'*

Some plant species that are not invasive in their native range become invasive when introduced into new areas, where there are presumed to be fewer natural enemies or other constraints upon them. This experience is a basis for the idea of 'ecological release'. Outside the native range, co-evolved pests and pathogens are likely to be absent, potentially 'releasing' the plants from such controls on their abundance and spread (Mitchell and Power, 2003). Analogous 'release' is suspected if a plant acquires transgenic (or non-transgenic) virus resistance genes as a result of hybridisation (e.g. Butler and Reichhardt, 1999; Ellstrand, 2003; Ellstrand *et al.*, 1999; Raybould and Gray, 1993; Scheffler and Dale, 1994).

2. *Where Is the Harm?*

B. rapa and *B. oleracea* can hybridise with *B. napus* under field conditions (Wilkinson *et al.*, 2000) and, because *B. napus* is cultivated commercially in many countries (Heritage, 2003), there is a potential for introgression of genes from transgenic oilseed rape. Using a combination of remote sensing, field surveys, molecular genetics and mathematical modelling, Wilkinson *et al.* (2003) estimated that in the United Kingdom (largely in England) c. 50,000 *B. napus* × *B. rapa* hybrid plants form annually.

3. *Exposure*

Risk assessment protocols call for an assessment of exposure. This issue has been addressed repeatedly over the last 15 years, although not necessarily under that label (Chèvre *et al.*, 2004; Gray and Raybould, 1999; Scheffler and Dale, 1994). *B. nigra* and *B. napus* will hybridise when pollinated manually, but no spontaneous hybrids have been recorded in the field. On this basis, the likelihood of any hybridisation between *B. nigra* and a virus-tolerant *B. napus* under field conditions is small. Exactly how small may be 'nice to know' but is not essential for this purpose.

B. oleracea and *B. napus* will form hybrids when pollinated manually (Scheffler and Dale, 1994) and there are unpublished and unconfirmed records of spontaneous hybrid formation in the field. Accordingly, hybrids between transgenic *B. napus* and *B. oleracea* could occur provided the *B. napus* was grown 'sufficiently close' to populations of wild *B. oleracea*. The exact measure of the 'critical' distance is again not essential information, but numerous determinations/predictions have been made (e.g. Ellstrand, 2003; Meagher and Vassiliadis, 2003).

Some may argue that forced, laboratory-based, hybridisation is unrealistic, but there is no need for realism in this assessment; such hybridisation provides a 'worst case' measure that is particularly appropriate for this purpose.

To investigate the scale of the hazard in a virological context, experiments were designed to assess whether ecological release could occur if a TuMV resistance gene introgressed into a population of *B. nigra, B. oleracea* or *B. rapa*. Only essential data relevant to the risk assessment are given in a later section.

When *B. nigra* seedlings from four wild populations were challenged with TuMV by sap inoculation using an isolate of the virus from a neighbouring plant of wild *B. oleracea*, 17 of 18 seedlings inoculated became infected. A further 22 seedlings were exposed to aphids that had recently fed on TuMV-infected brassicas and 20 of these became infected. All infected seedlings developed systemic necrosis within 10 days and were dead 2–3 weeks after inoculation (Thurston *et al.*, 2001). In essence, these data show that at least some populations of wild *B. nigra* are highly susceptible and sensitive to isolates of TuMV from their vicinity and hence ecological release is possible. However, TuMV 'naturally' infects *B. nigra* only rarely in coastal regions of England (Raybould *et al.*, 2003). Thus, in this location the hazard from any introgression of a TuMV resistance gene seems to be low. Also, when combined with exposure data, it seems reasonable that the risk of ecological release of TuMV-resistant *B. nigra* would be negligible. If correct (i.e. based on a reasonable index), there is therefore no requirement for large and costly field exposure tests to assess this system further.

When *B. oleracea* was tested similarly by sap inoculation, none of the inoculated and infected seedlings died but the concentration of TuMV in infected seedlings differed many-fold (Raybould *et al.*, 2000). Thus, although qualitatively different from *B. nigra*, the thrust of the experience was that at least some genotypes of wild *B. oleracea* are infectible and sensitive to TuMV. Therefore, ecological release of *B. oleracea* would be possible, provided that some other factor such as space was not limiting.

Tests on wild *B. rapa* revealed a broadly similar picture, although, in this species, infected plants did not die but they were severely stunted when manually inoculated with TuMV (Pallett *et al.*, 2002). Thus, on this preliminary basis, at least some populations of wild *B. rapa* are susceptible and sensitive to TuMV and hence ecological release is possible. No TuMV infections were found in the field-grown *B. rapa* at one site in southern England. However, such modest experience alone has little significance because the possibility cannot be excluded that an isolate of this virus from elsewhere is more pathogenic and that infection with it would result in rapid death in competition with neighbours or directly. In any event, if situations are revealed in which the absence of TuMV in a population of field-grown *Brassica* species is actually due to infection followed by rapid death of individuals from a population, then resource-intensive demographic studies will be necessary.

The most common attitude in the ecological community is that, when investigating possible harm to the environment, much more than a minimum set of data should be sought. This has provided the justification, as almost a first resort, for resource-consuming field trials. Risk assessments that involve field trials are site-specific and time-consuming. Any extra field tests that delay the exploitation of a technological opportunity also incur costs (Cross, 1996). Thus, the approach to risk assessment outlined here seems promising as a preliminary screen and may have value in the context of different wild plant relatives of crops and viruses. Having regard to the economic importance of grasses and the fact that some small grain cereals have been obtained with transgenic luteoviral genes that may impact on fitness, there is scope for building on, for example, the study of Malmstrom *et al.* (2005a,b) to assess specific impacts. There is certainly opportunity for testing the possible environmental impact of conventional genes conferring virus resistance/tolerance similarly—even though this is not required under statute.

X. Conclusion

Wild vascular plants are the principal progenitors of species now being cultivated for food, fodder, fibre and ornament. They also provide sources for a diverse range of products including pharmaceuticals and genes that provide many traits used to breed improved cultivated plants. Such traits include resistance and tolerance against pests and pathogens. The economically most significant herbaceous plant species that provide human dietary staples and stockfeed have often been selected and interbred to an extent that now distinguishes them greatly from their land race and wild ancestors. By contrast, in general, forest trees, ornamental trees and shrubs have been much less highly selected since their primary collection in wild sites. Substantial diversity in original collection sites, such as wildflowers in their natural settings, provides valuable income from ecotourism. Wildflowers are also becoming increasingly popular for use as cut flowers, in home gardens, and for roadside plantings and parks. However, biodiversity is seriously threatened in many regions, with loss of wild plant species a growing concern necessitating a major focus on conservation measures both nationally and internationally. With the continually expanding volume and greater rapidity of international trade in plants and plant products, greater frequency of introduction of invasive viruses that threaten plant biodiversity is inevitable. Therefore, the need to study viruses of wild plants in their natural ecosystems and develop strategies to protect plant biodiversity from them is becoming increasingly urgent.

Viral genomes consist of dynamic populations of mutable molecules with evolutionary relationships that are often as much reticulate as tree-like. Viruses from wild plant populations provide useful data on evolution that cannot be obtained otherwise. However, scientists who publish studies on long-established (selected) laboratory virus cultures sometimes forget this natural diversity. Novelty gives a 'fast track' to publication and the overused word 'unique' probably helps in this process! Utility and relevance to the real world may be implied, but these concepts are rarely prominent and the prevalence of a phenomenon is not routinely stated. Indeed, many records are based on unrepresentative single isolations! To some extent this criticism is lessened for wild plant studies because the precise determination of virus frequencies in them is more difficult than in crops. When surveys of wild plants are done, they tend to be based on semi-automated serological or nucleic acid hybridisation systems that are undoubtedly sensitive, but also are very selective and discriminating. Consequently, the data reviewed here are biased towards detection of the expected rather than the unexpected. Furthermore, since bioassays are no longer done routinely, unknown viruses are missed and pathological variation and its causes, as with the presence of extra-genomic nucleic acid species such as satellites and viroids, remain unrecognised.

It is notable that wild plants contain an appreciable number (even a disproportionate abundance) of viruses with properties that distinguish them from the viruses found in cultivated plants. Properties uncovered in viruses infecting wild plants may be exploitable for human benefit, viruses from algae offering particular opportunities for gene 'mining' as well as biological control. Predictive models and decision-support systems that help rationalise the use of insecticides to lessen virus spread are available in some pathosystems (e.g. Thackray and Jones, 2003; Thackray *et al.*, 2004). Knowledge about the co-existence strategies used by viruses and their hosts in naturally regenerating wild plant communities is fragmentary but, when more is known, new predictive approaches may be developed for such purposes. Systematic surveys and even modest prevalence data concerning viruses in wild plants are relatively uncommon. Typically, such knowledge was obtained as a result of searches for wild reservoirs of viruses (or their vectors) that threatened economically important cultivated species. In other instances, the testing, collection and challenge of wild species in their natural environments was driven by a wish to find sources of virus resistance or tolerance (Cooper and Jones, 1983) for use by plant breeders. When plants in their natural environment are infected with viruses with which they co-evolved rather than ones derived from new encounters, it is commonplace for the infected individuals to show little, if any, obvious abnormality. Consequently, ecologists have tended to overlook both the possibility of their presence and their impacts on plant community structure, genetic diversity, or even gross species diversity, even though slight change in one species may drive dramatic change in the balance between species (e.g. Jones and Nicholas, 1998).

Until about 20 years ago, ecologists and plant virologists rarely if ever worked together. This intellectual isolation was encouraged to break down when funding agencies realised that the field release of genetically modified crops containing virus-derived sequences might be imminent and that more environmental knowledge was necessary for informed decisions to be made. Specific funds were allocated and a few key targets investigated but, despite new information concerning gene flow from crops to their wild relatives and knowledge concerning virus occurrence and impacts in a few wild species, much critical information is still lacking. Thus, the association of visible damage with diminished plant survival, reproduction and establishment is lacking in situations where no seedling recruitment and survival data are collected. Whole life cycle studies are now recognised as pivotal to assessment of potential environmental harm. However, they are seriously hampered because many food plants are naturally perennials (or biennials) grown as annuals, for example, in nature *B. oleracea*, grown for cauliflower, broccoli, cabbage, Brussels sprouts, etc., has a lifespan measured in decades. Wild woody perennials and some perennial grasses are so long lived as to closely approximate to immortal! Consequently, researchers focus on annuals and hope that their experiences will be more generally relevant. Increasingly, the conceptual and mathematical models that ecologists use to describe interactions involving natural populations and communities are seen to enable inferences to be made efficiently in other settings. However, the predictive utility of models depends on appropriate parameterisation and validation with real data that should come from wild plants that are regenerating naturally.

References

Adlerz, W. C. (1972a). *Melothria pendula* plants infected with watermelon mosaic virus 1 as a source of inoculum for cucurbits in Collier County, Florida. *J. Econ. Entomol.* **65:**1303–1306.

Adlerz, W. C. (1972b). *Momordica charantia* as a source of watermelon mosaic virus 1 for cucurbit crops in Palm Beach County, Florida. *Plant Dis. Reptr.* **56:**563–564.

Adlerz, W. C. (1981). Weed hosts of aphid-borne viruses of vegetable crops in Florida. *In* "Pests, Pathogens and Vegetation" (J. M. Thresh, ed.), pp. 467–478. Pitman Press, London.

Ajayi, O., and Dewar, A. M. (1983). The effect of barley yellow dwarf virus on field populations of the cereal aphids, *Sitobion avenae* and *Metopolophium dirhodum*. *Ann. Appl. Biol.* **103:**1–11.

Anderson, C. W. (1959). A study of field sources and spread of five viruses of peppers in central Florida. *Phytopathology* **49:**97–101.

Anderson, P. K., Cunningham, A. A., Patel, N. G., Morales, F. J., Epstein, P. R., and Daszak, P. (2004). Emerging infectious diseases of plants: Pathogen pollution, climate change and agrotechnology drivers. *Trends Ecol. Evol.* **19:**535–544.

Balcombe, D. C., Saunders, G. R., Bevan, M. W., Mayo, M.A, and Harrison, B. D. (1986). Expression of biologically active viral satellite RNA from the nuclear genome of transformed plants. *Nature* **321:**446–449.

Barker, H. (1989). Specificity of the effect of sap-transmissible viruses in increasing the accumulation of lutcovirus in co-infected plants. *Ann. Appl. Biol.* **115:**71–78.

Begon, M., and Harper, J. L. (1989). "Ecology: Individuals, Populations and Communities." Blackwells, Oxford.

Bem, F., and Murant, A. F. (1979a). Transmission and differentiation of six viruses infecting hogweed (*Heracleum sphondylium*) in Scotland. *Ann. Appl. Biol.* **92:**237–242.

Bem, F., and Murant, A. F. (1979b). Host range, purification and serological properties of heracleum latent virus. *Ann. Appl. Biol.* **92:**243–256.

Bennett, C. W. (1952). Origin and distribution of new or little known virus diseases. *Plant Dis. Reptr. Suppl.* **211:**43–46.

Bertioli, D. J., Hayle, A., and Cooper, J. I. (1993). A new virus isolated from an ash tree with die-back. *J. Phytopathol.* **139:**367–372.

Bonsquet, P. A. (1917). Wild vegetation as a source of curly-top infection of sugar beets. *J. Econ. Ent.* **10:**392–397.

Bos, L. (1981). Wild plants in the ecology of virus diseases. *In* "Plant Diseases and Vectors: Ecology and Epidemiology" (K. Maramarosch and K. F. Harris, eds.), pp. 1–28. Academic Press, New York.

Bos, L. (1992). New plant virus problems in developing countries: A corollary of agricultural modernisation. *Adv. Virus Res.* **38:**349–407.

Bos, L., Huijberts, N., Huttinga, H., and Maat, D. Z. (1983). Further characterization of dandelion yellow mosaic virus from lettuce and dandelion. *Neth. J. Plant Pathol.* **89:**207–222.

Brasier, C. M. (1983). Oak tree mortality in Iberia. *Nature* **360:**539.

Brasier, C. M. (1986). The population biology of Dutch elm disease: Its principal features and some implications for other host-pathogen systems. *Adv. Plant Pathol.* **5:**55–118.

Brasier, C. M. (2000). The rise of hybrid fungi. *Nature* **405:**134–135.

Broadbent, L. (1964). Control of plant virus diseases. *In* "Plant Virology" (M. K. Corbett and H. D. Sisler, eds.), pp. 330–364. University of Florida Press, Gainsville.

Broadbent, L., Cornford, C. E., Hull, R., and Tinsley, T. W. (1949). Overwintering of aphids, especially *Myzus persicae* (Sulzer), in root clamps. *Ann. Appl. Biol.* **36:**513–524.

Brown, J. K. (2002). The molecular epidemiology of begomoviruses. *In* "Plant Viruses as Molecular Pathogens" (J. A. Kahn and J. Dijkstra, eds.), pp. 279–315. The Haworth Press, Oxford.

Brown, J. K., and Bird, J. (1992). Whitefly transmitted geminiviruses in the Americas and the Caribbean basin: Past and present. *Plant Dis.* **76:**220–225.

Brown, R. M. (1972). Algal viruses. *Adv. Virus Res.* **17**:243–277.

Brunt, A. A. (1970). *Cacao yellow mosaic virus. CMI/AAB Descr. Pl. Viruses*, No. 11, p. 4.

Brunt, A. A., and Stace-Smith, R. (1978). Some hosts, properties and possible affinities of a labile virus from *Hypochoeris radicata* (Compositae). *Ann. Appl. Biol.* **90**:205–214.

Brunt, A. A., Crabtree, K., and Gibbs, A. (1990). "Viruses of Tropical Plants," p. 707. CAB International, Wallingford, Oxon.

Brunt, A. A., Crabtree, K., Dallwitz, M. J., Gibbs, A. J., and Watson, L. (1996). "Viruses of Plants. Descriptions and Lists from the VIDE Database," p. 1484. CAB International, Wallingford, Oxon.

Buddenhagen, I. W. (1977). Resistance and vulnerability of tropical crops in relation to their evolution and breeding. *Ann. N. Y. Acad. Sci.* **287**:309–326.

Butler, D., and Reichhardt, A. (1999). Long-term effect of GM crops serves up food for thought. *Nature* **398**:654–656.

Buzayan, J. M., Gerlach, W. L., and Breuning, G. (1986). Satellite tobacco ringspot virus RNA: A subset of the RNA sequence is sufficient for autocatalytic processing. *Proc. Natl. Acad. Sci. USA* **83**:8859–8862.

Cadman, C. H. (1961). Raspberry viruses and virus diseases in Britain. *Hort. Res.* **1**:1–52.

Chen, P., Buss, G. R., Roane, C. W., and Tolin, S. A. (1994). Inheritance in soybean of resistant and necrotic reactions to soybean mosaic virus strains. *Crop Sci.* **34**:414–422.

Cheng, Y., and Jones, R. A. C. (1999). Distribution and incidence of the necrotic and non-necrotic strains of bean yellow mosaic virus in wild and crop lupins. *Austral. J. Agric. Res.* **50**:589–599.

Cheng, Y., Jones, R. A. C., and Thackray, D. J. (2002). Deploying strain specific hyper-sensitive resistance to diminish temporal virus spread. *Ann. Appl. Biol.* **140**:69–79.

Chèvre, A. M., Ammitzboll, H., Breckling, B., Dietz-Pfeilstetter, A., Eber, F., Fargue, A., Gomez-Campo, C., Jenczewski, E., Jørgensen, R., Lavigne, C., Meier, M. S., den Nijs, H. C. M., *et al.* (2004). A review on interspecific gene flow from oilseed rape to wild relatives. *In* "Introgression from Genetically Modified Plants into Wild Relatives" (H. C. M. den Nijs, D. Bartsch, and J. Sweet, eds.), pp. 235–251. CAB International, Wallingford, UK.

Cho, J. J., Mau, R. F. L., Gonsalves, D., and Mitchell, W. C. (1986). Reservoir weed hosts of tomato spotted wilt virus. *Plant Dis.* **70**:1014–1017.

Choi, G. H., and Nuss, D. L. (1992). Hypovirulence of chestnut blight fungus conferred by an infectious viral cDNA. *Science* **257**:800–803.

Cockerham, G. (1970). Genetical studies on resistance to potato viruses X and Y. *Heredity* **25**:309–348.

Cohen, S., Kern, J., Harpaz, I., and Bar-Joseph, R. (1988). Epidemiological studies of the tomato yellow leaf curl virus (TYLCV) in the Jordan Valley, Israel. *Phytoparasitica* **16**:259–270.

Cooper, J. I. (1971). The distribution in Scotland of tobacco rattle virus and its nematode vectors in relation to soil type. *Plant Pathol.* **20**:51–58.

Cooper, J. I. (1979). "Virus Diseases of Trees and Shrubs," p. 74. Institute of Terrestrial Ecology, Cambridge.

Cooper, J. I. (1993). "Virus Diseases of Trees and Shrubs," 2nd Ed., p. 205. Chapman and Hall, London.

Cooper, J. I., and Edwards, M. L. (1981). The distribution of poplar mosaic virus in hybrid poplars and virus detection by ELISA. *Ann. Appl. Biol.* **99**:53–61.

Cooper, J. I., and Harrison, B. D. (1973a). The role of weed hosts and the distribution and activity of vector nematodes in the ecology of tobacco rattle virus. *Ann. Appl. Biol.* **73**:53–66.

Cooper, J. I., and Harrison, B. D. (1973b). Distribution of potato mop-top virus in Scotland in relation to soil and climate. *Plant Pathol.* **22:**73–78.
Cooper, J. I., and Jones, A. T. (1983). Responses of plants to viruses: Proposals for the use of terms. *Phytopathology* **73:**127–128.
Cooper, J. I., and Massalski, P. R. (1984). Viruses and virus-like diseases affecting *Betula* spp. *Proc. Roy. Soc. Edin.* **85B:**183–195.
Cooper, J. I., and Raybould, A. F. (1997). Transgenes for stress tolerance: Consequences for weed evolution. *In* "Proceedings of Brighton Crop Protection Conference," pp. 265–272.
Cooper, J. I., and Tinsley, T. W. (1978). Some epidemiological consequences of drastic ecosystem changes accompanying exploitation of tropical rainforest. *La terre et la vie* **32:**221–240.
Cooper, J. I., and Walsh, J. A. (2003). Genetic modification of disease resistance, viral pathogens. *In* "Encyclopedia of Applied Plant Sciences" (B. Thomas, D. Murphy, and B. Murray, eds.), pp. 257–262. Academic Press, London, UK.
Cooper, J. I., Edwards, M. L., Arnold, M. A., and Massalski, P. R. (1983). A tobravirus that invades *Fraxinus mariesii* in the United Kingdom. *Plant Pathol.* **32:**469–472.
Cooper, J. I., Massalski, P. R., and Edwards, M. L. (1984). *Cherry leaf roll virus* in the female gametophyte and seed of birch and its relevance to vertical virus transmission. *Ann. Appl. Biol.* **105:**55–64.
Coutts, B. A., and Jones, R. A. C. (2000). Viruses infecting canola (*Brassica napus*) in south-west Australia: Incidence, distribution, spread and infection reservoir in wild radish (*Raphanus raphanistrum*). *Aust. J. Agric. Res.* **51:**925–936.
Coutts, B. A., and Jones, R. A. C. (2005). Incidence and distribution of viruses infecting cucurbit crops in the Northern Territory and Western Australia. *Aust. J. Agric. Res.* **56:**847–858.
Creamer, R., and Falk, B. W. (1990). Direct detection of transcapsidated barley yellow dwarf luteovirus in doubly infected plants. *J. Gen. Virol.* **71:**211–217.
Cross, F. B. (1996). Paradoxical perils of the precautionary principle. *Washington Lee Law Rev.* **53:**851–925.
Dawe, A. L., and Nuss, D. L. (2001). Hypovirulence and chestnut blight: Exploiting viruses to understand and modulate fungal pathogenesis. *Ann. Rev. Genet.* **35:**1–29.
Dawkins, R. (1978). Replicator selection and the extended phenotype. *Z. Tierpsycol.* **47:**61–76.
Dawkins, R. (1999). "The Extended Phenotype," Revised edition, p. 314. Oxford University Press.
De Miranda, J. R., Munoz, M., Wu, R., and Espinoza, A. M. (2001). Phylogenetic position of a novel tenuivirus from the grass *Urochloa plantaginea*. *Virus Genes* **22:**329–333.
Diener, T. O. (1996). Origin and evolution of viroids and viroidlike satellite RNAs. *Virus Genes* **11:**119–131.
Diener, T. O. (2001). The viroid: Biological oddity or evolutionary fossil. *Adv. Virus Res.* **57:**137–184.
Dinant, S., Blaise, F., Kusiak, C., Astier-Manifacier, S., and Albouy, J. (1993). Heterologous resistance to potato virus Y in transgenic tobacco plants expressing the coat protein of lettuce mosaic potyvirus. *Phytopathology* **83:**818–824.
Dinant, S., Maisonneuve, B., Albouy, J., Chupeau, Y., Chupeau, M.-C., Bellec, Y., Gaudefroy, F., Kusiak, C., Souche, S., Robaglia, C., and Lot, H. (1997). Coat protein gene-mediated protection in *Lactuca sativa* against lettuce mosaic potyvirus strains. *Mol. Breeding* **3:**75–86.
Donson, J., and Hull, R. (1983). Physical mapping and molecular cloning of caulimovirus DNA. *J. Gen. Virol.* **64:**2281–2288.

Doolittle, S. P. (1954). The use of wild *Lycopersicum* species for tomato disease control. *Phytopathology* **44:**409–414.
Duffus, J. E. (1963). Incidence of beet virus diseases in relation to overwintering beet-fields. *Plant Dis. Reptr.* **47:**428–431.
Duffus, J. E. (1971). Role of weeds in the incidence of virus diseases. *Annu. Rev. Phytopathol.* **9:**319–340.
Duffus, J. E., Zink, F. W., and Bardin, R. (1970). Natural occurrence of sowthistle yellow vein virus on lettuce. *Phytopathology* **60:**1383–1384.
Dunn, J. A., and Kirkley, J. (1966). Studies on the aphid, *Cavariella aegopodii* Scop. II., on secondary hosts other than carrot. *Ann. Appl. Biol.* **58:**213–217.
Ebrahim-Nesbat, F., and Heitefuss, R. (1989). Isolierung eines Potyviruses aus erkrankter Fichte im Bayerischen Wald. *Eur. J. Forest Pathol.* **19:**222–230.
Edwards, M. L., Cooper, J. I., Massalski, P. R., and Green, B. (1985). Some properties of virus-like agent in *Brachypodium sylvaticum* in the United Kingdom. *Plant Pathol.* **34:**95–104.
Ellstrand, N. C. (2003). "Dangerous Liaisons? When Cultivated Plants Mate with Their Wild Relatives," p. 244. The John Hopkins press, Baltimore.
Ellstrand, N. C., Prentice, H. C., and Hancock, J. F. (1999). Gene flow and introgression from domesticated plants into their wild relatives. *Annu. Rev. Ecol. Syst.* **30:**539–563.
Elnagar, S., and Murant, A. F. (1976a). Relations of the semi-persistent viruses, parsnip yellow fleck and Anthriscus yellows, with their vector, *Cavariella aegopodii*. *Ann. Appl. Biol.* **84:**153–167.
Elnagar, S., and Murant, A. F. (1976b). The role of the helper virus, Anthriscus yellows in the transmission of parsnip yellow fleck virus by the aphid *Cavariella aegopodii*. *Ann. Appl. Biol.* **84:**169–181.
Elnagar, S., and Murant, A. F. (1978). Relations of carrot red leaf and carrot mottle viruses with their aphid vector, *Cavariella aegopodii*. *Ann. Appl. Biol.* **89:**237–244.
Ewald, P. W. (1983). Host-parasite relations, vectors and the evolution of disease severity. *Microbiol. Rev.* **55:**586–620.
Fauquet, C. M., Mayo, M. A., Maniloff, J., Desselberger, U., and Ball, L. A. (2005). Virus Taxonomy. Eighth Report of the International Committee on Taxonomy of Viruses. Elsevier Academic Press, Amsterdam.
Fine, P. E. M. (1975). Vectors and vertical transmission: An epidemiological prespective. *Ann. N.Y. Acad. Sci.* **266:**173–194.
Finlay, K. W. (1952). Inheritance of spotted wilt resistance in the tomato. I. Identification of strains of the virus by the resistance or susceptibility of tomato species. *Australian J. Sci. Res.* **5:**303–314.
Finlay, K. W. (1953). Inheritance of spotted wilt resistance in the tomato. II. Five genes controlling spotted wilt resistance in four tomato types. *Australian J. Sci. Res.* **6:**153–163.
Fraile, A., Escriu, F., Aranda, M. A., Malpica, J. M., Gibbs, A. J., and Garcia-Arenal, F. (1997). A century of tobamovirus evolution in an Australian population of *Nicotiana glauca*. *J. Virol.* **71:**8316–8320.
Francki, R. I. B., Grivell, J. C., and Gibb, K. S. (1986). Isolation of velvet tobacco mottle virus capable of replication with and without a viroid-like RNA. *Virology* **148:**381–384.
Freitag, J. H., and Severin, H. H. P. (1945). Transmission of celery yellow spot virus by the honeysuckle aphid, *Rhopalosiphum conii* (Dvd.). *Hilgardia* **15:**375–386.
Friess, N., and Maillet, J. (1996). Influence of cucumber mosaic virus infection on the intraspecific competitive ability and fitness of purslane (*Portulaca oleraccea*). *New Phytologist* **132:**103–111.

Frohlich, D., Torres-Jerez, I., Bedford, I. D., Markham, P. G., and Brown, J. K. (1999). A phylogeographic analysis of the *Bemisia tabaci* species complex based on mitochondrial DNA markers. *Mol. Ecol.* **8:**1593–1602.

Fulton, J. P. (1969). Transmission of tobacco ringspot virus to the roots of a conifer by a nematode. *Phytopathology* **59:**236.

Gerlach, W. L., Llewellyn, D., and Haseloff, J. (1987). Construction of a plant disease resistance gene from the satellite RNA of tobacco ringspot virus. *Nature* **328:**802–805.

Ghabrial, S. A. (1998). Origin, adaptation and evolutionary pathways of fungal viruses. *Virus Genes* **16:**119–131.

Gibbs, A. J. (1978). Kennedya yellow mosaic virus. *CMI/AAB Descr. Pl. Viruses*, No. 139, p. 4.

Gibbs, A. J. (1980). A plant virus that partially protects its wild legume host against herbivores. *Intervirology* **13:**42–47.

Gibbs, A., and Guy, P. (1979). How long have there been viruses in Australia? *Australasian Plant Pathol.* **8:**41–42.

Gibbs, A. J., Hecht-Poinar, E., and Woods, R. D. (1966). Some properties of three related viruses: Andean potato latent, Dulcamara mottle and *Ononis* yellow mosaic. *J. Gen. Microbiol.* **44:**177–193.

Gibbs, J. N., Liese, W., and Pinon, J. (1984). Oak wilt for Europe? *Outlook Agric.* **13:**203–207.

Gibbs, N. J. (1978). Development of the Dutch elm disease epidemic in southern England 1971–1976. *Ann. Appl. Biol.* **88:**219–228.

Givnich, T. J. (1990). Leaf mottling: Relation to growth form and leaf phenology and possible role in camouflage. *Funct. Ecol.* **4:**463–474.

Gosh, S. K. (1981). Weed hosts of rice necrosis mosaic virus. *Plant Dis.* **65:**602–603.

Gosh, S. K. (1982). Growth promotion in plants by rice necrosis mosaic virus. *Planta* **155:**193–198.

Gotlieb, A. R. (1975). Apple mosaic virus infecting yellow birch in Vermont. *Proc. Am. Phytopathol. Soc.* **2:**97.

Grafen, A. (1990). Biological signals as handicaps. *J. Theor. Biol.* **144:**517–546.

Gray, A. J., and Raybould, A. F. (1999). Genetically Modified Organisms Research Report No. 15. Environmental Risks of Herbicide-Tolerant Oilseed Rape: A Review of the PGS Hybrid Oilseed Rape, p. 59. Department of the Environment, Transport and the Regions, London.

Greber, R. S., and Randles, J. W. (1986). Solanum nodiflorum mottle virus. *CMI/AAB Descr. Pl. Viruses,* No. 318, p. 5.

Guy, P. L., and Gibbs, A. J. (1981). A tymovirus of *Cardamine* sp. from Alpine Australia. *Australasian Plant Pathol.* **10:**12–13.

Guy, P. L., and Gibbs, A. J. (1985). Further studies on turnip yellow mosaic tymovirus isolates from an endemic Australian *Cardamine*. *Plant Pathol.* **34:**532–544.

Hamilton, W. (1964a). The genetical evolution of social behaviour I. *J. Theor. Biol.* **7:**1–16.

Hamilton, W. (1964b). The genetical evolution of social behaviour II. *J. Theor. Biol.* **7:**17–32.

Hamilton, W. (1972). Altruism and related phenomena, mainly in social insects. *Annu. Rev. Ecol. Syst.* **3:**193–232.

Hamilton, W. D., Axelrod, R., and Tanase, R. (1990). Sexual reproduction as an adaptation to resist parasites (a review). *Proc. Natl. Acad. Sci. USA* **87:**3566–3573.

Hammond, J. (1982). *Plantago* as a host of economically important viruses. *Adv. Virus Res.* **27:**103–138.

Hansen, J. A., Nyland, G., McElroy, F. D., and Stace-Smith, R. (1974). Origin, cause, host range and spread of cherry raspleaf disease in North America. *Phytopathology* **64:**721–727.

Hardcastle, T., and Gotlieb, A. R. (1980). An enzyme-linked immunosorbent assay for the detection of apple mosaic virus in yellow birch. *Can. J. For. Res.* **10:**278–283.

Hardy, G. E. St. J., Colquhoun, I. J., Shearer, B. L., and Tommerup, I. (2001). The impact and control of *Phytophthora cinnamomi* in native (natural) and rehabilitated forest ecosystems in Western Australia. *For. Snow Landsc. Res.* **76:**337–343.

Harlan, J. R. (1965). The possible role of weedy races in the evolution of crop plants. *Euphytica* **14:**1–6.

Harlan, J. R. (1971). Agricultural origins: Centres and non centres. *Science* **174:**468–474.

Harlan, J. R. (1981). Ecological settings for the emergence of agriculture. *In* "Pests, Pathogens and Vegetation" (J. M. Thresh, ed.), pp. 3–22. Pitman Press, London.

Harper, J. L. (1977). "Population Biology of Plants," p. 892. Academic Press, London.

Harper, J. L. (1988). An apophasis of plant population biology. *In* "Plant Population Ecology" (A. J. Davy, M. J. Hutchings, and A. R. Watkinson, eds.), pp. 435–452. Blackwell Scientific Publications, Oxford.

Harper, J. L. (1990). Pests pathogens and plant communities: An introduction. *In* "Pests, Pathogens and Plant Communities" (J. J. Burdon and S. R. Leather, eds.), pp. 3–14. Blackwell Scientific Publications, Oxford.

Harper, G., Hull, R., Lockhart, B., and Olszewski, N. (2002). Viral sequences integrated into plant genomes. *Annu. Rev. Phytopathol.* **40:**119–136.

Harrington, R., Dewar, A. M., and George, B. (1989). Forecasting the incidence of virus yellows in sugar beet in England. *Ann. Appl. Biol.* **114:**459–469.

Harrison, B. D. (1964). Infection of gymnosperms with nematode-transmitted viruses of flowering plants. *Virology* **24:**228–229.

Harrison, B. D. (1981). Plant virus ecology: Ingredients, interactions and environmental influences. *Ann. Appl. Biol.* **99:**195–209.

Hassel, M. P., Comins, H. N., and May, R. M. (1991). Spacial structure and chaos in insect population dynamics. *Nature* **353:**255–258.

Hawkes, J. R., and Jones, R. A. C. (2005). Incidence and distribution of *Barley yellow dwarf virus* and *Cereal yellow dwarf virus* in over-summering grasses in a Mediterranean-type environment. *Aust. J. Agric. Res.* **56:**257–270.

Heagle, A. S. (1973). Interactions between air pollutants and plant parasites. *Annu. Rev. Phytopathol.* **11:**365–388.

Heiniger, U., and Rigling, D. (1994). Biological control of chestnut blight in Europe. *Annu. Rev. Phytopathol.* **32:**581–599.

Hemida, S. K., and Murant, A. F. (1989). Host ranges and serological properties of eight isolates of parsnip yellow fleck virus belonging to the two major serotypes. *Ann. Appl. Biol.* **114:**101–109.

Heritage, J. (2003). Will GM rapeseed cut the mustard? *Science* **302:**401–403.

Herms, D. A., and Mattson, W. J. (1992). The dilemma of plants to grow or to defend. *Q. Rev. Biol.* **67:**283–335.

Hibino, H. (1996). Biology and epidemiology of rice viruses. *Annu. Rev. Phytopathol.* **34:**249–274.

Hill, R. A., and Sendashonga, C. (2003). General principles for risk assessment of living modified organisms: Lessons from chemical assessment. *Environ. Biosafety Res.* **2:**81–88.

Hillman, B. I., and Suzuki, N. (2004). Viruses of the chestnut blight fungus, *Cryphonectria parasitica*. *Adv. Virus Res.* **63:**423–472.

Hollings, M. (1978). Mycoviruses: Viruses that infect fungi. *Adv. Virus Res.* **22:**1–53.
Holmes, F. O. (1950). Indications of a New-World origin of tobacco-mosaic virus. *Phytopathology* **41:**341–349.
Hong, Y., Cole, T. E., Brasier, C. M., and Buck, K. W. (1998). Evolutionary relationships among putative RNA-dependent RNA polymerases encoded by a mitochondrial virus-like RNA in the Dutch elm disease fungus, *Ophiostoma novo-ulmi*, by other viruses and virus-like RNAs and by the *Arabidopsis* mitochondrial genome. *Virology* **246:**158–169.
Hull, R. (1952). Control of virus yellows in sugar beet seed crops. *J. Roy. Agric. Soc.* **113:**86–102.
Hull, R. (1968). A virus disease of Hart's tongue fern. *Virology* **35:**333–335.
Idris, A. M., Bird, J., and Brown, J. K. (1999). First report of a bean-infecting begomovirus from *Macroptilium lathyroides* in Puerto Rico. *Plant Dis.* **83:**1071.
Jacobi, V., and Castello, J. D. (1991). Isolation of tomato mosaic virus from waters draining forest stands in New York State. *Phytopathology* **81:**1112–1117.
Jacobi, V., Castello, J. D., and Flachmann, M. (1992). Isolation of tomato mosaic virus from red spruce. *Plant Dis.* **76:**518–522.
Jan, F. J., Fagoaga, C., Pang, S. Z., and Gonsalves, D. (2000). A single transgene derived from two distinct viruses confers multi-virus resistance in transgenic plants through homology-dependent gene silencing. *J. Gen. Virol.* **81:**2103–2109.
Jankowitsch, J., Mette, M. F., van der Winden, J., Matzke, M. A., and Matzke, A. J. M. (1999). Integrated pararetroviral sequences define a unique class of dispersed repetitive DNA in plants. *Proc. Natl. Acad. Sci. USA* **96:**13241–13246.
Johns, L. S. (1982). Purification and partial characterization of a carlavirus from *Taraxacum officinale*. *Phytopathology* **72:**1239–1242.
Jones, A. T., and Mayo, M. A. (1984). Satellite nature of the viroid-like RNA-2 of *Solanum nodaflorum mottle virus* and the ability of other plant viruses to support the replication of viroid-like RNA. *J. Gen. Virol.* **65:**1713–1721.
Jones, A. T., McElroy, F. D., and Brown, D. J. F. (1981). Tests for transmission of cherry leaf roll virus using *Longidorus, Paralongidorus* and *Xiphinema* nematodes. *Ann. Appl. Biol.* **99:**143–150.
Jones, R. A. C. (1981). The ecology of viruses infecting wild and cultivated potatoes in the Andean region of South America. *In* "Pests, Pathogens and Vegetation" (J. M. Thresh, ed.), pp. 89–107. Pitman, London.
Jones, R. A. C. (1985). Further studies on resistance-breaking strains of potato virus X. *Plant Pathol.* **34:**182–189.
Jones, R. A. C. (1990). Strain group specific and virus specific hypersensitive reactions to infection with potyviruses in potato cultivars. *Ann. Appl. Biol.* **117:**93–105.
Jones, R. A. C. (2001). Developing integrated disease management strategies against non-persistently aphid-borne viruses: A model program. *Integr. Pest Manage. Rev.* **6:**5–46.
Jones, R. A. C. (2004). Using epidemiological information to develop effective integrated virus disease management strategies. *Virus Res.* **100:**5–30.
Jones, R. A. C. (2005). Patterns of spread of two non-persistently aphid-borne viruses in lupin stands under four different infection scenarios. *Ann. Appl. Biol.* **146:**337–350.
Jones, R. A. C., Coutts, B. A., and Cheng, Y. (2003). Yield limiting potential of necrotic and non-necrotic strains of bean yellow mosaic virus in narrow-leafed lupin (*Lupinus angustifolius*). *Aust. J. Agric. Res.* **54:**849–859.
Jones, R. A. C., and Fribourg, C. E. (1979). Host plant reactions, some properties and serology of wild potato mosaic virus. *Phytopathology* **69:**446–449.

Jones, R. A. C., and Latham, L. J. (1996). Natural resistance to cucumber mosaic virus in lupin species. *Ann. Appl. Biol.* **129:**523–542.

Jones, R. A. C., and Nicholas, D. A. (1998). Impact of an insidious virus disease in the legume component on the balance between species within self-regenerating annual pasture. *J. Agric. Sci., Cambridge* **131:**155–170.

Jones, R. A. C., and Smith, L. J. (2005). Inheritance of hypersensitive resistance to *Bean yellow mosaic virus* in narrow-leafed lupin (*Lupinus angustifolius*). *Ann. Appl. Biol.* **146:**539–543.

Jorgensen, R., and Anderson, B. (1994). Spontaneous hybridisation between oilseed rape (*Brassica napus*) and weedy *B. campestris* (Brassicaceae): A risk of growing genetically modified oilseed rape. *Am. J. Bot.* **81:**1620–1626.

Kassanis, B. (1947). Studies on dandelion yellow mosaic and other virus diseases of lettuce. *Ann. Appl. Biol.* **34:**412–421.

Kelley, S. E. (1993). Viruses and the advantage of sex in *Anthoxanthum odoratum*: A review. *Plant Species Biol.* **8:**217–223.

Kelley, S. E. (1994). Viral pathogens and the advantage of sex in the perennial grass *Anthoxanthum odoratum*. *Proc. Roy. Soc. London B* **346:**295–302.

Krantz, H. D., Miks, D., Siegler, M. L., Capesius, I., Sensen, C. W., and Huss, V. A. (1995). The origin of land plants: Phylogenetic relationships among charophytes, bryophytes and vascular plants inferred from complete small subunit ribosomal RNA gene sequences. *J. Mol. Evol.* **41:**74–78.

Kusunoki, M., Hanada, K., Iwaki, M., Chang, M. V., Doi, Y., and Yora, K. (1986). Cycas necrotic stunt virus, a new member of nepoviruses found in *Cycas revoluta*; host range, purification, serology and some properties. *Ann. Phytopathol. Soc. Jpn.* **52:**302–311.

Latham, L. J., and Jones, R. A. C. (1997). Occurrence of tomato spotted wilt tospovirus in weeds, native flora and horticultural crops. *Aust. J. Agric. Res.* **48:**359–369.

Latham, L. J., and Jones, R. A. C. (2003). Incidence of *Celery mosaic virus* in celery crops in south-west Australia, and its management using a "celery-free period." *Australasian Plant Pathol.* **32:**527–531.

Latham, L. J., and Jones, R. A. C. (2004). Carrot virus Y: Symptoms, losses, incidence, epidemiology and control. *Virus Res.* **100:**89–99.

Lehmann, P., Walsh, J. A., Jenner, C. E., Kozubek, E., and Greenland, A. (1996). Genetically engineered protection against turnip mosaic virus infection in transgenic oilseed rape (*Brassica napus* var. *oleifera*). *J. Appl. Gen.* **37A:**118–121.

Leppik, E. E. (1970). Gene centres of plants as sources of disease resistance. *Annu. Rev. Phytopathol.* **8:**323–344.

Lin, N. S., Lin, W. V., Kiang, T., and Chang, T. Y. (1981). Investigation and study of bamboo witches' broom in Taiwan. *Q. J. Chin. For.* **14:**135–148.

Lister, R. M., and Murant, A. F. (1967). Seed-transmission of nematode-borne viruses. *Ann. Appl. Biol.* **59:**49–62.

Liu, Y. C., Durrett, R., and Milgroom, M. G. (1999). A spatially-structured stochastic model to simulate heterogenous transmission of viruses in fungal populations. *Ecol. Model.* **127:**291–301.

Liu, Y. Y., and Cooper, J. I. (1994). Satellites of plant viruses. *Rev. Plant Pathol.* **73:**37–387.

Liu, Y. Y., Cooper, J. I., Edwards, M. L., and Hellen, C. U. T. (1991). A satellite RNA of arabis mosaic nepovirus and its pathological impact. *Ann. Appl. Biol.* **118:**577–587.

Lively, C. M., Craddock, C., and Vrijenhoek, R. C. (1990). Red queen hypothesis supported by parasitism in sexual and clonal fish. *Nature* **344:**864–866.

Lovisolo, O., Hull, R., and Rosler, O. (2003). Coevolution of viruses with hosts and vectors and possible paleontology. *Adv. Virus Res.* **62:**325–379.
Luisoni, E., Milne, R. G., Accotto, G. P., and Boccardo, G. (1987). Cryptic viruses in hop trefoil (*Medicago lupulina*) and their relationships to other cryptic viruses in legumes. *Intervirology* **28:**144–156.
Ma, G., Chen, P., Buss, G. R., and Tolin, S. A. (1995). Genetic characteristics of two genes for resistance to soybean mosaic virus in PI486355 soybean. *Theor. Appl. Genet.* **91:**907–914.
MacClement, W. D., and Richards, M. G. (1956). Virus in wild plants. *Can. J. Bot.* **34:**793–799.
Mack, R. N., Simbeloff, D., Lonsdale, M., Evans, H., Cloute, M., and Bazzazf, F. A. (2000). Biotic invasions: Causes epidemiology, global consequences, and control. *Ecol. Appl.* **10:**689–710.
MacKenzie, S. (1985). Reciprocal transplantation to study local speciation and the measurement of components of fitness. PhD Thesis. University College of North Wales.
Madriz, J., de Miranda, J. R., Cabezas, E., Oliva, M., Hernandez, M., and Espinosa, A. M. (1998). Echinochloa hoja blanca virus and rice hoja blanca virus occupy distinct ecological niches. *J. Phytopathol.* **146:**305–308.
Magyarosy, A. C., and Mittler, T. E. (1987). Aphid feeding rates on healthy and beet curly top virus-infected plants. *Phytoparasitica* **15:**335–338.
Malmstrom, C. M., Hughes, C. C., Newton, L. A., and Stoner, C. J. (2005a). Virus infection in remnant native bunchgrasses from invaded California grasslands. *New Phytologist* **168:**217–230.
Malmstrom, C. M., McCulloch, A. J., Newton, L. A., Johnson, H. J., and Borer, E. T. (2005b). Invasive annual grasses indirectly increase virus incidence in California bunchgrasses. *Oecologia* **145:**153–164.
Mansoor, S., Khan, S., Bashir, A., Saeed, M., Zafar, Y., Malik, K., Briddon, R. W., Stanley, J., and Markham, P. G. (1999). Identification of a novel circular single-stranded DNA associated with cotton leaf curl disease in Pakistan. *Virology* **259:**190–199.
Maroon-Lango, C. J., Li, R., Mock, R. G., and Hammond, J. (2005). Molecular characterization of a new flexivirus in ryegrass and its relationship to the genera *Allexivirus, Carlavirus, Foveavirus* and *Potexvirus* as well as unassigned members of the family *Flexiviridae*. Abstract of the International Congress of Virology, July 24, 2005, p. 63.
Martinez-Soriano, J. P., Galindo-Alonso, J., Maroon, C. J. M., Yucel, I., Smith, D. R., and Diener, T. O. (1996). Mexican papita viroid: Putative ancestor of crop viroids. *Proc. Natl. Acad. Sci. USA* **93:**9397–9401.
Maskell, L. C., Raybould, A. F., Cooper, J. I., Edwards, M.-L., and Gray, A. J. (1999). Effects of turnip mosaic virus and turnip yellow mosaic virus on the survival, growth and reproduction of wild cabbage (*Brassica oleracea*). *Ann. Appl. Biol.* **135:**401–407.
McDougall, K. L., Hardy, G. E. S., and Hobbs, R. J. (2002). Distribution of *Phytophthora cinnamomi* in the northern jarrah (*Eucalyptus marginata*) forest of Western Australia in relation to dieback age and topography. *Aust. J. Bot.* **50:**107–114.
McKirdy, S. J., and Jones, R. A. C. (1993). Occurrence of barley yellow dwarf virus serotypes MAV and RMV in over-summering grasses. *Aust. J. Agric. Res.* **44:**1195–1209.
McKirdy, S. J., and Jones, R. A. C. (1994a). Infection of alternative hosts associated with narrow-leafed lupin (*Lupinus angustifolius*) and subterranean clover (*Trifolium subterraneum*) by cucumber mosaic virus and its persistence between growing seasons. *Aust. J. Agric. Res.* **45:**1035–1049.
McKirdy, S. J., and Jones, R. A. C. (1994b). Infection of alternative hosts associated with annual medics (*Medicago* spp.) by alfalfa mosaic virus and its persistence between growing seasons. *Aust. J. Agric. Res.* **45:**1413–1426.

McKirdy, S. J., and Jones, R. A. C. (1995). Bean yellow mosaic potyvirus infection of alternative hosts associated with subterranean clover (*Trifolium subterraneum*) and narrow-leafed lupins (*Lupinus angustifolius*): Field screening procedure, relative susceptibility/resistance rankings, seed transmission and persistence between growing seasons. *Aust. J. Agric. Res.* **46:**135–152.

McKirdy, S. J., Coutts, B. A., and Jones, R. A. C. (1994). Occurrence of bean yellow mosaic virus in subterranean clover pastures and perennial native legumes. *Aust. J. Agric. Res.* **45:**183–194.

Meagher, T. R., and Vassiliades, C. (2003). Spatial geometry determines gene flow in plant populations. *In* "Genes in the Environment" (R. S. Hails, J. E. Beringer, and H. C. J. Godfray, eds.), pp. 76–90. Blackwells Scientific Publications, Oxford.

Milgroom, M. G., and Cortesi, P. (2004). Biological control of chestnut blight with hypovirulence: A critical analysis. *Annu. Rev. Phytopathol.* **42:**311–338.

Mitchell, C. E., and Power, A. G. (2003). Release of invasive plants from fungal and viral pathogens. *Nature* **421:**625–627.

Muller, D. G., and Stache, B. (1992). Worldwide occurrence of virus-infections in filamentous marine brown algae. *Helgolander Meersunters* **46:**1–8.

Muller, D. G., Kawai, H., Stache, B., and Lanka (1990). A virus infection in the marine brown alga *Ectocarpus siliculosus* (Phaeophyceae). *Bot. Acta* **103:**72–82.

Muller, D. G., Kapp, M., and Knippers, R. (1996). Viruses in marine brown algae. *Adv. Virus Res.* **50:**49–67.

Murant, A. F. (1990). Dependence of groundnut rosette virus on its satellite RNA as well as on groundnut rosette assistor luteovirus for transmission by *Aphis craccivora*. *J. Gen. Virol.* **71:**2163–2166.

Murant, A. F., and Goold, R. A. (1968). Purification, properties and transmission of parsnip yellow fleck, a semi-persistent, aphid-borne virus. *Ann. Appl. Biol.* **62:**123–137.

Murant, A. F., and Lister, R. M. (1967). Seed-transmission in the ecology of nematode-borne viruses. *Ann. Appl. Biol.* **59:**63–76.

Murant, A. F., and Mayo, M. A. (1982). Satellites of plant viruses. *Ann. Rev. Phytopathol.* **20:**49–70.

Naylor, M., Godfray, H. C. J., Pallet, D. W., Tristem, M., Reeves, J. P., and Cooper, J. I. (2003). Mutualistic interactions amongst viruses? *In* "Genes in the Environment" (R. S. Hails, J. E. Beringer, and H. C. J. Godfray, eds.), pp. 205–225. Blackwells Scientific Press, Oxford.

Nee, S. (2000). Mutualism, parasitism and competition in the evolution of coviruses. *Phil. Trans. Roy. Soc. Lond. Ser. B.* **355:**1607–1613.

Nelson, M. R., Yoshgimura, M. A., and Tremaine, J. H. (1975). Saguaro cactus virus. *CMI/AAB Descr. Plant Viruses* **148:**4.

Nuss, D. L. (1996). Using hypoviruses to probe and perturb signal transduction processes underlying fungal pathogenesis. *Plant Cell* **8:**1846–1853.

Pallett, D. W., Thurston, M. I., Cortina-Borja, M., Edwards, M.-L., Alexander, M., Mitchell, E., Raybould, A. F., and Cooper, J. I. (2002). The incidence of viruses in wild *Brassica rapa ssp. sylvestris* in southern England. *Ann. Appl. Biol.* **141:**163–170.

Pathipanawat, W., Jones, R. A. C., and Sivasithamparam, K. (1996). Resistance to alfalfa mosaic virus in button medic (*Medicago oribicularis*). *Aust. J. Agric. Res.* **47:**1157–1167.

Peters, R. H. (1991). "A Critique for Ecology," p. 366. Cambridge University Press, Cambridge.

Piemeisel, R. L. (1954). Replacement and control; changes in vegetation in relation to control of pests and diseases. *Bot. Rev.* **20:**1–32.

Pimentel, D., McNair, S., Janecka, J., Wightman, J., Simmonds, C., O'Connell, C., Wong, E., Russel, L., Zern, J., Aquino, T., and Tsomondo, T. (2001). Economic and environmental threats of alien plant, animal, and microbe invasions. *Agric. Ecosyst. Environ.* **84:**1–20.

Pinon, J. (1979). Origine et principaux caracteres des souches francaises d'Hypoxylon mammatum. *Eur. J. Forest Pathol.* **9:**129–142.

Pirone, T. P., and Blanc, S. (1996). Helper-dependent vector transmission of plant viruses. *Annu. Rev. Phytopathol.* **34:**227–247.

Posnette, A. F., Robertson, N. F., and Todd, J. Mc A. (1950). Virus diseases of cacao in West Africa. V. Alternative host plants. *Ann. Appl. Biol.* **37:**229–241.

Pound, G. S. (1946). Control of virus diseases of cabbage seed plants in Western Washington by plant bed isolation. *Phytopathology* **36:**1035–1039.

Preisig, O., Moleleki, N., Smit, W. A., Wingfield, B. D., and Wingfield, M. J. (2000). A novel RNA mycovirus in a hypovirulent isolate of a plant pathogen *Diaporthe ambigua*. *J. Gen. Virol.* **81:**3107–3114.

Preston, C. D., Pearman, D. A., and Dines, T. D. (2002). "New Atlas of the British & Irish Flora," p. 910. Oxford University Press, Oxford, UK.

Querci, M., Owens, R. A., Bartolini, I., Lazarte, V., and Salazar, L. F. (1997). Evidence for heterologous encapsidation of potato spindle tuber viroid in particles of potato leaf roll virus. *J. Gen. Virol.* **78:**1207–1211.

Quiot, J. B., Marchoux, G., Douine, L., and Vigoroux, A. (1979). Ecologie et epidemiologie du virus de la mosaique du concombre dans le Sud-Est de la France. V. Rôle des especes spontanées dans la conservation du virus. *Ann. Phytopathol.* **11:**325–348.

Randles, J. W., Harrison, B. D., and Roberts, I. M. (1976). Nicotiana velutina mosaic virus: Purification, properties and affinities with other rod-shaped viruses. *Ann. Appl. Biol.* **84:**193–204.

Randles, J. W., Davies, C., Hatta, T., Gould, A. R., and Francki, R. I. B. (1981). Studies on encapsidated viroid-like RNA. I. Characterization of velvet tobacco mottle virus. *Virology* **108:**111–122.

Raybould, A. F. (2005). Assessing the environmental risks of transgenic volunteer weeds. *In* "Crop Ferality and Volunteerism" (J. Gressel, ed.), pp. 389–401. CRC Press, Boca Raton, Florida.

Raybould, A. F., and Cooper, J. I. (2005). Tiered tests to assess the environmental risk of fitness changes in hybrids between transgenic crops and wild relatives: The example of virus resistant *Brassica napus*. *Environ. Biosafety Res.***4:**127–140.

Raybould, A. F., and Gray, A. J. (1993). Genetically modified crops and hybridization with wild relatives: A UK perspective. *J. Appl. Ecol.* **30:**199–219.

Raybould, A. F., and Wilkinson, M. J. (2005). Assessing the environmental risks of gene flow from genetically modified crops to wild relatives. *In* "Gene Flow from GM Plants" (G. M. Poppy and M. J. Wilkinson, eds.), pp. 169–185. Blackwell Publishing, Oxford.

Raybould, A. F., Edwards, M.-L., Clarke, R. T., Pallett, D., and Cooper, J. I. (2000). Heritable variation for the control of turnip mosaic virus and cauliflower mosaic virus replication in wild cabbage. *Beiträge zur Züchtungsforschung – Bundesanstalt für Züchtungsforschung an Kulturpflanzen*. *In* "Proceedings of the 7th Aschersleben Symposium: New Aspects of Resistance Research on Cultivated Plants," 4–8.

Raybould, A. F., Alexander, M. J., Mitchell, E., Thurston, M. I., Pallett, D. W., Hunter, P., Walsh, J. A., Edwards, M.-L., Jones, A. M. E., Moyes, C. L., Gray, A. J., and Cooper, J. I. (2003). The ecology of *Turnip mosaic virus* in populations of wild *Brassica* species. *In* "Genes in the Environment" (R. S. Hails, J. E. Beringer, and H. C. J. Godfray, eds.), pp. 226–224. Blackwell Scientific Press, Oxford.

Rochow, W. F. (1970). Barley yellow dwarf virus: Phenotypic mixing and vector specificity. *Science* **7:**875–878.

Roosinck, M. J., Sleat, D., and Palukaitis, P. (1992). Satellite RNAs of plant viruses: Structures and biological effects. *Microbiol. Rev.* **56:**265–279.

Sakimura, K. (1953). Potato virus Y in Hawaii. *Phytopathology* **44:**217–218.

Sanger, M., Passmore, B., Falk, B. W., Bruening, G., Ding, B., and Lucas, W. J. (1994). Symptom severity of beet western yellows strain ST9 is conferred by the associated RNA and is not associated with virus release from the phloem. *Virology* **200:**48–55.

Scheffler, J. A., and Dale, P. J. (1994). Opportunities for gene transfer from transgenic oilseed rape (*Brassica napus*) to related species. *Transgenic Res.* **3:**263–278.

Severin, H. H. P. (1934). Weed host range and overwintering of curly top virus. *Hilgardia* **8:**263–280.

Severin, H. P. (1919). Investigations of the beet leafhopper (*Eutettix tenella* Baker) in California. *J. Econ. Ent.* **12:**312–326.

Shattuck, V. I. (1992). The biology, epidemiology and control of turnip mosaic virus. *Plant Breed. Rev.* **14:**199–238.

Shepherd, R. J., Fulton, J. P., and Wakeman, R. J. (1969). Properties of a virus causing pokeweed mosaic. *Phytopathology* **59:**219–222.

Shukla, D. D., Tosic, M., Jilka, J., Ford, R. E., Toler, R. W., and Langham, M. A. C. (1989). Taxonomy of potyviruses infecting maize, sorghum and sugar cane in Australia and the United States as determined by reactivities of polyclonal antibodies directed towards virus-specific N-termini of coat proteins. *Phytopathology* **79:**223–229.

Shukla, D. D., Ward, C. W., and Brunt, A. A. (1994). *The Potyviridae*. CAB International, Wallingford, Oxford. p. 516.

Silander, J. A. (1985). Microevolution in clonal plants. *In* "Population Biology and Evolution of Clonal Organisms" (J. B. C. Jackson, L. W. Buss, and R. E. Cook, eds.), pp. 107–152. Yale University Press, New Haven.

Simons, J. N. (1957). Effects of insecticides and physical barriers on field spread of pepper veinbanding mosaic virus. *Phytopathology* **47:**139–145.

Simons, J. N., Conover, R. A., and Walter, J. M. (1956). Correlation of occurrence of potato virus Y with areas of potato production in Florida. *Plant Dis. Reptr.* **40:**531–533.

Sinclair, W. A., Iuli, R. J., Dyer, R. J., Marshall, P. T., Matteoni, J. A., Hibben, C. R., Standosz, G. R., and Burns, B. S. (1990). Ash yellows; geographic range and association with decline of white ash. *Plant Dis.* **74:**604–607.

Skoric, D., Krajacic, M., and Stefanac, Z. (1997). Cucumovirus with a satellite-like RNA isolated from *Robinia pseudoacacia* L. *Periodicum Biologorum* **99:**125–128.

Skotnicki, A., Gibbs, A., and Wrigley, N. G. (1976). Further studies on *Chara corallina* virus. *Virology* **75:**457–468.

Skotnicki, M. L., Mackenzie, A. M., Torronen, M., Brunt, A. A., and Gibbs, A. J. (1992). Cardamine chlorotic fleck virus, a new carmovirus from the Australian alps. *Australasian Plant Pathol.* **21:**120–122.

Skotnicki, M. L., Selkirk, P. M., Kitajima, E., McBride, T. P., Shaw, J., and Mackenzie, A. (2003). The first subantartctic plant virus report: *Stilbocarpa mosaic bacilliform badnavirus* (SMBV) from Macquarie Island. *Polar Biol.* **26:**1–7.

Slykhuis, J. T. (1955). *Aceria tulipae* Kiefer (Acarnia: Eriophyidae) in relation to the spread of wheat streak mosaic. *Phytopathology* **45:**116–128.

Smith, K. M. (1972). "A Text Book of Plant Virus Diseases," 3rd Ed., p. 684. Longman Group Ltd., London.

Snyder, W. C., and Rich, S. (1942). Mosaic of celery caused by the virus of alfalfa mosaic. *Phytopathology* **32:**537–539.

Solomon-Blackburn, R. M., and Barker, H. (2001). A review of host major-gene resistance to potato viruses X, Y, A and V in potato: Genes, genetics and mapped locations. *Heredity* **86:**8–16.

Stevens, M., Patron, N. J., Dolby, C. A., Weekes, R., Hallswporth, P. B., Lemaire, O., and Smith, H. G. (2005a). Distribution and properties of geographically distinct isolates of sugar beet yellowing viruses. *Plant Pathol.* **54:**100–107.

Stevens, M., Freeman, B., Liu, H.-Y., Herrbach, E., and Lemaire, O. (2005b). Beet poleroviruses: Close friends or distant relatives? *Mol. Plant Pathol.* **6:**1–9.

Stevenson, W. R., and Hagedorn, D. J. (1973). Further studies on seed transmission of pea-seed-borne mosaic virus in *Pisum sativum. Plant Dis. Reptr.* **57:**248–252.

Stobbs, L. W., Van Driel, L., Whybourne, K., Carlson, C., Tulloch, M., and Van Lier, J. (2005). Distribution of plum pox virus in residential sites, commercial nurseries, and native plant species in the Niagra region, Ontario, Canada. *Plant Dis.* **89:**822–827.

Stone, W. J. H., and Nelson, M. R. (1966). Alfalfa mosaic (calico) of lettuce. *Plant Dis. Reptr.* **50:**629–631.

Strauss, E. E., Laksman, D. K., and Tavantzis, S. M. (2000). Molecular characterisation of the genome of a partitivirus from the basidiomycete *Rhizoctonia solani. J. Gen. Virol.* **81:**549–555.

Stubbs, L. L., Guy, J. A. D., and Stubbs, K. L. (1963). Control of lettuce necrotic yellows virus disease by the destruction of common sowthistle (*Sonchus oleraceus*). *Aust. J. Exp. Agr. Anim. Husb.* **3:**215–218.

Su, S. P., and Tsai, L. S. (1983). The resistance of Paulownia against witches' broom disease. *Q. J. Chin. For.* **16:**187–202.

Tarutani, K., Nagasaki, K., and Yamaguici, M. (2000). Viral impacts on total abundance and clonal composition of the harmful bloom-forming phytoplankton *Hetersigma akashiwo. Appl. Environ. Microbiol.* **66:**4916–4920.

Tavantzis, S. (ed.) (2002). "Molecular Biology of Double Stranded RNA: Concepts and Applications in Agriculture, Forestry and Medicine." CRC Press, Boca Raton, Florida.

Thackray, D. J., and Jones, R. A. C. (2003). Forecasting aphid outbreaks and epidemics of *Barley yellow dwarf virus*: A decision support system for a Mediterranean-type climate. *Australasian Plant Pathol.* **32:**438 (Abstr.).

Thackray, D. J., Diggle, A. J., Berlandier, F. A., and Jones, R. A. C. (2004). Forecasting aphid outbreaks and epidemics of *Cucumber mosaic virus* in lupin crops in a Mediterranean-type environment. *Virus Res.* **100:**67–82.

Thackray, D. J., Jones, R. A. C., Bwye, A. M., and Coutts, B. A. (2000). Further studies on the effects of insecticides on aphid vector numbers and spread of cucumber mosaic virus in narrow-leafed lupins (*Lupinus angustifolius*). *Crop Protec.* **19:**121–139.

Thackray, D. J., Smith, L. J., Cheng, Y., Perry, J. N., and Jones, R. A. C. (2002). Effect of strain-specific hypersensitive resistance on spatial patterns of virus spread. *Ann. Appl. Biol.* **141:**45–59.

Thomas, P. R. (1970). Host status of some plants for *Xiphinema diversicaudatum* (Micol.) and their susceptibility to viruses transmitted by this species. *Ann. Appl. Biol.* **65:**169–178.

Thottaphilly, G., van Lent, J. M. W., Rossel, H. W., and Sehgal, O. P. (1992). Rottboellia yellow mottle virus, a new sobemovirus affecting *Rottboellia cochinchinensis* (itch grass) in Nigeria. *Ann. Appl. Biol.* **120:**405–415.

Thresh, J. M. (1980a). An ecological approach to the epidemiology of plant virus diseases. *In* "Comparative Epidemiology" (J. Palti and J. Kranz, eds.), pp. 57–70. Pudoc, Wageningen.

Thresh, J. M. (1980b). The origins and epidemiology of some important plant virus diseases. *Appl. Biol.* **5:**1–65.

Thresh, J. M. (ed.) (1981). "Pests, Pathogens and Vegetation," p. 517. Pitman, Boston.

Thresh, J. M. (1982). Cropping practices and virus spread. *Annu. Rev. Phytopathol.* **20:**193–218.

Thresh, J. M. (2003). Control of plant virus diseases in sub-Saharan Africa: The possibility and feasibility of an integrated approach. *African Crop Sci. J.* **11:**199–223.

Thurston, M. I., Pallett, D. W., Cortina-Borja, M., Edwards, M.-L., Raybould, A. F., and Cooper, J. I. (2001). The incidence of viruses in wild *Brassica nigra* in Dorset (UK). *Ann. Appl. Biol.* **139:**277–284.

Tien, P., and Wu, G. (1991). Satellite RNA for biological control of plant disease. *Adv. Virus Dis.* **39:**321–339.

Tinsley, T. W. (1971). The ecology of cocoa viruses. 1. The role of wild hosts in the incidence of swollen shoot virus in West Africa. *J. Appl. Ecol.* **8:**491–495.

Todd, J. Mc A. (1951). An indigenous source of swollen shoot disease of cacao. *Nature (London)* **167:**952–953.

Toko, H. V., and Bruehl, G. W. (1959). Some host and vector relationships of strains of the barley yellow dwarf virus. *Phytopathology* **49:**343–347.

Tomlinson, J. A., Carter, A. L., Dale, W. T., and Simpson, C. J. (1970). Weed plants as sources of cucumber mosaic virus. *Ann. Appl. Biol.* **66:**11–16.

Valkonen, J. P. T. (1994). Natural genes and mechanisms for resistance to viruses in cultivated and wild potato species (*Solanum* spp.). *Plant Breeding* **112:**1–16.

Valkonen, J. P. T., Jones, R. A. C., Slack, S. A., and Watanabe, K. N. (1996). Resistance specificities to viruses in potato: Standardisation of nomenclature. *Plant Breeding* **115:**433–438.

Valverde, R. A. (1985). Spring beauty latent virus: A new member of the bromovirus group. *Phytopathology* **75:**395–398.

Van Dijk, P., and Bos, L. (1989). Survey and biological differentiation of viruses of wild and cultivated Umbelliferae of the Netherlands. *Neth. J. Plant Pathol.* **95** (Suppl. 2):1–34.

Van Etten, J. L., and Meints, R. H. (1999). Giant viruses infecting algae. *Annu. Rev. Microbiol.* **53:**447–494.

Van Etten, J. L., Lane, L. C., and Meints, R. H. (1991). Viruses and virus-like particles of eukaryotic algae. *Microbiol. Rev.* **55:**586–620.

Van Etten, J. L., Graves, M. V., Muller, D. G., Boland, W., and Delaroque, N. (2002). Phycodnaviridae-large DNA algal viruses. *Arch. Virol.* **147:**1479–1516.

Van Regenmortel, M. H. V. (1972). Wild cucumber mosaic virus. *CMI/AAB Descr. Plant Viruses*, No. 105, p. 4.

Vederevskaja, T. D., Kegler, H., Gruntzig, M., and Bauer, E. (1985). Zur Bedeutung von Unkrautern als Reservoire des Scharka-Virus (plum pox virus). *Archiv für Phytopathologie und Pflantzenschutz* **21:**409–410.

Wallis, R. L. (1967). Green peach aphids and the spread of beet western yellows in the Northwest. *J. Econ. Ent.* **60:**313–315.

Watson, L., and Gibbs, A. J. (1974). Taxonomic patterns in the host ranges of viruses among grasses, and suggestions on generic sampling for host range studies. *Ann. Appl. Biol.* **77:**23–32.

Watson, M. A., Hull, R., Blencowe, J. W., and Hamlyn, B. G. M. (1951). The spread of beet yellows and beet mosaic viruses in the sugar beet root crop. I. Field observations on the virus diseases of sugar beet and their vectors *Myzus persicae* Sulz. and *Aphis fabae* Koch. *Ann Appl. Biol.* **38:**743–764.

Watson, M. A., Heathcote, G. D., Lauckner, F. B., and Sowray, P. A. (1975). The use of weather data and counts of aphids in the field to predict the incidence of yellowing viruses of sugar-beet crops in England in relation to the use of insecticides. *Ann. Appl. Biol.* **81:**181–198.

Watson, M. T., and Falk, B. W. (1994). Ecological and epidemiological factors affecting carrot motley dwarf development in carrots grown in the Salinas Valley of California. *Plant Dis.* **78:**477–481.

Wei, C. Z., Osaki, H., Iwanami, T., Matsumoto, N., and Ohtsu, Y. (2003). Molecular characterization of dsRNA segments 2 and 5 and electron microscopy of a novel reovirus from a hypovirulent isolate, W370, of the plant pathogen *Rosellinia nacatrix*. *J. Gen. Virol.* **84:**2431–2437.

Wen, F., and Lister, R. M. (1991). Heterologous encapsidation in mixed infections among four isolates of barley yellow dwarf virus. *J. Gen. Virol.* **72:**2217–2223.

Werker, A. R., Dewar, A. M., and Harrington, R. (1998). Modelling the incidence of virus yellows in sugar beet in the UK in relation to numbers of migrating *Myzus persicae*. *J. Appl. Ecol.* **35:**811–818.

Wilkinson, M. J., Davenport, I. J., Charters, Y. M., Jones, A. E., Allainguillaume, J., Butler, H. T., Mason, D. C., and Raybould, A. F. (2000). A direct regional scale estimate of transgene movement from genetically modified oilseed rape to its wild progenitors. *Mol. Ecol.* **9:**983–991.

Wilkinson, M. J., Elliot, L. J., Allainguillaume, J., Shaw, M. W., Norris, C., Welters, R., Alexander, M., Sweet, J., and Mason, D. C. (2003). Hybridization between *Brassica napus* and *B. rapa* on a national scale in the United Kingdom. *Science* **302:**457–459.

Wilson, W. H., Tarran, G. A., Schroeder, D., Cox, M., Oke, J., and Malin, G. (2002). Isolation of viruses responsible for the demise of an *Emiliania huxleyi* bloom in the English Channel. *J. Mar. Biol. Ass. UK* **82:**369–377.

Yahara, T., and Oyama, K. (1993). Effects of virus infection on demographic traits of an agamosoermous population of *Eupatorium chinense* (Asteraceae). *Oecologia* **96:**310–315.

Yarwood, C. E. (1959). Virus increase in seedling roots. *Phytopathology* **49:**220–223.

ADVANCES IN VIRUS RESEARCH, VOL 67

GENETIC DIVERSITY OF PLANT VIRUS POPULATIONS: TOWARDS HYPOTHESIS TESTING IN MOLECULAR EPIDEMIOLOGY

B. Moury, C. Desbiez, M. Jacquemond, and H. Lecoq

INRA Avignon, Station de Pathologie Végétale, Domaine St Maurice
BP94 84143 Montfavet cedex, France

Over the last two decades, analysis of the variation in the genome of plant viruses has revealed a level of diversity previously unsuspected. Using molecular techniques, new groups of viruses have been described, including recombinant or reassortant viruses. Characterization of the molecular diversity within plant virus populations has also been the aim of an increasing number of studies. However, to date, most of these studies described plant virus populations and did not analyse their behaviour. More precise quantitative estimates have been obtained of population dynamics and genetics parameters, such as genetic drift and positive selection, for some plant viruses. These kinds of approaches may allow tests of epidemiological hypotheses concerning the respective roles of plant hosts, vectors, physical environment or agricultural practices on virus populations and to identify which of these factors may have major influences on epidemics and the emergence of virus variants.

0065-3527/06 $35.00
DOI: 10.1016/S0065-3527(06)67002-4

I. Introduction

Viruses share several properties that distinguish them from other living entities. Their mutation rates (i.e. the number of misincorporations per nucleotide in their genome and per round of replication) are 10^4–10^5 times larger than for prokaryotes or eukaryotes (Drake and Holland, 1999). Also, their population size (the actual number of virus particles in infected plants or animals) is several orders of magnitude larger than that of any other micro- or macroorganism and their generation times are short. For example, a single tobacco leaf could contain 10^{11}–10^{12} particles of *Tobacco mosaic virus* (TMV) (Harrison, 1956; Malpica *et al.*, 2002). A consequence of these characteristics is that it is likely to be quite complicated to describe adequately the diversity within virus populations. Variation among virus isolates was initially revealed by differences in biological or structural properties such as their ability to infect particular plant species or genotypes, to induce particular symptoms, to be transmitted by particular vectors or to interact with specific antibodies. However, it was shown that these traits are usually related to a very limited number of nucleotide changes in the virus genome (Blanc *et al.*, 1998; Harrison, 2002; Hébrard *et al.*, 2005; Perry *et al.*, 1998) and only rarely allow quantitative measures of the virus genetic variability. In addition, most of these traits are not neutral in regard to the evolution and dynamics of virus populations, but often are a target of selection by host or vector and consequently not always suitable to analyse the structure of these populations. The development of simpler and cheaper molecular techniques allowing the characterization of DNA or RNA virus genomes has partially circumvented these major obstacles. From the taxonomic viewpoint, this has led to a large increase in the number of plant virus species and genera that have been distinguished and also to the establishment of quantitative criteria to delimit different species (Adams *et al.*, 2004, 2005; Fauquet and Stanley, 2005; van Regenmortel *et al.*, 2000). In contrast, a better knowledge of virus genetic diversity and the development of molecular tools allowing the characterization and accurate tracing of virus populations could provide a major contribution to plant virus epidemiology. Molecular epidemiology is already well developed for animal and human viruses (Vandamme, 2002), while it still remains an almost unexplored field for plant viruses. It could provide more accurate answers to important questions such as where virus infections come from, how they spread and the nature of the major forces that drive their evolution. It could help to reassess, for instance, the actual role of reservoirs in local epidemics and the relative importance of local and long distance virus spread. It would also help to gain a better understanding of the interactions between different virus populations, when spreading in a cultivated crop. This chapter presents results obtained on plant virus diversity and population genetics. We illustrate how the analysis of the patterns of genome variation could be a 'window' through which to measure the role of the different evolutionary forces that shape plant virus populations. We also emphasize that analysis of the representativeness of samples could help compare the diversity in different plant virus populations. Throughout this chapter, simple approaches are presented that can be used to assess statistically the role of different mechanisms affecting plant virus populations.

II. Methods of Measuring the Diversity of Plant Virus Populations

Analysis of the diversity of plant virus populations has been performed using different technical approaches, applied to samples collected via different strategies and of widely different sizes. The analytical technique, the choice of the genome or protein target and the sampling strategy should be adapted to the particular goals of each study. The diversity of populations can be estimated by different criteria (García-Arenal *et al.*, 2001; Nei, 1987), which provide different levels of information: (1) the number of different haplotypes present in the population (richness), (2) the frequency of each haplotype in the population and (3) the genetic distance within and between haplotypes. A haplotype will be defined as a group of virus isolates which possess a particular allele at a locus (or a particular set of alleles at separate loci). An allele is most often determined by a particular profile of DNA fragments visualized through electrophoresis, or by genomic sequences that exhibit only few nucleotide changes (see Section II.A). Ideally, an accurate estimation of these three parameters would be required to provide a precise view of the viral population and each of them depends to different extents on the quality of the method used to determine haplotypes and on the sampling strategy (see Morris *et al.*, 2002, for a critical review of studies of the microbial diversity, excluding viruses).

Different techniques have been applied to characterize plant virus populations, mainly monoclonal antibodies, restriction fragment length polymorphism (RFLP) (Vigne *et al.*, 2004), single-strand conformation polymorphism (SSCP) (d'Urso *et al.*, 2003; Lin *et al.*, 2004), ribonuclease protection assay of a labelled cRNA probe (RPA) (Bonnet *et al.*, 2005; Fraile *et al.*, 1996, 1997), heteroduplex mobility assay (HMA) (Berry *et al.*, 2001; Lin *et al.*, 2000), ribonuclease T1 fingerprint (Rodríguez-Cerezo *et al.*, 1989) and nucleotide sequence analyses of genome fragments or whole genomes (Tomimura *et al.*, 2003). Different sampling strategies have been applied to plant virus populations to determine their variability. To reach valid conclusions about populations by induction from samples, statistical procedures typically assume that the samples are obtained in a random fashion. Many sampling strategies in plant virology focus on symptomatic plants, which could induce a bias in the representativeness of the sampled population. Very few publications mention either hierarchical sampling (d'Urso *et al.*, 2003) or collection of samples along a transect (Colvin *et al.*, 2004; Legg and Ogwal, 1998).

A. *Number of Haplotypes*

Estimating the number of haplotypes in populations requires that the analytical technique used to characterize the virus is sufficiently discriminating, that is, it can reveal sufficient polymorphism and that the target genome region contains enough informative sites. The discrimination power differs across techniques. For RFLP, SSCP, RPA or HMA, the number of haplotypes can usually be defined as the number of different profiles of DNA fragments visualized through electrophoresis. For nucleotide sequences of parts of the virus genome, or ribonuclease T1 fingerprints, which provide a high-discrimination power, the definition of haplotypes is less trivial since almost all samples have different sequences, provided that a relatively large region of the genome is examined. Consequently, the total number of haplotypes tends to increase linearly with the number of samples and this number is, therefore, insufficiently informative to compare populations. An alternative approach is to build a sequence alignment, to calculate pairwise nucleotide diversity between them and to group in the same haplotype samples that differ by less than a nucleotide identity threshold (1–5%, given that the 10% threshold is the species boundary for many plant viruses) (Fauquet and Stanley, 2005; van Regenmortel *et al.*, 2000). Ideally, the chosen threshold should ensure that the between-haplotype distances are larger than those within haplotype.

Among these haplotypes, the occurrence of recombinant or reassortant viruses can be revealed. Recombination is the process by which segments of genetic information are exchanged between the nucleotide strands of different genetic variants during replication, whereas reassortment is the genetic exchange of entire genomic molecules between variants of viruses that have a segmented genome, illustrated by antigenic shift in *Human influenza A virus* in animals or *Cucumber mosaic virus* (CMV) in plants (Fraile *et al.*, 1997). Sequence data are particularly powerful to detect the location of recombination sites and to attribute a confidence value to the recombinant or reassortant status of a virus sample. Many methods and diverse softwares have been developed for these purposes (Posada and Crandall, 2001). In some cases, virus recombinant artifacts could be revealed when plants are infected simultaneously with several isolates of a virus or different closely related viruses. *In vitro* recombination during reverse transcription-polymerase chain reaction (RT-PCR) amplification is well documented and can be caused either by the reverse transcriptase during cDNA synthesis (Negroni and Buc, 2001) or the *Taq* polymerase (Bradley and Hillis, 1997). The recombinants generated *in vitro* between two cucumoviruses or two strains of CMV are the same as those identified in co-infected plants (Fernandez-Delmond *et al.*, 2004). Also, selective amplification of one or the other virus variants, depending on the oligonucleotide primers used for each genomic segment, cannot be excluded (Alma Bracho *et al.*, 2004). If virus recombinants or reassortants are confirmed as new strains or species in independent samples, they can be useful to trace epidemics (Glasa *et al.*, 2004), or to understand the biological importance of recombination and reassortment (Bonnet *et al.*, 2005; Desbiez and Lecoq, 2004; Fraile *et al.*, 1997).

There have been very few attempts to estimate the actual number of haplotypes in viral populations. This problem relates to the analysis of the representativeness or exhaustivity of sampling. To our knowledge, among the publications presenting information concerning plant virus diversity, none has estimated the representativeness of their samples relative to the overall population, while this is more commonplace for diversity studies of macroorganisms. The utility of various statistical approaches for assessing the richness of microbial populations for which the true diversity is unknown has been reviewed (Hughes *et al.*, 2001). The relationship between the number of haplotypes observed and sampling effort gives information about the total diversity of the sampled population. This pattern can be visualized by 'accumulation curves', where the cumulative number of observed haplotypes is plotted against sample size (Fig. 1). Different populations show different shapes of curves, depending on differences in relative abundances of haplotypes in the sampled population. The downward concavity of the curve increases with the extent of sampling of the population and curves eventually reach a horizontal asymptote at the actual population richness. Comparing richness and sampling effort between studies and even between different treatments within a study is rendered difficult by the fact that the different treatments have usually been sampled unequally. Certain statistical approaches allow comparisons when sample sizes are unequal. The most promising approaches are: (1) rarefaction and (2) non-parametric richness estimators (Hughes *et al.*, 2001). Rarefaction curves result from averaging randomizations of the observed accumulation curves (Heck *et al.*, 1975) and allow comparison of the observed richness among samples through the calculation of the variance around the repeated randomizations. They cannot, however, provide measures of confidence concerning the actual richness of the population. In contrast, non-parametric richness estimators provide estimates of, and confidence intervals around, the extrapolated actual haplotype richness. Several of these estimators were adapted from mark–release–recapture statistics for estimating the size of animal populations. These non-parametric estimators consider the proportion of species that have been observed before ('recaptured') to those that are observed only once (Colwell and Coddington, 1994). In a very diverse population, the probability that a haplotype will be observed more than once will be low and most haplotypes will be represented by a single individual in a sample. In a less diverse population, the probability that a haplotype will be observed more than once will be higher and many haplotypes will be observed several times in a sample. The Chao1 estimator, for example, estimates haplotype richness as

$$\mathrm{S}_{\mathrm{Chao1}} = \mathrm{S}_{\mathrm{obs}} + n_1^2/(2 \times n_2) \text{ if } n_2 \neq 0;$$
$$\mathrm{S}_{\mathrm{Chao1}} = \mathrm{S}_{\mathrm{obs}} + n_1 \times (n_1 - 1)/2 \text{ if } n_2 = 0,$$

where $\mathrm{S}_{\mathrm{obs}}$ is the number of observed haplotypes, n_1 is the number of singletons (haplotypes observed once) and n_2 is the number of doubletons (haplotypes observed twice) (Chao, 1984). Other non-parametric estimators are described in Colwell and Coddington (1994). The precision of Chao1, that is, the variance of richness estimate to be expected from multiple samples, can be estimated by bootstrap simulation (Chao, 1984), or by a closed-form solution for the variance of $\mathrm{S}_{\mathrm{Chao1}}$ (Chao, 1987), which allows calculation of confidence intervals (CIs) around $\mathrm{S}_{\mathrm{Chao1}}$ and comparisons of predictions of total richness between samples. To investigate the potential of these approaches, we applied them to different plant virus data sets. The first set was taken from Vigne *et al.* (2004) and consisted of samples of grapevine infected by *Grapevine fanleaf virus* (GFLV), a virus transmitted by soil-borne nematodes, that had been collected from plants grafted onto transgenic or non-transgenic rootstocks. The initial aim of this study was to investigate a putative role of the transgenes in contributing to recombinant viruses, but the data are used to compare the diversity of viruses in both kinds of plants. Nine virus haplotypes were observed in the transgenic plants versus eight haplotypes in non-transgenic ones, which were sampled less exhaustively. Rarefaction curves (Fig. 1) presented quite a different shape, suggesting that there would be a high probability that newly collected non-transgenic grapevine samples would fall into RFLP groups not described, whereas the nine groups described for the collection of 117 transgenic grapevine samples represent a value much closer to the asymptote. This was also supported by the Chao1 estimate of actual numbers of haplotypes (9.0 for transgenic vs 11.0 for non-transgenic grapevines, Table I). However, the 95% CIs of both estimates overlap slightly and it cannot be concluded that there is a significant difference. If the number of samples collected from non-transgenic grapevines in the close vicinity of the experimental field plots ('outsiders') is added, the estimated diversity of GFLV collected from non-transgenic grapevines increases slightly (S_{Chao1} = 15.5) and the 95% CIs between transgenic and

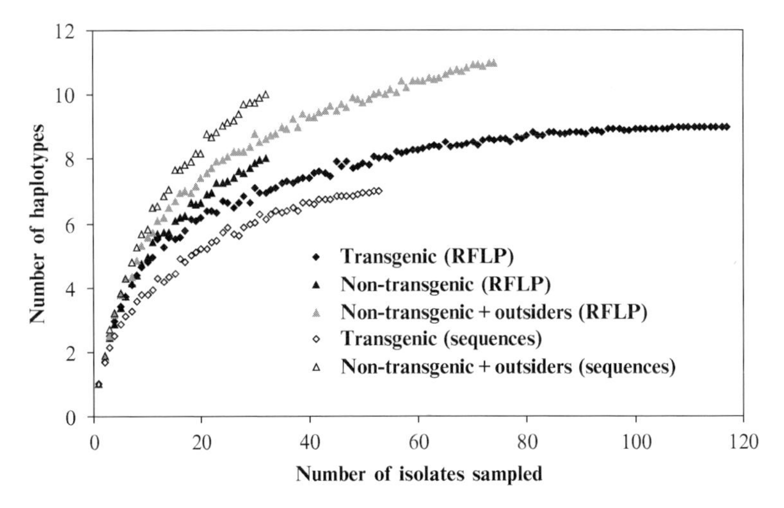

FIG 1. Observed haplotype richness of *Grapevine fanleaf virus* samples collected in fields of transgenic or non-transgenic grapevine versus sample size (data from Vigne *et al.*, 2004). The number of haplotypes observed for a given sample size (accumulation curve) was averaged over 100 simulations. Diversity was revealed either from RFLP or sequence analyses of the coat-protein coding region (haplotypes were groups of sequences distant by less than 3.6% nucleotide difference) and was also studied for non-transgenic grapevines near experimental fields (outsiders).

TABLE I

ESTIMATIONS OF PLANT VIRUS POPULATION RICHNESS (ACTUAL NUMBER OF HAPLOTYPES) IN THREE CASE STUDIES

Publication/ virus	Virus population	Genome region	Analytical method	N[a]	n[b]	S_{Chao1}[c]	$CI_{95\%}$[d]
Vigne *et al.*, 2004/ GFLV	Transgenic grapevines	CP	RFLP	117	9	9.0	*6.0–10.0*[e]
	Non-transgenic grapevines	CP	RFLP	32	8	11.0	8.4–32.9
	Non-transgenic grapevines + outsiders	CP	RFLP	74	11	15.5	*11.5–51.9*
	Transgenic grapevines	CP	Sequence	53	7	7.3	7.0–11.7
	Non-transgenic grapevines + outsiders	CP	Sequence	32	10	11.5	10.2–22.9
Ooi and Yahara, 1999/ TLCV	Asexual *Eupatorium makinoi*	C4	SSCP	36	3	3.0	*3.0–3.0*[e]
	Sexual *E. glehni*	C4	SSCP	42	6	7.0	*6.1–19.7*
Lin *et al.*, 2004/CMV	α[f]	2b	SSCP	63	6	8.0	6.2–28.1
	β	2b	SSCP	18	11	19.2	12.6–53.4
	α	MP	SSCP	63	12	24.3	14.2–80.2
	β	MP	SSCP	18	12	52.5	18.3–273.1
	α	3′ NTR of RNA 3	SSCP	63	7	11.5	7.5–47.9
	β	3′ NTR of RNA 3	SSCP	18	11	27.0	14.0–95.8

[a] Number of samples.

[b] Observed number of haplotypes. The 12th RFLP group in Vigne *et al.* (2004) that is composed of variable profiles due to mixed infections was omitted.

[c] Non-parametric estimator of actual richness of the population (Chao, 1984). Other parametric and non-parametric richness estimators, as implemented in the SPADE software (Chao and Shen, 2003), lead to similar results (data not shown).

[d] 95% confidence interval around S_{Chao1} obtained with the SPADE software. $CIs_{95\%}$, which do not overlap between treatments of an experiment, suggesting significant difference between their actual diversity, are in italics.

[e] The variance formula in Chao (1987) was not valid in these cases and so the bootstrap method (1000 simulations with the R software; Ihaka and Gentleman, 1996) was used to estimate $CI_{95\%}$ (Chao, 1984).

[f] α and β populations correspond to isolates collected from cucurbit plants in two growing seasons and in a single location, and to isolates collected from various hosts over a decade in different California locations, respectively.

Abbreviations: GFLV, *Grapevine fanleaf virus*; CMV, *Cucumber mosaic virus*; TLCV, *Tobacco leaf curl virus*; CP, coat protein; MP, movement protein; NTR, non-translated region; RFLP, restriction fragment length polymorphism; SSCP, single-strand conformation polymorphism.

non-transgenic plants no longer overlap (Table I). This indicates a higher diversity of viruses collected from non-transgenic grapevines. This could be due to the impact of the resistance of some transgenic grapevine genotypes towards some GFLV variants. Selection of virulent variants by these transgenic grapevines could have reduced the richness in the virus population, a hypothesis that is worth testing. It is notable that richness estimators obtained from sequence data of the entire coat-protein (CP) coding region, where haplotypes were defined as groups in which sequences differed by less than 3.6% (Vigne *et al.*, 2004), were very close to the previous ones, despite the fact that sequences represented a much higher information level than RFLPs (RFLPs covered 0.4–3.0% of the CP cistron (Vigne *et al.*, 2004)) and the samples analysed with the different methods were largely independent. Another interesting issue is to estimate the sample size that would be necessary and sufficient to detect a significant difference between the richness estimates of transgenic versus non-transgenic grapevines. The size of the 95% CIs could be extrapolated for larger sample sizes in the case of transgenic or non-transgenic plants including outsiders (Fig. 2, RFLP data). For non-transgenic plants excluding outsiders, no extrapolation was possible due to insufficient sampling (data not shown). A both curve fits well with the CIs of S_{Chao1} estimates for both samples and reveals that a total of *c.* 200 samples would be necessary to distinguish significantly two populations with Chao1 estimates that differ by two units, as do the transgenic and non-transgenic populations.

A second data set we used to explore statistical comparisons of diversity was that of Ooi and Yahara (1999). They collected *Tobacco leaf curl virus* from two populations of *Eupatorium* spp., which differed in their reproduction regimes. By SSCP analysis of a similar number of samples, they observed six versus three haplotypes in the two different plant populations (Table I). From this and additional data, they concluded that a higher virus diversity existed in the first population. Analysis of this distribution by non-parametric richness estimators (Table I) corroborate their assumptions since 95% CIs of these estimators in the two groups do not overlap.

The last data set we examined consisted of CMV isolates collected in two different virus populations: (1) group α contained isolates collected from cucurbit plants in two growing seasons and in a single location in California and (2) group β contained isolates collected from various hosts over a decade in different California locations (Lin *et al.*, 2004). Since about three times more α than β samples were characterized and similar numbers of SSCP haplotypes were observed (Table I), direct

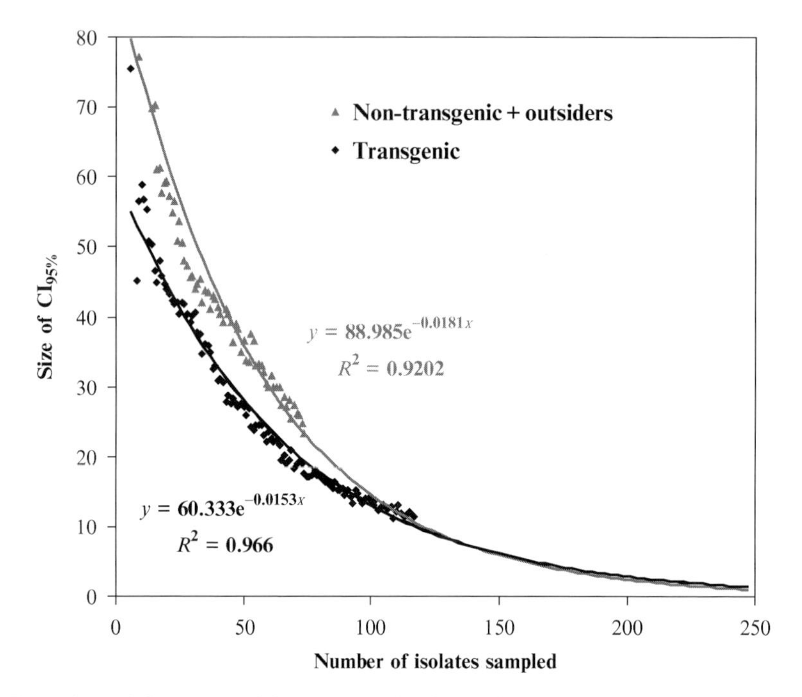

FIG 2. Size of the 95% confidence intervals (CIs) of Chao1 estimate (upper limit of the CI-lower limit of the CI) (Chao, 1984) for *Grapevine fanleaf virus* samples collected in transgenic or non-transgenic grapevines, as sample size increases (Vigne *et al.*, 2004; RFLP data). Values are averages of 1000 simulations using the R software (Ihaka and Gentleman, 1996).

richness comparisons are difficult. Because of the different geographic regions and timespans represented for the β population, a larger number of haplotypes is to be expected in this population. Non-parametric richness estimators suggest that two to three times more haplotypes could exist in the β population, but 95% CIs are largely overlapping. These large CIs are probably due to the many singletons in the populations and illustrate the difficulties in selecting an appropriate sampling size to analyse plant virus populations.

B. *Frequency Distribution of Haplotypes*

To estimate the frequency of haplotypes in a population requires a more exhaustive and accurate sampling than simply estimating the number of haplotypes, since it should represent the entire population with limited biases. The frequency of haplotypes relates to the structure of populations. The definition of structure adopted here is that of Morris *et al.* (2002), that is, a quantitative measure of the relative abundance of the different haplotypes characterized that ideally includes an estimation of the variance associated with that structure. An estimate of the structure of plant virus populations facilitates more fundamental studies concerning the dynamics of these populations and the impact of external factors on them.

Distributions of haplotypes between plant virus populations are most often compared by contingency χ^2 tests (Arboleda and Azzam, 2000; Azzam *et al.*, 2000; Bonnet *et al.*, 2005; Fraile *et al.*, 1997; McNeil *et al.*, 1996). Such analyses have revealed significant differentiation among plant virus populations according to situation (host plants, time, geographical locations, ecological or agricultural conditions) (García-Arenal *et al.*, 2001). However, in χ^2 analyses of contingency tables, it is recommended that expected frequencies should exceed 1.0 and no more than 20% be less than 5.0 (Cochran, 1954; Roscoe and Byars, 1971). Alternatively, Yates' continuity correction for small samples (Yates, 1934) or other tests (Mantel, 1963; Woolf, 1955) should be employed. If a 2×2 contingency table has insufficiently large frequencies for a χ^2 analysis, then the Fisher exact test (Fisher, 1958) is an advisable alternative. These basic considerations are useful for many virus data sets comprising many singletons or doubletons and departing from required hypotheses for χ^2 analyses. Monte Carlo simulations can be used to compare estimates of exact confidence intervals around distributions.

When multiple areas are sampled, haplotype distributions between populations can be compared with the F statistic (d'Urso *et al.*, 2003; Weir and Cockerham, 1984) (see Section IV.B).

C. Within- and Between-Population Genetic Distances

Quantitative estimates of the diversity between or within haplotypes can be obtained from RFLP, ribonuclease T1 fingerprint and from nucleotide sequence analyses. HMA, RPA and SSCP analyses, in contrast, do not allow direct measurement of genetic distances since the variation detected by these methods is not necessarily correlated with the number of nucleotide differences. For SSCP and HMA, the sensitivity of the methods depends on the size of the analysed fragment and, sometimes, on the location of the mutations in the fragment (Sheffield *et al.*, 1993). However, sequencing of cDNA fragments that share the same SSCP profiles showed that their genetic diversity was low (Desbiez, C., and Lecoq, H., unpublished data; Rubio *et al.*, 2001, Vives *et al.*, 2002). Two statistical parameters are used frequently to quantify the diversity at the nucleotide level: (1) the mean pairwise nucleotide differences (summary statistic π) and (2) the number of polymorphic sites (summary statistic θ) among sequences of the sampled isolates (Nei, 1987). These estimates are the basis of Tajima's D test of neutrality (Tajima, 1989), which considers that under the neutral mutation model and in a panmictic population $\pi = \theta$. Nucleotide diversities can also be partitioned in non-synonymous (amino acid changing) and synonymous (silent) sites (Li, 1993; Pamilo and Bianchi, 1993), providing an estimate of the selective pressure on the target genome region (see section IV.B) and being the basis of the McDonald and Kreitman (1991) test for positive selection, which predicts that under neutral evolution the ratio of non-synonymous to synonymous changes estimated within a species does not differ significantly from that between two species. Distance estimates between haplotypes can differ markedly according to the genome region examined (Aleman-Verdaguer *et al.*, 1997; García-Arenal *et al.*, 2001). Particular genome regions of plant viruses can exhibit extremely low variation (Pfosser and Baumann, 2002; Schneider and Roossinck, 2001), which can give the false impression of a lack of variation of the virus population. When no evaluation of the representativeness of the genome region under study is available, it is advisable to examine several regions to account for the genetic distances among virus populations. Estimation of distances between plant virus populations has notably clarified the taxonomy and classification of viruses by comparing more precisely the intra- and inter-species variation (Adams et al., 2004; Fauquet and Stanley, 2005; Shukla and Ward, 1989).

Secondary or composite indices can also be defined, which confound richness and the relative abundance and/or identity among haplotypes, and can be compared through statistical tests (Nei and Tajima, 1981; Shannon and Weaver, 1949; Simpson, 1949).

III. Observed Diversity and Structure of Plant Virus Populations

Plant virus population diversity and structure have been reviewed by García-Arenal *et al.* (2001, 2003) and only major trends will be summarized here. The main characteristic of the diversity observed among plant virus populations is a great stability over time (García-Arenal *et al.*, 2001), as examplified with the few nucleotide differences shown between *Tobacco mild green mosaic virus* or *Wound tumor virus* populations collected in the same area over long time periods (up to 100 years), or between geographically isolated populations that diverged 13,000–14,000 years ago (Keese *et al.*, 1989). This contrasts with some fast-evolving animal viruses such as *Human influenza A virus, Hepatitis C virus, Foot-and-mouth disease virus* or *Human immunodeficiency virus 1* (Gorman *et al.*, 1990; Perelson *et al.*, 1996). This difference could be attributed to the selection pressure on animal viruses imposed by the immune system of the host. Other animal viruses, nevertheless, show a greater stability (Grenfell *et al.*, 2004; Jenkins *et al.*, 2002). Estimates of the pairwise nucleotide sequence diversity among plant virus populations are usually low (usually $\pi < 0.20$; Chare and Holmes, 2004; García-Arenal *et al.*, 2001). Some plant viruses, however, seem to show a higher level of polymorphism (Teycheney *et al.*, 2005). As emphasized by García-Arenal *et al.* (2003), there is no evident correlation between plant virus diversity and any particular characteristic of the plant virus biology or life cycle.

The most common composition of plant virus populations is a differing number of haplotypes, separated only by small genetic distances that exhibit an L-shaped rank-abundance curve, that is, one or a few major haplotypes plus many minor haplotypes (García-Arenal *et al.*, 2001). This abundance curve shape is observed classically for micro- or macroorganisms. A consequence of such distributions is that, except when the major haplotypes are not the same between populations (Fraile *et al.*, 1997; Ooi and Yahara, 1999; Skotnicki *et al.*, 1996), it could be difficult to reveal significant differences between haplotype frequencies in different populations, whereas richness differences could be revealed (see section II.A for GFLV). In addition, direct comparison of haplotype distribution between populations could be difficult statistically due to the numerous singletons or doubletons in the population.

Analysis of the structure of plant virus populations requires estimating and comparing their diversity at different levels, for example, at different spatial scales (plant, field, region), in different hosts and different years. Hierarchical sampling or randomized complete block studies could be extremely valuable for this purpose and to study factors affecting plant virus differentiation, but this is rarely used. When multiple factors vary (plant genotype, geographical locations), the influence of these factors can be assessed by hierarchically partitioning the sampled population among these factors and determining the contribution of each separate factor to the observed diversity (d'Urso *et al.*, 2003).

Differentiation of plant virus populations according to various factors was shown in a few cases (García-Arenal *et al.*, 2001). Viruses which spread only over short distances by contact between plants, soil-borne vectors or air-borne vectors of limited mobility may be expected to exhibit a geographically differentiated structure, as has been shown in some cases (Fargette *et al.*, 2004; Fraile *et al.*, 1996). In contrast, viruses transmitted efficiently over long distances would be expected to be less spatially structured, as with many insect-borne viruses (Albiach-Marti *et al.*, 2000; Azzam *et al.*, 2000; Desbiez *et al.*, 2002; McNeil *et al.*, 1996).

IV. Inferring the Processes that Determine the Genetic Structure and Evolution of Virus Populations

Inferring virus population dynamic processes (mutation, recombination, genetic drift, natural selection, population growth or decline, migration) from diversity studies should be possible in the frame of classical population genetics (Moya *et al.*, 2000). This is essential to understand the role of the various internal or external factors acting on virus populations, but is rendered difficult by confusion between the different forces operating. However, more or less simplified models allow estimation of some of these population genetics parameters. The diversity observed in plant virus populations could be compared to that expected under the different models, allowing estimation of the occurrence and intensity of different evolutionary forces (d'Urso *et al.*, 2003; Moya *et al.*, 1993; Stenger *et al.*, 2002).

A. Mutation Rates, Recombination Frequencies and Genetic Drift

Mutation rates, recombination frequencies and genetic drift are difficult to estimate from natural diversity studies since it is not easy to estimate them independently of each other and independently of the action of natural selection. Thus, they were evaluated preferentially in experiments where the environment and initial virus population were controlled (but see Moya *et al.*, 1993). There is very little information on mutation rates or recombination frequencies in plant viruses, in part due to the difficulties of estimating generation times. Spontaneous mutation rates of TMV under minimum selection from the host plant were estimated at 0.10–0.13 mutations per genome per replication cycle, a value similar to that of animal viruses (Malpica *et al.*, 2002). An experimental estimation of plant virus recombination rates was made for *Cauliflower mosaic virus*, a DNA virus, during the systemic invasion of the host. It was estimated at 2×10^{-5} to 4×10^{-4} recombinations per nucleotide per replication cycle, which is a relatively high value (Froissart *et al.*, 2005). Frequent recombination was also observed for *Brome mosaic virus*, an RNA virus, in small groups of cells around local hypersensitive lesions (Bruyere *et al.*, 2000).

Genetic drift occurs at any step where molecules or pathways necessary for the virus life cycle are in insufficient quantities with regard to the total virus population. In theory, genetic drift can be associated with virus replication, accumulation, cell-to-cell or systemic movement and plant-to-plant dissemination. The effects of drift and selection can be difficult to distinguish, since both occur at the same steps in the virus life cycle and their effect on population diversity can be similar. Both phenomena usually reduce the within-population diversity, but can increase that between populations. However, the effects of drift are stochastic and unpredictable individually, whereas the effects of selection should be reproducible for identical environments and initial composition of the virus population. Genetic drift during plant invasion was estimated independently for three unrelated RNA viruses infecting different hosts and similar results were obtained for the different systems (French and Stenger, 2003; Li and Roossinck, 2004; Sacristán *et al.*, 2003). The effective population size, that is, the number of individual viruses giving rise to the entire virus population in the plant, was estimated to be about 10 (French and Stenger, 2003; Sacristán *et al.*, 2003), which indicates the possibility of substantial genetic drift, when compared to the size of entire virus populations. Genetic drift is also probably intense during plant-to-plant transmission of viruses by vectors, but no precise estimates are available yet.

B. Positive Selection

Experimental evidence for selection was obtained from consistent selection of particular variants from heterogeneous viral populations (Ayme *et al.*, 2006; Chen *et al.*, 2002; Desbiez *et al.*, 2003; Kühne *et al.*, 2003; Liang *et al.*, 2002). Analytical evidence was also obtained from: (1) consistent differentiation of plant virus populations according to the environment (host plant, geographical locations...) (Bonnet *et al.*, 2005; d'Urso *et al.*, 2003; Mastari *et al.*, 1998; Skotnicki *et al.*, 1996), (2) comparison of observed haplotype distributions with those expected under neutrality (d'Urso *et al.*, 2003; Moya *et al.*, 1993) and analysis of nucleotide and amino acid substitutions from sequence alignments, either from (3) covariation patterns (Altschuh *et al.*, 1987; Liang *et al.*, 2002;

Simmonds and Smith, 1999) or (4) measures of amino acid replacement rates (Hurst, 2002; Moury, 2004; Moury *et al.*, 2002; Schirmer *et al.*, 2005; Yang and Bielawski, 2000). Quantitative measures of natural selection acting on plant viruses were obtained mostly by method (4).

The observed haplotype distribution can be compared to that expected for a strictly neutrally evolving population (Tajima, 1989), or models including other evolutionary forces such as selection (Otto, 2000). The observed haplotype distribution also allows calculation of the average heterozygosity of the population (i.e. the probability that two individual randomly-selected samples belong to different haplotypes). This observed heterozygosity value can also be compared to the expected heterozygosity in a model of neutral evolution (Watterson, 1978). Both methods have been applied only occasionally to the study of plant virus populations (d'Urso *et al.*, 2003; Stenger *et al.*, 2002). It was suggested that polymorphism-based neutrality tests that evaluate single locus are extremely sensitive to demographic events such as population expansion (Nielsen, 2001; Przeworski, 2002). Consequently, it is recommended that nucleotide variability should be screened at multiple loci in order to disentangle the effects of demography and selection. Ideally, these loci should be unlinked to avoid genetic 'hitch-hiking' effects on polymorphism, since in this situation the effects of selection are generally expected to be locus-specific, whereas demographic processes will have relatively uniform effects across the genome. Plant viruses possess small genomes and largely unknown patterns and frequencies of recombination and genetic hitch-hiking effects of selection on the polymorphism present on the whole genome can be expected. However, d'Urso *et al.* (2003) examined the heterozygosity values at two loci in the genome of *Citrus triteza virus* (CTV), namely the A and p20 segments, and showed that one of them departed more than the other from neutral expectations, suggesting selection acting on the former (A segment). With spatially separated populations, it is also possible to identify genome regions involved in adaptation by comparing relative levels of differentiation among multiple loci. A selective sweep will reduce the within-population diversity at the locally adaptive allele and increase the between-population diversity by selection against locally deleterious alleles that are introduced by migration. Simultaneous changes in the within- and between-population components of genetic diversity can be measured by Wright's (1951) F_{ST} statistic (Lewontin and Krakauer, 1973). D'Urso *et al.* (2003) used this to assess the genetic diversity between several geographically separated CTV populations and a much higher differentiation was observed in the p20 segment than in the A segment, which signifies a lower selection pressure. Loci that exhibit both reduced variability and a skew in the haplotype distribution compared to other loci, such as the A segment of CTV, are good candidates for selection because the two properties are not correlated under neutral models of evolution. These methods are therefore well-suited to identify candidate regions associated with beneficial mutations, which could be the target of further in-depth analyses.

Another approach is to use the ratio (ω) of non-synonymous to synonymous substitution rates to estimate the selective pressure at the protein level (Kimura, 1983). A value of $\omega > 1$ means that non-synonymous mutations offer fitness advantages to the protein and have higher fixation probabilities than synonymous mutations (diversifying or positive selection). In contrast, ω values close to 0 mean that the protein is essentially conserved at the amino acid level (purifying or negative selection) and $\omega = 1$ corresponds to neutral evolution. In many cases, calculation of ω as an average over all codons in the gene and over the entire evolutionary time that separates the sequences provides only limited information and impedes the detection of diversifying selection events. Many proteins appear to be under purifying selection most of the time (Li, 1997) and a large proportion of their amino acids is largely invariable (with ω close to 0) due to structural constraints. When knowledge of the functional domains of the protein is not available, or when only a few codons or a few lineages undergo diversifying selection, a better approach is to devise statistical models that allow for heterogeneous ω ratios among codons or lineages (Nielsen and Yang, 1998; Yang *et al.*, 2000). This can be achieved by maximum likelihood (Yang, 1997) or maximum parsimony (Suzuki and Gojobori, 1999) methods. Such analyses can provide valuable information on the mechanisms of adaptation of plant viruses, on the major constraints, either internal or external, that influence them and can allow hypotheses to be raised about plant virus epidemiological processes. Also, fruitful links can be established with biochemical analyses on function and structure of viral proteins. García-Arenal *et al.* (2001) have shown that negative selection predominates during evolution of plant viruses (average ω values were no more than 0.31, when measured on a series of proteins from different viruses). Although average ω values are low when calculated on entire genes or cistrons, this does not preclude that positive selection can act strongly on a small number of codons. Positive selection was detected on a few codon positions of some plant virus genome parts (Glasa *et al.*, 2002; Moury, 2004; Moury *et al.*, 2002, 2004; Schirmer *et al.*, 2005; Tsompana *et al.*, 2005). Constraints responsible for these variation patterns were suggested to be adaptation to the plant host or to aphid vectors. For *Beet necrotic yellow vein virus* (BNYVV), positive selection on one codon position was particularly high and its intensity comparable to that exerted on fast-evolving animal viruses such as *Measles virus* (Woelk *et al.*, 2002) or *Foot-and-mouth disease virus* (Fares *et al.*, 2001). This amino acid position was suggested to determine the ability to overcome partial resistance to BNYVV present in some beet cultivars (Schirmer *et al.*, 2005). Attempts to evaluate the relative intensity of the different evolutionary constraints on plant viruses are few. Chare and Holmes (2004) compared the selective pressure (average ω values) in the CP-coding region of plant viruses that were either vector-borne or transmitted by other routes. They found higher selective constraints in the former group, a characteristic shared by animal viruses (Woelk and Holmes, 2002), which may be explained by specific interactions between CPs and vector receptors necessary for transmission. In CMV, selection of different variants by different aphid vector populations or species was suggested to be responsible for positive selection patterns in the CP of the virus (Moury, 2004), as shown by mutagenesis experiments (Perry *et al.*, 1998), and because there was no signature of positive selection by plant host in that protein.

In another attempt to unravel the relative importance of various constraints in the evolution of plant viruses, we compared ω values in the ten cistrons of the genome of six potyviruses with contrasted biological properties (Table II) for which the complete genomes of a minimum of ten different isolates were available in data banks. The methodology was as in Moury *et al.* (2002), except that only models M8 and M7 were compared, since the main interest was in estimating average ω values and not in the precise detection of positive selection acting on particular codons (*cf.* Chare and Holmes, 2004; Woelk and Holmes, 2002). The average ω values were lower than 1.0 (between 0.01 and 0.32; Table III), confirming the predominance of negative selection. Significant differences were observed between the different potyviral proteins (Table III). The P1 protein was the least constrained for five of six potyviruses, with average ω values about twice as large as those of P3 and CP proteins, which ranked second. In contrast, the NIa protease, NIb, 6K1 and CI cistrons were the most constrained. Positive selection was significant on a small number of amino acid sites of the CP and VPg cistrons of *Potato virus Y*, of the P3 cistron of *Plum pox virus* and of the CP of *Sugarcane mosaic virus*, as shown with different data sets by Bousalem *et al.* (2003), Glasa *et al.* (2002) and Moury *et al.* (2002, 2004). Perhaps the most striking feature of these analyses is that ω estimates along the ten cistrons

TABLE II
CHARACTERISTICS OF SIX POTYVIRUSES FOR WHICH THE SELECTIVE PRESSURE ON GENOME EVOLUTION WAS ANALYSED

Virus	Plant host range[a]		Host characteristics[b]	Number of sequences
	Families	Genera		
PVY	1 (3) D	4	Annual/vegetative	18
TuMV	3 (15) D	7	Annual or biennial	30
PPV	1 (10) D	1	Perennial/vegetative	10
SCMV	1 (2) M	6	Perennial/vegetative	10
SMV	1 (5) D	1	Annual/vertical transmission	10
ZYMV	1 (6) D	4	Annual	13

[a] After Brunt *et al.* (1996). Experimental host range is between parentheses.
[b] Use of vegetative propagation and occurrence of vertical transmission are mentioned when frequent for some plant hosts.
Abbreviations: M, monocotyledonous; D, dicotyledonous; PVY, *Potato virus Y*; TuMV, *Turnip mosaic virus*; PPV, *Plum pox virus*; SCMV, *Sugarcane mosaic virus*; SMV, *Soybean mosaic virus*; ZYMV, *Zucchini yellow mosaic virus*.

TABLE III
AVERAGE ω VALUES[a] ESTIMATED FOR THE TEN PROTEINS EXPRESSED BY THE GENOMES OF SIX POTYVIRUSES

Virus	Protein									
	P1	HC-Pro	P3	6K1	CI	6K2	VPg	NIa- Pro	NIb	CP
PVY	0.27	0.08	0.13	0.08	0.03	0.11	*0.10*	0.06	0.05	*0.17*
TuMV	0.25	0.02	0.12	0.01	0.03	0.08	0.08	0.02	0.03	0.07
PPV	0.18	0.03	*0.13*	0.03	0.03	0.13	0.04	0.05	0.04	0.20
SCMV	0.15	0.02	0.06	0.05	0.01	0.04	0.05	0.02	0.03	*0.12*
SMV	0.27	0.11	0.14	0.02	0.12	0.01	0.04	0.02	0.06	0.04
ZYMV	0.32	0.12	0.10	0.07	0.08	0.10	0.04	0.02	0.05	0.08
Mean[b]	0.24**a**	0.06**cd**	0.11**b**	0.04**cd**	0.05**d**	0.08**bcd**	0.06**cd**	0.03**d**	0.04**d**	0.11**bc**

[a] Average values were obtained by the model retained by a likehood ratio test among models M7 and M8 in the codeml program of the PAML software (Yang, 1997). Values are in italics when positive selection was significant for a number of amino acid sites.
[b] Values followed by the same letters are not significantly different ($P > 0.05$) in a non-parametric Tukey-type multiple comparison test (Zar, 1984).
Abbreviations as in Table II.

were correlated between potyviruses (Table IV). Twelve of 15 Pearson's correlation coefficients between average ω values in the proteins of these potyviruses were significantly higher than zero ($P < 0.05$). Ten of 15 Pearson's correlation coefficients between average amino acid distances, but only 2 of 15 Pearson's correlation coefficients between average nucleotide diversity indices in the cistrons of these potyviruses, were significantly higher than zero (Table IV). The mean Pearson's correlation coefficients between ω values of the potyviruses were significantly higher than that of nucleotide diversities ($P < 0.05$), but only marginally higher than that of amino acid diversities

TABLE IV

PEARSON'S CORRELATION COEFFICIENTS OF THE AVERAGE RATIO BETWEEN NON-SYNONYMOUS AND SYNONYMOUS SUBSTITUTION RATES (ω, CALCULATED WITH PAML, YANG, 1997) OR PAIRWISE GENETIC DISTANCES (P-DISTANCES CALCULATED ON NUCLEOTIDE OR AMINO ACIDS USING MEGA3, KUMAR *ET AL.*, 2004) IN THE TEN CISTRONS OF SIX POTYVIRUSES

Compared viruses	Nucleotide diversity (p-distance)	Amino acid diversity (p-distance)	Amino acid replacement rate (ω)
PVY vs TuMV	0.53	**0.74**	***0.90***
PVY vs PPV	-0.30	0.56	***0.83***
PVY vs SCMV	***0.86***	**0.95**	***0.95***
PVY vs SMV	0.01	0.43	**0.65**
PVY vs ZYMV	-0.15	**0.83**	**0.85**
TuMV vs PPV	-0.33	***0.71***	***0.70***
TuMV vs SCMV	**0.67**	*0.81*	**0.80**
TuMV vs SMV	-0.20	0.48	**0.79**
TuMV vs ZYMV	-0.44	***0.75***	**0.88**
PPV vs SCMV	-0.07	***0.64***	***0.85***
PPV vs SMV	0.55	0.11	0.35
PPV vs ZYMV	0.08	***0.70***	0.55
SCMV vs SMV	0.20	0.43	0.50
SCMV vs ZYMV	-0.12	**0.85**	**0.71**
SMV vs ZYMV	0.50	**0.64**	**0.86**
Mean[a]	0.12**a**	0.64**b**	0.74**b**

[a] Values followed by the same letters are not significantly different ($P > 0.05$) in a non-parametric Tukey-type multiple comparison test (Zar, 1984).

Abbreviations as in Table II.

Values significantly different from zero ($P < 0.05$) are in bold. Cases corresponding to Spearman's rank correlation coefficients that are significantly different from zero are in italics.

($P = 0.10$). Since these potyviruses have different host ranges, infect plants with different biological properties (annual, biennial or perennial, vegetatively or seed-propagated, monocotyledonous or dicotyledonous) under different climatic conditions and are transmitted by different vector populations, different selective constraints and different intensities among these constraints would be expected to act on them. The fact that most ω values along their genome are significantly correlated, suggests that internal constraints, that is selection pressures associated with the maintenance of functional or structural features between their proteins, predominate in potyvirus evolution and could explain why positive selection signatures in their genome are only sporadic. Molecular demonstration of such internal constraints has been obtained, for example, in the case of the structure of nucleocapsids of tobamoviruses (Altschuh *et al.*, 1987), in the compatibility of movement and coat proteins for cell-to-cell movement of CMV (Salánki *et al.*, 2004) or in the maintenance of RNA structures involved in replication (Argüello-Astorga *et al.*, 1994; Bacher *et al.*, 1994) and/or movement (Choi *et al.*, 2005).

C. Adaptative Role of Recombination

Most early studies on recombination relied on laboratory isolates deficient in important traits in the viral cycle and thus submitted to a high selection pressure for emergence (Gal-On *et al.*, 1998; Greene and Allison, 1994). The fitness of such recombinants in natural conditions was seldom studied. *In vitro* intraspecific or interspecific recombinants were also obtained for several viruses (Briddon *et al.*, 1990; Tobias *et al.*, 2001). Such recombinants frequently had a poor fitness, but occasionally a recombinant or reassortant could present a higher accumulation or more severe symptoms than parental genomes in controlled experiments on a few experimental hosts (Ding *et al.*, 1996; Paalme *et al.*, 2004). In natural conditions, recombination is usually inferred from sequence analyses, but its biological impact is not known (Revers *et al.*, 1996; Rubio *et al.*, 2001; Tomimura *et al.*, 2004). Different kinds of constraints have been shown that determine the appearance or fitness of recombinants. In experimental conditions, recombination 'hotspots' associated with AU-rich sequences and/or local double-stranded structures were characterized for some RNA viruses (Nagy and Bujarski, 1997; Shapka and Nagy, 2004) in relation to the mechanisms involved in recombination, whereas homologous recombination was not observed in highly structured regions (Alejska *et al.*, 2005). Homologous recombination between CMV and the related species *Tomato aspermy virus* (TAV) in conditions of minimum selection pressure occurred at, or near regions of high-sequence similarity between the parental strains or viruses (de Wispelaere *et al.*, 2005). Once they appeared, the fate of recombinants was suggested to depend on the complexity of interactions involving the recombinant region and the genome background, as was shown with *Maize streak virus* (MSV) *in vitro* recombinants (Martin *et al.*, 2005). The relative fitness of MSV recombinants was correlated positively with the nucleotide similarity of the exchanged fragments and negatively with the complexity of interactions involving the exchanged region.

If these constraints play a major role in the emergence of virus recombinants, the distribution of recombination breakpoints of naturally occurring recombinants should depart from random. We tested the pattern of recombination events along the genomes of natural recombinant virus strains for complete spatial randomness (CSR) in a similar way as done classically for spatial point patterns (Diggle, 1983; Manly, 1997) (Table V). Three parameters were tested for CSR: (1) the distance between breakpoints, (2) the distance to the nearest neighbouring breakpoint, which emphasizes the importance of small distances and (3) the distance of any site of the genome to the nearest breakpoint, which emphasizes the importance of empty space (Diggle, 1983). For intraspecific potyvirus (*Turnip mosaic virus* (TuMV) and *Potato virus Y* (PVY)) or *Mastrevirus* (MSV) recombination, the data sets available did not reveal a significant clustering of breakpoints (Table V) and suggest a near-random emergence of recombinants. In contrast, interspecific recombinants between the two cucumoviruses CMV and TAV, obtained in the laboratory from co-infected plants, clearly showed significant departure from random (Table V). This has been attributed to an absence of breakpoints in the CP cistron, either due to a lack of similarity between viruses in that region or lack of functionality of the recombined CP (de Wispelaere *et al.*, 2005).

Different roles have been attributed to recombination: it may help create high-fitness variants and/or may purge virus populations from accumulation of deleterious mutations (Worobey and Holmes, 1999). A comparison of the frequency of recombinants (and/or reassortants) and parental genotypes on a large scale and during several years was performed with CMV and revealed that the majority of recombinants and reassortants remained only minor components in viral populations, suggesting a poor relative fitness, although one particular recombinant could reach a higher frequence (Bonnet *et al.*, 2005; Fraile *et al.*, 1997). To our knowledge, the fitness of emergent plant

TABLE V
TEST OF COMPLETE SPATIAL RANDOMNESS OF RECOMBINATION BREAKPOINTS FROM PLANT VIRUS POPULATIONS

Data set	Number of sequences	Number of recombination breakpoints[a]	Tests of spatial randomness[b]	Source data
TuMV complete genomes	38	15	Not rejected ($P > 5\%$)	Tomimura *et al.* (2003)
TuMV concatenated partial sequences	142	23	Not rejected ($P > 5\%$)	Tomimura *et al.* (2004)
PVY complete and partial sequences	>18[c]	12	Not rejected ($P > 5\%$)	Moury *et al.* (2002) and unpublished
MSV complete genomes	26	21	Not rejected ($P > 5\%$)	Martin *et al.* (2005)
CMV × TAV RNA 3 sequences[d]	195	38	(1), (2), (3) rejected ($P < 2.5\%$)	de Wispelaere *et al.* (2005)

[a] When several isolates showed recombination breakpoints at the same genome position exchanging regions of the same parental groups they were not considered as independent.

[b] The distance between breakpoints (1), the distance to the nearest neighbouring breakpoint (2) and the distance of any site of the genome to the nearest breakpoint (3) were compared to results obtained from 1000 simulations under complete randomness (Diggle, 1983) using the R software (Ihaka and Gentleman, 1996).

[c] The number of complete sequences is indicated. Additional partial sequences available in data banks were analysed, leading to an underrepresentation of some genome regions which was not taken into account in our simulations.

[d] Recombinants obtained in plants artificially inoculated with both viruses.

Abbreviations: TuMV, *Turnip mosaic virus*; PVY, *Potato virus Y*; MSV, *Maize streak virus*; CMV, *Cucumber mosaic virus*; TAV, *Tomato aspermy virus*.

virus recombinants has never been compared to that of the parental populations. Another issue is that a correlation between particular biological properties and the recombinant status of a virus strain does not imply that the recombination event was the direct cause of these biological properties. Similar necrotic symptoms in potato tubers were induced by multi-recombinant PVY strains in Europe (Glais et al., 2002) and by non-recombinant strains in North America (Nie and Singh, 2003). Further studies are, therefore, needed to unravel the epidemiological significance of recombination or reassortment in plant virus populations.

D. Population Size Variation

A fundamental result of the coalescent theory in population genetics is the finding of a relationship between coalescent time (i.e. the time at which two alleles share their most common ancestor) and population size (Kingman, 1982). Two nucleotide sequences drawn at random from a small population have a higher probability of having a more recent coalescence (i.e. fewer substitution differences) than do two sequences drawn at random from a large population. Thus a change in population size over time will leave a 'signature' in the pattern of nucleotide substitutions among individuals within a population that will depend on the direction (growth or decline) and date (ancient or recent) of this change. Different methods have been developed to reveal these demographic patterns, each involving a particular set of assumptions (reviewed in Emerson *et al.*, 2001). They were mainly applied to animal viruses (Grassly *et al.*, 1999; Ong *et al.*, 1996, 1997; Pybus *et al.*, 1999, 2000), the required number of samples being quite prohibitive. Stenger *et al.* (2002) used such models to analyse the observed distribution of *Wheat streak mosaic virus* (WSMV) haplotypes, but none could explain the large number of singletons in the distribution. This could result from the complex action of different evolutionary forces that are difficult to disentangle.

E. Migration

Given their diverse means of dispersal (persistent or non-persistent transmission by air-borne insect vectors, transmission by soil-borne vectors such as fungi or nematodes, contact or wound transmission, vertical seed transmission, dispersal in infected plant material by propagation or trade), extremely diverse patterns of virus migration are to be expected. However, for almost all plant viruses, the frequence versus distance distribution of migration events is unknown and for several reasons. Dynamics of vectors are useful to predict epidemics, especially in the case of viruses transmitted persistently (Fabre *et al.*, 2003; Werker *et al.*, 1998) and insect movement can be estimated both at field and regional scales (Tatchell, 1990). However, the viruliferous status of vectors is seldom determined (Fabre *et al.*, 2005) and remains extremely difficult to assess for transient, non-persistent viruses–vector interactions (Olmos *et al.*, 2005). Moreover, detecting viruses in a vector does not imply that the virus will be transmitted or even that the detected virus is in a transmissible state (Dedryver *et al.*, 2005; Gray and Gildow, 2003). Inferring migration events from observed plant infection patterns is also rendered difficult by the lack of information on the virus variability, either to identify independent migration events or to separate primary and secondary infections. Consequently, few studies have examined virus variability at a local scale (field or region) with the aim of assessing virus spread (Lecoq *et al.*, 2005; Teycheney *et al.*, 2005).

Studies on emerging virus diseases can also provide important information on 'founder effects' following a virus introduction into a new environment. Situations can be quite diverse, depending on the virus. For CTV in Italy, it appeared that within each site in which the virus was found, populations were very homogeneous since only one haplotype was detected by SSCP (Davino *et al.*, 2005). Similarly, for *Tomato yellow leaf curl Sardinia virus* (TYLCSV) and *Cucurbit yellow stunting disorder virus* (CYSDV), a high genetic stability has been observed in southern Spain during the 8 years after their introduction (Marco and Aranda, 2005; Sanchez-Campos *et al.*, 2002). In contrast, important molecular, serological and biological diversities were observed between *Zucchini yellow mosaic virus* (ZYMV) isolates collected during a 6-year period after its first introduction in Martinique (Desbiez *et al.*, 2002). Such differences could be due to more diverse environmental conditions, host variability or intrinsic factors due to the virus biology. Similar molecular approaches could be applied to identify potent virus sources. Another important question is to know whether homogeneous or diverse virus populations are spreading in the field during an epidemic. Biological tests have shown that different CMV virus strains occur in a single field and even in the same individual plant (Quiot *et al.*, 1979). By contrast, it is generally thought that cross-protection between related virus strains should limit mixed infections. However, cross-protection efficiency seems to be less effective between genetically distant viruses (Lecoq and Raccah, 2001). Specific monoclonal antibodies have been used to compare the spread of a specific *Soybean mosaic virus* (SMV) G5 strain released from a point source to the spread of exogenous isolates (non-G5) originating from the local environment in order to obtain spatial maps of virus spread (Nutter *et al.*, 1998). Also, different virus populations can be observed in different years in the same geographical area (Lecoq *et al.*, 2005; McNeil *et al.*, 1996), but the contribution of migration patterns or vector population dynamics remains to be established.

When a spatial structure of virus diversity is observed, this suggests a lack of intense gene flow (Fargette *et al.*, 2004; Fraile *et al.*, 1996; Lecoq *et al.*, 2005). Also, minimal distances for which spatially structured populations are observed may indicate distance thresholds above which virus migrations are drastically reduced. For *Rice tungro spherical virus* and *Rice tungro bacilliform virus*, no structure was found for populations sampled up to 200 km apart, whereas populations collected thousands of kilometres apart, in the Philippines and Indonesia, differentiated (Arboleda and Azzam, 2000; Azzam *et al.*, 2000). At the field scale, during the ZYMV survey in Martinique (Desbiez *et al.*, 2002), ZYMV was detected in an isolated location in an old self-sown squash and in a nearby young zucchini squash field. Nine isolates from the cultivated field were identical at the molecular and serological levels, but differed significantly from the isolate from the old volunteer squash, suggesting that this neighbouring plant was not the original virus source for the epidemic in the cultivated field. A similar observation was made for *Watermelon mosaic virus* in south-eastern France: virus populations found in 2001 in weeds (Shepherd's purse, henbit or fumitory) differed from those in a nearby cultivated squash field (Lecoq *et al.*, 2005).

Assumptions about virus migration between geographical locations have been made, when virus populations separated by large distances showed high similarities, especially when new viruses or variants are being introduced into a region. Eventually, such data could allow epidemics to be traced back to their origin. Examples of such studies concern the introduction of CTV in south Italian citrus orchards (Davino *et al.*, 2005), emergence of CMV subgroup IB in Europe (Bonnet *et al.*, 2005; Gallitelli, 2000), ZYMV introduction in French West Indies (Desbiez *et al.*, 2002), *Papaya ringspot virus* (PRSV) introductions in France (Lecoq *et al.*, 2003), TYLCV introductions into the Caribbean and subsequent spread (Bird *et al.*, 2001; Polston *et al.*, 1999; Salati *et al.*, 2002; Sinisterra *et al.*, 2000) and dispersal of recombinant *Plum pox virus* isolates (Glasa *et al.*, 2005). However, separate virus populations can be similar in terms of genetic distance either due to intense gene flow (migration) between them or recent common ancestry, and the two causes could be difficult to distinguish. In no case was a confidence level attributed to these assumptions, which would be highly desirable because of the influences of the level of polymorphism in the sequenced regions and of the representativeness of the putative progenitor populations. To validate that an epidemic resulted from an introduced virus population would require a rejection of the hypothesis that it was the consequence of *in situ* virus diversification. Moreover, identifying a particular geographic region as the source of a recent virus introduction necessitates a refutation of the hypothesis that all other surrounding regions would be the actual source.

Phylogeny reconstruction has allowed hypotheses to be raised about origins of viral populations. Often, phylogenetic groups that branched closest to the root node of the tree were inferred to be progenitors of other groups. For instance, a Colombian group of *Sugarcane yellow leaf virus* (Moonan and Mirkov, 2002), a west European group of TuMV (Tomimura *et al.*, 2003), an Indian group of PRSV (Bateson *et al.*, 2002), an Asian-Oceanic group of *Yam mild mosaic virus* (Bousalem *et al.*, 2003) and East African populations of *Rice yellow mottle virus* (Fargette *et al.*, 2004) were considered progenitors or 'ancestral' to other groups. However, as emphasized (Crisp and Cook, 2005), intuitive assertion of ancestry from phylogenetic trees is likely to lead to errors. Semantically, if character states can be ancestral (plesiomorphic) or derived (apomorphic), these concepts do not apply to sister-groups (i.e. groups that bifurcated from a common node in the tree) that survived to date since these sister-groups, by definition, diverged simultaneously from their most recent common ancestor. Thus, extant sister-groups cannot be qualified as ancestral, basal or early diverging (Crisp and Cook, 2005). More importantly, equally probable (i.e. equally costly considering the total number of events) scenarios that involve episodes of diversification and population splitting (by plant host or geographical isolation) but do not necessarily involve dispersal from one area to the other often can be constructed to interpret the tree structure (Crisp and Cook, 2005). Therefore, in many instances, independent information is needed in conjunction with the phylogenetic tree to reconstruct ancestry, such as historical (Bousalem *et al.*, 2003) or recombination (Tomimura *et al.*, 2003) events.

F. Quasi-Species Nature of Plant Viruses

The structure of virus populations within an infected plant is reminiscent of that of quasi-species, where a stationary mutant distribution of infinite size is centered around one or several master sequence(s) and surrounded by a 'swarm' of more or less distant variants (Eigen, 1971). However, whether plant virus populations do behave as quasi-species is still controversial. The quasi-species model implies several key dynamic properties: high mutation rates and extremely large population sizes, which could be satisfied by plant viruses, but also a lack of stochastic changes in the population structure and an influence of selection forces on the entire virus population instead of on individual variants. Although the quasi-species model fits well with properties of bacteriophage or animal virus populations (Domingo, 2002; Novella, 2003), extremely narrow genetic 'bottlenecks' during plant infection and probably also during plant-to-plant transmission could allow unpredictable stochastic variations in plant virus population structure, incompatible with the quasi-species concept (French and Stenger, 2005).

V. Conclusions

The development of genome exploration techniques and analytical methods for studying population genetic should allow a more precise understanding of plant virus epidemiology through the study of genomic diversity within virus populations. Ideally, an experiment should be designed so as to estimate the variance associated with undesired variation sources and to optimize the ability to measure or differentiate the processes at play. This strategy has, however, several inconveniences: (1) it relies on regular epidemics and for many viruses, it might not be desirable to introduce laboratory virus strains in the field; (2) it depends on natural virus diversity, which should be sufficient to obtain adequate information; (3) the scale at which epidemiological processes operate (and thus the size of the experiment) is not always known and (4) these experiments only explore a limited fraction of natural epidemiological situations. An easier and less costly approach is to collect virus samples in different epidemiological situations and most studies of plant virus diversity have been performed in such a context. Consequently, few environmental parameters are controlled and analysis is rendered more complex. However, based on knowledge of plant virus diversity and structure, several guidelines can be provided to optimize molecular epidemiology:

1. The sampling strategy should be adapted to the goal of the study. *A priori* estimations of the power of a sampling design can be calculated for given statistical methods and given preliminary results about the observed diversity. Hierarchical sampling and hierarchical partitioning of samples among variation factors can be useful to study the structure of virus populations and the factors that differentiate them (d'Urso *et al.*, 2003). Analysis of the representativeness of sampling can provide information about the unexplored diversity and can help compare the diversity in populations with unequal sampling effort.

2. Concerning the genome target, diversity levels, evolutionary patterns and/or functional roles of different regions in the virus genome are privileged information to choose targets appropriate to a particular study (e.g. neutral or adaptive). With no *a priori* knowledge about the distribution of variability across the virus genome, or to what parts of the genome selective pressures are exerted, multi-locus approaches may be most appropriate (d'Urso *et al.*, 2003; Moreno *et al.*, 2004).

3. The choice of the laboratory method used to reveal polymorphism depends on the virus diversity in the target region and on the statistical method used to estimate and compare viral diversity. Some statistical methods may have an optimal power at intermediate levels of polymorphism. For instance, when too many singleton haplotypes are characterized, the Chao1 CIs are large and the χ^2 test is not appropriate.

4. The choice of the diversity indices to be estimated and/or compared depends on the goal of the study and the data obtained (haplotype distributions, genetic distances, nucleotide sequences). Richness estimators could be useful to assess the role of agricultural practices, environmental factors or ecological situations on the virus biodiversity (i.e. actual number of haplotypes). Haplotype distributions and the associated diversity indices are most suited to comparing the structure of populations and identifying factors that differentiate them. They could also be used to test for departure from neutral mutation models, but distinction between different evolutionary forces is impossible with most plant virus data sets (Stenger *et al.*, 2002). Sequence data combined with phylogenetic methods are most appropriate to quantify the effects of different evolutionary forces owing to theoretical developments.

Acknowledgments

The authors gratefully acknowledge Dr. Joël Chadœuf for the programme of CSR analysis and Drs. Cindy Morris, Thierry Candresse, Frédéric Fabre and Joël Chadœuf for their comments on the manuscript.

References

Adams, M. J., Antoniw, J. F., Bar-Joseph, M., Brunt, A. A., Candresse, T., Foster, G. D., Martelli, G. P., Milne, R. G., and Fauquet, C. M. (2004). The new plant virus family *Flexiviridae* and assessment of molecular criteria for species demarcation. *Arch. Virol.* **149:**1045–1060.

Adams, M. J., Antoniw, J. F., and Fauquet, C. (2005). Molecular criteria for genus and species discrimination within the family *Potyviridae*. *Arch. Virol.* **150:**459–479.

Albiach-Martí, M. R., Mawassi, M., Gowda, S., Satyanarayana, T., Hilf, M. E., Shanker, S., Almira, E. C., Vives, M. C., López, C., Guerri, J., Flores, R., Moreno, P., *et al.* (2000). Sequences of citrus tristeza virus separated in time and space are essentially identical. *J. Virol.* **74:**6856–6865.

Alejska, M., Figlerowicz, M., Malinowska, N., Urbanowicz, A., and Figlerowicz, M. (2005). A universal BMV-based RNA recombination system: How to search for general rules in RNA recombination. *Nucleic Acids Res.* **33:**e105.

Aleman-Verdaguer, M.-E., Goudou-Urbino, C., Dubern, J., Beachy, R. N., and Fauquet, C. (1997). Analysis of the sequence diversity of the P1, HC, P3, NIb and CP genomic regions of several yam mosaic potyvirus isolates: Implications for the intraspecies molecular diversity of potyviruses. *J. Gen. Virol.* **78:**1253–1264.

Alma Bracho, M., García-Robles, I., Jiménez, N., Torres-Puente, M., Moya, A., and González-Candelas, F. (2004). Effect of oligonucleotide primers in determining viral variability within hosts. *Virology Journal* **1:**13.

Altschuh, D., Lesk, A. M., Bloomer, A. C., and Klug, A. (1987). Correlation of co-ordinated amino acid substitutions with function in viruses related to tobacco mosaic virus. *J. Mol. Biol.* **193:**693–707.

Arboleda, M., and Azzam, O. (2000). Inter- and intra-site diversity of natural field populations of rice tungro bacilliform virus in the Philippines. *Arch. Virol.* **145:**275–289.

Argüello-Astorga, G., Herrera-Estrella, L., and Rivera-Bustamante, R. (1994). Experimental and theoretical definition of geminivirus origin of replication. *Plant Mol. Biol.* **26:**553–556.

Ayme, V., Souche, S., Caranta, C., Jacquemond, M., Chadœuf, J., Palloix, A., and Moury, B. (2006). Different mutations in the genome-linked protein VPg of *Potato virus Y* confer virulence on the $pvr2^3$ resistance in pepper. *Mol. Plant-Microbe Interact.* **19:**557–563.

Azzam, O., Yambao, M. L. M., Muhsin, M., McNally, K. L., and Umadhay, K. M. L. (2000). Genetic diversity of rice tungro spherical virus in tungro-endemic provinces of the Philippines and Indonesia. *Arch. Virol.* **145:**1183–1197.

Bacher, J. W., Warkentin, D., Ramsdell, D., and Hancock, F. (1994). Selection versus recombination: What is maintaining identity in the 3′ termini of blueberry leaf mottle nepovirus RNA1 and RNA2? *J. Gen. Virol.* **75:**2133–2137.

Bateson, M. F., Lines, R. E., Revill, P., Chaleeprom, W., Ha, C. V., Gibbs, A. J., and Dale, J. L. (2002). On the evolution and molecular epidemiology of the potyvirus *Papaya ringspot virus*. *J. Gen. Virol.* **83:**2575–2585.

Berry, S., and Rey, M. E. C. (2001). Differentiation of cassava-infecting begomoviruses using heteroduplex mobility assays. *J. Virol. Methods* **92:**151–163.

Bird, J., Idriss, A. M., Rogan, D., and Brown, J. K. (2001). Introduction of the exotic *Tomato yellow leaf curl virus*-Israel in tomato in Puerto Rico. *Plant Dis.* **85:**1028.

Blanc, S., Ammar, E. D., Garcia-Lampasona, S., Dolja, V. V., Llave, C., Baker, J., and Pirone, T. P. (1998). Mutations in the potyvirus helper component protein: Effects on interactions with virions and aphid stylets. *J. Gen. Virol.* **79:**3119–3122.

Bonnet, J., Fraile, A., Sacristán, S., Malpica, J. M., and García-Arenal, F. (2005). Role of recombination in the evolution of natural populations of *Cucumber mosaic virus*, a tripartite RNA plant virus. *Virology* **332:**359–368.

Bousalem, M., Dallot, S., Fuji, S., and Natsuaki, K. T. (2003). Origin, world-wide dispersion, bio-geographical diversification, radiation and recombination: An evolutionary history of *Yam mild mosaic virus* (YMMV). *Infect. Genet. Evol.* **3:**189–206.

Bradley, R. D., and Hillis, D. M. (1997). Recombinant DNA sequences generated by PCR amplification. *Mol. Biol. Evol.* **14:**592–593.

Briddon, R. W., Pinner, M. S., Stanley, J., and Markham, P. G. (1990). Geminivirus coat protein gene replacement alters insect specificity. *Virology* **177:**85–94.

Brunt, A., Crabtree, K., Dallwitz, M., Gibbs, A., and Watson, L. (1996). "Viruses of Plants: Descriptions and Lists from the VIDE Database." C. A. B. International, UK.

Bruyere, A., Wantroba, M., Flasinski, S., Dzianott, A., and Bujarski, J. J. (2000). Frequent homologous recombination events between molecules of one RNA component in a multipartite RNA virus. *J. Virol.* **74:**4214–4219.

Chao, A. (1984). Non-parametric estimation of the number of classes in a population. *Scand. J. Stat.* **11:**265–270.

Chao, A. (1987). Estimating the population size for capture-recapture data with unequal catchability. *Biometrics* **43:**783–791.

Chao, A., and Shen, T.-J. (2003). Program SPADE (Species Prediction And Diversity Estimation). Program and User's Guide published at http://chao.stat.nthu.edu.tw.

Chare, E. R., and Holmes, E. C. (2004). Selection pressures in the capsid genes of plant RNA viruses reflect mode of transmission. *J. Gen. Virol.* **85:**3149–3157.

Chen, Y.-K., Goldbach, R., and Prins, M. (2002). Inter- and intramolecular recombinations in the *Cucumber mosaic virus* genome related to adaptations to Alstroemeria. *J. Virol.* **76:**4119–4124.

Choi, I.-R., Horken, K. M., Stenger, D. C., and French, R. (2005). An internal RNA element in the P3 cistron of Wheat streak mosaic virus revealed by synonymous mutations that affect both movement and replication. *J. Gen. Virol.* **86:**2605–2614.

Cochran, W. G. (1954). Some methods for strengthening the common χ^2 tests. *Biometrics* **10:**417–451.

Colvin, J., Omongo, C. A., Maruthi, M. N., Otim-Nape, G. W., and Thresh, J. M. (2004). Dual begomovirus infections and high *Bemisia tabaci* populations: Two factors driving the spread of a cassava mosaic disease pandemic. *Plant Pathol.* **53:**577–584.

Colwell, R. K., and Coddington, J. A. (1994). Estimating terrestrial biodiversity through extrapolation. *Phil. Trans. R. Soc. London B* **345:**101–118.

Crisp, M. D., and Cook, L. G. (2005). Do early branching lineages signify ancestral traits? *Trends Ecol. Evol.* **20:**122–128.

Davino, S., Rubio, L., and Davino, M. (2005). Molecular analysis suggests that recent *Citrus tristeza virus* outbreaks in Italy were originated by at least two independent introductions. *Eur. J. Plant Pathol.* **111:**289–293.

de Wispelaere, M., Gaubert, S., Trouilloud, S., Belin, C., and Tepfer, M. (2005). A map of the diversity of RNA3 recombinants appearing in pants infected with *Cucumber mosaic virus* and *Tomato aspermy virus*. *Virology* **331:**117–127.

Dedryver, C. A., Riault, G., Tanguy, S., Le Gallic, J. F., Trottet, M., and Jacquot, E. (2005). Intra-specific variation and inheritance of BYDV-PAV transmission in the aphid *Sitobion avenae*. *Eur. J. Plant Pathol.* **111:**341–354.

Desbiez, C., and Lecoq, H. (2004). The nucleotide sequence of *Watermelon mosaic virus* (WMV, *Potyvirus*) reveals interspecific recombination between two related potyviruses in the 5′ part of the genome. *Arch. Virol.* **149:**1619–1632.

Desbiez, C., Wipf-Scheibel, C., and Lecoq, H. (2002). Biological and serological variability, evolution and molecular epidemiology of *Zucchini yellow mosaic virus* (ZYMV, *Potyvirus*) with special reference to Caribbean islands. *Virus Res.* **85:**5–16.

Desbiez, C., Gal-On, A., Girard, M., Wipf-Scheibel, C., and Lecoq, H. (2003). Increase in *Zucchini yellow mosaic virus* symptom severity in tolerant zucchini cultivars is related to a point mutation in P3 protein and is associated with a loss of relative fitness on susceptible plants. *Phytopathology* **93:**1478–1484.

Diggle, P. J. (1983). "Statistical Analysis of Spatial Point Patterns," p. 148. Academic Press, London.

Ding, S.-W., Shi, B.-J., Li, W.-X., and Symons, R. H. (1996). An interspecies hybrid RNA virus is significantly more virulent than either parental virus. *Proc. Natl. Acad. Sci., USA* **93:**7470–7474.

Domingo, E. (2002). Quasispecies theory in virology. *J. Virol.* **76:**463–465.

Drake, J. W., and Holland, J. J. (1999). Mutation rates among RNA viruses. *Proc. Natl. Acad. Sci. USA* **96:**13910–13913.

d'Urso, F., Sambade, A., Moya, A., Guerri, J., and Moreno, P. (2003). Variation of haplotype distributions of two genomic regions of *Citrus tristeza virus* populations from eastern Spain. *Mol. Ecol.* **12:**517–526.

Eigen, M. (1971). Self-organization of matter and the evolution of biological macromolecules. *Naturwissenschaften* **58:**465–523.

Emerson, B. C., Paradis, E., and Thébaud, C. (2001). Revealing the demographic histories of species using DNA sequences. *Trends Ecol. Evol.* **16:**707–716.

Fabre, F., Dedryver, C. A., Leterrier, J. L., and Plantegenest, M. (2003). Aphid abundance on cereals in autumn predicts yield losses caused by BYDV. *Phytopathology* **93:**1217–1222.

Fabre, F., Mieuzet, L., Plantegenest, M., Dedryver, C. A., Leterrier, J. L., and Jacquot, E. (2005). Spatial and temporal patterns in the frequency of *Rhopalosiphum padi* bearing barley yellow dwarf viruses and its epidemiological significance. *Agr. Ecosyst. Environ.* **106:**49–55.

Fares, M. A., Moya, A., Escarmis, C., Baranowski, E., Domingo, E., and Barrio, E. (2001). Evidence for positive selection in the capsid protein-coding region of the foot-and-mouth disease virus (FMDV) subjected to experimental passage regimens. *Mol. Biol. Evol.* **18:**10–21.

Fargette, D., Pinel, A., Abubakar, Z., Traoré, O., Brugidou, C., Fatogoma, S., Hébrard, E., Choisy, M., Séré, Y., Fauquet, C., and Konaté, G. (2004). Inferring the evolutionary history of *Rice yellow mottle virus* from genomic, phylogenetic, and phylogeographic studies. *J. Virol.* **78:**3252–3261.

Fauquet, C. M., and Stanley, J. (2005). Revising the way we conceive and name viruses below the species level: A review of geminivirus taxonomy calls for new standardized isolate descriptors. *Arch. Virol.* **150:**2151–2179.

Fernandez-Delmond, I., Pierrugues, O., de Wispelaere, M., Guilbaud, L., Gaubert, S., Divéki, Z., Godon, C., Tepfer, M., and Jacquemond, M. (2004). A novel strategy for creating recombinant infectious RNA virus genomes. *J. Virol. Methods* **121:**247–257.

Fisher, R. A. (1958). "Statistical Methods for Research Workers," 13th Ed. Hafner, New York.

Fraile, A., Malpica, J. M., Aranda, M. A., Rodríguez-Cerezo, E., and García-Arenal, F. (1996). Genetic diversity in tobacco mild green mosaic tobamovirus infecting the wild plant *Nicotiana glauca*. *Virology* **223:**148–155.

Fraile, A., Alonso-Prados, J. L., Aranda, M. A., Bernal, J. J., Malpica, J. M., and García-Arenal, F. (1997). Genetic exchange by recombination or reassortment is infrequent in natural populations of a tripartite RNA plant virus. *J. Virol.* **71:**934–940.

French, R., and Stenger, D. C. (2003). Evolution of *Wheat streak mosaic virus*: Dynamics of population growth within plants may explain limited variation. *Ann. Rev. Phytopathol.* **41:**199–214.

French, R., and Stenger, D. C. (2005). Population structure within lineages of *Wheat streak mosaic virus* derived from a common founding event exhibits stochastic variation inconsistent with the deterministic quasi-species model. *Virology* **343:**179–189.

Froissart, R., Roze, D., Uzest, M., Galibert, L., Blanc, S., and Michalakis, Y. (2005). Recombination every day: Abundant recombination in a virus during a single multicellular host infection. *PLoS Biol.* **3:**e89.

Gallitelli, D. (2000). The ecology of *Cucumber mosaic virus* and sustainable agriculture. *Virus Res.* **71:**9–21.

Gal-On, A., Meiri, E., Raccah, B., and Gaba, V. (1998). Recombination of engineered defective RNA species produces infective potyvirus *in planta*. *J. Virol.* **72:**5268–5270.

García-Arenal, F., Fraile, A., and Malpica, J. M. (2001). Variability and genetic structure of plant virus populations. *Ann. Rev. Phytopathol.* **39:**157–186.

García-Arenal, F., Fraile, A., and Malpica, J. M. (2003). Variation and evolution of plant virus populations. *Int. Microbiol.* **6:**225–232.

Glais, L., Tribodet, M., and Kerlan, C. (2002). Genetic variability in *Potato potyvirus Y* (PVY): Evidence that $PVY^{N}W$ and PVY^{NTN} variants are single to multiple recombinants between PVY^{O} and PVY^{N} isolates. *Arch. Virol.* **147:**363–378.

Glasa, M., Marie-Jeanne, V., Moury, B., Kúdela, O., and Quiot, J.-B. (2002). Molecular variability of the P3–$6K_1$ genomic region among geographically and biologically distinct isolates of *Plum pox virus*. *Arch. Virol.* **147:**563–575.

Glasa, M., Palkovics, L., Kominek, P., Labonne, G., Pittnerova, S., Kudela, O., Candresse, T., and Subr, Z. (2004). Geographically and temporally distant natural recombinant isolates of *Plum pox virus* (PPV) are genetically very similar and form a unique PPV subgroup. *J. Gen. Virol.* **85:**2671–2681.

Glasa, M., Paunovic, S., Jevremovic, D., Myrta, A., Pittnerová, S., and Candresse, T. (2005). Analysis of recombinant *Plum pox virus* (PPV) isolates from Serbia confirms genetic homogeneity and supports a regional origin for the PPV-Rec subgroup. *Arch. Virol.* **150:**2051–2060.

Gorman, O. T., Bean, W. J., Kawaoka, Y, and Webster, R. G. (1990). Evolution of the nucleoprotein gene of influenza A virus. *J. Virol.* **64:**1487–1497.

Grassly, N. C., Harvey, P. H., and Holmes, E. C. (1999). Population dynamics of HIV-1 inferred from gene sequences. *Genetics* **151:**427–438.

Gray, S. M., and Gildow, E. (2003). Luteovirus-aphid interactions. *Annu. Rev. Phytopathol.* **41:**539–566.

Greene, A. E., and Allison, R. F. (1994). Recombination between viral RNA and transgenic plant transcripts. *Science* **263:**1423–1425.

Grenfell, B. T., Pybus, O. G., Gog, J. R., Wood, J. L. N., Daly, J. M., Mumford, J. A., and Holmes, E. C. (2004). Unifying the epidemiological and evolutionary dynamics of pathogens. *Science* **303:**327–332.

Harrison, B. D. (1956). The infectivity of extracts made from leaves at intervals after inoculation with viruses. *J. Gen. Microbiol.* **15:**210–220.

Harrison, B. D. (2002). Virus variation in relation to resistance breaking in plants. *Euphytica* **124:**181–192.

Hébrard, E., Pinel-Galzi, A., Catherinot, V., Labesse, G., Brugidou, C., and Fargette, D. (2005). Internal point mutations of the capsid modify the serotype of *Rice yellow mottle virus*. *J. Virol.* **79:**4407–4414.

Heck, K. L., Belle, G. V., and Simberloff, D. (1975). Explicit calculation of the rarefaction diversity measurement and the determination of sufficient sample size. *Ecology* **56:**1459–1461.

Hughes, J. B., Hellmann, J. J., Ricketts, T. H., and Bohannan, B. J. M. (2001). Counting the uncountable: Statistical approaches to estimating microbial diversity. *Appl. Envir. Microbiol.* **67:**4399–4406.

Hurst, L. D. (2002). The Ka/Ks ratio: Diagnosing the form of sequence evolution. *Trends Ecol. Evol.* **18:**486–487.

Ihaka, R., and Gentleman, R. (1996). R: A language for data analysis and graphics. *J. Comput. Graph. Stat.* **58:**299–314.

Jenkins, G. M., Rambaut, A., Pybus, O. G., and Holmes, E. C. (2002). Rates of molecular evolution in RNA viruses: A quantitative phylogenetic analysis. *J. Mol. Evol.* **54:**156–165.

Keese, P., McKenzie, A., and Gibbs, A. J. (1989). Nucleotide sequence of an Australian isolate of turnip yellow mosaic tymovirus. *Virology* **172:**536–546.

Kimura, M. (1983). "The Neutral Theory of Molecular Evolution." Cambridge University Press, Cambridge, UK.

Kingman, J. C. (1982). On the genealogy of large populations. *J. Appl. Prob.* **19:**27–43.

Kühne, T., Shi, N., Proeseler, G., Adams, M. J., and Kanyuka, K. (2003). The ability of a bymovirus to overcome the *rym4*-mediated resistance in barley correlates with a codon change in the VPg coding region on RNA1. *J. Gen. Virol.* **84:**2853–2859.

Kumar, S., Tamura, K., and Nei, M. (2004). MEGA3: Integrated software for molecular evolutionary genetics analysis and sequence alignment. *Brief. Bioinform.* **5:**150–163.

Lecoq, H., and Raccah, B. (2001). Cross-protection: Interactions between strains exploited to control plant virus diseases. *In* "Biotic Interactions in Plant Pathogen Associations" (M. J. Jeger and N. J. Spence, eds.), pp. 177–192. CAB International, Wallingford, UK.

Lecoq, H., Desbiez, C., Wipf-Scheibel, C., and Girard, M. (2003). Potential involvement of melon fruit in long-distance dissemination of cucurbits potyviruses. *Plant Dis.* **87:**955–959.

Lecoq, H., Desbiez, C., Wipf-Scheibel, C., Costa, C., and Girard, M. (2005). Molecular epidemiology of *Watermelon mosaic virus* (WMV, *Potyvirus*) in cucurbits: From simple to complex patterns. IX International Plant Virus Epidemiology Symposium, Lima, Peru, April 4–7, 2005, Abstract, p. 53.

Legg, J. P., and Ogwal, S. (1998). Changes in the incidence of African cassava mosaic geminivirus and the abundance of its whitefly vector along south-north transects in Uganda. *J. Appl. Entomol.* **122:**169–178.

Lewontin, R. C., and Krakauer, J. (1973). Distribution of gene frequency as a test of the theory of the selective neutrality of polymorphisms. *Genetics* **74:**175–195.

Li, H., and Roossinck, M. J. (2004). Genetic bottlenecks reduce population variation in an experimental RNA virus population. *J. Virol.* **78:**10582–10587.

Li, W.-H. (1993). Unbiased estimation of the rates of synonymous and nonsynonymous substitution. *J. Mol. Evol.* **36:**96–99.

Li, W.-H. (1997). "Molecular Evolution." Sinauer Associates, Sunderland, Massachusetts.

Liang, X.-Z., Lee, B. T. K., and Wong, S.-M. (2002). Covariation in the capsid protein of *Hibiscus chlorotic ringspot virus* induced by serial passaging in a host that restricts movement leads to avirulence in its systemic host. *J. Virol.* **76:**12320–12324.

Lin, H.-X., Rubio, L., Smythe, A. B., and Falk, B. W. (2004). Molecular population genetics of *Cucumber mosaic virus* in California: Evidence for founder effects and reassortment. *J. Virol.* **78:**6666–6675.

Lin, S.-S., Hou, R. F., and Yeh, S.-D. (2000). Heteroduplex mobility and sequence analyses for assessment of variability of *Zucchini yellow mosaic virus*. *Phytopathology* **90:**228–235.

Malpica, J. M., Fraile, A., Moreno, I., Obies, C. I., Drake, J. W., and García-Arenal, F. (2002). The rate and character of spontaneous mutation in an RNA virus. *Genetics* **162:**1505–1511.

Manly, B. J. F. (1997). "Randomization, Bootstrap and Monte Carlo Methods in Biology," p. 399. Chapman and Hall, London.

Mantel, N. (1963). Chi-square test with one degree of freedom: Extension of the Mantel Haenszel procedure. *J. Am. Stat. Assoc.* **58:**690–700.

Marco, C. F., and Aranda, M. A. (2005). Genetic diversity of a natural population of *Cucurbit yellow stunting disorder virus*. *J. Gen. Virol.* **86:**815–822.

Martin, D. P., van der Walt, E., Posada, D., and Rybicki, E. P. (2005). The evolutionary value of recombination is constrained by genome modularity. *PLoS Genet.* **1:**e51.

Mastari, J., Lapierre, H., and Dessens, J. T. (1998). Asymmetrical distribution of barley yellow dwarf virus PAV variants between host plant species. *Phytopathology* **88:**818–821.

McDonald, J. H., and Kreitman, M. (1991). Adaptive protein evolution at the Adh locus in *Drosophila*. *Nature* **351:**652–654.

McNeil, J. E., French, R., Hein, G. L., Baezinger, P. S., and Eskridge, K. M. (1996). Characterization of genetic variability among natural populations of wheat streak mosaic virus. *Phytopathology* **86:**1222–1227.

Moonan, F., and Mirkov, T. E. (2002). Analyses of genotypic diversity among North, South, and Central American isolates of *Sugarcane yellow leaf virus*: Evidence for Colombian origins and for intraspecific spatial phylogenetic variation. *J. Virol.* **76:**1339–1348.

Moreno, I. M., Malpica, J. M., Díaz-Pendon, J. A., Moriones, E., Fraile, A., and García-Arenal, F. (2004). Variability and genetic structure of the population of watermelon mosaic virus infecting melon in Spain. *Virology* **318:**451–460.

Morris, C. E., Bardin, M., Berge, O., Frey-Klett, P., Fromin, N., Girardin, H., Guinebretière, M.-H., Lebaron, P., Thiéry, J. M., and Troussellier, M. (2002). Microbial biodiversity: Approaches to experimental design and hypothesis testing in primary scientific literature from 1975 to 1999. *Microbiol. Mol. Biol. R.* **66:**592–616.

Moury, B. (2004). Differential selection of genes of *Cucumber mosaic virus* subgroups. *Mol. Biol. Evol.* **21:**1602–1611.

Moury, B., Morel, C., Johansen, E., and Jacquemond, M. (2002). Evidence for diversifying selection in potato virus Y and in the coat protein of other potyviruses. *J. Gen. Virol.* **83:**2563–2573.

Moury, B., Morel, C., Johansen, E., Guilbaud, L., Souche, S., Ayme, V., Caranta, C., Palloix, A., and Jacquemond, M. (2004). Mutations in *Potato virus Y* genome-linked protein determine virulence toward recessive resistances in *Capsicum annuum* and *Lycopersicon hirsutum*. *Mol. Plant Microbe Interact.* **17:**322–329.

Moya, A., Rodríguez-Cerezo, E., and García-Arenal, F. (1993). Genetic structure of natural populations of the plant RNA virus tobacco mild green mosaic virus. *Mol. Biol. Evol.* **10:**449–456.

Moya, A., Elena, S. F., Bracho, A., Miralles, R., and Barrio, E. (2000). The evolution of RNA viruses: A population genetics view. *Proc. Natl. Acad. Sci., USA* **97:**6967–6973.

Nagy, P. D., and Bujarski, J. J. (1997). Engineering of homologous recombination hotspots with AU-rich sequences in brome mosaic virus. *J. Virol.* **71:**3799–3810.

Negroni, M., and Buc, H. (2001). Mechanisms of retroviral recombination. *Annu. Rev. Genet.* **35:**275–302.

Nei, M. (1987). "Molecular Evolutionary Genetics," p. 512. Columbia University Press, New York.

Nei, M., and Tajima, F. (1981). DNA polymorphisms detectable by restriction endonucleases. *Genetics* **97:**145–163.

Nie, X. Z., and Singh, R. P. (2003). Evolution of North American PVY^{NTN} Strain Tu 660 from local PVY^{N} by mutation rather than recombination. *Virus Genes* **26:**39–47.

Nielsen, R. (2001). Statistical tests of selective neutrality in the age of genomics. *Heredity* **86:**641–647.

Nielsen, R., and Yang, Z. (1998). Likelihood models for detecting positively selected amino acid sites and application to the HIV-1 envelope gene. *Genetics* **148:**929–936.

Novella, I. S. (2003). Contributions of vesicular stomatitis virus to the understanding of RNA virus evolution. *Curr. Opin. Microbiol.* **6:**399–405.

Nutter, F. W., Schultz, P. M., and Hill, J. H. (1998). Quantification of within-field spread of soybean mosaic virus in soybean using strain-specific monoclonal antibodies. *Phytopathology* **88:**895–901.

Olmos, A., Bertolini, E., Gil, M., and Cambra, M. (2005). Real-time assay for quantitative detection of non-persistently transmitted *Plum pox virus* RNA targets in single aphids. *J. Virol. Methods* **128:**151–155.

Ong, C. K., Nee, S., Rambaut, A., and Harvey, P. H. (1996). Inferring the population history of an epidemic from a phylogenetic tree. *J. Theor. Biol.* **182:**173–178.

Ong, C. K., Nee, S., Rambaut, A., Bernard, H.-U., and Harvey, P. H. (1997). Elucidating the population histories and transmission dynamics of papillomaviruses using phylogenetic trees. *J. Mol. Evol.* **44:**199–206.

Ooi, K., and Yahara, T. (1999). Genetic variation of geminiviruses: Comparison between sexual and asexual host populations. *Mol. Ecol.* **8:**89–97.

Otto, S. (2000). Detecting the form of selection from DNA sequence data. *Trends Genet.* **16:**526–529.

Paalme, V., Gammelgard, E., Järvekülg, L., and Valkonen, J. P. T. (2004). *In vitro* recombinants of two nearly identical potyviral isolates express novel virulence and symptom phenotypes in plants. *J. Gen. Virol.* **85:**739–747.

Pamilo, P., and Bianchi, N. O. (1993). Evolution of the Zfx and Zfy genes: Rates and interdependence between the genes. *Mol. Biol. Evol.* **10:**271–281.

Perelson, A. S., Neumann, A. U., Markowitz, M., Leonard, J. M., and Ho, D. D. (1996). HIV-1 dynamics *in vivo*: Virion clearance rate, infected cell life-span, and viral generation time. *Science* **271:**1582–1586.

Perry, K. L., Zhang, L., and Palukaitis, P. (1998). Amino acid changes in the coat protein of cucumber mosaic virus differentially affect transmission by the aphids *Myzus persicae* and *Aphis gossypii*. *Virology* **242:**204–210.

Pfosser, M. F., and Baumann, H. (2002). Phylogeny and geographical differentiation of *Zucchini yellow mosaic virus* isolates (Potyviridae) based on molecular analysis of the coat protein and part of the cytoplasmic inclusion protein genes. *Arch. Virol.* **147:**1599–1609.

Polston, J. E., McGovern, R. J., and Brown, L. G. (1999). Introduction of *Tomato yellow leaf curl virus* in Florida and implications for the spread of this and other geminiviruses of tomato. *Plant Dis.* **83:**984–988.

Posada, D., and Crandall, K. A. (2001). Evaluation of methods for detecting recombination from DNA sequences: Computer simulations. *Proc. Natl. Acad. Sci. USA* **98:**13757–13762.

Przeworski, M. (2002). The signature of positive selection at randomly chosen loci. *Genetics* **160:**1179–1189.

Pybus, O. G., Holmes, E. C., and Harvey, P. H. (1999). The mid-depth method and HIV-1: A practical approach for testing hypotheses of viral epidemic history. *Mol. Biol. Evol.* **16:**953–959.

Pybus, O. G., Rambaut, A., and Harvey, P. H. (2000). An integrated framework for the inference of viral population history from reconstructed genealogies. *Genetics* **155:**1429–1437.

Quiot, J.-B., Devergne, J.-C., Cardin, L., Verbrugghe, M., Marchoux, G., and Labonne, G. (1979). Ecologie et épidémiologie du virus de la mosaïque du concombre dans le Sud-Est de la France VII. Répartition de deux types de populations virales dans les cultures sensibles. *Annu. Rev. Phytopathol.* **11:**359–373.

Revers, F., Le Gall, O., Candresse, T., Le Romancer, M., and Dunez, J. (1996). Frequent occurrence of recombinant potyvirus isolates. *J. Gen. Virol.* **77:**1953–1965.

Rodríguez-Cerezo, E., Moya, A., and García-Arenal, F. (1989). Variability and evolution of the plant RNA virus pepper mild mottle virus. *J. Virol.* **63:**2198–2203.

Roscoe, J. T., and Byars, J. A. (1971). Sample size restraints commonly imposed on the use of chi-square statistic. *J. Amer. Statist. Assoc.* **66:**755–759.

Rubio, L., Ayllon, M. A., Kong, P., Fernandez, A., Polek, M., Guerri, J., Moreno, P., and Falk, B. W. (2001). Genetic variation of *Citrus tristeza virus* isolates from California and Spain: Evidence for mixed infections and recombination. *J. Virol.* **75:**8054–8062.

Sacristán, S., Malpica, J. M., Fraile, A., and García-Arenal, F. (2003). Estimation of population bottlenecks during systemic movement of *Tobacco mosaic virus* in tobacco plants. *J. Virol.* **77:**9906–9911.

Salánki, K., Gellért, Á., Huppert, E., Náray-Szabó, G., and Balázs, E. (2004). Compatibility of the movement protein and the coat protein of cucumoviruses is required for cell-to-cell movement. *J. Gen. Virol.* **85:**1039–1048.

Salati, R., Nahkla, M. K., Rojas, M. R., Guzman, P., Jaquez, J., Maxwell, D. P., and Gilbertson, R. L. (2002). *Tomato yellow leaf curl virus* in the Dominican Republic: Characterization of an infectious clone, virus monitoring in whiteflies, and identification of reservoir hosts. *Phytopathology* **92:**487–496.

Sanchez-Campos, S., Diaz, J. A., Monci, F., Bejarano, E. R., Reina, J., Navas-Castillo, J., Aranda, M. A., and Moriones, E. (2002). High genetic stability of the begomovirus *Tomato yellow leaf curl Sardinia virus* in southern Spain over a 8-year period. *Phytopathology* **92:**842–849.

Schirmer, A., Link, D., Cognat, V., Moury, B., Beuve, M., Meunier, A., Bragard, C., Gilmer, D., and Lemaire, O. (2005). Phylogenetic analysis of isolates of beet necrotic yellow vein virus collected worldwide. *J. Gen. Virol.* **86:**2897–2911.

Schneider, W. L., and Roossinck, M. J. (2001). Genetic diversity in RNA virus quasispecies is controlled by host-virus interactions. *J. Virol.* **75:**6566–6571.

Shannon, C. E., and Weaver, W. (1949). "The Mathematical Theory of Communications." University of Illinois Press, Urbana.

Shapka, N., and Nagy, P. D. (2004). The AU-rich RNA recombination hot spot sequence of *Brome mosaic virus* is functional in tombusviruses: Implications for the mechanism of RNA recombination. *J. Virol.* **78:**2288–2300.

Sheffield, V. C., Beck, J. S., Kwitek, A. E., Sandstrom, D. W., and Stone, E. M. (1993). The sensitivity of single-strand conformation polymorphism analysis for the detection of single base substitutions. *Genomics* **16:**325–332.

Shukla, D. D., and Ward, C. W. (1989). Structure of potyvirus coat protein and its applications in the taxonomy of the potyvirus group. *Advances in Virus Res.* **36:**273–314.

Simmonds, P., and Smith, D. B. (1999). Structural constraints on RNA virus evolution. *J. Virol.* **73:**5787–5794.

Simpson, E. H. (1949). Measurement of diversity. *Nature* **163:**688.

Sinisterra, X., Patte, C. P., Siewnath, S., and Polston, J. (2000). Identification of *Tomato yellow leaf curl virus-Is* in the Bahamas. *Plant Dis.* **84:**592.

Skotnicki, M. L., Mackenzie, A. M., and Gibbs, A. J. (1996). Genetic variation in populations of kennedya yellow mosaic tymovirus. *Arch. Virol.* **141:**99–110.

Stenger, D. C., Seifers, D. L., and French, R. (2002). Patterns of polymorphism in *Wheat streak mosaic virus*: Sequence space explored by a clade of closely related viral genotypes rivals that between the most divergent strains. *Virology* **302:**58–70.

Suzuki, Y., and Gojobori, T. (1999). A method for detecting positive selection at single amino acid sites. *Mol. Biol. Evol.* **16:**1315–1328.

Tajima, F. (1989). Statistical method for testing the neutral mutation hypothesis by DNA polymorphism. *Genetics* **123:**597–601.

Tatchell, G. M. (1990). Monitoring and forecasting aphid problems. *In* "Aphid-Plant Interactions: Populations to Molecules" (D. C. Peters, J. A. Webster, and C. S. Chlouber, eds.), pp. 215–231. Oklahoma State University Press, Stillwater, Oklahoma, USA.

Teycheney, P.-Y., Laboureau, N., Iskra-Caruana, M.-L., and Candresse, T. (2005). High genetic variability and evidence for plant-to-plant transfer of *Banana mild mosaic virus*. *J. Gen. Virol.* **86:**3179–3187.

Tobias, I., Palkovics, L., Tzekova, L., and Balazs, E. (2001). Replacement of the coat protein gene of plum pox potyvirus with that of zucchini yellow mosaic potyvirus: Characterization of the hybrid potyvirus. *Virus Res.* **76:**9–16.

Tomimura, K., Gibbs, A. J., Jenner, C. E., Walsh, J. A., and Ohshima, K. (2003). The phylogeny of *Turnip mosaic virus*: Comparisons of 38 genomic sequences reveal a Eurasian origin and a recent "emergence" in East Asia. *Mol. Ecol.* **12:**2099–2111.

Tomimura, K., Spak, J., Katis, N., Jenner, C. E., Walsh, J. A., Gibbs, A., and Ohshima, K. (2004). Comparisons of genetic structure of populations of Turnip mosaic virus in West and East Eurasia. *Virology* **330:**408–423.

Tsompana, M., Abad, J., Purugganan, M., and Moyer, W. (2005). The molecular population genetics of the *Tomato spotted wilt virus* (TSWV) genome. *Mol. Ecol.* **14:**53–66.

van Regenmortel, M. H. V., Fauquet, C. M., Bishop, D. H. L., Carstens, E. B., Estes, M. K., Lemon, S. M., Maniloff, J., Mayo, M. A., McGeoch, D. J., Pringle, C. R., and Wickner, R. B. (eds.) (2000). Virus Taxonomy. Seventh Report of the International Committee on Taxonomy of Viruses. Academic Press, San Diego, California.

Vandamme, A. M. (2002). Virus evolution and molecular epidemiology. *Virus Res.* **85:**1–3.

Vigne, E., Komar, V., and Fuchs, M. (2004). Field safety assessment of recombination in transgenic grapevines expressing the coat protein gene of *Grapevine fanleaf virus*. *Transgenic Res.* **13:**165–179.

Vives, M. C., Rubio, L., Galipienso, L., Navarro, L., Moreno, P., and Guerri, J. (2002). Low genetic variation between isolates of Citrus blotch virus from different host species and of different geographical origins. *J. Gen. Virol.* **83:**2587–2591.

Watterson, G. (1978). The homozygosity test of neutrality. *Genetics* **88:**405–417.

Weir, B. S., and Cockerham, C. C. (1984). Estimating F-statistics for the analysis of population structure. *Evolution* **38:**1358–1370.

Werker, A. R., Dewar, A. M., and Harrington, R. (1998). Modelling the incidence of virus yellows in sugar beet in the UK in relation to numbers of *Myzus persicae*. *J. Appl. Ecol.* **35:**811–818.

Woelk, C. H., and Holmes, E. C. (2002). Reduced positive selection in vector-borne RNA viruses. *Mol. Biol. Evol.* **19:**2333–2336.

Woelk, C. H., Pybus, O. G., Jin, L., Brown, D. W., and Holmes, E. C. (2002). Increased positive selection pressure in persistent (SSPE) versus acute measles virus infections. *J. Gen. Virol.* **83:**1419–1430.

Woolf, B. (1955). On estimating the relation between blood group and disease. *Ann. Hum. Genet.* **19:**251–253.

Worobey, M., and Holmes, E. C. (1999). Evolutionary aspects of recombination in RNA viruses. *J. Gen. Virol.* **80:**2535–2543.

Wright, S. (1951). The genetical structure of populations. *Ann. Eugen.* **15:**323–354.

Yang, Z. (1997). PAML: A program package for phylogenetic analysis by maximum likelihood. *Comp. Appl. Biosci.* **13:**555–556.

Yang, Z., and Bielawski, J. P. (2000). Statistical methods for detecting molecular adaptation. *Trends Ecol. Evol.* **15:**496–503.

Yang, Z., Nielsen, R., Goldman, N., and Krabbe Pedersen, A.-M. (2000). Codon-substitution models for heterogeneous selection pressure at amino acid sites. *Genetics* **155:**431–449.

Yates, F. (1934). Contingency tables involving small numbers and the χ^2 test. *J. Roy. Stat. Soc.* **1**(Suppl.)**:**217–235.

Zar, J. H. (1984). "Biostatistical Analysis." Prentice-Hall, Inc., Englewood Cliffs, New Jersey.

ADVANCES IN VIRUS RESEARCH, VOL 67

PLANT VIRUS EPIDEMIOLOGY: THE CONCEPT OF HOST GENETIC VULNERABILITY

J. M. Thresh

Natural Resources Institute, University of Greenwich, Chatham Maritime
Kent ME4 4TB, United Kingdom

There are many reports in the plant pathology literature of diseases that were first reported or became prevalent following the introduction of a particular cultivar or group of closely related cultivars. This has led to the concept of host genetic vulnerability, which is considered here. Numerous examples of vulnerability are discussed and they relate to virus diseases of a wide range of tropical, subtropical and temperate crops, including some of great economic importance. The adoption of vulnerable cultivars is seen as one of the factors 'driving' epidemics by disrupting the equilibria to be expected between pathogens and their hosts. Such an interpretation is consistent with the observation that many examples of genetic vulnerability relate to cultivars that have been introduced to new areas where they were severely affected by viruses or virus strains not encountered previously.

I. Introduction

It has long been known that some crop cultivars are affected particularly severely by disease and there are many examples in the plant pathology literature. However, this topic received greatly increased

0065-3527/06 $35.00
DOI: 10.1016/S0065-3527(06)67003-6

attention following the 1970 epidemic of southern corn leaf blight in the United States. This was caused by *Colletotrichum maydis* (*Cochlio-*Cochliobolus heterostrophus) and associated with the introduction of hybrid maize developed by using male-sterile lines later found to have introduced extreme blight susceptibility (Moore, 1970). The occurrence of such a damaging epidemic in a country that has so many plant pathologists and a long and distinguished history of crop research led to considerable debate on what became known as 'genetic vulnerability'. One important outcome was the influential proceedings of a special conference on the topic arranged by the New York Academy of Sciences (Day, 1977). Particular concern was expressed at the decrease in genetic diversity that was occurring in many crops as a consequence of an increasing emphasis on a limited number of cultivars that were often of similar genetic background.

Plant virologists made only a limited contribution to the debate, either at the time or subsequently. However, host genetic vulnerability has been reported in studies on plant virus diseases, as described previously (Thresh, 1980, 1990). These and other examples are discussed here to show the importance of vulnerability and how it can be regarded as the underlying cause of some of the most damaging epidemics that have occurred.

II. Examples of Host Genetic Vulnerability

A. *Cassava Diseases*

1. *Cassava Mosaic Disease*

Cassava mosaic disease (CMD) has been known since the first reports from what is now Tanzania in 1894. It soon became apparent from observations in West Africa and elsewhere that some cultivars were more severely damaged by the disease than others and that farmers avoided serious losses by adopting ones that were relatively unaffected. These findings have been substantiated by subsequent observations in many African countries and especially during the 1990s epidemic in Uganda.

The onset and progress of the epidemic in Uganda has been described and discussed in detail (Otim-Nape and Thresh, 2006; Otim-Nape *et al.*, 2000; Legg *et al.*, this volume, pp. 355–418). The first reports were from a localized part of Luwero district where there was at the time a heavy reliance on the local cassava landrace Senyonjo, which was very severely affected. It then became apparent from surveys and field observations that the epidemic was progressing into other areas and that some districts were much more affected than others.

In Kumi and some of the other worst-affected districts, the local landrace Ebwanateraka predominated in virtually all plantings in 1990–1992 and few other cultivars were being grown (Table I). Elsewhere, there was much greater diversity, many cultivars were being grown and Ebwanateraka was absent or scarcely represented. Ebwanateraka had been selected and adopted some years previously by Ugandan farmers at a time when CMD was unimportant and caused little or no obvious damage. It was high-yielding, cropped early and the tuberous roots were of excellent quality. For these reasons, Ebwanateraka was adopted throughout large areas of Uganda to produce roots for local consumption and for sale in urban areas and also for export to the neighbouring countries of Sudan and Kenya.

TABLE I

THE NUMBER OF CULTIVARS THAT PREDOMINATED IN ONE OR MORE OF THE PLANTINGS ASSESSED IN EACH OF FOUR SURVEYS OF 12 DISTRICTS OF UGANDA BETWEEN 1990–1992 AND 2003[a]

District	1990–1992[b]	1994[b]	1997[b]	2003[b]
Iganga	0/1 (E)[c]	0/1 (E)[c]	3/5 (E)[c]	8/18 (–)[c]
Kamuli	0/4 (E)	0/4 (E)	2/4 (–)	5/17 (–)
Kumi	0/1 (E)	2/7 (–)	3/4 (–)	4/5 (–)
Luwero	0/1 (E)	0/8 (–)	1/6 (–)	9/19 (–)
Soroti	0/1 (E)	2/5 (–)	2/3 (–)	3/5 (–)
Tororo	0/3 (E)	0/6 (–)	2/4 (E)	15/25 (–)
Hoima	0/13 (–)	0/18 (–)	0/8 (–)	2/20 (–)
Kibaale	0/11 (–)	0/9 (–)	0/12 (–)	1/11 (–)
Masaka	0/13 (–)	0/7 (–)	0/5 (–)	6/16 (–)
Masindi	0/8 (–)	0/8 (–)	1/9 (–)	1/15 (–)
Mubende	0/11 (–)	0/10 (–)	0/14 (–)	1/21 (–)
Nebbi	0/13 (–)	0/13 (–)	–	1/13 (–)

[a] Data for Iganga and five other districts in which the landrace cultivar Ebwanateraka predominated in 1990–1992 and for six other districts in which Ebwanateraka did not predominate and where there was much greater diversity (Bua *et al.*, 2005; Otim-Nape *et al.*, 1998, 2001).

[b] The number of CMD-resistant cultivars that predominated in one or more of the plantings surveyed as a fraction of the total number of cultivars recorded during each of the four surveys. Note the first appearance of CMD-resistant cultivars in 1994 and their increasing importance in subsequent surveys.

[c] Districts in which Ebwanateraka did (E) or did not (–) predominate.

Ebwanateraka proved to be extremely vulnerable to the severe epidemic form of CMD that progressed through much of Uganda in the 1990s and losses were devastating (Otim-Nape *et al.*, 2000). All but the most recently affected plants were so stunted that they produced little or no yield of tuberous roots and suitable stem cuttings were not available for further plantings. Many rural households suffered an almost total loss of income, food shortages occurred and famine-related deaths were reported to a subsequent Presidential Committee of Inquiry. The hardship was particularly severe following the 1993–1994 drought as cassava had previously been available as a famine reserve to be used at times of scarcity. In the almost complete absence of cassava, farmers had no option but to resort to sweet potato or other crops and commodity prices soared.

The situation in the worst-affected areas was so serious that food relief supplies had to be provided and various governmental and non-governmental projects issued cuttings of CMD-resistant and other cassava cultivars. Details of these rehabilitation projects are available elsewhere (Otim-Nape *et al.*, 2000) and the changeover in cultivars that has occurred over the last decade is apparent from the results of comprehensive farm surveys made in 1990–1992, 1994 and 2003 (Table I). There have also been additional surveys in some areas and up to 25 cultivars are now grown in some districts where previously only Ebwanateraka predominated. Some of the introduced cultivars were selected and released by the National Cassava Programme as resistant to CMD. Others were landraces that had been selected locally by farmers or introduced from elsewhere in Uganda and found to be somewhat resistant to or tolerant of infection. Production has been restored and the crisis has been overcome.

The situation was completely different in many parts of west and southwest Uganda where cassava production was relatively unimportant. Many cultivars were being grown and farmers had less reliance on cassava for food security and income. In many individual plantings, numerous cultivars were being grown and the onset of the epidemic did not have very drastic affects. Some cultivars were severely damaged and farmers could be seen uprooting the worst-affected plants whilst retaining those least affected. Farmers were soon able to adapt, even though they had little or no knowledge of CMD and only limited access to agricultural extension services. This was a notable example of the benefits of crop genetic diversity in providing farmers with at least some degree of resilience in being able to adapt to arthropod pests, diseases, or environmental hazards.

There is much additional evidence of farmers being able to adjust in this way from experience in Uganda and other parts of Africa following the first appearance of cassava mealybug (*Phenacoccus manihoti*), cassava green mite (*Mononychellus tanajoa*) and cassava bacterial blight (*Xanthomonas axonopodis* pv. *manihoti*) (Nweke *et al.*, 1994). This is apparent from the comprehensive 'COSCA' survey of cassava in several African countries. It reported a substantial turnover in the cultivars being grown as some were abandoned and others were introduced. The changes were attributed to the need to overcome the damaging effects of arthropod pests, diseases and unfavourable environmental factors (Nweke, 1994).

2. Cassava Brown Streak Disease

Cassava brown streak disease (CBSD) was first reported in what is now Tanzania in 1936 (Storey, 1936). It was found later in other coastal areas of eastern and southern Africa and also further inland in Malawi, Uganda (Nichols, 1950) and more recently in Democratic Republic of Congo (Mahungu *et al.*, 2003). An important feature of CBSD is that the symptoms are very variable in type and severity (Nichols, 1950). Only some of the many landraces being grown in the affected areas develop stem necrosis, dieback and necrosis of the tuberous roots in addition to leaf discolouration. Farmers in some areas have to some extent adjusted to the problem by discarding the most vulnerable landraces and adopting relatively tolerant ones that express only inconspicuous leaf symptoms. Moreover, farmers tend to select the most vigorous plants that are unaffected by necrosis to provide stem cuttings for further plantings. Another 'coping strategy' is to sacrifice some yield and harvest early before the root necrosis has had time to develop (Hillocks *et al.*, 2001). These practices have been adopted intuitively by farmers who have little or no knowledge of plant pathology and may be unaware that CBSD is an infectious disease that is caused by a whitefly-borne virus (Maruthi *et al.*, 2005).

The effects of CBSD have been particularly devastating in coastal areas of the Nampula region of northern Mozambique. Soils there are generally poor and the population is heavily dependent on cassava, which is one of the few subsistence crops that can be grown in the harsh environment. The occurrence of CBSD was not recognized until the 1990s (Hillocks *et al.*, 2002) and was associated with the release of the cultivar 'Calamidade' in rehabilitation programmes mounted after the devastation caused by the severe cyclone in 1994. The origin of Calamidade is not known and it was introduced with the best of intentions as a means of restoring production after the catastrophe. However, it proved to be extremely vulnerable and in a recent survey 96% of the 4000 plants examined expressed root necrosis and the mean symptom severity score was 3.5 on a scale of 1 (no necrosis) to 5 (most severe necrosis). Calamidade is now being replaced by relatively tolerant cultivars and it is estimated that the discounted benefits of doing this will be US$ 15–20 million (McSween, S., personal communication).

Early government-funded attempts were made in Tanzania to breed CBSD-resistant cultivars (Jennings, 1960) and some of these have been adopted by farmers and given local names. Moreover, recently there have been introductions of clonal selections and also of seed provided by the International Institute of Tropical Agriculture (IITA), Nigeria. CBSD is not known to occur in West Africa and some of the introductions were extremely vulnerable to infection when grown in evaluation trials in coastal areas of Tanzania, Mozambique and Kenya. One of the introductions was the IITA clone TMS (Tropical *Manihot* Series) 42025, which developed very severe symptoms.

A recent development has been the appearance and spread of a severe form of CBSD in Uganda (Legg *et al.*, this volume, pp. 355–418). Two Nigerian landraces (TME (Tropical *Manihot esculenta*) 14 and TME 204) and some of the seedling progenies derived from them were extremely vulnerable to infection and developed conspicuous leaf symptoms and root necrosis. This has caused serious concern because much of the TME material has good growth characteristics and contains the dominant and highly effective *CMD-2* gene for resistance to cassava mosaic viruses (Akano *et al.*, 2002). Consequently, the TME clones and their derivatives have been distributed widely in Uganda and neighbouring countries following the occurrence of the mosaic pandemic (Legg *et al.*, this volume, pp. 355–418). The full consequences of this development and the future of TME-derived genotypes in areas affected by or at risk from CBSD are not yet apparent.

B. *Sugarcane Diseases*

1. *Sugarcane Mosaic Disease*

The history of sugarcane production in Louisiana, United States, provides a striking example of the way in which a virus disease can have a drastic and continuing influence on the type of cultivar grown and on the breeding practices adopted.

Sugarcane mosaic disease, later shown to be caused by an aphid-borne virus, was first reported in 1892. It has since been found in virtually all countries where sugarcane is grown. The first reports from Louisiana were in 1919 when the main cultivars being grown were traditional 'noble' canes (*Saccharum officinarum*) of tropical origin that had been grown successfully for many years. The noble canes proved to be extremely vulnerable to mosaic and infection soon became widespread through the dissemination of infected planting material and spread by aphid vectors (Summers *et al.*, 1948). Production declined from an average of 200,000 tonnes per year to only 47,000 tonnes in 1926, when losses were estimated to be US$ 100 million (Brandes and Sartoris, 1936). In the 1920s, it seemed that sugarcane production in Louisiana might have to be abandoned. Many farmers were declared bankrupt and sugar producing companies went out of business. However, the situation changed when 'P.O.J.' cultivars derived from crosses between *S. officinarum* and *S. barberi* were introduced from the Proefstation, East Java. These cultivars were much more tolerant of infection than the noble canes, which were almost entirely replaced between 1924 and 1929. The P.O.J. cultivars were then largely superseded by Co.281 and Co.290. These were mosaic-tolerant cultivars from Coimbatore, India, derived from *S. officinarum* × *S. spontaneum* × *S. barberi* crosses.

The next important development was the release in 1930 of the first of a sequence of locally-produced 'C.P.' cultivars that were selected at Canal Point, Florida, for their resistance to mosaic. By 1955, C.P. selections accounted for 95% of the total area and led to a big decrease in the incidence of mosaic. This was no longer regarded as a serious problem and in 1954 a decision was taken to release an introduced cultivar from Coimbatore (N.Co.310). It was high-yielding and cold-resistant, but susceptible to mosaic and somewhat tolerant. This led to a marked deterioration in the disease situation. Moreover, strain H of *Sugarcane mosaic virus* was first reported in 1956 and shown to infect C.P.22–101 and other widely-grown cultivars that had previously been highly resistant (Abbott, 1958).

By 1960, only 12% of the Florida area was planted with cultivars that were still rated as resistant. Infection again became prevalent and several widely-grown cultivars had to be totally or partially replaced. It was also necessary to reintroduce roguing as a control measure in attempts to ensure the availability of healthy stocks of vegetative planting material (Abbott, 1962).

Roguing is inconvenient, expensive and not always effective, especially in areas where infection is prevalent or where virus-tolerant cultivars are grown. Several of these are still widely planted because it has been difficult to breed virus-resistant cultivars of acceptable yield and quality. A particular problem has been that *S. spontaneum*, which was the main source of resistance used in the original C.P. cultivars, is susceptible to strain H (Abbott and Todd, 1962). These factors explain why mosaic has continued to cause losses, albeit much less than those reported in the 1920s before the first of the resistant cultivars was introduced (Breaux, 1985).

2. *Sugarcane Fiji Leaf Gall Disease*

Additional ways in which disease problems are closely related to the vulnerability of the cultivars being grown are illustrated by experience with sugarcane Fiji leaf gall disease (FLGD) (previously referred to as Fiji disease) in Bundaberg and other parts of Queensland, Australia. This disease is caused by a planthopper-borne virus that first attracted attention in Australia in the 1920s when existing cultivars were being replaced by P.O.J.2878, which was less susceptible to a prevalent bacterial gumming disease. The new 'wonder cane' had been selected in Indonesia and proved to be highly susceptible to FLGD. This caused continuing losses despite a well-organized control campaign, involving inspection and eradication and the provision of healthy planting material (Egan and Toohey, 1977).

The official Queensland figures for numbers of sugarcane plants eradicated in attempts to control FLGD underestimate the magnitude of the losses. This is because they exclude data for the worst-affected fields that were abandoned before counts were made. Nevertheless, they reveal a peak of infection during the 1944–1945 season. This was followed by a decline as P.O.J.2878 was replaced by a less susceptible cultivar and eventually by resistant ones. Losses became insignificant until there was an upsurge of FLGD beginning in the late 1960s, following a big increase in vector populations and the introduction of inherently higher-yielding cultivars that were less resistant to infection (Egan, 1976).

Particular problems were encountered with the widely-grown N.Co.310 from India, which is prone to attack by the planthopper vector and susceptible to infection with FLGD. Infected plants of N.Co.310 are slow to develop obvious symptoms and difficult to detect by the plant health inspectors who carry out routine surveys. This decreased the effectiveness of eradication measures and by 1976 it was estimated that there were 10 million infected stocks in the Bundaberg area alone. Virtually all farms were affected and thousands of hectares had to be ploughed out prematurely. In some localities it became difficult to produce adequate supplies of healthy planting material and for any real solution to the problem it was necessary to replace N.Co.310 by less susceptible types (Fig. 1).

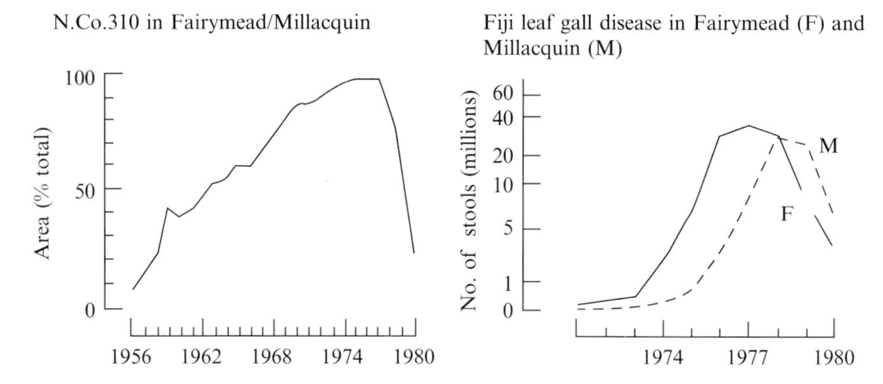

FIG 1. The cultivation of sugarcane cv. N.Co.310 in the Fairymead and Millacquin localities of Bundaberg, Queensland: 1956–1980 (left) and the incidence of sugarcane Fiji leaf gall disease in the Fairymead (F) and Millacquin (M) localities: 1972–1981 (right). Data from Egan (1976) and unpublished. [Note the use of a logarithmic scale to accommodate the wide range of incidence values.] Figure 1 is reproduced from Thresh (1990) by kind permission of the publisher Springer, Netherlands.

N.Co.310 was removed from the list of approved cultivars in much of the Bundaberg area at the end of 1978 and from the remaining localities in 1980. This was a crucial development in bringing the epidemic under control. A similar policy was also effective in preventing the losses that are otherwise likely to have occurred in the Plane Creek area of Queensland where FLGD was first reported in 1982 and immediate steps were taken to replace N.Co.310. From these experiences it is apparent why the disease has been referred to as 'N.Co.310 disease' (Ryan, 1988).

C. *Tree Crop Diseases*

1. *Cacao Swollen Shoot Disease*

Cacao swollen shoot disease (CSSD) has featured in previous reviews on 'catastrophic' and 'threatening' diseases (Klinkowski, 1970; Thurston, 1973). It was first reported in Gold Coast (now Ghana) in 1936, although there is anecdotal evidence that it occurred earlier in the 1920s. This was little more than 20 years after cacao was first introduced to the West African mainland. Virtually all the initial plantings were of the Amelonado cultivar, which originated in South America. Production in Ghana soon expanded rapidly and at first the cultivar seemed ideally suited to the lowland rainforest areas. Moreover, the beans produced were of the type and quality that were required by chocolate manufacturers in Europe and North America. Accordingly, no attempt was made to extend the range of germplasm available by making further introductions from South or Central America until CSSD appeared and the full magnitude of the CSSD problem was recognized in the early 1940s.

It was then apparent that Amelonado trees were extremely vulnerable and they were killed within 1–2 years of infection with the virulent 'New Juaben' (IA) strain of the mealybug-transmitted virus that causes CSSD (Brunt, 1975; Crowdy and Posnette, 1947). This strain predominated in the Eastern Region of Ghana, which was at the time the most important cacao-growing area in the world. Strains of similar type also occurred in adjacent parts of Côte d'Ivoire, which explains the very severe losses that have occurred in these areas. Ampofo (1997) estimated that 193 million trees had been removed in Ghana alone in attempts to contain the spread of CSSD and millions more were killed by the disease before they could be removed by the authorities responsible for the 'cutting out' campaign. Serious losses have also occurred in Côte d'Ivoire, Nigeria, Togo and elsewhere in West Africa, where Amelonado was the usual cultivar being grown. Overall, the losses have been immense, mainly due to the destruction of infected trees. Even if such trees survive, they produce substandard yields and are prone to attack by cacao mirids (Thresh, 1960). Moreover, attempts to control CSSD have been costly and for many years the expense of the cutting out campaign dominated the budget of the Ghana Department of Agriculture. In addition, personnel and resources were diverted from other cacao improvement projects or other means of enhancing agricultural production. CSSD control measures and cacao cultivation have been abandoned in large areas of Eastern Region and first Ashanti/Brong Ahafo and more recently Western Region became the main areas of production in Ghana.

Cacao is so crucial to the economy of Ghana and to the well-being of the rural communities that in 1943, at the height of the Second World War (1939–1945), scarce resources were allocated for additional cacao genotypes to be introduced (Posnette, 1951). This was to extend the range of material available for a virus resistance breeding programme. The introductions were acquired in Trinidad from the collection assembled there of material obtained during previous expeditions to the Amazon Basin of South America where cacao originated. Seeds were transported to alleviate quarantine problems. They were then established mainly in Ghana, although some were planted in Nigeria.

By 1948, the introduced trees were producing seed and progenies were screened for resistance to *Cacao swollen shoot virus* (Posnette and Todd, 1951). Several of the introductions were found to be considerably more resistant to infection by the mealybug vectors than the equivalent Amelonado controls. The introductions were also relatively tolerant of infection and developed inconspicuous symptoms. An additional feature of some of the introductions was that they were early-bearing, vigorous and easier to establish than the Amelonado grown previously. Moreover, the beans produced were acceptable to manufacturers for processing and in 1954 seed from selected Amazon trees was released for use by farmers (Knight and Rogers, 1955). There was a particular demand where replanting was being done to replace trees destroyed by CSSD, or those removed during control operations. Selected Upper Amazon types were also interbred and some were crossed with Amelonado or Trinitario types to produce vigorous hybrids. In addition, Upper Amazons have been used in breeding programmes to produce progenies that have even greater resistance to virus infection (Thresh *et al.*, 1988).

The material derived from the original 1944 introductions and others made since have made a substantial contribution to the continued production of cacao in Ghana. This is because of their superior agronomic attributes and resistance/tolerance to CSSD. Consequently, they are likely to make an even greater contribution in future. It is already apparent that the disease would have had much less impact if other genotypes or a wider range of introductions had been made at the outset of cacao production in the 1880s. Nevertheless, large areas of Amelonado plantings remain. They are at severe risk of infection and it will be many years before a complete changeover to other less vulnerable cultivars has been achieved.

2. *Coconut Diseases*

The virus disease now known as coconut foliar decay (CFD) is limited to Vanuatu in the Pacific. It was not known until additional cultivars were introduced to the islands in the 1960s as part of a coconut improvement programme to replace the ageing plantations of local cultivars (Hanold *et al.*, 2003). The introduced 'Malayan Red Dwarf' was particularly susceptible to infection and began to show symptoms within 18 months of planting. There was up to 90% mortality within 10 years in areas where the local 'Vanuatu tall' was unaffected (Calvez *et al.*, 1980).

In subsequent studies, it was established that CFD is caused by a seemingly indigenous planthopper-borne virus that infects local cultivars but has little or no effect on their growth. It is estimated that about half the introduced germplasm seedlings were destroyed by CFD.

This has severely affected the scope for crop improvement by restricting the range of cultivars that can be grown successfully in Vanuatu. Those recommended and now being used are less productive than the hybrids being grown elsewhere in the Pacific Region where CFD is not known to occur. Moreover, germplasm from Vanuatu cannot be distributed safely due to the possibility of disseminating CFD virus or its vectors with seed. This has the effect of eliminating a potential source of export earnings for the country (Persley, 1992).

Elsewhere in the Pacific Region, a viroid causes the very severe disease of coconut known as 'cadang-cadang'. This affects parts of the Philippines, where Zelazny *et al.* (1982) reported that *c.* 30 million trees had been killed in southern Luzon alone. It is notable that the 'LAOT population' is the predominant type of coconut being grown in the affected areas and it was one of the most susceptible cultivars assessed in screening tests and by natural exposure (Rodriguez, 2003).

3. *Citrus Tristeza Disease*

The main areas of citrus production are in regions where the crop has been introduced. There has been much movement of bud-wood and rooted plants and this accounts for the presence of several important viruses in virtually all countries with major commercial plantings (McClean, 1957). Particular problems have been encountered with the aphid-borne *Citrus tristeza virus*. When this virus spreads to sweet orange trees (*Citrus sinensis*) growing on sour orange rootstocks (*C. aurantium*), it causes a lethal decline (Bennett and Costa, 1949). Some other stock/scion combinations are much less sensitive to infection. Consequently, they have been widely used in South Africa where tristeza appears to have been present for many years. The use of sour orange rootstocks had to be abandoned there, despite their resistance to *Phytophthora* root rot (Webber, 1943).

The reasons for the failure of sour orange as a rootstock and the prevalence of infection in South Africa were not apparent until investigations were initiated into the virus first identified as the cause of the disease known in Brazil as tristeza (Meneghini, 1946) and the similar 'quick decline' disease of citrus in California (Fawcett and Wallace, 1946). The extensive plantings of sweet orange on sour orange rootstocks had previously been grown successfully in these areas for many years, suggesting that tristeza virus had been introduced relatively recently, or that it had only just begun to spread.

Tristeza has caused enormous losses in South America where it was first recorded in Argentina in 1930 and subsequently in Brazil (1937), Uruguay (1940), Venezuela (1950) and Paraguay. In the worst-affected areas, almost all the vulnerable trees on sour orange rootstocks were killed within a few years, or plantations became so worthless that they were abandoned. By 1949, an estimated 6 million trees had been destroyed in the São Paulo state of Brazil alone, and this amounted to 75% of all the orange trees present. The oriental citrus aphid *Toxoptera citricidus* was shown to be a highly efficient vector. Moreover, spread between regions was facilitated by the distribution of large quantities of plant material, including tolerant varieties and stock/scion combinations (Bennett and Costa, 1949).

There is considerable information on the early history of citrus-growing in Argentina, including detailed records of the introduction and wide distribution of two large shipments of nursery material from South Africa between 1927 and 1930. Tristeza could have been introduced at this time and it is known that the imported scion-wood failed to develop when grafted onto sour orange rootstocks of local origin, although it grew normally on other varieties. Certainly the disease was already widespread when first discovered, causing the death of many trees in each of the main areas of production (DuCharme *et al.*, 1951).

Losses have continued and became even more severe as *T. citricidus* spread to new areas. In recent decades it has progressed northwards in South America and into the Caribbean region and also into Mexico and the southern states of the United States (Yokomi *et al.*, 1994). Moreover, *T. citricidus* has been reported recently in Spain and is expected to cause even greater losses than those encountered previously (Cambra, M., personal communication). In all these areas, many of the trees are being grown on sour orange rootstocks and it is inevitable that they will be severely damaged.

4. *Temperate Fruit Crop Diseases*

Plum pox virus infects plum, peach and other deciduous fruit crops. It has caused serious damage in many European countries since it was first seen in Bulgaria in 1915 (Atanasoff, 1932). Moreover, the virus has been reported recently in North and South America and also in Japan. The losses experienced in Europe would have been less severe but for the vulnerability of some of the most popular cultivars that were being grown. Pozegača in the Balkans, Hauszwetsche in Germany and Victoria in the United Kingdom are all plum cultivars of this type. Their fruits develop conspicuous symptoms of necrosis and blemishes that render the crops virtually worthless.

Problems have also arisen with apple and pear due to the extreme vulnerability of particular cultivars. For example, several types of quince were found to be unsuitable for use as rootstocks for pear in England because of apparent growth incompatibility effects (Hatton, 1928). These were shown later to be due to the extreme sensitivity of the quinces to viruses that were prevalent at the time in tolerant pear cultivars (Cropley, 1967).

The existence or prevalence of several other important viruses of some other pome fruit crops was not appreciated until serious disease problems were encountered after attempts to introduce new rootstock or scion cultivars. This is illustrated by experience in North America where the use of Virginia Crab and Spy 227 as apple rootstocks or interstocks has been restricted by their extreme sensitivity to viruses that were widespread but latent in many commercial varieties (Gardner *et al.*, 1946; Tukey and Brase, 1943). Similarly, apple chat fruit and rubbery wood diseases were not known until a sensitive cultivar (Lord Lambourne) was introduced and revealed the prevalence of infection in rootstocks and other apple varieties (Luckwill and Crowdy, 1950).

D. Cereal Diseases

1. Maize Rough Dwarf Disease

Maize was introduced from the New World to the Old over 400 years ago and has been cultivated extensively in Italy and other Mediterranean countries where maize rough dwarf disease (MRDD) occurs. However, the disease has a very recent history and its first appearance was closely associated with the introduction of high-yielding American hybrid varieties soon after the Second World War (Harpaz, 1972).

Trials of the new varieties started in Italy in 1946 and the first commercial plantings totalling 1500 ha were in 1948, when MRDD was first recorded. It attracted little attention until 1949, when outbreaks were so serious as to threaten the whole future of hybrid varieties. These were conspicuously more susceptible than those grown previously and 90% infection was reported in one area of northern Italy where local varieties were virtually unaffected (Grancini, 1962; Trebbi, 1950).

There was a similar sequence of events in Israel where hybrid varieties were first released in 1952. Plantings totalled 7000 ha by 1957, when MRDD was first recorded. In 1958, major outbreaks occurred in all regions where hybrids had been introduced. Up to 75% infection occurred in plantings along the coastal plains where there was severe dwarfing and much premature death. Infection was only 5–7% in local open-pollinated dent varieties grown under similar conditions.

MRDD provides a striking example of the way in which new and unexpected problems can occur when long-established cultivars are replaced. The disease does not occur in North America and the hybrids released to Mediterranean countries had not previously been exposed to infection. Consequently, it is hardly surprising that they were severely affected by comparison with established local cultivars. These were so resistant or tolerant that infection was entirely overlooked until susceptible genotypes were introduced.

2. Rice Tungro Disease

The first introduction of modern cultivars to a region is frequently accompanied by major changes in traditional cropping practices. This makes it difficult to assess whether any immediate increase in the prevalence of arthropod pests or diseases is due to the exceptional vulnerability of the new genotypes, or to other causes. The situation can be extremely complex, as illustrated by the history of rice tungro disease in South-East Asia.

What appears to be tungro has been known in some Asian countries for many years under various local names (Sõgawa, 1976). These include *'mantak'*, which has been recognized in Indonesia since 1869 and *'penyakit merah'*, first reported in Malaysia in 1934. The causal viruses and their leafhopper vectors (*Nephotettix* spp.) received little detailed attention until 1963, when serious damage occurred in experimental plantings at the International Rice Research Institute (IRRI) in the Philippines (Rivera and Ou, 1965). Serious epidemics occurred subsequently in the 1960s and 1970s in the Philippines and in several other South-East Asian countries. This was attributed to the widespread use of certain IRRI cultivars (Bos, 1992; Buddenhagen, 1977), although this is but one of several interrelated factors responsible.

The high-yielding cultivars that were introduced and led to the 'green revolution' in rice production differed from the many traditional ones grown previously in being heavy-tillering, photo-insensitive and of short growing season and stature (Khush, 1977). These features facilitated intensive methods of cultivation using close spacing and artificial fertilizers. There was also increased use of irrigation to extend the natural growing season and so permit successive crops to be grown in close or even overlapping sequence. Such conditions favour the build-up of weeds and also arthropod pests and diseases, including some that seldom damaged old landrace cultivars grown by traditional methods. Some of the new varieties soon proved to be vulnerable. For example, the IRRI varieties IR5, IR8 and especially IR22 were severely affected in the Philippines following their release between

1968 and 1970. In Malaysia, unprecedented losses occurred in a traditional long-season variety following an early and heavy influx of leafhopper vectors from earlier plantings of the new short-season improved types (Lim, 1972). Tungro was also reported for the first time in India and became prevalent in certain areas immediately after the use of the improved cultivar T(N)-1 and its derivatives since 1964. These varieties proved to be highly susceptible to leafhoppers and planthoppers, which previously had been only sporadic and unimportant pests of rice in India (Kulshreshtha *et al.*, 1970).

The high-yielding cultivars soon accounted for almost all plantings. Over large areas a few modern types replaced the traditional ones grown previously. By 1983, IR36 was planted on an estimated 11 million hectares and became the most widely grown of all crop cultivars. There were also changes in cultural practices and these were associated with greatly increased losses due to arthropod pests and diseases. This posed problems for the plant breeders and others concerned with crop improvement. It was appreciated that there was a need to avoid undue reliance on a few widely-grown cultivars of similar genetic background. It was also apparent that cultivars were required that were more resistant to tungro and its leafhopper vectors than those grown previously (Khush, 1977). Moreover, the difficulty of achieving this became clear when outbreaks of tungro in the Philippines, Indonesia and elsewhere were associated with the 'breakdown' of IR36 and other initially vector-resistant cultivars as the vector populations adapted and overcame the resistance being deployed (Dahal *et al.*, 1990). It was also realized that the widespread use of insecticides to control the leafhopper vectors was undesirable on health grounds and also because it was not readily compatible with the need to decrease the use of insecticides as a means of enhancing the natural enemies of rice brown planthopper (*Nilaparvata lugens*) and other insect pests (Kenmore *et al.*, 1984).

Rice is hugely important for the nutrition and well-being of a large proportion of the human population in South-East Asia. Consequently, changes in methods of crop production and the apparent increase in problems due to arthropod pests and diseases, including tungro, have received much attention. The need to avoid undue reliance on chemicals and on cultivars having a type of resistance that is vulnerable to breakdown has been appreciated. It has also been realized that losses due to tungro can be decreased by avoiding unduly vulnerable varieties, by careful choice of sowing date and by planting crops synchronously after a break in production to decrease the carry-over of inoculum (Azzam and Chancellor, 2002).

Considerable success has been achieved in this way in decreasing the losses due to tungro and virus-resistant cultivars are now available. However, progress has been undermined by farmers continuing to grow popular cultivars of high quality despite their inherent susceptibility to tungro, or after a breakdown of their resistance to the leafhopper vector. Vulnerable cultivars have been retained in part because the occurrence of tungro is so unpredictable and sporadic that they can be grown successfully and profitably in some seasons when the use of resistant cultivars would have brought no benefits. For example, the susceptible high quality cv. IR64 is still widely grown in the Philippines, even though at times it is severely affected by tungro. This has led to experiments on mixing IR64 with the resistant 'Matatag 9' to provide at least some degree of protection and the incidence of tungro in IR64 in the mixture was halved compared with that in a pure stand of the cultivar (Leung *et al.*, 2003). Another approach is to plant vulnerable cultivars mainly at times of year when the experience is that tungro is unlikely to be prevalent. These examples illustrate the ways in which farmers can and do exploit different means of enabling them to continue growing vulnerable cultivars and why these are sometimes so difficult to displace.

3. *Rice Waika Disease*

The history of rice waika disease in Kyushu, Japan, is closely associated with the use of one particular cultivar (Kiritani, 1983). The disease was first reported in 1967 and shown to be caused by a leafhopper-borne virus that was identified later as *Rice tungro spherical virus*. The disease-affected area in Kyushu increased to a maximum in 1973 and then declined to zero by 1977 (Table II). The prevalence of infection was closely linked with the adoption of cv. Reiho which was introduced in 1968 and by 1971 accounted for more than half the total area of rice grown. The area of Reiho then decreased when it became apparent that it was particularly vulnerable to waika disease and the cultivar was withdrawn in 1977.

4. *Rice Yellow Mottle Disease*

Rice yellow mottle disease (RYMD) in Africa resembles rice tungro in Asia in being indigenous to the region in which it occurs. Moreover, both diseases were first described soon after an intensification of production and the adoption of new cultivars. The first reports of RYMD were from Kisumu district of Kenya, around the shore of Lake Victoria (Bakker, 1970). The disease was soon recognized as a potential threat

TABLE II
THE OCCURRENCE OF RICE WAIKA DISEASE IN JAPAN AND CULTIVATION OF RICE CV. REIHO: 1966–1977[a]

Year	Area affected (ha)	Cv. Reiho (% total area)
1966	0	0
1967	2	0
1968	50	0
1969	150	5.7
1970	23	27.1
1971	1647	51.9
1972	11,297	49.8
1973	24,825	36.8
1974	612	26.0
1975	24	22.0
1976	17	17.1
1977	0	0

[a] Adapted from Kiritani (1983).

to rice in the major irrigation projects being developed nearby. The limited areas of rice grown previously had been mainly in small, seasonal rain-fed plantings that provided little opportunity for damaging outbreaks to develop.

Irrigation allowed continuous cropping and an extensive growth of grasses, weeds and self-sown rice plants that persisted through the dry season. These conditions facilitated the perennation and build-up of *Rice yellow mottle virus* and its beetle vectors and RYMD soon became prevalent. There has been a similar situation in many other parts of Africa, including Madagascar and the Zanzibar islands. In Kenya and some other countries the first reports of RYMD were associated with the introduction of 'improved' high-yielding cultivars of Asian rice (*Oryza sativa* subsp. *indica*) that proved to be vulnerable to infection. They had been introduced to replace the locally selected rices grown previously. These were relatively resistant to infection, which explains why RYMD had been inconspicuous and overlooked previously.

Some of the worst epidemics of RYMD have been in the Republic of Niger, where the irrigated area increased from 571 ha in 1974 to 8500 ha in 1986 (John *et al.*, 1988; Reckhaus and Adamou, 1986). The disease was not noted until 1982, yet it was prevalent in almost all the irrigated areas in 1984 and 1985. The main cultivar grown was an introduced lowland type from Asia—IR 1529-680-3. Large areas were devastated in 1984 when the overall incidence of infection exceeded 25%.

5. Sorghum Mosaic and Red Leaf Disease

The aphid-borne *Maize dwarf mosaic virus* causes a leaf mosaic, reddening and necrosis of sorghum in United States and elsewhere. Severe epidemics affected Texas and the Great Plains region in 1967 and Arizona in 1968. A major factor contributing to the epidemics was that most of the commercial sorghum genotypes being grown were Redlan × Caprock hybrids. Both parents were shown to be highly susceptible and they develop the very damaging red leaf symptom (Toler, 1985).

E. Hop Mosaic Disease

The many different hop cultivars grown in England can be divided into two distinct groups, according to their reaction to *Hop mosaic virus* (HMV). Cultivars of the Golding group produce cones that attract premium prices because of their high quality for use in brewing. When infected, Goldings develop conspicuous leaf mosaic symptoms and usually die or grow so badly that they have to be replaced. All the other cultivars grown are extremely tolerant to HMV and fail to produce symptoms, even though until recent years all established clones were infected throughout. Both types of cultivar are grown commercially because of their distinct quality attributes in meeting the requirements of brewers.

HMV is aphid-borne and serious outbreaks are likely to occur in Goldings whenever they are planted close to sources of infection. These are other female cultivars, or tolerant male plants when used inadvertently as pollinators. This became apparent at an early stage of the investigations which began soon after mosaic was first recorded in England between 1907 and 1910 (Keyworth, 1947; Mackenzie *et al.*, 1929; Salmon, 1923). It was established that losses could be avoided or kept to acceptable levels by growing Goldings with appropriately sensitive males at isolated sites or on entirely separate farms. Such measures are effective, but not always convenient for growers to adopt. Moreover, it is expensive for nurserymen to grow Goldings away from other cultivars in order to avoid any risk of mosaic disease and to meet the requirements of official plant health certification schemes. Another development has been that changes in market demand induced some growers of Goldings to plant tolerants for the first time on their farms and heavy losses occurred (Thresh, 1979).

There is a similar problem in Germany where mosaic appeared at isolated sites in the Hersbruck area of Bavaria soon after tolerant cultivars were introduced from the main hop-growing areas of the Hallertau. Losses also occurred in the traditional Hersbruck cultivar when it was introduced to the Hallertau (Thresh, 1979). There is no detailed information on the incidence of mosaic in Hallertau cultivars, but they could have been responsible for introducing HMV to England. Some of the first outbreaks ever recorded were in collections of cultivars and at a farm trial near recently introduced German material (Mackenzie *et al.*, 1929).

Hop growers are not unique in growing both virus-sensitive and tolerant cultivars despite the disease problems that can arise and the control measures that must be adopted to avoid the losses becoming unacceptable. For example, the strawberry cv. Royal Sovereign was retained for many years in United Kingdom for its high-quality fruit and despite its vulnerability to aphid-borne viruses. This necessitated the use of virus-free stocks of planting material and isolation from cv. Climax and other tolerant cultivars being grown widely at the time. There was a similar situation with the sensitive high-quality raspberry cv. Lloyd George and the relatively virus-tolerant cultivars being grown (Jones, 1976).

F. Other Diseases

1. Lettuce Disease

Turnip mosaic virus (TuMV) is transmitted non-persistently by aphids and causes prevalent diseases of cruciferous crops in many parts of the world. It was not reported in lettuce until 1966, when damaging outbreaks occurred in California (Zink and Duffus, 1969). They were restricted to a crisp-head cultivar 'Calmar', which was being grown widely because of its resistance to the downy mildew fungus (*Bremia lactucae*). In comprehensive tests on a wide range of genotypes all seven downy mildew-resistant cultivars of crisp-headed type were found to be susceptible to TuMV, whereas all others were immune (Table III).

There was no segregation within populations of each cultivar and an examination of the available pedigrees suggested that mosaic susceptibility was introduced to lettuce from two mildew-resistant lines of a closely related wild species (*Lactuca serriola*). Different sources of resistance had been used in developing the mildew-resistant cultivars of butterhead and other types that were immune to mosaic.

TABLE III
THE REACTION OF DOWNY MILDEW-RESISTANT AND MILDEW-SUSCEPTIBLE LETTUCE CULTIVARS OF DIFFERENT TYPE TO *TURNIP MOSAIC VIRUS*[a]

Lettuce type	Mildew-resistant[b]	Mildew-susceptible[b]
Crisp-head	7/7	0/40
Butterhead	0/4	0/11
Leaf	0/2	0/7
Cos	0/1	0/3
Latin	–	0/2
Stem	–	0/2

[a] Zink and Duffus, 1969.
[b] The number of cultivars of each type that were susceptible to *Turnip mosaic virus* as a proportion of the total number tested.

The apparent linkage between the genes for mildew resistance and mosaic susceptibility in crisp-headed cultivars was confirmed subsequently (Zink and Duffus, 1970). This showed that both reactions were due to dominant genes designated *Tu* and *Dm*. There was some crossing over and the linkage was broken with relative ease, indicating that it would not be difficult to select crisp-headed varieties resistant to both mildew and mosaic.

2. *Potato Diseases*

There are early reports of the vulnerability of the American potato cv. Russet Burbank. When plants are infected with *Potato leafroll virus*, the tubers produced develop a dark discolouration termed net necrosis (Gilbert, 1928). This renders the tubers unfit for the fresh market in the United States and for all but the least profitable of processed products (Thomas *et al.*, 2000). Losses can be kept to acceptable levels only by adopting costly control measures. These involve planting healthy stocks of 'seed' tubers and the use of insecticides to control the aphid vectors. Nevertheless, Russet Burbank is still widely grown because of its favourable attributes and it remains the most important cultivar in western United States.

Vulnerable cultivars of potato have also been reported in Europe where different strains of *Potato virus Y* (PVY) can cause serious losses. These became particularly severe in the UK cv. Record during the 1980s and 1990s, at a time when it was the most widely-grown cultivar used to produce potatoes for processing into crisps. High incidences of the necrotic strain PVY^N (mean 58.5%) were recorded in an assessment of ware crops, even where these had been grown from certified stocks of seed tubers (Barker, 1994). Such rapid deterioration was attributed to the spread of PVY^N from 'volunteer' tubers remaining in the ground from previous crops (Jones *et al.*, 1996). It was also suspected that current season infection with PVY^N was associated with a condition known as 'sugar spot', which is a processing problem that occurs when crisps are being fried. The losses caused by PVY, together with the association with sugar spot, contributed to a rapid decline in the cultivation of cv. Record and it has now been largely replaced by other less vulnerable cultivars.

Potato growers have also experienced problems with PVY in many other parts of Europe due to an increased prevalence of 'NTN' strains causing tuber necrotic ringspot disease (Kus, 1994). Strains of this type were first identified in Slovenia in 1988 and soon largely destroyed 'seed' potato production in the country. Necrotic ringspot was regarded as the most devastating disease of potato encountered in Slovenia during the twentieth century (Kus, 1995). Many local cultivars were destroyed, including cv. Igor which had been the most popular. A transgenic resistant form of cv. Igor was developed (Racman *et al.*, 2001), but it has not been adopted commercially and potato production in Slovenia has been restored by introducing other less vulnerable cultivars.

3. *Other Diseases*

There are many other reports of particularly susceptible cultivars, including several early examples. One of these is the sugarcane cv. Uba in which streak disease was first reported (Storey, 1925). The cultivar was so susceptible in South Africa that it was replaced between 1933 and 1935 and streak then became a rare disease (Bock and Bailey, 1989).

In United States, *Soil-borne wheat mosaic virus* caused severe losses when a susceptible cultivar was first grown at a site where the fungus-borne virus had not previously attracted any attention (Koehler *et al.*, 1952). Later, *Red clover necrotic mosaic virus* was first reported in England from trial plantings incorporating some of the newly introduced tetraploid cultivars of early-flowering broad red clover (Bowen and Plumb, 1979). Hungaropoly and three other cultivars of similar genetic background were highly susceptible compared with many of the diploid types grown previously. The tetraploids were also at risk because they persisted for relatively long periods and so provided increased opportunities for infection to occur.

Unexpected problems were encountered in the United Kingdom and elsewhere in Europe when attempts were made to increase the productivity of Brussels sprout crops by introducing F_1 hybrid cultivars (Tomlinson and Ward, 1981). These had been developed by crossing selected inbred lines and were first grown extensively in the United Kingdom in 1974. Many of these crops developed yellow mosaic symptoms that were particularly severe in the F_1 hybrid Fasolt. The symptoms were shown to be caused by dual infection with two aphid-borne viruses (*Cauliflower mosaic virus* and *Turnip mosaic virus*). These had occurred previously in brassica crops, but were regarded as unimportant in Brussels sprout. It was established that the inbred parents used to develop the first hybrids differed considerably in their response to infection and that the most sensitive should be avoided to prevent any further release of vulnerable genotypes.

An additional example of vulnerability became apparent in studies on the very damaging epidemic of cotton leaf curl disease that affected large areas of Pakistan in the 1990s and spread to adjoining parts of India (Briddon and Markham, 2000). The disease had been noted in the late 1960s and for the next 20 years was regarded as no more than a local nuisance. Infection then became prevalent and it was noted that once a region was infected the severity of the symptoms depended largely on the cultivar grown. S12, which had been released in 1988, was particularly susceptible to infection. This was of great consequence because the cultivar had become very popular and accounted for up to 46% of the total area of cotton grown during the early years of the epidemic. Other cultivars were relatively tolerant and they have been used to maintain production until resistant cultivars become available.

Earlier, there had been a similar sequence of events in Sudan where the cotton leaf curl problem had been overcome by switching from susceptible cultivars derived from *Gossypium barbadense* to other more tolerant types (Nour and Nour, 1964). The vulnerability of *G. barbadense* cultivars was also apparent in India where the first report of cotton leaf curl disease was in an introduction of this type from Sudan (Varma and Malathi, 2003).

Elsewhere in India in the late 1970s exotic accessions of cowpea from West Africa were affected by the indigenous *Mungbean yellow mosaic India virus* (MYMIV), which had not previously been a problem in Indian cowpeas. By 1984, MYMIV had become the most severe constraint to cowpea production in India, affecting even well-established local cultivars that had been grown successfully for more than 20 years (Varma and Malathi, 2003).

III. Discussion

A. *Significance of Host Genetic Vulnerability*

The foregoing sections provide information on host vulnerability to 30 virus diseases of a diverse range of tropical, subtropical and temperate crops. Many of these are of great economic importance and influence the livelihoods and well-being of millions of farmers and their families in some of the most impoverished regions of the world. The examples are listed by host, but other categorization is possible. It is particularly notable that several examples relate to the way in which the introduction of a cultivar, or small group of related cultivars, has led to the occurrence and identification of a previously unrecognized disease caused by a virus that was already established in the environment:

Sugarcane streak in Africa (Section II.F.3)
Cacao swollen shoot in West Africa (Section II.C.1)
Coconut foliar decay in Vanuatu (Section II.C.2)
Maize rough dwarf in the Mediterranean (Section II.D.1)
Rice tungro and waika in Asia (Section II.D.2/3)
Rice yellow mottle in Africa (Section II.D.4).

The introduction of other vulnerable cultivars has led to the greatly increased prevalence of diseases that occurred previously but were considered to be of little or no importance:

Sugarcane Fiji leaf gall disease in Queensland (Section II.B.2)
Yellow mosaic disease of Brussels sprout in United Kingdom (Section II.F.3)
Cotton leaf curl disease in India (Section II.F.3).

All these examples can be regarded as the outcome of 'new encounters' [*sensu* Buddenhagen (1977)] between susceptible hosts and viruses that were either unrecognized or unimportant. This is but one of several ways in which long-established equilibria between viruses and their hosts can be disrupted and lead to disease problems. Other disruptive influences are the introduction of entirely new crops or viruses, or of particularly damaging strains of viruses that already occur.

It is somewhat simplistic and misleading to attribute the occurrence and widespread distribution of damaging epidemics to a single underlying cause. Diseases are unlikely to inflict severe losses unless susceptible hosts encounter virulent virus strains in circumstances and seasons permitting many plants to be infected and severely damaged. Inevitably, weather, cropping practices and genetic features of host and pathogen are involved in all major epidemics and cannot be considered in isolation. This is apparent from experience with rice tungro (Section II.D.2), rice yellow mottle (Section II.D.4), cacao swollen shoot (Section II.C.1) and several of the other diseases considered in the foregoing sections. Nevertheless, there can be advantages in stressing the overriding importance of particular factors that 'drive' epidemics. Such drivers can be climatic, biotic or associated with features of crop husbandry. This emphasises the diversity of the diseases encountered, which can become prevalent for very different reasons, as discussed previously (Thresh, 1980).

From this perspective, the extreme vulnerability of particular cultivars can be regarded as one of the crucial driving factors leading to damaging outbreaks. Many of the diseases discussed in the previous sections would have been unrecognized or unimportant if vulnerable cultivars had not been grown. This emphasises how human intervention can greatly influence the incidence and prevalence of crop diseases. Many other features of crop husbandry have similarly deleterious effects and this led the plant breeder, N. W. Simmonds (1962), to write '*disease patterns are to a great extent a product of plant breeding and agricultural practices*'. The veracity of this comment has since become ever more apparent (Bos, 1992; Thresh, 1982).

Obvious questions arise. Why are vulnerable cultivars introduced and why are they sometimes adopted on a very large scale if the consequences can be so undesirable? There is no single answer to these questions, but it is important for virologists to appreciate that the main concern of farmers and others involved in crop production is to produce crops that are marketable and profitable and not necessarily free or largely free of disease. This is sometimes overlooked by pathologists whose primary concern is plant health. There are certainly many examples of diseased crops being more profitable than healthy ones and of vulnerable cultivars being preferred to others that are more resistant.

One reason why vulnerable crops are introduced and adopted is that the disease encountered was either unknown or unimportant at the time the introductions were made. Moreover, it is evident that in many instances the cultivars introduced had such favourable growth, yield, quality or other attributes that they were soon in great demand from growers, consumers and processors. In these circumstances, there was no compelling reason why their adoption should be delayed whilst further evaluations were done, even if these were possible. This is exemplified by experience with cacao (Section II.C.1), which was introduced to West Africa more than 40 years before swollen shoot disease was recognized. Furthermore, it was largely fortuitous that the initial introductions were mainly of the Amelonado type. This seemed to be so well adapted to lowland rainforest conditions in West Africa that no further introductions were made until swollen shoot occurred and the need for virus-resistant cultivars became apparent.

The situation was completely different in the latter part of the twentieth century when attempts were made to extend the range of cassava genotypes available in sub-Saharan Africa where the crop had been introduced by the Portuguese in the sixteenth century (Carter *et al.*, 1992). Additional material was introduced in the 1970s and later from South/Central America where the crop originated. The introductions were established at the International Institute of Tropical Agriculture, Nigeria, where CMD has become endemic. The disease is not known in the neotropics and the exotic material soon became severely affected (Porto *et al.*, 1994). It was then apparent that such cultivars would be inappropriate for use in Africa until some form of resistance to mosaic had been introduced by further breeding. Moreover, it was appreciated that mosaic-resistant lines from Africa should be utilized in South/Central American cassava breeding programmes to produce locally-adapted cultivars for use should the need arise.

These conclusions would have been drawn less readily or not at all if the introductions had been grown only at a site where mosaic was absent or unimportant. This is apparent from experience in Uganda (Section II.A.1) where the vulnerable Ebwanateraka was adopted by many farmers at a time and in areas where mosaic was known or assumed to be unimportant. Moreover, the merits of the mosaic-resistant TME genotypes of cassava became apparent in areas and at times when cassava brown streak was absent or unimportant and their adoption may have to be reassessed following the recent outbreak of the disease in Uganda (Section II.A.2). The situation was similar in the Queensland sugarcane-growing areas of Australia when N.Co.310 was introduced and adopted widely at a time when FLGD was seemingly unimportant (Section II.B.2). However, exotic cultivars such as N.Co.310 may have such outstanding merits that farmers, researchers and processors are reluctant to forego the advantages to be gained from their use and may be prepared to overlook or ignore seemingly minor defects. There may also be an incentive to disseminate new cultivars more widely for disaster relief programmes, commercial profit, to gain scientific prestige, to claim plant breeders' rights or to meet the insatiable demands of horticulturalists for novelty. This partially explains why some cultivars are grown far beyond the areas for which they were bred and where they encounter damaging pathogens to which they have not been exposed previously.

From the examples considered here and other experiences, there are powerful incentives for a more enlightened attitude to the development, selection and adoption of new cultivars. This was stressed previously by Buddenhagen (1977), but there is little indication that his prescient advice and recommendations have been followed. It is particularly important to be aware of the risks posed by exotic introductions and these should not be grown widely until they have been tested thoroughly at many sites and in different seasons. Even if this is done, problems can arise because statistically improbable events are difficult to detect in a limited number of small-scale trials and yet become mathematical certainties when introductions are released and adopted widely.

There are great benefits to be gained by decreasing the likelihood of problems arising because severe losses are sustained when outbreaks occur, control measures become necessary and widely-grown cultivars have to be replaced. Cacao producers in Ghana are still adversely affected by the initial reliance of their forebears on Amelonado. Sixty years after the first reports of swollen shoot disease, this cultivar still accounts for a substantial proportion of all plantings (Section II.C.1). Citrus is another tree crop in which it is difficult, slow and expensive to achieve a changeover in the cultivars grown and there may be disadvantages in doing so (Section II.C.3). Sour orange is still widely used as a rootstock in many regions and has advantages in providing resistance to *Phytophthora* root rot disease. However, trees on sour orange rootstocks are at serious risk from the effects of tristeza disease, especially in regions being invaded by particularly virulent strains of the causal virus and/or the most efficient vector aphid (*T. citricidus*).

Problems were also experienced in achieving the changeover in cultivars that became necessary with sugarcane in Queensland during the 1970s (Section II.B.2) and with cassava in Uganda (Section II.A.1). Both crops are propagated vegetatively and large amounts of bulky planting material are required for new plantings. Moreover, it is necessary to maintain continuity of production to meet the requirements of consumers or processors. With both crops the time-scale required to change the cultivars grown was in years and losses continued during the changeover period. In addition, valuable genotypes have been destroyed or discarded from breeding populations because of coconut foliar decay (CFD) disease (Section II.C.2). Moreover, it is considered that 10–15 years were required to recover the lost genetic potential in the south Queensland sugarcane breeding programme after the most recent epidemic of Fiji leaf gall disease (Smith, 2000). Comparable losses were encountered due to CFD disease in Vanuatu (Section II.C.2), which has greatly restricted the opportunity for crop improvement (Persley, 1992).

B. *Significance of Host Genetic Diversity*

Considerable attention has been given to the hazards associated with an undue reliance on a narrow range of cultivars and there has been a corresponding emphasis on the benefits of genetic diversity. Moreover, comparisons have been made between the relative uniformity of modern agricultural practices and the intra- and inter-specific diversity which is such a notable feature of many natural plant communities and traditional farming systems (Day, 1977). Modern practices include the use of a restricted range of cultivars, usually grown without intercrops, in large homogeneous stands and often in close or overlapping succession at the same sites. These innovations have contributed to disease problems and led to the realization that biodiversity benefits productivity by providing at least some degree of resilience in withstanding the effects of arthropod pests and pathogens, or unfavourable environmental conditions. Conversely, uniformity is regarded as an unsatisfactory but largely inevitable feature of modern agriculture that can enhance the losses caused by pests and diseases.

There has been considerable debate on these issues and it has been suggested that the losses caused by pests and diseases could be decreased and crop productivity improved by developing modern cropping practices that have some of the features and resilience of earlier more diverse systems. Virologists have made only a limited contribution to the discussion and there is little information on the consequences of either inter- or intra-specific diversity. Nevertheless, there are several reasons why the use of crop mixtures can be expected to decrease the opportunity for viruses to be spread by vectors or other means, by:

- an increase in the mean separation distance between susceptible plants within the stand;
- one or more of the component crops grown in the mixture camouflaging the susceptibles;
- repelling or impeding the movement of vectors;
- effects on crop microclimate;
- non-susceptible crops intercepting inoculum that could otherwise lead to further spread.

Evidence for some these effects has been obtained in intercropping studies on several pathosystems in the tropics (Thresh, this volume, pp. 245–295) and for grass/legume mixtures in temperate regions (Brink and McLaughlin, 1990). However, the information available is totally inadequate in relation to the importance of intercropping in many regions and the scope for using non-host intercrops to protect vulnerable genotypes.

The effects of intra-specific diversity can be of great importance, as illustrated by experience with cassava during the epidemic of mosaic in Uganda (Section II.A.1). Farmers were soon able to adjust to the problem in areas where there was considerable intra-specific diversity, but not elsewhere. There has also been some adjustment to the problem of CBSD in parts of eastern and southern Africa (Section II.A.2). The situation is likely to be similar with other tropical crops, including common bean (Sperling and Leovinsohn, 1993), sweet potato (Aritua *et al.*, 1998), yams, rice, cowpea and maize.

There is only limited quantitative evidence from field trials on the effects of intra-specific diversity on virus spread within crop stands. In one of the first studies with a leafhopper-borne virus of maize, a resistant cultivar had no effect on the incidence of infection in a susceptible one when sole and mixed stands were compared (Power, 1987). The apparent lack of effect was attributed to the effects of the resistant cultivar in enhancing the movement of the vectors within mixed stands (Power, 1988). In similar studies on two aphid-borne viruses, resistant cultivars decreased spread to susceptible ones. This was demonstrated with *Barley yellow dwarf virus* in oats (Power, 1991) and with *Soybean mosaic virus* in soybean (Irwin and Kampmeier, 1989). In the latter experiment, the incidence of infection in a susceptible isoline was inversely proportional to the percentage of resistant plants in the blend. Subsequently, Sserubombwe *et al.* (2001) recorded a significant decrease in the incidence of CMD in a susceptible local landrace when intercropped with resistant or partially resistant cultivars that were not adversely affected. The beneficial effects of diversity recorded in these and other experiments with wheat (Hariri *et al.*, 2001) and rice (Leung *et al.*, 2003) are likely to occur more widely and justify increased attention. This is apparent from the more extensive studies of plant pathologists on fungal and bacterial diseases (Browning and Frey, 1969; Wolfe, 1985). These have led to an appreciation of the merits of multilines and mixtures as a means of avoiding an undue reliance on chemicals or other methods of disease control (Leung *et al.*, 2003; Mundt, 2002; Smithson and Lenné, 1996).

C. *Resistance Breeding*

The situation of farmers in developed countries is very different from those elsewhere because of the generally limited range of cultivars being grown. This restricts the opportunity to overcome disease problems by exploiting intra- and inter-specific diversity. However, farmers in developed countries usually have the advantage of access to specially bred cultivars produced by commercial companies or research establishments. These are able to utilize the latest scientific techniques and innovations and the wide range of genetic material available in collections and gene banks.

Two basic approaches have been distinguished. 'Negative' selection refers to the situation in which vulnerable cultivars are discarded in favour of those that are damaged less severely and continue to crop satisfactorily. In contrast, 'positive' selection involves the deliberate selection of highly resistant cultivars identified in rigorous screening programmes. Negative selection is practised intentionally or inadvertently by farmers, extensionists and researchers in selecting and recommending the most appropriate cultivars to adopt. An inevitable consequence is that there is a trend towards some form of tolerance and symptoms are mild or absent, even though virus is present. This is a particular feature of vegetatively-propagated crops in which there has been an opportunity for mutual adaptation between viruses and their hosts. The examples given here (Section II.C.4) relate to deciduous fruits, but many other horticultural crops, ornamentals and potato demonstrate similar trends.

There is a long history of breeding for host-plant resistance by positive selection as a means of overcoming virus disease problems. Examples are the successful programmes mounted early in the twentieth century in the United States to combat sugarcane mosaic (Summers *et al.*, 1948) and sugarbeet curly top diseases (Coons, 1949, 1953). The enormous losses caused by these diseases provided a powerful incentive to researchers and adequate funds were allocated to mount vigorous programmes of research and development. Moreover, these programmes have been sustained to develop the succession of cultivars needed to meet the changing requirements of producers and processors. It has also been necessary to overcome new problems as they arise due to the emergence of new strains of virus that are more damaging than those present previously, or able to overcome the resistance being deployed.

Such substantial commitments of funds and resources are not always justified and must be considered in relation to the advantages and cost implications of other approaches to disease control and crop improvement. Nevertheless, the need for resistant cultivars can be compelling, as apparent from experience in Europe and North America following the establishment of rhizomania disease (Asher *et al.*, 2002). This is caused by a fungus-borne virus that is not readily controlled by other means. For many other diseases resistant varieties are not such a high priority, especially if outbreaks occur sporadically and unpredictably. Furthermore, it may be difficult to combine adequate levels of resistance to the full range of pests and diseases encountered with the quality and other favourable attributes required, or if a penalty in yield or quality is incurred for doing so. Additional problems arise when there are commercial or other pressures for cultivars to be distributed and used far beyond the area in which they originated.

An important development has been in the ability to transform plants by genetic manipulation. This provides an opportunity to overcome the problem of vulnerability by introducing genes for resistance to vulnerable cultivars that are in other respects highly satisfactory and suitable for adoption. The potato cultivars Russet Burbank in United States and Igor in Europe have already been transformed (Section II.F.2) and virus-resistant cultivars of papaya have been bred by transforming parental lines (Gonsalves, this volume, pp. 317–354). Such biotechnological approaches are likely to be applicable more widely, but there are formidable problems to be overcome before transgenics are accepted for general use. This is apparent from experience in developing virus-resistant transgenic papayas which are one of only two such crops to have been commercialized in United States (Gonsalves, this volume, pp. 317–354). Consequently, vulnerable cultivars remain at risk, considerable losses occur and expenses are incurred in adopting control measures to prevent or decrease the incidence of infection. This is done by adopting one or more of the approaches discussed elsewhere in this volume (Jones, this volume, pp. 205–244; Thresh, this volume, pp. 245–295). The use of resistant cultivars, transgenics or other crop species to protect vulnerable cultivars within mixed stands is of particular interest and avoids the undesirable effects of pesticides. Much can also be achieved by adjusting cropping practices to decrease the risks of infection associated with recent trends in crop production. Nevertheless, it seems inevitable that there will be further instances of genetic vulnerability. The likelihood of this occurring could be decreased considerably by utilizing the experience gained to date and presented here.

Acknowledgments

Acknowledgements are due to many colleagues and correspondents for helpful advice, and to H. Barker (Scottish Crop Research Institute) and S. Gerrish (British Potato Council) for information on potato and potato viruses.

References

Abbott, E. V. (1958). Strains of sugarcane mosaic virus in Louisiana. *Sugar Bull. New Orleans* **37:**49–51.

Abbott, E. V. (1962). Problems in sugarcane disease control in Louisiana. *Proc. Int. Soc. Sugar Cane Technol.* **11:**739–742.

Abbott, E. V., and Todd, E. (1962). Mosaic in clones of *Saccharum spontaneum* and in Kassoer. *Proc. Int. Soc. Sugar Cane Technol.* **11:**753–755.

Akano, A. O., Dixon, A. G. O., Mba, C., Barrera, E., and Fregene, M. (2002). Genetic mapping of a dominant gene conferring resistance to cassava mosaic disease. *Theor. Appl. Gen.* **105:**521–525.

Ampofo, S. T. (1997). The current swollen shoot virus disease situation in Ghana. *In* "Proceedings 1st International Cocoa Pests and Diseases Seminar," pp. 175–178. Ghana, 1995.

Aritua, V., Adipala, E., Carey, E. E., and Gibson, R. W. (1998). The incidence of sweet potato virus disease and virus resistance of sweet potato grown in Uganda. *Ann. Appl. Biol.* **132:**399–411.

Asher, M. J. C., Chwarszczynska, D. M., and Leaman, M. (2002). The evaluation of rhizomania resistant sugar beet for the UK. *Ann. Appl. Biol.* **141:**101–109.

Atanasoff, D. (1932). Plum pox. A new virus disease. Yearbook, Faculty of Agriculture, University of Sofia. **11:**49–69 (In Bulgarian).

Azzam, O., and Chancellor, T. C. B. (2002). The biology, epidemiology and management of rice tungro disease in Asia. *Plant Dis.* **86:**88–100.

Bakker, W. (1970). Rice yellow mottle, a mechanically transmissible virus disease of rice in Kenya. *Neth. J. Plant Pathol.* **76:**53–63.

Barker, H. (1994). Incidence of potato virus Y infection in seed and ware tubers of the potato cv. Record. *Ann. Appl. Biol.* **124:**179–183.

Bennett, C. W., and Costa, A. S. (1949). Tristeza disease of citrus. *J. Agric. Res.* **78:**207–237.

Bock, K. R., and Bailey, R. A. (1989). Streak. *In* "Diseases of Sugarcane: Major Diseases" (C. Ricaud, B. T. Egan, A. G. Gillespie, Jr., and C. G. Hughes, eds.), pp. 323–332. Elsevier Science, Netherlands.

Bos, L. (1992). New plant virus problems in developing countries: A corollary of agricultural modernization. *Adv. Virus Res.* **38:**349–407.

Bowen, R., and Plumb, R. T. (1979). The occurrence and effects of red clover necrotic mosaic virus in red clover *(Trifolium pratense)*. *Ann. Appl. Biol.* **91:**227–236.

Brandes, E. W., and Sartoris, G. B. (1936). Sugarcane: Its origin and improvement. *USDA Year Book*, pp. 561–623. Washington, DC.

Breaux, R. D. (1985). Controlling diseases of sugarcane in Louisiana by breeding, potentialities and realities. *J. Am. Soc. Sugar Cane Technol.* **4:**58–61.

Briddon, R. W., and Markham, P. G. (2000). Cotton leaf curl virus disease. *Virus Research* **71:**151–159.

Brink, G. E., and McLaughlin, M. R. (1990). Influence of seeding rate and interplanting with tall fescue on virus infection of white clover. *Plant Dis.* **74:**51–53.
Browning, J. A., and Frey, K. J. (1969). Multiline cultivars as a means of disease control. *Annu. Rev. Phytopathol.* **7:**355–382.
Brunt, A. A. (1975). The effects of cocoa swollen-shoot virus on the growth and yield of Amelonado and Amazon cocoa *(Theobroma cacao)* in Ghana. *Ann. Appl. Biol.* **80:**169–180.
Bua, A., Sserubombwe, W. S., Alicai, T., Baguma, Y. K., Omongo, C. A., Akullo, D., Tumwesigye, S., Apok, A., and Thresh, J. M. (2005). The incidence and severity of cassava mosaic virus disease and the varieties of cassava grown in Uganda: 2003. *Roots* (in press).
Buddenhagen, I. W. (1977). Resistance and vulnerability of tropical crops in relation to their evolution and breeding. *Ann. NY Acad. Sci.* **287:**309–326.
Calvez, C., Renard, J. L., and Marty, G. (1980). Tolerance of the hybrid coconut Local X Rennell to New Hebrides disease. *Oléagineaux* **35:**443–451.
Carter, S. E., Fresco, L. O., Jones, P. G., and Fairbairn, J. N. (1992). "An Atlas of Cassava in Africa: Historical, Agroecological and Demographic Aspects of Crop Distribution," Centro Internacional de Agricultura Tropical, Cali, Colombia.
Coons, G. H. (1949). The sugar beet: Product of science. *Sci. Mon.* **68:**149–164.
Coons, G. H. (1953). Disease resistance breeding of sugar beets: 1918–1952. *Phytopathology* **43:**297–303.
Cropley, R. (1967). Decline and death of pear on quince rootstocks caused by virus infection. *J. Hort. Sci.* **42:**113–115.
Crowdy, S. H., and Posnette, A. F. (1947). Virus diseases of cacao in West Africa. II. Cross-immunity experiments with viruses 1A, 1B and 1C. *Ann. Appl. Biol.* **34:**403–411.
Dahal, G., Hibino, H., Cabunagan, R. C., Tiongco, E. R., Flores, Z. M., and Aguiero, V. M. (1990). Changes in cultivar reaction to tungro due to changes in 'virulence' of the leafhopper vector. *Phytopathology* **80:**659–665.
Day, P. R. (ed.) (1977). The genetic basis of epidemics in agriculture. *Ann. NY Acad. Sci.* **287:**1–400.
DuCharme, E. P., Knorr, L. C., and Speroni, H. A. (1951). Observations on the spread of tristeza in Argentina. *Citrus Mag.* **13:**10–14.
Egan, B. T. (1976). The fall and rise of Fiji disease in southern Queensland. *Proc. 43rd Conf. Qd. Soc. Sugar. Cane Technol.* **43:**73–77.
Egan, B. T., and Toohey, C. L. (1977). The Bundaberg approved plant source scheme. *Proc. 44th Conf. Qd. Soc. Sugar Cane Technol.* **44:**55–59.
Fawcett, H. S., and Wallace, J. M. (1946). Evidence of the virus nature of quick decline. *Calif. Citorg.* **32:**88–89.
Gardner, F. E., Marth, P. C., and Magness, J. R. (1946). Lethal effects of certain apple scions on spy 227 stock. *Proc. Am. Soc. Hortic. Sci.* **48:**195–199.
Gilbert, A. H. (1928). Net necrosis of Irish potato tubers. *Vt. Agric. Exp. Stn. Bull.* **289:**36.
Grancini, P. (1962). Ulteriori notizie sul nanismo ruvido del mais. *Maydica* **17:**17–25.
Hanold, D., Morin, J. P., Labouisse, J. P., and Randles, J. W. (2003). Coconut and other palm trees. Foliar decay disease in Vanuatu. *In* "Virus and Virus-like Diseases of Major Crops in Developing Countries" (G. Loebenstein and G. Thottappilly, eds.), pp. 583–596. Kluwer Academic, London.
Hariri, D., Fouchard, M., and Prud'homme, H. (2001). Incidence of *Soil-borne wheat mosaic virus* in mixtures of susceptible and resistant wheat cultivars. *Eur. J. Plant Pathol.* **107:**625–631.
Harpaz, I. (1972). "Maize Rough Dwarf," 251 pp. Israel Universities Press, Jerusalem.

Hatton, R. G. (1928). The behaviour of certain pears on various quince rootstocks. *J. Pomol.* **7:**216–233.

Hillocks, R. J., Raya, M. D., Mtunda, K., and Kiozia, H. (2001). Effects of brown streak virus disease on yield and quality of cassava in Tanzania. *J. Phytopathol.* **149:**389–394.

Hillocks, R. J., Thresh, J. M., Tomas, J., Botao, M., Macia, R., and Zavier, R. (2002). Cassava brown streak disease in northern Mozambique. *Int. J. Pest Manag.* **48:**178–182.

Irwin, M. E., and Kampmeier, G. E. (1989). Vector behaviour, environmental stimuli, and the dynamics of plant virus epidemics. *In* "Spatial Components of Plant Virus Disease Epidemics" (M. J. Jeger, ed.), pp. 14–39. Prentice Hall, New Jersey.

Jennings, D. L. (1960). Observations on virus diseases of cassava in resistant and susceptible varieties. II. Brown streak disease. *Emp. J. Exp. Agric.* **28:**261–270.

John, V. T., Thottappilly, G., Masajo, T. M., and Reckhaus, P. M. (1988). Rice yellow mottle virus disease in Africa. International Conference of Plant Pathology, Japan, 20–27 August 1988, Unpublished contribution.

Jones, A. T. (1976). The effect of resistance to *Amphorophora rubi* in raspberry *(Rubus idaeus)* on the spread of aphid-borne viruses. *Ann. Appl. Biol.* **82:**503–570.

Jones, D. A. C., Woodford, J. A. T., Main, S. C., Pallett, D., and Barker, H. (1996). The role of volunteer potatoes in the spread of potato virus Y^N in ware crops of cv. Record. *Ann. Appl. Biol.* **129:**471–478.

Kenmore, P. E., Cariño, F. O., Perez, C. A., Dych, V. A., and Gutierrez, A. P. (1984). Population regulation of the rice brown planthopper (*Nilaparvata lugens* Stål) within rice fields in the Philippines. *J. Plant Prot. Trop.* **1:**19–37.

Keyworth, W. G. (1947). Mosaic disease of the hop: A study of tolerant and sensitive varieties. *Rep. E. Malling Res. Stn.* 1946, pp. 142–148.

Khush, G. S. (1977). Breeding for resistance in rice. *Ann. NY Acad. Sci.* **287:**296–308.

Kiritani, K. (1983). Changes in cropping practices and the incidence of hopper-borne diseases of rice in Japan. *In* "Plant Virus Epidemiology" (R. T. Plumb and J. M. Thresh, eds.), pp. 239–247. Blackwell Scientific, Oxford.

Klinkowski, M. (1970). Catastrophic plant diseases. *Annu. Rev. Plant Pathol.* **8:**37–60.

Knight, R. L., and Rogers, H. H. (1955). Recent introductions to West Africa of *Theobroma cacao* and related species. I. A review of the first ten years. *Emp. J. Exp. Agric.* **23:**113–125.

Koehler, B., Bever, W. M., and Bonnett, O. T. (1952). Soil-borne wheat mosaic. *Bull. 556, Ill. Agric. Exp. Stn.*, pp. 566–599.

Kulshreshtha, J. P., Kalode, M. B., Prakasa Rao, P. S., Misra, B. C., and Varma, A. (1970). High yielding varieties and the resulting changes in the pattern of rice pests in India. *Oryza* **7:**61–64.

Kus, M. (1994). Krompir (Potato) Ljubljana. ČZP Kmečki glas, Slovenia.

Kus, M. (1995). The epidemic of the tuber necrotic ringspot strain of potato virus Y (PVYNTN) and its effects on potato crops in Slovenia. *Proc. 9th EAPR Virology Section Meeting Bled.*, pp. 159–160.

Leung, H., Zhu, Y., Revilla-Molina, I., Fan, J. X., Chen, H., Pangga, I., Vera Cruz, G., and Mew, T. W. (2003). Using genetic diversity to achieve sustainable rice disease management. *Plant Dis.* **87:**1156–1169.

Lim, G. S. (1972). Studies on penyakit merah disease of rice. III. Factors contributing to an epidemic in North Krian, Malaysia. *Malays. Agric. J.* **48:**278–294.

Luckwill, L. C., and Crowdy, S. H. (1950). Virus diseases of fruit trees. II. Observations on rubbery wood, chat fruit and mosaic in apples. *Ann. Rep. Agric. Hort. Res. Stn. Long Ashton, Univ. Bristol 1949*, pp. 68–79.

Mackenzie, D., Salmon, E. S., Ware, W., and Williams, R. (1929). The mosaic disease of the hop II. Grafting experiments. *Ann. Appl. Biol.* **16:**359–381.

Mahungu, N. M., Bidiaka, M., Tata, H., Lukombo, S., and N'luta, S. (2003). Cassava brown streak disease-like symptoms in Democratic Republic of Congo. *Roots* **8**(2)**:**8–10.

Maruthi, M. N., Hillocks, R. J., Mtunda, K., Raya, M. D., Muhanna, M., Kiozia, H., Rekha, A. R., Colvin, J., and Thresh, J. M. (2005). Transmission of *Cassava brown streak virus* by *Bemisia tabaci* (Gennadius). *J. Phytopathol.* **153:**307–312.

McClean, A. P. D. (1957). Virus infections in citrus trees. *Plant. Prot. Bull. FAO* **5:**133–141.

Meneghini, M. (1946). Sôbre a natureza e transmissibilidade da doença 'Tristeza' dos citrus. *O Biológico* **12:**285–287.

Moore, W. F. (1970). Origin and spread of southern corn leaf blight in 1970. *Pl. Dis. Reptr.* **54:**1104–1108.

Mundt, C. C. (2002). Use of multiline cultivars and cultivar mixtures for disease management. *Annu. Rev. Phytopathol.* **40:**381–410.

Nichols, R. F. W. (1950). The brown streak disease of cassava: Distribution, climatic effects and diagnostic symptoms. *East Afr. Agric. J.* **15:**154–160.

Nour, M. A., and Nour, J. J. (1964). Identification, transmission and host range of leaf curl infecting cotton in Sudan. *Empire Cotton Growers' Rev.* **41:**27–37.

Nweke, F. I. (1994). Farm level practices relevant to cassava plant protection. *Afr. Crop Sci. J.* **2:**563–582.

Nweke, F. I., Dixon, A. G. O., Asiedu, R., and Folayan, S. A. (1994). "Cassava Varietal Needs of Farmers and the Potential for Production Growth in Africa." Collaborative Study of Cassava in Africa. Working Paper 10, IITA, Ibadan, Nigeria.

Otim-Nape, G. W., and Thresh, J. M. (2006). The current pandemic of cassava mosaic virus disease in Uganda. *In* "The Epidemiology of Plant Diseases" (B. M. Cooke, D. Gareth Jones, and B. Kay, eds.), 2nd Ed., pp. 521–550. Kluwer Academic, London, UK.

Otim-Nape, G. W., Thresh, J. M., and Shaw, M. W. (1998). The incidence and severity of cassava mosaic virus disease in Uganda: 1990–92. *Trop. Sci.* **38:**25–37.

Otim-Nape, G. W., Bua, A., Thresh, J. M., Baguma, Y., Ogwal, S., Ssemakula, G. N., Acola, G., Byabakama, B. A., Colvin, J., Cooter, R. J., and Martin, A. (2000). "The Current Pandemic of Cassava Mosaic Virus Disease in East Africa and its Control." NARO/NRI.DFID, Chatham, UK.

Otim-Nape, G. W., Alicai, T., and Thresh, J. M. (2001). Changes in the incidence and severity of cassava mosaic virus disease, varietal diversity and cassava production in Uganda. *Ann. Appl. Biol.* **138:**313–327.

Persley, G. (1992). Replanting the tree of life. Towards an international agenda for coconut palm research, TAC, CGIAR, FAO, Rome, Italy.

Porto, M. C. M., Asiedu, R., Dixon, A., and Hahn, S. K. (1994). An agroecologically-orientated introduction of cassava germplasm from Latin America into Africa. *In* "Tropical Root Crops in a Developing Economy" (F. Ofori and S. K. Hahn, eds.), pp. 118–129. *Proc. 9th Symp.* ISTRC, Accra, Ghana, 1991.

Posnette, A. F. (1951). Progeny trials with cacao in the Gold Coast. *Emp. J. Expt. Agric.* **19:**242–252.

Posnette, A. F., and Todd, J. McA. (1951). Virus diseases of cacao in West Africa. VIII. The search for virus-resistant cacao. *Ann. Appl. Biol.* **38:**785–800.

Power, A. G. (1987). Plant community diversity, herbivore movement; and an insect-transmitted disease of maize. *Ecology* **68:**1658–1669.

Power, A. G. (1988). Leafhopper response to genetically diverse host plant stands. *Entomol. Exp. et Applic.* **49:**213–219.

Power, A. G. (1991). Virus spread and vector dynamics in genetically diverse plant populations. *Ecology* **72:**232–241.
Racman, D. S., Mcgeachy, K., Reavy, B., Štrukelj, B., Žel, J., and Barker, H. (2001). Strong resistance to potato tuber necrotic ringspot disease in potato induced by transformation with coat protein gene sequences from an NTN isolate of *Potato virus Y*. *Ann. Appl. Biol.* **139:**269–275.
Reckhaus, P. M., and Adamou, I. (1986). Rice diseases and their economic importance in Niger. *FAO Plant. Prot. Bull.* **34:**77–82.
Rivera, C. T., and Ou, S. H. (1965). Leafhopper transmission of 'tungro' disease of rice. *Plant. Dis. Reptr.* **49:**127–131.
Rodriguez, M. J. B. (2003). Coconut and other palm trees: Viroid diseases. *In* "Virus and Virus-like Diseases of Major Crops in Developing Countries" (G. Loebenstein and G. Thottappilly, eds.), pp. 567–579. Kluwer Academic, London.
Ryan, C. C. (1988). Epidemiology and control of Fiji disease virus of sugarcane. *Adv. Dis. Vector Res.* **5:**163–176.
Salmon, E. S. (1923). The 'mosaic' disease of the hop. *J. Minist. Agric. Fish.* **29:**927–934.
Simmonds, N. W. (1962). Variability in crop plants, its use and conservation. *Biol. Rev.* **37:**422–465.
Smith, G. R. (2000). Fiji disease. *In* "A Guide to Sugarcane Diseases" (P. Rott, R. Bailey, J. C. Comstock, B. Croft, and S. Saumtally, eds.), pp. 239–244. CIRAD/ISSCT, Montpellier, France.
Smithson, J. B., and Lenné, J. M. (1996). Varietal mixtures: A viable strategy for sustainable productivity in subsistence agriculture. *Ann. Appl. Biol.* **128:**127–158.
Sõgawa, K. (1976). Rice tungro virus and its vectors in tropical Asia. *Rev. Plant. Prot. Res.* **9:**21–46.
Sperling, L., and Loevinsohn, M. (1993). The dynamics of adoption, distribution and mortality of bean varieties among small farmers in Rwanda. *Ag. Systems* **41:**441–453.
Sserubombwe, W. S., Thresh, J. M., Otim-Nape, G. W., and Osiru, D. O. S. (2001). Progress of cassava mosaic virus disease and whitefly vector populations in single and mixed stands of four cassava varieties grown under epidemic conditions in Uganda. *Ann. Appl. Biol.* **138:**161–170.
Storey, H. H. (1925). Streak disease of sugar-cane. *S. Afr. Dept. Agric. Bull.* 39.
Storey, H. H. (1936). Virus disease of East African plants VI. A progress report on studies of the disease of cassava. *East Afr. Agric. J.* **2:**34–39.
Summers, E. M., Brandes, E. W., and Rands, R. D. (1948). "Mosaic of Sugarcane in the United States with Special Reference to Strains of the Virus." USDA Tech. Bull. 955, Washington, DC.
Thomas, P. E., Lawson, E. C., Zalewski, J. C., Reed, G. L., and Kaniewski, W. K. (2000). Extreme resistance to *Potato leafroll virus* in potato cv. Russet Burbank mediated by the viral replicase gene. *Virus Res.* **71:**49–62.
Thresh, J. M. (1960). Capsids as a factor influencing the effect of swollen shoot disease on cacao in Nigeria. *Emp. J. Expt. Ag.* **28:**193–200.
Thresh, J. M. (1979). Hop-growing in Germany. *Newsl. Fed. Brit. Plant Pathologists* **2:**47–48.
Thresh, J. M. (1980). The origins and epidemiology of some important plant virus diseases. *Appl. Biol.* **5:**1–65.
Thresh, J. M. (1982). Cropping practices and virus spread. *Annu. Rev. Phytopathol.* **20:**193–218.

Thresh, J. M. (1990). Plant virus epidemiology: The battle of the genes. *In* "Recognition and Response in Plant-Virus Interactions" (R. S. S. Fraser, ed.), NATO ASI Series Vol. H 41, pp. 93–121. Springer-Verlag, Berlin.
Thresh, J. M., Owusu, G. K., Boamah, A., and Lockwood, G. (1988). Ghanaian cocoa varieties and swollen shoot virus. *Crop Prot.* **7:**219–231.
Thurston, H. D. (1973). Threatening plant diseases. *Annu. Rev. Phytopathol.* **11:**27–52.
Toler, R. W. (1985). Maize dwarf mosaic, the most important virus disease of sorghum. *Plant Dis.* **69:**1011–1015.
Tomlinson, J. A., and Ward, C. M. (1981). The reactions of some Brussels sprout F_1 hybrids and inbreds to cauliflower mosaic and turnip mosaic viruses. *Ann. Appl. Biol.* **97:**205–212.
Trebbi, T. (1950). Il nanismo del mais in Provincia di Brescia nel 1949. *Notiz. Mal. Piante* **8:**13–16.
Tukey, H. B., and Brase, K. D. (1943). An uncongeniality of the McIntosh apple when top-worked onto Virginia Crab. *Proc. Am. Soc. Hort. Sci.* **43:**139–142.
Varma, A., and Malathi, V. G. (2003). Emerging geminivirus problems: A serious threat to crop production. *Ann. Appl. Biol.* **142:**145–164.
Webber, H. J. (1943). The 'tristeza' disease of sour-orange rootstock. *Proc. Am. Soci. Hort. Sci.* **43:**160–168.
Wolfe, M. S. (1985). The current status and prospects of multiline cultivars and variety mixtures for disease resistance. *Annu. Rev. Phytopath.* **23:**251–274.
Yokomi, R. K., Lastra, R., Stoetzel, M. B., Damsteegt, V. D., Lee, R. F., Garnsey, S. M., Gottwald, T. R., Roch-Peña, M. A., and Niblett, C. L. (1994). Establishment of the brown citrus aphid (Homoptera: Aphididae) in Central America and the Caribbean basin and transmission of citrus tristeza virus. *J. Econ. Entomol.* **87:**1078–1085.
Zelazny, B., Randles, J. W., Boccardo, G., and Imperial, J. S. (1982). The viroid nature of the cadang-cadang disease of coconut palm. *Scientia Filipinas* **2:**46–63.
Zink, F. W., and Duffus, J. E. (1969). Relationship of turnip mosaic virus susceptibility and downy mildew *(Bremia lactucae)* resistance in lettuce. *J. Am. Soc. Hort. Sci.* **94:**403–407.
Zink, F. W., and Duffus, J. E. (1970). Linkage of turnip mosaic virus susceptibility and downy mildew *(Bremia lactucae)* resistance in lettuce. *J. Am. Soc. Hort. Sci.* **95:**420–422.

ADVANCES IN VIRUS RESEARCH, VOL 67

HISTORY AND CURRENT DISTRIBUTION OF BEGOMOVIRUSES IN LATIN AMERICA

Francisco J. Morales

International Centre for Tropical Agriculture, AA 6713, Cali, Colombia

Viruses transmitted by the whitefly *Bemisia tabaci* were first studied in Brazil in the mid-1930s. The viruses associated with the 'infectious chlorosis of the *Malvaceae*' in Brazil were later shown to infect cultivated plants such as common bean, tomato and cotton. The first disease reported to be of epidemiological and economic importance in Brazil was bean golden mosaic. This disease soon became widespread in Brazil, the Caribbean region and northwestern Mexico. However, the causal viruses were later shown to be distinct species. Of these, *Bean golden yellow mosaic virus* became the type species of the genus *Begomovirus*. Tomato was the second crop to be attacked by different begomoviruses, particularly in Mexico and the Caribbean region, including Venezuela. *Tomato yellow mosaic virus*, originally identified in Venezuela, disseminated rapidly in the Caribbean region and, recently, in Colombia, South America. Pepper and chilli (*Capsicum* spp.) have also been severely affected by begomoviruses in Latin America. Other crops affected include: cotton, tobacco, cucurbits, potato, soybean, cotton and some fruit crops. Genetic resistance to begomoviruses has only

0065-3527/06 $35.00
DOI: 10.1016/S0065-3527(06)67004-8

been exploited in common bean in Latin America, and so crop production in whitefly-affected regions has depended heavily on insecticides, with adverse environmental and biological consequences. Begomoviruses in Latin America are predominantly found in tropical wet/dry regions, but can also cause damage in sub-tropical dry/humid climates, provided that the dry season lasts at least 4 months with average precipitation not exceeding 80 mm.

I. Introduction

Latin America (Fig. 1) has the highest incidence and diversity of whitefly-transmitted geminiviruses (genus: *Begomovirus*; family: *Geminiviridae*) in the world. Consequently, this region suffers major yield losses due to the year-round cultivation of susceptible crops and constant attack of these viruses and their vector, the whitefly *Bemisia tabaci* (Morales and Anderson, 2001). Begomoviruses affect many important food and industrial crops in Latin America, such as common bean, cucurbits, tomato and peppers, from northern Mexico to Argentina, including the entire Caribbean region. The presence of *B. tabaci* is an essential condition for the occurrence of begomovirus outbreaks: the higher the populations of the vector, the greater the impact of these viruses on susceptible crops. Fortunately, *B. tabaci* does not tolerate well temperatures below 17°C or the abundant rainfall of the wet Amazon region. *B. tabaci* is not an important pest or virus vector below 30° of latitude south, which includes most of Argentina, Uruguay and Chile in South America, due to the harsh winter conditions in this temperate region. In Latin America, temperature is also a function of altitude, even along the equator, throughout the year. Thus, regions above 1200 m of altitude are generally free of begomoviruses of economic importance. However, the 'El Niño' phenomenon and apparent global warming have occasionally allowed *B. tabaci* and some begomoviruses to affect crops up to 1700 m of altitude (Morales, F. J., unpublished information). Latin America also includes some of the rainiest regions (3500–8000 mm/yr) in the world, namely the Pacific coast of Colombia, Panama, and Costa Rica. The Gulf coast of Central America and Mexico also receives abundant precipitations from June until October. These humid regions are usually planted with crops, such as bananas, African oil palm and cocoa, which are not attacked by whiteflies or begomoviruses.

FIG 1. Map of Latin America showing location of the equator and both Tropics (Cancer and Capricorn).

A. Ecology of Begomoviruses in Latin America

Begomoviruses affect several food and industrial crops in over 22 countries of Latin America, a region with extreme agro-ecosystems ranging from deserts to tropical rainforests, and hot lowland coastal areas to freezing highlands above 3000 m. Latin America has most of its territory within the Tropics of Cancer and Capricorn, where temperatures in the lowlands and mid-altitude regions are adequate for whitefly development throughout the entire year. The area affected by *B. tabaci* and begomoviruses in Latin America, extends at least 10° north and south of the Tropics, up to the US–Mexico border and down to southern Brazil and northern Argentina (Morales and Anderson, 2001).

Considering the economic importance of whiteflies as pests and vectors of plant viruses in the Tropics, the Tropical Whitefly IPM Project (TWFP) was created in 1996 as a collaborative effort of various international and national agricultural research institutions (Anderson and Morales, 2005; www.tropicalwhiteflyipmproject.cgiar.org). During Phase I of the TWFP, *B. tabaci* and begomovirus samples were collected from 303 geo-referenced locations in Latin America, including all countries where the whitefly and the viruses it transmits are considered serious constraints to food production. These data were analysed using 'FloraMap', a computer programme developed to predict the distribution of plants or potential target sites for pests (Jones *et al.*, 1999). This software offers a cluster analysis function tool that condensed the multiple climate types identified into six distinct climate clusters. Cluster 1 included the whitefly/begomovirus regions of northern Mexico and Argentina that lie outside the Tropics and some marginal tropical regions nearby. Cluster 2 grouped some mid-altitude regions in the Andean region of South America and southern Brazil. Cluster 3 included Mesoamerica, the Cuban highlands and eastern/western Brazil. Cluster 4 grouped the Caribbean and Bolivian lowlands, and the Mato Grosso and Acre states of Brazil. Cluster 5 included climates of the Pacific coast of Mexico and Central America, and the Brazilian Cerrados (Goias, Minas Gerais, Brasilia, Piaui). Cluster 6 considered the hot lowlands of the Caribbean coast, parts of the Amazon and coastal Brazil. Although these clusters are in quite diverse climates, they all have a well-defined dry season lasting at least 4 months, with average rainfall under 80 mm/month. The temperature during the dry season was not a critical factor, but the mean monthly temperature of the hottest month should not be below 21 °C for *B. tabaci* to thrive (Morales and Jones, 2004).

A begomovirus/*B. tabaci* distribution map based on the climate probability model used by FloraMap, showed agricultural regions (>30% cultivated land) not yet affected by begomoviruses in central–eastern Argentina, southwestern Uruguay, western Ecuador, northern Peru and Central Chile. There are no reports of begomoviruses in Chile yet, but recent surveys have detected the presence of begomoviruses affecting tomato north and south of Lima, Peru and in northern Uruguay (Morales, F. J., unpublished). A recent survey conducted in Argentina (Truol *et al.,* 2005) revealed the presence of biotype B of *B. tabaci* in the departments of Cordoba and Buenos Aires, as predicted by FloraMap.

A modified Koeppen climate classification used to analyse the 303 collection points, showed that 55% of the samples collected were from tropical wet/dry (Aw) regions; 22% were located in tropical and subtropical dry/humid climates (Bw/Cw). The remaining 23% were in the wet equatorial and trade wind littoral climates (Af/Am). Thus, begomoviruses in Latin America can affect crops in different climates, as long as there are agro-ecosystems with a relatively warm and prolonged dry season and suitable reproductive hosts for *B. tabaci.* The transformation of semi-desert areas into irrigated districts for agricultural purposes has been a common and desirable undertaking in Latin America in recent decades. Unfortunately, these environments are very conducive to the development of high populations of *B. tabaci* and diseases induced by begomoviruses.

B. History of Begomoviruses in Latin America

Research on plant viruses transmitted by the *B. tabaci* in Latin America was initiated in the mid-1930s by an enthusiastic group of Brazilian and foreign scientists working in Brazil. These scientists were captivated by the investigations conducted by Baur (1904) in Germany on the 'infectious variegation' of *Abutilon* spp., after noticing the similarities of this disease with the widespread variegations of several malvaceous plant species in Brazil (Costa, 1955; Silberschmidt, 1943, 1948; Silberschmidt and Tommasi, 1955). In the early 1900s, Baur observed that the 'infectious variegation' of *Abutilon striatum* (syn. thompsonii), a tropical ornamental species introduced to Europe from the neo-tropics, could be reproduced by grafting malvaceous species with vascular tissue from *A. striatum.* Baur (1904) referred to the symptoms observed on the grafted malvaceous hosts as 'infectious chlorosis of the *Malvaceae*' and speculated on the possible existence of a vector of the causal agent in the neo-tropical centre of origin of *Abutilon* spp.

In 1946, Orlando and Silberschmidt (1946) published on the transmission of the causal agent of the 'infectious chlorosis of the *Malvaceae*' by *B. tabaci*. Baur's mechanical transmission experiments were reproduced in Brazil (Costa and Carvalho, 1960a) by transferring *Abutilon mosaic virus* (AbMV) from *A. striatum* to the weed *Sida micrantha* through grafting. The latter plant species was then used as source of inoculum to infect other malvaceous species, such as cotton (*Gossypium hirsutum*), *Malva parviflora, M. rotundifolia* and *Sida rhombifolia.* The assumption that AbMV was the causal virus of these diseases (Costa and Carvalho, 1960a,b) has since persisted, even though Baur (1904) and Flores and Silberschmidt (1967) had reserved the term 'abutilon mosaic' for the variegation of *Abutilon* spp., and 'infectious chlorosis of *Malvaceae*' for the symptoms observed on other malvaceous hosts.

Puerto Rico, in the Caribbean region, was another 'centre' of research in Latin America on what later became known as begomoviruses. For example, variegated plants of a *Sida* sp. were seen in this island in the early 1930s (Cook, 1931). In the 1950s, Julio Bird initiated a series of investigations on the aetiology of several diseases of wild and cultivated plants infected by viruses transmitted by *B. tabaci*. He referred to these viruses as 'rugaceous', based on the pioneering classification of Holmes (1949). *Jatropha gossypifolia* (Euphorbiaceae) is a weed commonly found in Puerto Rico affected by mosaic (Bird, 1957). The causal virus was shown to differ from a begomovirus of *Euphorbia prunifolia* described previously in Brazil (Costa and Bennett, 1950), even though both viruses were shown to be potential pathogens of common bean (*Phaseolus vulgaris*) and other legumes.

Bird (1958) also conducted research on the 'infectious chlorosis' of *Sida carpinifolia*, and concluded that there were many similarities in the studies conducted in Puerto Rico and Brazil (Orlando and Silberschmidt, 1946) on the disease known as 'infectious chlorosis of *Malvaceae*'. Bird and co-workers (1975) described other rugaceous diseases, such as the mosaics of *Merremia quinquefolia* (Convolvulaceae), *Jacquemontia tamnifolia* (Convolvulaceae), *Euphorbia prunifolia* (Euphorbiaceae) and *Rhynchosia minima* (Leguminosae).

C. *Plant Hosts of Begomoviruses in Latin America*

The main crops affected by begomoviruses in Latin America have been common bean (*P. vulgaris*), tomato (*Lycopersicon esculentum*) and sweet and hot peppers (*Capsicum* spp.). Common bean is by far the most affected crop, having over 4 million ha exposed to different begomoviruses capable of inflicting total losses (Morales and Anderson, 2001). The area planted to tomato (*c.* 290,000 ha) and peppers (*c.* 195,000 ha) in Latin America, is relatively small, but the economic losses caused by different begomoviruses in these crops largely surpass the economic losses reported for common bean in the entire region. This is partly due to the high value of vegetables and, also, the cultivation of begomovirus-resistant common bean varieties as virus-resistant varieties are not available for vegetables in Latin America. These crops and the main begomoviruses that affect them in Latin America are dealt with in the following sections, but are not the only plant species affected by begomoviruses in the region. Cotton (*Gossypium* spp.), tobacco (*Nicotiana* spp.), soybean (*Glycine max*), various native and introduced cucurbits and some fruit crops, have been affected by begomoviruses in past and recent years and are also mentioned in this chapter.

The role of wild hosts in the evolution and epidemiology of begomoviruses continues to be a controversial topic of research and discussion among virologists in industrialised and developing nations alike. Different begomoviruses of *Sida* spp. affected by the 'infectious chlorosis of *Malvaceae*', have been characterised in the last decade at the molecular level (Frischmuth *et al.*, 1997; Höfer *et al.*, 1997; Lima *et al.*, 2002; Rampersad and Umaharan, 2003). These studies have shown much genetic relatedness among these and other begomoviruses of cultivated species (e.g., *Bean dwarf mosaic virus*, *Tomato mottle virus* (ToMoV), and so forth) in the Americas. Recently, Jovel *et al.* (2004) in Brazil reported on the molecular characterisation of a 1977 virus isolate from *Sida micrantha,* and showed that the original sample selected contained two different begomoviruses related to the original AbMV isolate from the West Indies. Moreover, their phylogenetic analysis showed that AbMV is very closely related to *Sida golden mosaic virus* from Florida.

In a brief study conducted at the International Centre for Tropical Agriculture (CIAT) in Palmira, Colombia, the wild species *Aspilia tenella* (Compositae), *Desmodium uncinatum, Macroptilium lathyroides, Rhynchosia minima* (Leguminosae), *Malva* sp., *Malvastrum* sp., *Sida rhombifolia* (Malvaceae), *Euphorbia prunifolia* (Euphorbiaceae), *Melochia villosa* (Sterculiaceae) and *Pavonia* sp. (Boraginaceae) were used as sources of unknown begomoviruses (detected by ELISA using broad-spectrum monoclonal antibodies against bi-partite begomoviruses available at CIAT) for biological transmission tests with *B. tabaci.*

The results obtained showed that all of the tested wild-hosts contained begomoviruses able to infect common bean. The begomovirus(es) present in the *M. lathyroides* and *Rhynchosia minima* also infected pigeon pea (*Cajanus cajan*) and the wild legumes tested. The other begomoviruses only infected *P. vulgaris* and their respective plant hosts (Morales *et al.*, 2000). Thus, the role and importance of wild hosts in the survival, evolution and epidemiology of begomoviruses cannot be ignored, considering the perennial nature and resilience of most wild plants in the Tropics.

D. The Whitefly Vector: Bemisia tabaci

The importance, distribution and incidence of begomoviruses in Latin America is directly associated to the distribution and population dynamics of their insect vector, the whitefly species *B. tabaci*. Its geographic distribution depends largely on climatic conditions that favour the reproduction of the species: usually warm temperatures, moderate relative humidity and relatively low-to-moderate rainfall. These conditions are found commonly in tropical America, at altitudes that range from sea level to over 1200 m of altitude. The ecology of *B. tabaci* in Latin America has been studied by Morales and Jones (2004). Besides favourable weather conditions, the existence of a suitable reproductive host is another critical factor responsible for the rapid build-up of *B. tabaci* populations. Cultivated plant species used by this species to complete its life cycle are termed 'reproductive hosts', and their importance is directly proportional to the area they occupy in a given agro-ecosystem. The rapid increase in the area planted with soybean in Brazil, from over 1 to 15 million ha in the last three decades, has been the main factor responsible for the increasing whitefly and begomovirus problems that this country has suffered since the mid-1970s (Morales and Anderson, 2001).

Hosts differ in their capacity to support whitefly reproduction and some whitefly populations prefer certain plant species. Bird proposed the existence of 'races' of *B. tabaci* in Puerto Rico (Bird, 1957; Bird and Sánchez, 1971) based on the host and virus transmission specificity exhibited by *B. tabaci*. However, most taxonomists do not accept this proposal, because this whitefly species shows marked host preference according to the reproductive hosts available in a given region. Flores and Silberschmidt (1958) reared *B. tabaci* on different hosts and concluded that there were 'ecological biotypes' rather than 'races'. Nevertheless, some *B. tabaci* populations and biotypes seem to have 'evolved' with certain plant hosts and viruses, such as the *B. tabaci* genotypes, associated with cassava and cassava mosaic viruses in Africa (Sseruwagi *et al.*, 2005).

One of the most important factors responsible for the increasing incidence of whiteflies and begomoviruses in Latin America was the introduction of the B biotype of *B. tabaci* on ornamental plants in the early 1990s and its subsequent, rapid dissemination under natural conditions (Morales and Anderson, 2001). This new biotype exhibits a radically different behaviour from that of the 'original' biotype (A), being far more polyphagous and aggressive in terms of its fecundity and adaptation to different hosts and environments (Barro, 1995). Biotype B also induces physiological disorders of considerable economic importance that were previously unknown in the agricultural regions invaded by this new pest. These disorders include 'silver leaf' of squash, 'irregular ripening' of tomato (Costa and Brown, 1991; Ramírez *et al.*, 1998), and severe chlorosis of some fruits produced by some of the plant species (e.g., common bean and some cucurbits) affected by large populations of the new biotype (Morales, F. J., personal observation). These significant differences and other biological considerations have led some scientists to consider this biotype to be a new whitefly species: *Bemisia argentifolii* (Bellows *et al.*, 1994; Perring *et al.*, 1993). Currently, the proposal is to treat *B. tabaci* as a 'species-complex' (Martin *et al.*, 2000; Perring, 2001).

The highly polyphagous nature of the B biotype of *B. tabaci* has facilitated the adaptation of this new pest to a changing agricultural environment characterised by the transition of monoculture (e.g., maize or common bean in a farm or region) to mixed cropping systems (e.g., maize, common bean, tomato, pepper and peanut in a farm or region) in Latin America. The high-reproductive capacity of the new biotype also makes it a serious pest even in the absence of begomoviruses. Populations exceeding 2000 whitefly individuals per trifoliolated leaf have been recorded for common bean, causing severe cases of 'sooty mould' and, ultimately, plant death. As a vector, the new biotype may not be as efficient as the original biotype in transmitting begomoviruses (Duffus *et al.*, 1992), but the much larger populations of the B biotype offset by far the differences in transmission efficiency observed between the A and B biotypes. Moreover, as the B biotype begins to encounter and transmit begomoviruses, its transmission efficiency increases significantly with time (Morales, F., and Cuellar, Maria H., unpublished data). Finally, the B biotype has displaced the A biotype from important agricultural regions of Latin America where it has appeared (Morales *et al.*, 2005) and it has also displaced the second most important whitefly pest species, *Trialeurodes vaporariorum*, from mid-altitude (950–1600 m) agricultural regions where the original *B. tabaci* biotype could not thrive (Rodriguez *et al.*, 2005). Occasionally, the B biotype has been observed to transmit begomoviruses at altitudes over 1500 m above sea level in the central highlands of Colombia (Morales, F. J., unpublished information).

II. Begomoviruses as Pathogens of Cultivated Plant Species

A. *Cotton*

The family Malvaceae includes approximately 100 genera and 1700 species; of which the genera *Abutilon, Malva* and *Sida* account for over 400 species of predominantly neotropical origin (Fryxell, 1997). Considering the biological importance of malvaceous plant species in the perpetuation and possible evolution of several begomoviruses in Latin America, it is not surprising to learn that cotton (*Gossypium* spp.) was one of the first malvaceous crops of economic importance to be affected by a mosaic caused by a whitefly-transmitted virus (Costa, 1937). Costa demonstrated that the infection of cotton with the *Abutilon infectious variegation virus* in Brazil, depends on the transmission of the virus from malvaceous weeds (e.g., *Sida micrantha*) to cotton by *B. tabaci*, whereas transmission of the virus from infected to healthy cotton in the field was 'negligible' (Costa, 1955). The first report of *B. tabaci* as a vector of plant viruses in Mexico was associated with cotton production in the Valley of Mexicali in the early 1950s (Brown and Nelson, 1987).

Cotton (*Gossypium hirsutum* and *G. barbadense*) is native of the Americas, later exploited by the European conquerors as a plantation and export crop for the European market. Various wars in Europe, Asia and the United States increased the global demand for cotton and thus encouraged the independent Latin American nations to increase the area planted to this crop. It is still uncertain when *B. tabaci* was introduced in the Americas from the Old World, but the colonial (16–19th centuries) and republican (19–20th centuries) cotton plantations present throughout the Americas were a suitable host to this insect. A high demand for cotton in the 1940s promoted a substantial increase in the area planted to cotton in the Pacific lowlands of Central America, from a few thousand to over 300,000 ha. Following the Second World War (1939–1945), the advent and subsequent intensive use of insecticides on cotton, rapidly elevated *B. tabaci* to a prominent pest status in Latin America. In 1964, a whitefly-borne virus of cotton that appeared in Guatemala halted cotton cultivation. By 1990, cotton production in the Central American Pacific region had been reduced 90% due to the high costs associated with whitefly control (Spillari, 1994). Cotton was also severely affected by *B. tabaci* in northwestern Mexico and southwestern United States, in the first quarter of the 20th century. In the early 1950s, 'leaf crumple' was detected in this region and in Guatemala as an economically important disease of cotton (Brown *et al.*, 2002). The bipartite genome of *Cotton leaf crumple virus* (CLCrV) suggests that this begomovirus may have originated by recombination (Idris and Brown, 2004) among ancestors of AbMV and *Squash leaf curl virus* (SLCV). The latter begomovirus attacks cucurbits and common bean in northwestern Mexico, as detailed later in this chapter. Cotton production in Latin America is not currently endangered by begomoviruses, but the potential for severe outbreaks is real. It is not known why cotton begomoviruses have not evolved into more pathogenic variants, considering they are probably the oldest whitefly-transmitted viruses reported to affect any cultivated plant species in Latin America. Perhaps the significant reduction (>70%) in the area used to plant cotton in Latin America from 1970 (5.9 million ha) to 2000 (1.7 million ha) (http://faostat.fao.org), may help explain this fact.

B. *Tobacco*

Tobacco (*Nicotiana tabacum*), another plant species native to the Americas, was originally considered to be the main host of *B. tabaci*, as its name suggests. However, tobacco has not been observed to be a preferred host of *B. tabaci* or the begomoviruses transmitted by this species in Latin America. In fact, tobacco plant residues are still used in Latin America as an effective insecticide against *B. tabaci* and early research in Brazil (Silberschmidt and Tommasi, 1956) showed that the begomoviruses causing the 'infectious chlorosis of *Malvaceae*' do not readily infect this solanaceous species. Nevertheless, *B. tabaci* can colonise tobacco in Latin America, albeit in relatively moderate numbers (TWFP, unpublished results), and has been reported to be a vector of opportunistic begomoviruses infecting tobacco in Brazil, Venezuela, Puerto Rico, Dominican Republic, Mexico, Guatemala and Colombia (Bird and Maramorosch, 1978; Costa and Forster, 1939; Morales and Anderson, 2001; Morales *et al.*, 2000; Paximadis *et al.*, 1999; Wolf *et al.*, 1949).

Symptoms induced by begomoviruses in tobacco are usually of the 'leaf curl' type. Affected tobacco plants may also show severe stunting and variegation; disease incidence may be significant (>30%) in some tobacco plantings (Morales *et al.*, 2000). Two begomoviruses isolated in the state of Chiapas, southern Mexico (Paximadis *et al.*, 1999) were very closely related to *Cabbage leaf curl virus* (MEX 15 isolate) or *Pepper golden mosaic virus* (PepGMV)(MEX 32 isolate). The Colombian tobacco begomovirus was closely related to a begomovirus of *Merremia* sp. from Puerto Rico. Coincidentally, the Puerto Rican begomovirus from *M. quinquefolia* had been observed to infect tobacco (Bird *et al.*, 1975). A begomovirus isolated in Cuba, tentatively named tobacco leaf rugose virus, is related to *Jatropha mosaic virus* from Puerto Rico (Domínguez *et al.*, 2002). These observations demonstrate that begomoviruses may also become important pathogens of tobacco in Latin America if whitefly populations continue to increase as a result of poorly managed mixed cropping systems. Begomoviruses are already important pathogens of tobacco in Africa, Asia and the Pacific region (Muniyappa, 1980; Paximadis *et al.*, 1999).

C. Tomato

The significant increase in tomato (*Lycopersicon esculentum*) production in Latin America in the 1970s and 1980s (almost double of that in the 1960s), was followed by an equally noticeable increase in the number of begomoviruses capable of attacking this crop in the region (Morales and Anderson, 2001). Tomato is native to the Americas, where it is widely cultivated even though most of the genetic improvement has taken place elsewhere (Morales, 2001).

The first disease of tomato associated with *B. tabaci* in Latin America was an 'infectious chlorosis' of tomato observed in Brazil (Flores *et al.*, 1960). It is not clear whether this disease was caused by begomoviruses associated with wild malvaceous hosts in Brazil, as suggested by the title of the paper: *Observations on the infectious chlorosis of malvaceous plants in field tomatoes*, or there were other begomoviruses of tomato involved, as suggested by Bird and Maramorosch (1978). These authors cite Costa (1974) and Costa *et al.* (1975) as claiming that there were six viral diseases of tomato in Brazil associated with *B. tabaci* and only tomato golden yellow mosaic was mechanically transmissible. Eventually, Matys *et al.* (1975) isolated and partially characterised *Tomato golden mosaic virus* (TGMV).

The second begomovirus disease of tomato in Latin America was described in Venezuela in 1963, as a 'yellowish mosaic' of tomato. By 1975, this disease had already disseminated in the main tomato-producing regions of Venezuela and was named tomato yellow mosaic (Lastra and Gil, 1981; Lastra and Uzcátegui, 1975). In 1981 and 1985,

Tomato yellow mosaic virus (ToYMV) was reported to infect potato (*Solanum tuberosum*) under field conditions in the state of Aragua, Venezuela (Debrot, 1981; Debrot and Centeno, 1985). These reports were ignored when a potato sample from the tomato yellow mosaic-affected region in Venezuela was later reported to contain 'a new geminivirus infecting potatoes' (Roberts *et al.*, 1986). As a result, several begomoviruses related to ToYMV, found affecting tomato in the Caribbean region (Engel *et al.*, 1998; Guzmán *et al.*, 1997; Polston *et al.*, 1998; Urbino *et al.*, 2004) are currently labelled potato yellow mosaic. An investigation conducted in 2000 with a Venezuelan ToYMV isolate preserved since the early 1980s, clearly demonstrated that this isolate had nucleotide and amino acid sequence identities of >95% with the corresponding genomic regions of *Potato yellow mosaic virus* (Morales *et al.*, 2001). Hence, the correct name of this begomovirus is ToYMV. It has been detected recently in two geographically distinct tomato-producing regions of Colombia, the municipality of Fusagasuga and the Cauca Valley (Morales and Martínez, unpublished results; Morales *et al.*, 2002). Despite this evidence, the *Geminiviridae* Study Group of the International Committee on Taxonomy of Viruses persists in labelling these tomato viruses as '*Potato yellow mosaic*', in their latest report (Fauquet *et al.*, 2005). Moreover, ToYMV has been eliminated in this report from the list of begomovirus species and its name has been given to a tentative begomovirus found in Brazil, for which there is not even a partial sequence. Some of the tomato begomoviruses currently found in the Cauca Valley of Colombia also show genetic affinities to *Bean dwarf mosaic virus* (BDMV), originally isolated in the same region (Morales *et al.*, 1990, 2002).

Northwestern Mexico is one of the main agricultural regions affected by *B. tabaci* and begomoviruses in Latin America (Morales and Anderson, 2001), perhaps because of the early association between the whitefly pest and cotton cultivation in this region. In 1971, a disease of tomato, referred to as *enchinamiento*, was described in the Valley of Culiacan, Sinaloa. The causal begomovirus, named *Chino del tomate virus*, has caused sporadic outbreaks in tomato plantings in the state of Sinaloa since 1976 (Brown and Nelson, 1988) and now occurs in the central states of Jalisco, San Luis Potosí and Guanajuato (Garzón-Tiznado *et al.*, 2002). The DNA A component of this virus shares notable similarities with the corresponding genomic component of AbMV and some begomoviruses of Malvaceae (www.danforthcenter.org). There was a proposal to re-name this virus as *Tomato leaf crumple virus*, but the *Geminiviridae* study group has chosen to retain the original name, *Chino del tomate virus* (Fauquet *et al.*, 2005). An alternative name would have been *Tomato chino virus*, considering that 'chino' is the name given by Mexican farmers to the characteristic 'foliar malformation' induced by the virus.

The expansion of non-traditional agricultural export crops in Latin America has had a significant impact on the emergence of new begomoviruses affecting tomato. Basically, small-scale farmers in developing countries have been suffering a gradual erosion of the income derived from traditional agricultural products, relative to the significant increase in the price of manufactured goods produced by industrialised nations, which includes agricultural inputs (Braun *et al.*, 1993; Hilhorst *et al.*, 1995). This situation has forced small-scale farmers to diversify their cropping systems, by introducing high-value (e.g., tomato, pepper) crops. Unfortunately, the availability of new hosts for *B. tabaci* and the begomoviruses it transmits, the increased use of pesticides, which results in insecticide-resistant whitefly populations and reduces bio-control agents together with the lack of technical assistance have resulted in significant yield losses in the case of non-traditional crops (Morales and Anderson, 2001).

One of the most unfortunate events in the history of begomoviruses in Latin America has been the introduction of the Old World *Tomato yellow leaf curl virus* (TYLCV). This irresponsible act apparently occurred following the introduction of infected tomato seedlings from Israel into the Dominican Republic (Polston *et al.*, 1994). The virus rapidly disseminated in the Hispaniola Island, causing the collapse of the tomato industry of Haiti and the Dominican Republic. Economic losses in the Dominican Republic alone were estimated at $50 million dollars in a single year. TYLCV has now disseminated in the Caribbean region, mainly to Cuba, Puerto Rico, Guadeloupe and Jamaica (Bird *et al.*, 2001; Gonzales and Valdez, 1995; McGlashan *et al*, 1994; Urbino and Tassius, 1999). In the late 1990s, TYLCV emerged in the Peninsula of Yucatan, Mexico (Ascencio-Ibañez *et al.*, 1999). This virus not only attacks tomato, but also several other crops, including peppers, tobacco, common bean, and cucurbits (Dalmon and Marchoux, 2000; Martinez-Zubiaur *et al.*, 2002, 2004).

Regarding the evolution of tomato begomoviruses in South America, a large-scale survey of tomato-growing regions in the Federal District and seven different states of Brazil (Rio de Janeiro, São Paulo, Minas Gerais, Bahia, Pernambuco, Paraiba and Rio Grande do Norte) revealed at least seven new begomovirus species (Ribeiro *et al.*, 2003). Two of the new species were tentatively named tomato chlorotic mottle virus and tomato rugose mosaic virus. These virus species were related to the begomoviruses that cause bean golden mosaic, tomato golden mosaic and tomato yellow vein streak in Brazil. The latter virus was first isolated from common bean (*P. vulgaris*) in Argentina (Morales and Anderson, 2001) but was later found affecting tomato in Brazil (Faria *et al.*, 1997). These findings, together with the close relationship between some tomato and common bean begomoviruses, suggests that the evolution and emergence of new begomoviruses is not necessarily associated to the long-term infection of plant species in the same or related families. The expansion of soybean into northern Brazil and the dry conditions that predominate in northeast Brazil has facilitated the emergence of different begomoviruses in this region (Morales and Anderson, 2001). Ceara and the Valley of São Francisco (Pernambuco and Bahia) are locations in northeast Brazil, where tomato crops have been affected by begomoviruses (Lima *et al.*, 2000, 2001).

In Venezuela, limited surveys conducted by the TWFP in 2001 revealed the existence of a new begomovirus species that differs from ToYMV in the state of Lara. The begomovirus detected showed the highest genomic similarity (93%) with *Merremia mosaic virus* (Morales and Geraud, unpublished data), originally isolated in Puerto Rico (Brown *et al.*, 2001b). ToYMV is still present in Venezuela and has recently been detected in the neighbouring tomato-growing regions of Colombia (Santander), the central highlands (Cundinamarca) and the Cauca Valley of Colombia (Morales *et al.*, 2002). This begomovirus is also giving rise to new begomoviruses of tomato in Colombia, through recombination with common bean begomoviruses (Morales and Martinez, unpublished data).

Begomoviruses are not yet important pathogens of tomato in other South American countries, but these pathogens are now emerging in Peru (Murayama *et al.*, 2005) and other countries located in the southern cone of South America. The distribution of begomoviruses in Latin America would be larger than currently documented, both in the horizontal (latitude) and vertical (altitude) geographical sense, if the numerous begomoviruses found in wild hosts in this region were characterised.

In the Lesser Antilles, tomato yellow mosaic continues to be the predominant disease in tomato-growing regions. Unfortunately, the misnomer 'potato yellow mosaic' still persists in the literature (Urbino *et al.*, 2004), even though the disease in both tomato and potato has been shown to be caused by ToYMV (Debrot, 1981; Morales *et al.*, 2001). In Cuba, tomato yellow leaf curl remains an economically important disease of tomato (Martinez-Zubiaur *et al.*, 2003). Additionally, two bipartite begomoviruses have been known to infect tomato in this island since 1997: *Tomato mottle Taino virus* (ToMoTV) and *Tomato mosaic Havana virus* (ToMHV) (Martinez-Zubiaur, 1998; Martinez-Zubiaur *et al.*, 1998; Ramos *et al.*, 1997). ToMoTV can form pseudo-recombinants with ToYMV (Ramos *et al.*, 2003) and infect potato in Cuba (Cordero *et al.*, 2003). In the Dominican Republic and Haiti, tomato yellow leaf curl is also the main biotic constraint to tomato production, although the introduction of commercial TYLCV-resistant cultivars has greatly contributed to the sustainable management of this virus problem (Morales, unpublished). In Puerto Rico, four begomoviruses affect tomato production: TYLCV, ToMoV, ToYMV and a begomovirus isolated from the weed *Merremia* sp. (Idris *et al.*, 1998). In Jamaica, TYLCV has been present since the early 1990s (McGlashan *et al.*, 1994) and a bipartite begomovirus, *Tomato dwarf leaf curl virus*, was isolated from single and doubly infected (TYLCV) tomato plants in the late 1990s (Roye *et al.*, 1999).

In Mexico, *Chino del tomate virus* has disseminated to other states, namely: Chiapas, Michoacan, Morelos, Tamaulipas and Baja California Sur (Hernandez, 1972; Holguin-Peña *et al.*, 2005; Montes-Belmont *et al.*, 1995; Torres-Pacheco *et al.*, 1996). In a national survey conducted by Dr Rafael Rivera-Bustamante of CINVESTAV (Centre for Research and Advanced Studies, Irapuato, Mexico) for the TWFP (Morales *et al.*, 2005b), *Pepper Huasteco yellow vein virus* (PHYVV) and PepGMV (formerly Texas pepper virus), were found to be pathogens of tomato in the states of Jalisco, Morelos, Hidalgo and San Luis Potosi, and Nayarit, Oaxaca and Hidalgo, respectively. Tomato plants doubly infected with PHYVV and PepGMV were found in some of the locations surveyed.

As mentioned earlier, ToYMV is now widely distributed in the Caribbean, both in its original and recombinant forms, particularly in the Lesser Antilles, Panama and Colombia (Engel *et al.*, 1998; Guzmán *et al.*, 1997; Morales and Martínez, unpublished results; Morales *et al.*, 2002; Polston *et al.*, 1998; Umaharan, *et al.*, 1998; Urbino *et al.*, 2004). A previous report on the detection of ToYMV in Costa Rica (Rosset *et al.*, 1990) has been corrected and the name of the Costa Rican begomovirus has been changed to *Tomato yellow mottle virus* (Nakhla *et al.*, 1994). In 1998, two other tomato-infecting begomoviruses were detected in Costa Rica (Karkashian *et al.*, 1998). A year later, *Tomato leaf curl Sinaloa virus* (ToLCSinV), a begomovirus originally described in northwestern Mexico (Brown *et al.*, 1993), was detected in Costa Rica (Idris *et al.*, 1999). ToLCSinV has also been detected in the neighbouring Nicaragua (Rojas *et al.*, 2000). The use of this name should be discouraged, because *Tomato leaf curl virus* is an

Old World begomovirus, not found in the neotropics. In Nicaragua, Rojas *et al.* (Rojas, 2005; Rojas *et al.*, 2005) reported on the existence of a begomovirus complex affecting tomato. These viruses include: ToLCSinV, *Tomato severe leaf curl virus* (ToSLCV), *Squash yellow mild mottle virus* (SYMMoV), *Euphorbia mosaic virus* (EuMV) and PepGMV. SYMMoV had previously been detected in the neighbouring country, Costa Rica, infecting cucurbits (Karkashian *et al.*, 2002), and ToSLCV is also known to occur in Honduras and Guatemala, the following two Central American countries north of Nicaragua (Maxwell *et al.*, 2002). PepGMV has been detected in Costa Rica (Lotrakul *et al.*, 2000) and Guatemala (Morales and Anderson, 2001) as well, and it was originally identified in Texas, United States (Stenger *et al.*, 1990). The dissemination of begomoviruses, such as PepGMV and ToLCSinV, from southwestern United States and northwestern Mexico to southern Central America, clearly illustrates the epidemiological potential of begomoviruses and their whitefly vector in agricultural regions covering millions of hectares of cultivated land. Apparently, the movement of begomoviruses in this region has also occurred from south to north, as ToSLCV and another begomovirus originally described in Central America (Maxwell *et al.*, 2002; Rojas *et al.*, 2000) as *Tomato mild mottle virus,* a name already given to a potyvirus-like virus in Africa (Walkey *et al.*, 1994), have been detected recently in Baja California Sur, Mexico (Holguin-Peña *et al.*, 2005). ToSLCV from Guatemala had been already detected in tomato fields in California, United States (Holguin-Peña *et al.*, 2003). Another begomovirus originally detected in Florida, United States (Abouzid *et al.*, 1992), ToMoV, has also spread south into Puerto Rico (Brown *et al.*, 1995) and west into the Yucatan Peninsula, Mexico (Garrido-Ramirez and Gilbertson, 1998).

*D. Peppers (*Capsicum *spp.)*

Sweet and hot pepper cultivars of the species *Capsicum annuum, C. frutescens* and *C. chinense* are severely attacked by begomoviruses in Mexico and Central America. *Pepper Huasteco yellow vein virus* (PHYVV = formerly pepper huasteco virus) was isolated in the late 1980s from Serrano pepper (*C. annuum*) in the Huasteca Plateau (southern Tamaulipas, northern Veracruz and eastern San Luis Potosi) of Mexico (Leal and Quintero, 1989). Hence, the correct name for this begomovirus should be *Pepper yellow vein Huasteco virus*, and its acronym should be (PepYVHV). This virus is an important pathogen of tomato, as described earlier, and in the national survey conducted by Dr Rafael Rivera-Bustamante of CINVESTAV for the TWFP, this virus was isolated from diseased hot peppers in the states of Campeche, Colima, Nayarit, Guanajuato, Veracruz, Morelos, Hidalgo, Queretaro and San Luis Potosi (Morales *et al.*, 2005). PHYVV has also been reported to infect pepper in the northwestern state of Sonora in Mexico (Ramirez-Arredondo *et al.*, 1998). PepGMV was first reported from Texas in 1987 (Stenger *et al.*, 1990) and was known as Texas pepper virus. PepGMV has since become an economically important pathogen of pepper and tomato in Mexico and Central America. In Mexico, PepGMV has been detected affecting hot peppers in Colima, Nayarit, Veracruz, Oaxaca, Aguascalientes, Morelos and San Luis Potosi (Morales *et al.*, 2005). This begomovirus also affects *C. annuum* in the Mexican state of Coahuila (Bravo-Luna *et al.*, 2000). *Sinaloa tomato leaf curl virus* was originally isolated from pepper and tomato in the state of Sinaloa, northwestern Mexico (Brown *et al.*, 1993), and it now occurs in Costa Rica (Idris *et al.*, 1999). *Pepper mild tigre virus* (PepMTV) was originally isolated from pepper in the state of Tamaulipas, Mexico (Brown *et al.*, 1989), but does not seem to be an economically important begomovirus elsewhere. This is still a tentative begomovirus (Fauquet *et al.*, 2005), but the name 'tigre' (tiger in Spanish) should not have an accent on the last 'e', as regularly included in the literature. PepGMV occurs in Central America, where it has been isolated from Tabasco pepper (*C. frutescens*) and Habanero pepper (*C. chinense*) (Lotrakul *et al.*, 2000).

In the Caribbean, *Tomato dwarf leaf curl virus* affects peppers in Jamaica (Roye *et al.*, 1999) and occurs in single or mixed infections with TYLCV. The latter virus was also found to infect peppers (*C. annuum*) in Cuba (Quiñones *et al.*, 2001) and Yucatan, Mexico (Ascencio-Ibañez, 1999). In Trinidad Tobago, at least two begomoviruses related to ToYMV and PHYVV have been isolated from *C. annuum* and *C. frutescens* (Umaharan *et al.*, 1998).

In South America, begomoviruses have been observed to affect sweet peppers (*C. annuum*) in northeastern Brazil (Lima *et al.*, 2001) and the Cauca Valley of Colombia. The latter begomovirus was found to be closely related to ToYMV (Morales and Martinez, unpublished information).

E. Potato

The first begomovirus shown to infect potato (*Solanum tuberosum*) in Latin America was ToYMV (Debrot, 1981; Debrot and Centeno, 1985). This virus was later erroneously named *Potato yellow mosaic virus* (Roberts *et al.*, 1986). This misnomer persists despite recent evidence confirming the identity of ToYMV as a major pathogen of tomato in the Caribbean region and an occasional pathogen of potato in Venezuela (Morales *et al.*, 2001). ToMoTV was detected in 1998 infecting potato in Cuba (Morales *et al.*, 2005a). This finding was confirmed recently in Cuba (Cordero *et al.*, 2003).

'Potato deforming mosaic' was first described in 1981 from the southeastern potato-growing region of Buenos Aires, Argentina (Delhey *et al.*, 1981). This disease was later reported in Rio Grande do Sul, Brazil (Daniels and Castro, 1985), and in 1992, it was associated with a begomovirus isolated from potato in the state of São Paulo, Brazil. The authors suggested that the causal begomovirus of potato deforming mosaic in Brazil differs from the begomovirus that causes tomato yellow mosaic (ex-potato yellow mosaic) in Venezuela (Vega *et al.*, 1992). A recent report (www.promedmail.org) informs that Brazilian scientists have demonstrated that 'potato deforming mosaic' is caused by a begomovirus described in 1997 as *Tomato yellow vein streak virus* (Faria *et al.*, 1997). This virus had also been isolated from common bean in northwestern Argentina in 1995 (Morales and Anderson, 2001), and, thus, should retain its original name, *Potato deforming mosaic virus*. A tri-segmented geminivirus-like pathogen of potato was associated with the disease 'Solanum apical leaf curling' in Peru (Hooker and Salazar, 1983), but the vector and the virus were not characterised further. The distribution and economic importance of this disease remain limited.

Currently, in Latin America, the most important viral disease of potato associated with a whitefly vector, is 'potato yellow vein'. However, this disease is caused by a crinivirus transmitted by the whitefly *Trialeurodes vaporariorum* (Salazar *et al.*, 2000).

F. Common Bean

In 1965, one of the most eminent of the Latin American virologists, Dr Alvaro Santos Costa, described three whitefly-transmitted diseases of common bean (*P. vulgaris*) in the state of São Paulo, Brazil (Costa, 1965). The first disease, 'bean crumpling', was attributed to the occasional infection of common bean plants by the EuMV, found in wild *Euphorbia prunifolia* near common bean fields. The second disease was 'bean mottle dwarf', later re-named 'bean dwarf mosaic' (Costa, 1975) and transmitted from *Sida* sp. to common bean by *B. tabaci*. The incidence of this disease was high in malvaceous weeds, but low (<5%) in common bean fields at the time. The third disease, 'bean golden mosaic', had been observed since 1961 in the state of São Paulo (Costa,

1965). Although Costa noted that this disease was 'not currently of sufficient economic importance', he pointed out its greater epidemiological potential as compared to the other two diseases of bean described. A decade later, Costa (1975) reported bean golden mosaic in the two main common bean-producing states of Brazil, São Paulo and Parana. He attributed the rapid dissemination of this disease to the noticeable increase in the population of *B. tabaci*, as a result of the exponential growth in the area planted with soybean, a reproductive host of this species. Soybean plantings in Brazil had increased from 1.3 million ha in the early 1970s to almost 6 million ha by 1975 (FAO, 1994) and bean golden mosaic had disseminated to the remaining common bean-growing states of Minas Gerais, Goias and Bahia (Costa, 1975).

In Central America, a 'golden mosaic' of common bean had been observed in the late 1960s from Guatemala to Panama (Gamez, 1970). In 1966, a 'yellow mottle' of common bean caused high-yield losses in the Pacific coastal lowlands of Guatemala, El Salvador and Nicaragua (Zaumeyer and Smith, 1966), where *B. tabaci* had previously become a pest of cotton in the 1950s (Spillari, 1994).

In the Caribbean region, the name 'bean golden yellow mosaic' was used for the first time in 1973 to describe a disease of common bean in Puerto Rico (Bird *et al.*, 1973). A 'golden mosaic' of common bean had also been reported in the Dominican Republic (Schieber, 1970) and Jamaica (Pierre, 1975) in 1970 and 1975, respectively. Similar symptoms were observed in Cuba in the early 1970s, particularly in the province of Velasco, where farmers referred to this disease as 'amachamiento' (female sterility), because the pods of affected common bean plants did not produce seeds or the seeds were underdeveloped. The emergence of this disease was associated with an increase in the population of *B. tabaci* (Blanco and Bencomo, 1978). In Haiti, Vakili (1973) made the first association between the presence of whiteflies and bean golden yellow mosaic in the lowlands. By 1978, Balthazar (1978) reported bean golden yellow mosaic in all major bean producing regions of the country. In Central America, other *Phaseolus* spp., such as *P. acutifolius, P. coccineus* and particularly *P. lunatus*, are commonly found affected by this disease under field conditions (Morales and Anderson, 2001).

In 1974, a 'yellow mosaic' was observed in common bean fields in the Valley of Culiacan, Sinaloa, northwestern Mexico (Lopez, 1974). The disease was later observed in the Valley of Mochis, Sinaloa, and the Valley of Santo Domingo, Baja California Sur (Salinas and Vázquez, 1979). The disease was referred to initially as 'bean golden mosaic', but a virus isolate collected in 1988 from affected common bean plants in the state of Sonora, was characterized molecularly as *Bean calico mosaic vius* (BCaMV) (Brown *et al.*, 1999; Loniello *et al.*, 1992), a distinct begomovirus related to SLCV, previously isolated from diseased squash in southwestern United States (Flock and Mayhew, 1981). During Phase I (1997–2000) of the TWFP, a survey conducted in northwestern Mexico: Los Mochis and Culiacan in the state of Sinaloa, and Etchojoa in the state of Sonora, detected three different types of begomoviruses in common bean, one identical to BCaMV, another type similar to BCaMV and a third type identical to SLCV (Morales *et al.*, 2005).

The 'soybean boom' in South America transcended the borders of Brazil, reaching Argentina in the late 1970s. The area planted to soybean in Argentina increased from 442,000 ha in 1975 to over 3 million ha by the mid-1980s (Morales and Anderson, 2001), causing a noticeable increase in the population of *B. tabaci* in the common bean-growing provinces of the northwest. By 1980, Argentina was already losing tens of thousands of hectares planted to common bean, due to a disease locally known as 'achaparramiento' (severe dwarfing). This disease was similar to the bean dwarf mosaic disease described originally from Brazil (Costa, 1965). BDMV was isolated and partially characterised in Colombia in the late 1980s (Morales *et al.*, 1990) based on its characteristic symptomatology. The Colombian isolate of BDMV was later characterised at the molecular level (Hidayat, 1990). Recently, two begomovirus isolates obtained from common bean plants showing bean dwarf mosaic-like symptoms in northwestern Argentina, were shown to be closely related to different begomoviruses isolated from *Sida* sp. in Latin America (Morales and Martinez, unpublished information), as Costa (1965) had concluded from biological transmission experiments four decades ago. Therefore, it is apparent that bean dwarf mosaic is caused by a complex of distinct begomoviruses transmitted by *B. tabaci* from *Sida* and other malvaceous species to common bean.

A different begomovirus was isolated and partially sequenced in 1995 from a common bean plant showing severe plant malformation in the province of Tucuman, Argentina, but the low incidence of this disease did not merit further research (Morales, unpublished data). This begomovirus was later isolated from tomato in the state of São Paulo, Brazil and named *Tomato yellow vein streak virus* (Faria *et al.*, 1997).

The most recent begomovirus shown to attack common bean in Latin America, in a consistent and economically important manner, was detected in the Cauca Valley of Colombia in 2002, following a prolonged dry spell that increased populations of the B biotype of

B. tabaci (Morales *et al.*, 2002). This bipartite begomovirus appears to be a recombinant between the A component of *Bean golden yellow mosaic virus* and the B component of ToYMV (Morales and Martínez, unpublished data; Morales *et al.*, 2002). The pathogenicity of some begomoviruses in both solanaceous crops and common bean has already been documented for PHYVV (Leal and Quintero, 1989) and TYLCV (Martinez-Zubiaur *et al.*, 2002; Navas-Castillo *et al.*, 2002).

G. Soybean

Soybean (*Glycine max*) is probably the most important reproductive host of *B. tabaci* in terms of the total area cultivated with this legume in Latin America (*c.* 39 million ha), even though there are other reproductive hosts that support higher whitefly populations per plant (Anderson *et al.*, 2005; Morales and Anderson, 2001).

The transmission of begomoviruses by *B. tabaci* to soybean was first observed in the early 1970s in Brazil. Affected soybean plants in the state of São Paulo showed leaf crinkling and plant dwarfing and the problem was associated with large populations of *B. tabaci* on soybean, cotton and common beans in 1972 (Costa *et al.*, 1973). However, the incidence and economic importance of begomoviruses affecting soybean in Brazil have remained low (Faria *et al.*, 2000). Recently, a begomovirus isolated from soybean in the state of Parana, Brazil, was shown to be related to *Sida micrantha mosaic virus* (Moreira *et al.*, 2005).

The presence of begomoviruses, affecting soybean in northwestern Argentina, has also been reported in the provinces of Salta and Tucuman (Laguna *et al.*, 2005; Pardina *et al.*, 1998) with incidences of up to 45% in some fields. A recent survey of several soybean fields conducted by the author in northwestern Argentina showed average virus incidences under 5%. No begomovirus was recovered from symptomatic soybean plants.

In the Cauca Valley of Colombia, a begomovirus was isolated from soybean plants affected by a conspicuous yellow mottle (Morales *et al.*, 2000). This begomovirus was closely related to *Cabbage leaf curl virus,* originally isolated in Florida, United States (Abouzid *et al.*, 1992). The incidence of this begomovirus in Colombia is very low.

It can be concluded here that soybean is a better whitefly than begomovirus host. Whether this situation will continue or begomoviruses might become a serious constraint to soybean production in the future, is not known. In Asia, *Mungbean yellow mosaic virus* (Gupta and

Keshwal, 2002) and *Soybean crinkle leaf virus* (Samretwanich *et al.*, 2001) are economically important begomoviruses of soybean.

H. Cucurbits

The family Cucurbitaceae includes two genera of American origin: *Cucurbita* and *Sechium*, both of which contain species cultivated in the Americas since agriculture was developed in this continent (Saade, 1995). The first major outbreaks of begomoviruses in the Americas were observed in southwestern United States and northwestern Mexico in 1976 (Flock and Mayhew, 1981) and 1981 (Dodds *et al.*, 1984; McCreight and Kishaba, 1991). A new disease referred to as 'squash leaf curl' affected squash (*Cucurbita maxima*) and other cucurbit species (*C. argyrosperma, C. pepo* and *Cucumis melo*), causing a characteristic upward curling and enations on the affected leaves. Squash leaf curl was shown initially to be caused by an apparent virus complex that included a pathogenic variant able to infect common bean (*P. vulgaris*). The causal virus was named SLCV and was characterised molecularly as a bipartite begomovirus (Lazarowitz and Lazdins, 1991). SLCV moved east into Arizona and, subsequently into Texas, where it was first observed (Isakeit and Robertson, 1994) infecting watermelon (*Citrullus lanatus*). SLCV was detected in the state of Sonora, northwestern Mexico in 1990 and 1991, and further south in the state of Sinaloa in 1992 (Ramírez *et al.*, 1995), affecting 'calabacita' (*Cucurbita pepo*). SLC-like symptoms had been observed previously in experimental plots located in Los Mochis and Culiacan, state of Sinaloa, in 1988 (Silva *et al.*, 1994).

Surveys conducted in Central America in 1999 by the TWFP, revealed the presence of begomoviruses in melon in the Valley of Zacapa, Guatemala, and in 'pipian' (*C. argyrosperma*) and 'ayote' (*C. moschata*) in El Salvador. The begomoviruses isolated from melon and 'pipian' had partial sequence identities of 85% between them, and sequence identities >80% when compared to SLCV (Morales, F. J., unpublished data). The begomovirus detected in melon in 1999 was later re-isolated by other scientists from the same location and named *Melon chlorotic leaf curl virus* (Brown *et al.*, 2001a). The begomovirus isolated from *C. argyrosperma* in El Salvador was also found later on in Costa Rica infecting squash, and was named squash yellow mild mottle (GenBank Accession AJ842151). This virus was detected recently in Nicaragua infecting cucurbits (Ala-Poikela *et al.*, 2005). However, the name given to this virus in Costa Rica appears in the GenBank as a synonym of *Melon chlorotic leaf curl virus*. In 1998, a begomovirus was isolated from melon plantings showing chlorosis and leaf rugosity in the Caribbean region of Colombia (department of Atlantico). This disease, referred to as 'melon chlorosis' (Morales *et al.*, 2000), was associated with the introduction of biotype B of *B. tabaci* to northern Colombia. The partial sequence obtained at that time showed a 94% identity with a fragment of the coat protein of SLCV and an even higher identity with *Melon chlorotic leaf curl virus*. A similar begomovirus named *Melon chlorotic mosaic virus* was reported recently from Venezuela. This begomovirus is also closely related (>80%) to SLCV (Ramirez *et al.*, 2004).

The capacity of SLCV to generate new pathogenic variants (currently considered as species) is likely to continue, as suggested from the numerous begomoviruses of cucurbits described in recent years, including *Cucurbit leaf curl virus* and *Cucurbit leaf crumple virus* found in southwestern United States. These viruses were shown to form viable re-assortants with related viruses in the SLCV cluster (Brown *et al.*, 2002). In the Caribbean, TYLCV was isolated from squash (*Cucurbita pepo*) in Cuba (Martinez-Zubiaur *et al.*, 2004).

The seemingly endless identification of 'new' begomovirus species possessing amino acid or nucleotide identities of 80–90% when compared to a 'parental' virus species, such as SLCV, suggests that the 90% sequence identity threshold used currently to differentiate begomovirus species should be reduced to 80%.

I. Fruit Crops

Fruit crops have not escaped infection by begomoviruses, particularly species in the Passifloraceae. Passionfruit (*Passiflora edulis*) has been attacked by begomoviruses in Puerto Rico (Brown *et al.*, 1993) and Brazil (Novaes *et al.*, 2003), causing mottling and 'little leaf mosaic', respectively. In the northern coast of Colombia, a plant and fruit malformation disease of giant granadilla (*Passiflora quadrangularis*) was shown to be caused by a begomovirus, using a broad spectrum monoclonal antibody that detects bipartite begomoviruses (Cancino *et al.*, 1995). In central Mexico, tomatillo (*Physalis ixocarpa*) was shown to be a host of PepYVHV (Torre-Almaraz *et al.*, 2002). The current popularity of the previously neglected tropical fruit species of Latin America will probably soon lead to an increase in the number of these species attacked by begomoviruses.

III. Current Situation and Outlook

Despite sporadic efforts to educate farmers on the most appropriate methods to manage whitefly pests and whitefly-transmitted viruses, and the millions of dollars spent on insecticides against whiteflies, these pests continue to cause significant yield losses and to affect more agricultural regions in Latin America. The main reason for the continuous expansion of these biotic problems is the lack of technical assistance in rural areas. Latin America is still devoting over 40% of its Gross Regional Product to pay the mounting external debt acquired since the 1970s. This forces governments to reduce their support to agricultural research down to an inoperative level of human and material resources. CIAT has not escaped the effects of the global economic crises. As a result of a dwindling supply of improved varieties produced by national and international agricultural research institutions, pathogens and pests continue to evolve and cause increasing levels of damage to basic food and cash crops due to the breakdown of previously resistant varieties and/or emergence of more aggressive pathogens and pests. An example of this situation is the continuous occurrence of outbreaks of new begomoviruses in Latin America in the last three decades and the ravages of the new B biotype of *B. tabaci* throughout Latin America. In the absence of a continuous effort to provide small- and medium-scale farmers with resistant cultivars to counteract old and new pathogens and pests, the complementary environmental and socio-economic research cannot possibly have any impact. This situation is considerably worse for plant genetic resources of high-value crops, such as tomato and peppers, on which little or no research is conducted in the region. Ironically, as Latin America is being forced to pay attention to environmental, socio-economic and health issues, millions of tonnes of highly toxic pesticides enter the environment and the food consumed by millions of people in developing nations, as desperate farmers do not find any other alternative to cope with the overwhelming problems posed by pests and diseases.

Agricultural policies in Latin America have focused more on the threat of Free Trade Agreements than on re-gaining the capacity to feed the people of this region and produce marketable surpluses of both basic and high-value commodities in a safe and profitable way. The continuous need of researchers to operate on special project funds available for a number of different concerns (not necessarily those of resource-poor farmers), hampers collaborative efforts to combat crop production problems caused by whiteflies and whitefly-borne viruses and, thus, minimise their impact on the poor rural sector of Latin America. More positively, Latin American scientists have greatly benefited from the notable advances that have taken place in the area of molecular biology in recent decades. Unfortunately, these advances have been achieved at the expense of basic crop production and crop improvement research and extension, instead of being complementary and promoting their integration.

The main biological threat remains the continuous invasion of new agricultural areas by biotype B of *B. tabaci*. This biotype has not only displaced the original A biotype from most of the areas it has invaded, but, as mentioned previously, it is also displacing *T. vaporariorum* from mid-altitude (900–1600 m) regions where the A biotype was not considered to be a pest. The B biotype is far more aggressive than biotype A in terms of the number of plant species attacked and the degree of damage caused. It also induces severe physiological problems, such as the irregular ripening of tomato and the intense chlorosis of snap beans, cucurbits and other horticultural crops. The B biotype of *B. tabaci* is not as efficient as the original biotype in transmitting begomoviruses, but its high populations make the B biotype a very effective virus vector. Tests conducted in several Central American and Caribbean countries (Morales *et al.*, 2005) using monoclonal antibodies prepared in the early 1990s against *Bean golden yellow mosaic virus* (Cancino *et al.*, 1995), have shown notable changes in the antigenic properties of this begomovirus, probably in response to the introduction of the B biotype in that decade. The broad host range of the B biotype is also responsible for the increasing emergence of recombinant begomoviruses capable of infecting different solanaceous species, such as tomato, peppers and potato, and also legumes, namely common bean (Morales *et al.*, 2002).

The TWFP is one of the most ambitious efforts to educate small- and medium-scale farmers throughout the Tropics on integrated whitefly management practices designed to develop sustainable and economically viable mixed cropping systems. The main thrust of the project is not to promote organic agriculture, but to make rational and safe use of chemical products with a view to re-establishing the biological balance of the highly disturbed agro-ecosystems that characterise whitefly-affected agricultural regions in the Tropics. Many well-intended IPM projects have been promoting IPM practices, such as barrier and trap crops, sticky traps, biological control, and repellents, to control whiteflies and whitefly-borne viruses, but these measures do not control high whitefly populations. Consequently, farmers become disappointed and increasingly reluctant to participate in IPM projects. It is quite evident that the basis of any viable begomovirus control project should be the development of resistant varieties of crops, but, as mentioned already, the breeding capacity and output in Latin America has actually been reduced in recent decades. Meanwhile, the TWFP is promoting strategies that exclude whitefly pests and vectors, such as cultural, physical and legal practices, emphasising at the same time the need to end the use and abuse of those insecticides to which both *Trialeurodes* and *Bemisia* spp. have developed resistance. Once these now highly disturbed agro-ecosystems have been restored to their original biological balance, all IPM measures known to control whitefly pests under experimental conditions will have the opportunity to make a significant contribution to the development of sustainable agricultural production systems in Latin America and other tropical regions affected by these pests.

Acknowledgments

The author is indebted to the Department for International Development (DFID) of the United Kingdom and to the International Centre for Tropical Agriculture (CIAT) for their support to the Tropical Whitefly IPM Project; and expresses his gratitude to the numerous scientists and research staff who have collaborated in the project from national agricultural research programmes and universities in Latin America, advanced research institutions abroad and international centres. However, the information and opinions expressed here do not reflect the mission or philosophy of the supporting or collaborating agencies and institutions and, therefore, are exclusively the responsibility of the author.

References

Abouzid, A. M., Hiebert, E., and Strandberg, J. (1992). Cloning, identification and partial sequencing of the genome components of a geminivirus infecting the *Brassicaceae*. *Phytopathology* **82:**1070.

Ala-Poikela, M., Svensson, E., Rojas, A., Horko, T., Paulin, T., Valkonen, J. P., and Kvarnheden, A. (2005). Genetic diversity and mixed infections of begomoviruses infecting tomato, pepper and cucurbit crops in Nicaragua. *Plant Pathol.* **54:**448–459.

Anderson, P. K., and Morales, F. J. (2005). "Whitefly and Whitefly-Borne Viruses in the Tropics: Building a Knowledge Base." CIAT Publication No. 341, Palmira, Colombia.

Anderson, P. K., Hamon, A., Hernandez, P., and Martin, J. (2005). Reproductive crop hosts of *Bemisia tabaci* (Gennadius) in Latin America and the Caribbean. *In* "Whitefly and Whitefly-Borne Viruses in the Tropics: Building a Knowledge Base" (P. K. Anderson and F. J. Morales, eds.), pp. 243–250. CIAT Publication No. 341, Palmira, Colombia.

Ascencio-Ibañez, J. T., Diaz-Plaza, R., Mendez-Lozano, J., Monsalve-Fonnegra, Z. I., Arguello-Astorga, G. R., and Rivera-Bustamante, R. F. (1999). First report of tomato yellow leaf curl geminivirus in Yucatan, Mexico. *Plant Dis.* **83:**1178.

Balthazar, S. (1978). Les viruses du haricot commun (*Phaseolus vulgaris*) en Haití. MAMV, Damien, Haití. p. 25.

Barro, P. J.de. (1995). *Bemisia tabaci* biotype B: A review of its biology, distribution and control, p. 58. CSIRO Australia Division of Entomology. Tech. Paper No. 36.

Baur, E. (1904). Über die infectiose chlorose der *Malvaceae*. *Kgl. Preuss. Akad. Wiss.*11–29.

Bellows, T. S., Perring, T. M., Gill, R. J., and Headrick, D. H. (1994). Description of a species of *Bemisia* (Homoptera: Aleyrodidae) infesting North American agriculture. *Ann. Entomol. Soc. Am.* **87:**195–206.

Bird, J. (1957). A whitefly-transmitted mosaic of *Jatropha gossypifolia*. Agric. Exp. Sta. Univ. P. R., Tech Paper 22, p. 35.

Bird, J. (1958). Infectious chlorosis of *Sida carpinifolia* in Puerto Rico. Agr. Exp. Sta. Univ. P. R., Tech Paper No. 26, p. 23.

Bird, J., and Sanchez, J. (1971). Whitefly-transmitted viruses in Puerto Rico. *J. Agric. Univ. P. R.* **55:**461–467.

Bird, J., Sanchez, J., and Vakili, N. G. (1973). Golden yellow mosaic of beans (*Phaseolus vulgaris*) in Puerto Rico. *Phytopathology* **63:**1435.

Bird, J., and Maramorosch, K. (1978). Viruses and virus diseases associated with whiteflies. *Adv. Virus Res.* **22:**55–110.

Bird, J., Sánchez, J., Rodríguez, R. L., and Juliá, F. J. (1975). Rugaceous (whitefly-transmitted) viruses in Puerto Rico. *In* "Tropical Diseases of Legumes" (J. Bird and K. Maramorosch, eds.), pp. 3–26. Academic Press, New York.

Bird, J., Idris, A. M., Rogan, D., and Brown, J. K. (2001). Introduction of the exotic *Totamo yellow leaf curl virus*-Israel in tomato to Puerto Rico. *Plant Dis.* **85:**1028.

Blanco, N., and Bencomo, I. (1978). Afluencia de la mosca blanca (*Bemisia tabaci*) vector del virus del mosaico dorado en plantaciones de frijol. *Cienc. Agric.* **2:**39–46.

Braun, J. von, Hotchkiss, D., and Imminck, M. (1993). Nontraditional export crops in traditional small-holder agriculture: Effects on production, consumption, and nutrition in Guatemala. *In* "Agriculture and Food Marketing in Developing Countries" (J. Abbott, ed.), pp. 378–387. CAB International, Wallingford, UK.

Bravo-Luna, L., Frias-Treviño, G. A., Sánchez-Valdes, V., and Garzon-Tiznado, J. A. (2000). Sources of inoculum and vectors of Texas pepper geminivirus of chilli (*Capsicum annuum*) in Ramos Arizpe, Coahuila, Mexico. *Rev. Mex. Fitopatol.* **18:**97–102.

Brown, J. K., and Nelson, M. R. (1987). Host range and vector and vector relationships of cotton leaf crumple virus. *Plant Dis.* **71:**522–524.

Brown, J. K., and Nelson, M. R. (1988). Transmission, host range, and virus-vector relationships of chino del tomate virus, a whitefly-transmitted geminivirus from Sinaloa, Mexico. *Plant Dis.* **72:**866–869.

Brown, J. K., Campodonico, O. P., and Nelson, M. R. (1989). A whitefly-transmitted geminivirus from pepper with tigre disease. *Plant Dis.* **76:**610.

Brown, J. K., Bird, J., and Fletcher, D. C. (1993). First report of passiflora leaf mottle disease caused by a whitefly-transmitted geminivirus in Puerto Rico. *Plant Dis.* **77:**1264.

Brown, J. K., Bird, J., Banks, G., Sosa, M., Kiesler, K., Cabrera, I., and Fornaris, G. (1995). First report of an epidemic in tomato caused by two whitefly-transmitted geminiviruses in Puerto Rico. *Plant Dis.* **79:**1250.

Brown, J. K., Ostrow, K. M., Idris, A. M., and Stenger, D. C. (1999). Biotic, molecular, and phylogenetic characterization of Bean calico mosaic virus, a distinct begomovirus species with affiliation in the Squash leaf curl virus cluster. *Phytopathology* **89:**273–280.

Brown, J. K., Idris, A. M., Rogan, D., Hussein, M. H., and Palmieri, M. (2001a). Melon chlorotic leaf curl virus, a new begomovirus associated with *Bemisia tabaci* infestations in Guatemala. *Plant Dis.* **85:**1027.

Brown, J. K., Idris, A. M., Torres-Jerez, I., Banks, G. K., and Wyatt, S. D. (2001b). The core region of the coat protein gene is highly useful for establishing the provisional identification and classification of begomoviruses. *Arch. Virol.* **146:**1581–1598.

Brown, J. K., Idris, A. M., Altieri, C., and Stenger, D. C. (2002). Emergence of a new cucurbit-infecting begomovirus species capable of forming viable reassortantants with related viruses in the *Squash leaf curl virus* cluster. *Phytopathology* **92:**734–742.

Cancino, M., Abouzid, A. M., Morales, F. J., Purcifull, D. E., Polston, J. E., and Hiebert, E. (1995). Generation and characterization of three monoclonal antibodies useful in detecting and distinguishing bean golden mosaic virus isolates. *Phytopathology* **85:**484–490.

Cook, M. T. (1931). New virus diseases of plants in Porto Rico. *J. Dep. Agric.* **15:**193–195.

Cordero, M., Ramos, P. L., Hernandez, L., Fernández, A. I., Echemendia, A. L., Peral, R., Gonzales, G., Garcia, D., Valdes, S., Esteves, A., and Hernández, K. (2003). Identification of *Tomato mottle Taino virus* strains in Cuban potato fields. *Phytoparasitica* **31:**478–489.

Costa, A. S. (1937). Nota sobre o mosaico do algodoeiro. *Rev. Agric. Piracicaba Bras.* **12:**453–470.

Costa, A. S. (1955). Studies on *Abutilon* mosaic in Brazil. *Phytopathol. Z.* **24:**97–112.

Costa, A. S. (1965). Three whitefly-transmitted diseases of beans in the State of São Paulo, Brazil. *FAO Plant Prot. Bull.* **13:**121–130.

Costa, A. S. (1974). Molestias do tomateiro no Brasil transmitidas pela mosca branca *Bemisia tabaci*. "VII Congresso Anual da Sociedade. Brasileira de Fitopatologia" Brasilia, D. F. Abstract.

Costa, A. S. (1975). Increase in the populational density of *Bemisia tabaci*, a threat of widespread virus infection of legume crops in Brazil. *In* "Tropical Diseases of Legumes" (J. Bird and K. Maramorosch, eds.), pp. 27–49. Academic Press, New York.

Costa, A. S., and Forster, R. (1939). Uma suspeita de virus do fumo (*Nicotiana tabacum* L.) semelhante a 'leaf curl' presente no estado de São Paulo. *J. Agron. Piracicaba* **2:**295–302.

Costa, A. S., and Bennett, C. W. (1950). Whitefly-transmitted mosaic of *Euphorbia prunifolia*. *Phytopathology* **40:**266–283.

Costa, A. S., and Carvalho, A. M. (1960a). Mechanical transmission and properties of the *Abutilon* mosaic virus. *Phytopathol. Z.* **37:**250–272.

Costa, A. S., and Carvalho, A. M. (1960b). Comparative studies between the *Abutilon* and *Euphorbia* mosaic virus. *Phytopathol. Z.* **38:**129–152.

Costa, H. S., and Brown, J. K. (1991). Variation in biological characteristics and in esterase patterns among populations of *Bemisia tabaci* Genn. and the association of one population with silverleaf symptom development. *Entomol. Exp. Appl.* **61:**211–219.

Costa, A. S., Costa, C. L., and Sauer, H. F. (1973). Outbreak of whiteflies on crops in Parana and Sao Paulo. *An. Soc. Entom. Bras.* **2:**20–30.

Costa, A. S., Oliveira, A. R., and Silva, D. M. (1975). *Transmisão mecânica do agente causal do mosaico dourado do tomateiro.* VIII Congr. An. Soc. Bras. Fitopatol., Mossoró, Brasil (Abstract).

Dalmon, A., and Marchoux, G. (2000). Which plants host *Tomato yellow leaf curl virus*? *Phytoma* **527:**14–17.
Daniels, G., and Castro, L. A. (1985). Ocorrencia do virus do mosaico deformante da batata no Rio Grande do Sul. *Fitopatol. Bras.* **10:**306.
Debrot, E. A. (1981). Natural infection of potatoes in Venezuela with the whitefly-transmitted mosaico amarillo del tomate virus. *In* "Proceedings International Workshop on Pathogens Transmitted by Whiteflies," p. 17. Keeble College, Oxford, UK (31 July 1981).
Debrot, E. A., and Centeno, F. (1985). Infección natural de la papa en Venezuela con el mosaico amarillo del tomate, un geminivirus transmitido por moscas blancas. *Agron. Trop.* **35:**125–138.
Delhey, R., Kiehr-Delhey, M., Heinze, K., and Calderoni, A. V. (1981). Symptoms and transmission of potato deforming mosaic in Argentina. *Potato Res.* **24:**123–133.
Dodds, J. A., Lee, J. G., Nameth, S. T., and Laemmlen, F. F. (1984). Aphid- and whitefly-transmitted cucurbit viruses in Imperial County, California. *Phytopathology* **74:**221–225.
Domínguez, M., Ramos, P. L., Echemendia, A. L., Peral, R., Crespo, J., Andino, V., Pujol, M., and Barroto, C. (2002). Molecular characterisation of tobacco leaf rugose virus, a new begomovirus infecting tobacco in Cuba. *Plant Dis.* **86:**1050.
Duffus, J. E., Cohen, S., and Liu, H. I. (1992). The sweet potato whitefly in western USA biotypes, plant interactions and virus epidemiology. *In* "Recent Advances in Vegetable Virus Research," pp. 76–77. 7th Conference ISHS Vegetable Virus Workshop Group, Athens, Greece, 12–16 July 1992.
Engel, M., Fernandez, O., Jeske, H., and Frischmuth, T. (1998). Molecular characterization of a new whitefly-transmissible bipartite geminivirus infecting tomato in Panama. *J. Gen. Virol.* **79:**2313–2317.
FAO. (1994). Anuario FAO de producción, p. 217. Roma, FAO.
Faria, J. C., Souza-Dias, J. A. C., Slack, S. A., and Maxwell, D. P. (1997). A new geminivirus associated with tomato in the State of São Paulo, Brazil. *Plant Dis.* **81:**423.
Faria, J. C., Bezerra, I. C., Zerbini, F. M., Ribeiro, S. G., and Lima, M. F. (2000). Current status of geminiviruses in Brazil. *Fitopatol. Bras.* **25:**125–137.
Fauquet, C. M., Mayo, M. A., Maniloff, J., Desselberger, U., and Ball, L. A. (2005). Virus Taxonomy: Eighth Report of the International Committee on Taxonomy of Viruses, p. 1259. Elsevier Academic Press, San Diego, CA.
Flock, R. A., and Mayhew, D. E. (1981). Squash leaf curl, a new disease of cucurbits in California. *Plant Dis.* **65:**75–76.
Flores, E., and Silberschmidt, K. (1958). Relations between insect and host plant in transmission experiments with 'infectious chlorosis' of *Malvaceae*. *An. Acad. Bras. Cien.* **30:**535–560.
Flores, E., and Silberschmidt, K. (1967). Contribution to the problem of insect and mechanical transmisión of infectious clorosis of *Malvaceae* and the disease displayed by *Abutilon thompsonii*. *Phytopath. Z.* **60:**181–195.
Flores, E., Silberschmidt, K., and Kramer, M. (1960). Observacões da clorose infecciosa das malváceas em tomateiros do campo. *Biológico* **26:**65–69.
Frischmuth, T., Engel, M., Lauster, S., and Jeske, H. (1997). Nucleotide sequence evidence to the occurrence of three distinct whitefly-transmitted, *Sida*-infecting bipartite geminiviruses in Central America. *J. Gen. Virol.* **78:**2675–2682.
Fryxell, P. A. (1997). The American genera of Malvaceae-II. *Brittonia* **49:**204–269.
Gamez, R. (1970). Los virus del frijol en Centroamérica. I. Transmisión por moscas blancas (*Bemisia tabaci* Genn.) y plantas hospedantes del virus del mosaico dorado. *Turrialba* **21:**22–27.

Garrido-Ramirez, E. R., and Gilbertson, R. L. (1998). First report of tomato mottle geminivirus infecting tomatoes in Yucatan, Mexico. *Plant Dis.* **82:**592.

Garzón-Tiznado, J. A., Acosta-García, G., Torres-Pacheco, I., Gonzales-Chavira, M., Rivera-Bustamante, R. F., Maya-Hernández, V., and Guevara-González, R. G. (2002). Presence of the geminiviruses *Pepper Huasteco virus* (PHV), Texas pepper virus-Tamaulipas, and Chino del tomate virus (CdTV) in the states of Guanajuato, Jalisco, and San Luis Potosí, Mexico. *Rev. Mex. Fitopatol.* **20:**45–52.

Gonzales, A. G., and Valdés, R. S. (1995). Virus del encrespamiento amarillo de las hojas del tomate (TYLCV) en Cuba. *CEIBA* **36:**103.

Gupta, K. N., and Keshwal, R. L. (2002). Studies on the events prior to epidemic of mungbean yellow mosaic virus on soybean. *Ann. Plant Prot. Sci.* **10:**118–120.

Guzmán, P., Arredondo, C. R., Emmatty, D., Portillo, R. J., and Gilbertson, R. L. (1997). Partial characterization of two whitefly-transmitted geminiviruses infecting tomatoes in Venezuela. *Plant Dis.* **81:**312.

Hernandez, R. F. (1972). Estudio sobre la mosca blanca en el estado de Morelos. *Agric. Tec. Mex.* **3:**165–170.

Hidayat, S. H. (1990). The construction and application of DNA probes for the detection of bean dwarf mosaic geminivirus. M. Sc. Thesis, Univ. Wisc., Madison, p. 53.

Hilhorst, T., Meerendonk, H., and Wit, T. (1995). New strategies to increase small farmers' benefits from export-oriented agriculture: Asparagus growers in Peru. *Bull. Royal Trop. Inst.* **338:**13–26.

Höfer, P., Engel, M., Jeske, H., and Frischmuth, T. (1997). Sequence of a new bipartite geminivirus isolated from the common weed *Sida rhombifolia* in Costa Rica. *J. Gen. Virol.* **78:**1785–1790.

Holguin-Peña, R. J., Vazquez-Juarez, R., and Rivera-Bustamante, R. (2003). First report of a geminivirus associated with leaf curl in Baja California Peninsula tomato fields. *Plant Dis.* **87:**1397.

Holguin-Peña, R. J., Vazquez-Juarez, R., and Rivera-Bustamante, R. F. (2005). A new begomovirus causes tomato leaf curl disease in Baja California Sur, Mexico. *Plant Dis.* **89:**341.

Holmes, F. O. (1949). "The Filterable Viruses," p. 1286. Williams and Wilkins Co., Baltimore.

Hooker, W. J., and Salazar, L. F. (1983). A new plant virus from the high jungle of the eastern Andes: Solanum apical leaf curling virus (SALCV). *Ann. Appl. Biol.* **103:**449–454.

Idris, A. M., and Brown, J. K. (2004). *Cotton leaf crumple virus* is a distinct western hemisphere begomovirus species with complex evolutionary relationships indicative of recombination and reassortment. *Phytopathology* **94:**1068–1074.

Idris, A. M., Lee, S. H., Lewis, E. A., Bird, J., and Brown, J. K. (1998). Three tomato-infecting begomoviruses from Puerto Rico. *Phytopathology* **88:**42.

Idris, A. M., Rivas-Platero, G., Torres-Jerez, I., and Brown, J. K. (1999). First report of Sinaloa tomato leaf curl geminivirus in Costa Rica. *Plant Dis.* **83:**303.

Isakeit, T., and Robertson, N. L. (1994). First report of *Squash leaf curl virus* on watermelon in Texas. *Plant Dis.* **78:**1010.

Jones, P. G., Gladkov, A., and Jones, A. L. (1999). FloraMap, a computer tool for predicting the distribution of maps and other organisms in the wild. Version 1. International Centre for Tropical Agriculture (CIAT), Cali, Colombia, p. 99.

Jovel, J., Reski, G., Rothenstein, D., Ringel, M., Frischmuth, T., and Jeske, H. (2004). *Sida micrantha* mosaic is associated with a complex infection of begomoviruses different from *Abutilon mosaic virus*. *Arch. Virol.* **149:**829–841.

Karkashian, J. P., Nakhla, M. K., Maxwell, D. P., and Ramirez, P (1998). Molecular characterization of tomato-infecting geminiviruses in Costa Rica. "International Workshop on *Bemisia* and Geminiviruses," p. 45. San Juan, Puerto Rico (7–12 June 1998).
Karkashian, J. P., Maxwell, D. P., and Ramirez, P. (2002). Squash yellow mottle geminivirus: A new cucurbit-infecting geminivirus from Costa Rica. *Phytopathology* **92:**125.
Laguna, I. G., Fiorona, M. A., Ploper, L. D., Galvez, M. R., and Rodriguez, P. E. (2005). Prospección de enfermedades virales del cultivo de soja en distintas areas de producción de Argentina. "XIII Congreso Latinoamericano de Fitopatologia," p. 564. Córdoba, Argentina (19–22 April 2005).
Lastra, J. R., and Gil, F. (1981). Ultrastructural host cell changes associated with tomato yellow mosaic. *Phytopathology* **71:**524–528.
Lastra, J. R., and Uzcátegui, R. C. (1975). Viruses affecting tomatoes in Venezuela. *Phytopath. Z.* **84:**253–258.
Lazarowitz, S. G., and Lazdins, I. B. (1991). Infectivity and complete nucleotide sequence of the cloned genomic components of a bipartite squash leaf curl geminivirus with a broad host range phenotype. *Virology* **180:**58–69.
Leal, R. A., and Quintero, S. (1989). Caracterización de una virosis del chile transmisible por mosquita blanca en la planicie Huasteca. *Rev. Mex. Fitopatol.* **7:**147–149.
Lima, G. S., Assunção, I. P., Resende, L. V., Ferreira, M. A., Viana, T. H., Gallindo, F. A., and Freitas, N. S. (2002). Detection of a begomovirus associated to weeds in the state of Pernambuco and partial molecular characterization of an isolate from *Sida rhombifolia*. *Summa Phytopathol.* **28:**353–356.
Lima, J. A., Goncalves, M. F., Oliveira, V. B., Torres, F. J., and Miranda, A. C. (2000). Serological and PCR detection of a begomovirus infecting tomato fields in Ibiapaba Mountain, Ceara. *Fitopatol. Bras.* **25:**104–108.
Lima, M. F., Bezerra, I. C., Ribeiro, S. G., and Avila, A. C. (2001). Distribution of geminiviruses in tomato and sweet pepper crops in twelve counties of the lower basin of the San Francisco Valley. *Fitopatol. Bras.* **26:**81–85.
Loniello, A. O., Martinez, R. T., Rojas, M. R., Gilbertson, R. L., Brown, J. K., and Maxwell, D. P. (1992). Molecular characterization of bean calico mosaic geminivirus. *Phytopathology* **82:**1149.
Lopez, G. H. (1974). Aumente sus rendimientos en frijol en el Valle de Culiacán. Circular No. 12, CIAS-INIA-SARH.
Lotrakul, P., Valverde, R. A., Torre, R., JeonGu, S., and Gomez, A. (2000). Occurrence of a strain of Texas pepper virus in Tabasco and Habanero pepper in Costa Rica. *Plant Dis.* **84:**168–172.
Martin, J. H., Mifsud, D., and Rapisarda, C. (2000). The whiteflies (Hemiptera: Aleyrodidae) of Europe and the Mediterranean basin. *Bull. Entom. Res.* **90:**407–448.
Martinez-Zubiaur, Y. (1998). Contribución al conocimiento de geminivirus que afectan el cultivo del tomate (*Lycopersicon esculentum* Mill) en Cuba. Doctoral Thesis, Instituto Superior de Ciencias Agropecuarias de la Habana, p. 39.
Martinez-Zubiaur, Y., Blas, C., Quiñones, M., Castellanos, C., Peralta, E., and Romero, J. (1998). Havana tomato virus, a new bipartite geminivirus infecting tomatoes in Cuba. *Arch. Virol.* **143:**1757–1772.
Martinez-Zubiaur, Y., Quiñónez, M., Fonseca, D., Potter, J., and Maxwell, D. P. (2002). First report of *Tomato yellow leaf curl virus* associated with beans, *Phaseolus vulgaris*, in Cuba. *Plant Dis.* **86:**814.
Martinez-Zubiaur, Y., Quiñónez, M., Fonseca, D., and Miranda, I. (2003). National survey of begomoviruses in tomato crops in Cuba. *Rev. Prot. Veg.* **18:**168–175.

Martinez-Zubiaur, Y., Fonseca, D., Quiñónez, M., and Palenzuela, I. (2004). Presence of tomato yellow leaf curl virus infecting squash (*Cucurbita pepo*) in Cuba. *Plant Dis.* **88:**572.

Matys, J. C., Silva, D. M., Oliveira, A. R., and Costa, A. S. (1975). Purificacão e morfologia do virus do mosaico dourado do tomateiro. *Summa Phytopathol.* **1:**267–274.

Maxwell, D. P., Nakhla, M. K., Maxwell, M. D., Ramirez, P., Karkashian, J. P., Roca, M. M., Roye, M., and Faria, J. C. (2002). Diversity of begomoviruses and their management in Latin America. *Phytopathology* **92:**127.

McCreight, J. D., and Kishaba, A. N. (1991). Reaction of cucurbit species to squash leaf curl virus and sweetpotato whitefly. *J. Am. Soc. Hortic. Sci.* **116:**137–141.

McGlashan, D., Polston, J. E., and Bois, D. (1994). Tomato yellow leaf curl geminivirus in Jamaica. *Plant Dis.* **78:**1219.

Montes-Belmont, R., Espin-García, S., Sosa-Hernandez, A., and Torres-Pacheco, R. (1995). Evaluación de extractos vegetales para el control de la virosis "chino del tomate" en dos regiones agroecologicas de Mexico. *Rev. Mex. Fitopatol.* **13:**111–116.

Morales, F. J. (2001). Conventional breeding for resistance to *Bemisia tabaci*-transmitted geminiviruses. *Crop Prot.* **20:**825–834.

Morales, F. J., and Anderson, P. K. (2001). The emergence and dissemination of whitefly-transmitted geminiviruses in Latin America. *Arch. Virol.* **146:**415–441.

Morales, F. J., and Jones, P. G. (2004). The ecology and epidemiology of whitefly-transmitted viruses in Latin America. *Virus Res.* **100:**57–65.

Morales, F., Niessen, A., Ramirez, B., and Castaño, M. (1990). Isolation and partial characterization of a geminivirus causing bean dwarf mosaic. *Phytopathology* **80:**96–101.

Morales, F. J., Muñoz, C., Castaño, M., and Velasco, A. C. (2000). Geminivirus transmitidos por mosca blanca en Colombia. *Fitopatol. Colomb.* **24:**95–98.

Morales, F. J., Lastra, R., Uzcategui, R. C. de, and Calvert, L. (2001). Potato yellow mosaic virus: A synonym of *Tomato yellow mosaic virus*. *Arch.Virol.* **146:**2249–2253.

Morales, F. J., Martinez, A. K., and Velasco, A. C. (2002). Nuevos brotes de begomovirus en Colombia. *Fitopatol. Colomb.* **26:**75–79.

Morales, F. J., Gonzales, G., Murguido, C., Echemendia, A., Martinez, Y., Hernández, Y., Faure, B., and Chailloux, M. (2005a). Cuba. *In* "Whitefly and Whitefly-Borne Viruses in the Tropics: Building a Knowledge Base" (P. K. Anderson and F. J. Morales, eds.), pp. 230–236. CIAT Publ. No. 341, Palmira, Colombia.

Morales, F. J., Rivera-Bustamante, R., Salinas, R., Torres-Pacheco, I., Diaz-Plaza, R., Aviles, W., and Ramirez, G. (2005b). Mexico. *In* "Whitefly and Whitefly-Borne Viruses in the Tropics: Building a Knowledge Base" (P. K. Anderson and F. J. Morales, eds.), pp. 177–187. CIAT Publication No. 341, Palmira, Colombia.

Moreira, A. G., Pereira, C. O., Andrade, E. C., and Zerbini, F. M. (2005). Caracterización molecular de dos begomovirus que afectan a la soja (*Glycine max*) y malezas asociadas, en Brasil. XIII Congreso Latinoamericano de Fitopatologia, Cordoba Argentina. Resumenes, p. 602.

Muniyappa, V. (1980). Viral diseases transmitted by whiteflies. *In* "Vectors of Plant Pathogens" (K. F. Harris and K. Maramorosch, eds.), pp. 39–85. Academic Press, New York.

Murayama, A., Aragon, L., and Fernández-Northcote, E. (2005). Nuevo begomovirus del grupo del Nuevo Mundo asociado al encrespamiento de la hoja del tomate en la costa del Peru. *In* "XIII Congreso Latinoamericano de Fitopatologia," p. 603. Cordoba, Argentina (19–22 April 2005).

Nakhla, M. K., Maxwell, M. D., Hidayat, S. H., Lange, D. R., Loniello, A. O., Rojas, M. R., Maxwell, D. P., Kitajima, E. W., Rojas, A., Anderson, P., and Gilbertson, R. L. (1994). Two geminiviruses associated with tomatoes in Central America. *Phytopathology* **84:**1155.

Navas-Castillo, J., Sanchez-Campos, S, Diaz, J. A., Saez-Alonso, E., and Moriones, E. (2002). Tomato yellow leaf curl virus-Is causes a novel disease of common bean and severe epidemics in tomato in Spain. *Plant Dis.* **83:**29–32.

Novaes, Q. S., Freitas-Astua, J., Auki, V. A., Kitajima, E. W., Camargo, L. E., and Rezende, J. A. (2003). Partial characterization of a bipartite begomovirus infecting yellow passion flower in Brazil. *Plant Pathol.* **52:**648–654.

Orlando, A., and Silberschmidt, K. (1946). Estudos sobre a disseminacão natural do virus da clorose infecciosa das malvaceas (Abutilon virus 1 Baur) e a sua relacão com o inseto-vetor *Bemisia tabaci* (Genn.) (Homoptera: Aleyrodidae). *Arq. Inst. Biol. São Paulo* **17:**136.

Pardina, P. E., Ploper, L. D., Truol, G. A., Hanada, K., Platero, G. R., Ramirez, P., Herrera, P. S., and Laguna, I. G. (1998). Detection of a geminivirus in soybean crops in northwestern Argentina. *Rev. Ind. Agric. Tucuman* **75:**51–56.

Paximadis, M., Idris, A. M., Torres-Jerez, I., Villarreal, A., Rey, M. E. C., and Brown, J. K. (1999). Characterisation of tobacco geminiviruses in the Old and New World. *Arch. Virol.* **144:**703–717.

Perring, T. M. (2001). The *Bemisia tabaci* species complex. *Crop Prot.* **20:**725–737.

Perring, T. M., Farrar, C. A., Bellows, T. S., Cooper, A. D., and Rodríguez, R. J. (1993). Evidence for a new species of whitefly: UCR findings and implications. *Calif. Agric.* **47:**7–8.

Pierre, R. E. (1975). Observations on the golden mosaic of bean (*Phaseolus vulgaris* L.). *In* "Tropical Diseases of Legumes" (J. Bird and K. Maramorosch, eds.), pp. 55–60. Academic Press, New York.

Polston, J. E., Bois, D., Serra, C. A., and Concepción, S. (1994). First report of a tomato yellow leaf curl-like geminivirus from tomato in the Western Hemisphere. *Plant Dis.* **78:**831.

Polston, J. E., Bois, D., Ano, G., Poliakoff, N., and Urbino, C. (1998). Occurrence of a new strain of potato yellow mosaic geminivirus infecting tomato in the Eastern Caribbean. *Plant Dis.* **82:**126.

Quiñones, M., Fonseca, D., Accotto, G. P., and Martinez, Y. (2001). Viral infections associated with the presence of begomovirus in pepper plants in Cuba. *Rev. Prot. Veg.* **16:**147–151.

Ramírez, J. A., Armenta, I., Delgadillo, F., and Rivera-Bustamante, R. F. (1995). Geminivirus transmitidos por mosquita blanca (*Bemisia tabaci* Gennn.) en los cultivos de chile y calabacita en el Valle de Mayo, Sonora, México. *Rev. Mex. Fitopatol.* **13:**100–105.

Ramírez-Arredondo, J. A., Cárdenas, I. A., Sánchez, F. D., and Garzón-Tiznado, J. A. (1998). Virus transmitted by whitefly (*Bemisia tabaci* Gennadius) in pepper and zucchini squash in the Mayo Valley, Sonora, Mexico. *Agric. Tec. Mex.* **24:**37–43.

Ramirez, P., Chicas, M., Salas, J., Maxwell, D., and Karkashian, J. (2004). Identificacion de un nuevo begomovirus en melon (*Cucumis melo* L.) en Lara, Venezuela. *Rev. MIP Agroecol.* **72:**22–30.

Ramos, P. L., Guerra, O., Peral, R., Oramas, P., Guevara, R. G., and Rivera-Bustamante, R. (1997). Taino tomato mottle virus, a new bipartite geminivirus from Cuba. *Plant Dis.* **81:**1095.

Ramos, P. L., Guevara, R. G., Peral, R., Ascencio-Ibañez, J. T., Polston, J. E., Arguello, G. R., Vega, J. C., and Rivera-Bustamante, R. F. (2003). Tomato mottle Taino virus pseudorecombines with PYMV but not with ToMoV: Implications for the delimitation of cis- and trans-acting replication specificity determinants. *Arch. Virol.* **148:**1697–1712.

Rampersad, S. N., and Umaharan, P. (2003). Detection of two bipartite geminiviruses infecting dicotyledonous weeds in Trinidad. *Plant Dis.* **87:**602.

Ribeiro, S. G., Ambrocevicius, L. P., Avila, A. C., Bezerra, I. C., Calegario, R. F., Fernandes, J. J., Lima, M. F., Mello, R. N., Rocha, H., and Zerbini, F. M. (2003). Distribution and genetic diversity of tomato-infecting begomoviruses in Brazil. *Arch. Virol.* **148:**281–295.

Roberts, E. J., Buck, K. W., and Coutts, R. (1986). A new geminivirus infecting potatoes in Venezuela. *Plant Dis.* **70:**603.

Rodriguez, I., Morales, H., Bueno, J. M., and Cardona, C. (2005). El biotipo B de *Bemisia tabaci* adquiere mayor importancia en el Valle del Cauca. *Rev. Col. Entomol.* **31:**21–28.

Rojas, A., Kvarnheden, A., and Valkonen, J. P. T. (2000). Geminiviruses infecting tomato crops in Nicaragua. *Plant Dis.* **84:**843–846.

Rojas, A. (2005). A complex of begomoviruses affecting tomato crops in Nicaragua. *In* "Proceedings IX International Plant Virus Epidemiology Symposium," p. 72. Lima, Peru (4–7 April 2005).

Rosset, P., Meneses, R., Lastra, R., and Gonzalez, W. (1990). Estimación de pérdidas e identificación del geminivirus transmitido al tomate por la mosca blanca *Bemisia tabaci* Genn. en Costa Rica. *Rev. MIP* **15:**24–34.

Roye, M. E., Wernecke, M. E., McLaughlin, W. A., Nakhla, M. K., and Maxwell, D. P. (1999). *Tomato dwarf leaf curl virus*, a new bipartite geminivirus associated with tomatoes and peppers in Jamaica and mixed infection with *Tomato yellow leaf curl virus*. *Plant Pathol.* **48:**370–378.

Saade, R. L. (1995). Estudios taxonómicos y ecogeográficos de las *Cucurbitaceae* latinoamericanas de importancia económica. *Inst. Int. Recur. Fitogen.* **28:**1(IPGRI).

Salazar, L. F., Müller, G., Querci, M., Zapata, J. L., and Owens, R. A. (2000). Potato yellow vein virus: Its host range, distribution in South America and identiication as a crinivirus transmitted by *Trialeurodes vaporarirum*. *Ann. Appl. Biol.* **137:**7–19.

Salinas, R. A., and Vázquez, G. M. (1979). El cultivo del frijol en el Valle de Santo Domingo, *B.C.S. INIA-CIAPAN-CAESTOD*, p. 8. Circular No. 88.

Samretwanich, K., Kittipakorn, K., Chiensombat, P., and Ikegami, M. (2001). Complete nucleotide sequence and genome organization of soybean crinkle leaf virus. *J. Phytopathol.* **149:**333–336.

Schieber, E. (1970). Enfermedades del frijol (*Phaseolus vulgaris*) en la República Dominicana. *Turrialba* **20:**20–23.

Silberschmidt, K. (1943). Estudos sobre a transmissão experimental da "clorose infecciosa" das malvaceas. *Arq. Inst. Biol. São Paulo* **14:**105–156.

Silberschmidt, K. M. (1948). Infectious chlorosis of *Phenax sonnerati*. *Phytopathology* **38:**395–398.

Silberschmidt, K., and Tommasi, C. R. (1955). Observações e estudos sobre especies de plantas suscetíveis à clorose infecciosa das malváceas. *An. Acad. Bras. Cienc.* **27:**195–214.

Silberschmidt, K., and Tommasi, L. R. (1956). A solanaceous host of the virus of infectious chlorosis of Malvaceae. *Ann. Appl. Biol.* **44:**161–165.

Silva, S., Rodriguez, R., Garzón-Tiznado, J. A., Delgadillo, F., and Cárdenas, E. (1994). Efecto de una variante del virus enrollamiento de la hoja de la calabaza (VEHC, Squash leaf curl virus, SqLCV) en genotipos de cucurbitaceas. *Rev. Mex. Fitopatol.* **12:**15–20.

Spillari, A. G. (1994). Problemática del complejo mosca blanca-virus en algodón en Centroamerica. Memorias III Taller Centroamericano y del Caribe sobre Mosca Blanca, pp. 23–38. Antigua Guatemala, Guatemala.

Sseruwagi, P., Brown, J. K., Maruthi, M. N., Colvin, J., Rey, M. E. C., and Legg, J. P. (2005). Diversity of *Bemisia tabaci* (Gennadius) (Hemiptera: Aleyrodidae) and significance to the epidemiology of whitefly-transmitted viruses in Uganda. *In* "Proceedings IX International Plant Virus Epidemiology Symposium." International Society for Plant Pathology, p. 78, Lima, Peru (4–7 April).

Stenger, D. C., Duffus, J. E., and Villalon, B. (1990). Biological and genomic properties of a geminivirus isolated from pepper. *Phytopathology* **80:**704–709.

Torre-Almaraz, R., Valverde, R., Mendez, J., Ibáñez, J. T., and Rivera-Bustamante, R. F. (2002). Preliminary characterization of a geminivirus in tomatillo (*Physalis ixocarpa* B.) in the central region of Mexico. *Agrociencia* **36:**471–481.

Torres-Pacheco, I., Garzón-Tiznado, J. A., Brown, J. K., and Rivera-Bustamante, R. F. (1996). Detection and distribution of geminiviruses in Mexico and the southern United States. *Phytopathology* **86:**1186–1192.

Truol, G., Corre, L. H., Vilarinho, M. R., and Laguna, I. G. (2005). Analisis de biotipos de moscas blancas vectoras de geminivirus en Argentina. "XIII Congreso Latinoamericano de Fitopatologia," p. 577. Cordoba, Argentina (19–22 April 2005).

Umaharan, P., Padidam, M., Phelps, R. H., Beachy, R. N., and Fauquet, C. M. (1998). Distribution and diversity of geminiviruses in Trinidad Tobago. *Phytopathology* **88:**1262–1268.

Urbino, C., and Tassius, K. (1999). First report of *Tomato yellow leaf curl virus* in tomato in Guadeloupe. *Plant Dis.* **35:**46–53.

Urbino, C., Polston, J. E., Patte, C. P., and Caruana, M. L. (2004). Characterization and genetic diversity of *Potato yellow mosaic virus* from the Caribbean. *Arch. Virol.* **149:**417–424.

Vakili, N. (1973). Bean (*Phaseolus vulgaris*) diseases encountered in Haitian lowlands. IICA, p. 9. Port-au-Prince, Haiti.

Vega, J., Winter, S., Pural, A., and Hamilton, R. I. (1992). Partial characterization of a whitefly-trasnmitted geminivirus from potato in Brazil. *Fitopatol. Bras.* **17:**167.

Walkey, D. G. A., Spence, N. J., Clay, C. M., and Miller, A. (1994). A potyvirus isolated from solanaceous hosts. *Plant Pathol.* **43:**931–937.

Wolf, F. A., Whitecomb, W. H., and Mooney, W. C. (1949). Leaf curl of tobacco in Venezuela. *J. Elisha Mitchel Sci. Soc.* **65:**38–47.

Zaumeyer, W. J., and Smith, F. F. (1966). Fourth report of the bean disease and insect survey in El Salvador. AID/USDA Tech. Assist, p. 13. Agreem., Beltsville, MD.

ADVANCES IN VIRUS RESEARCH, VOL 67

EVOLUTIONARY EPIDEMIOLOGY OF PLANT VIRUS DISEASE

M. J. Jeger,* S. E. Seal,[†] and F. Van den Bosch[‡]

*Division of Biology, Imperial College London, Wye Campus, Wye
Ashford TN25 5AH, United Kingdom
[†]Natural Resources Institute, University of Greenwich, Chatham Maritime
Kent ME4 4TB, United Kingdom
[‡]Biomathematics and Bioinformatics Division, Rothamsted Research
Harpenden AL5 2JQ, United Kingdom

Evolutionary epidemiology was first used as a term some 25 years ago in considering the predominantly clonal structure of field populations of many parasitic protozoa. The need for an approach to link population genetics and epidemiology has long been recognised for plant pathogens, including viruses, mostly in relation to the evolution

0065-3527/06 $35.00
DOI: 10.1016/S0065-3527(06)67005-X

of virulence. For viruses, most attention has been given to RNA viruses of vertebrates including humans, but even here there are many problems in clarifying how virus genetic variation is modulated by host response, transmission processes and epidemic dynamics. Most economically important plant viruses have arthropod vectors, which add a further level of complexity to evolutionary dynamics as vector ecology, behaviour and genetic variation must also be considered. Consequently, it has proved difficult to predict quantitatively the likely contribution of new cropping practices, or crop protection measures, such as vector control, host-plant resistance or cultural control, to emerging plant virus disease problems. Nevertheless, some important qualitative insights have been obtained through retrospective analysis of epidemics and through new modelling approaches.

I. Evolutionary Epidemiology

Evolutionary epidemiology is a term that is being used increasingly to describe the link between evolutionary biology and epidemiology. This chapter describes the origins of the term and the ways it has been used in various fields. We then examine its relevance to plant virus-evolution and epidemiology, and for a better understanding and management of virus disease in both crop and natural plant communities. Throughout this chapter we consider some of the theoretical underpinnings of evolutionary epidemiology and describe, where appropriate, work that has been done in the medical and veterinary arenas, notably with animal viruses and draw out the implications for plant virus epidemiology. The difficulties in combining approaches in population genetics and population dynamics have long been recognised as formidable (Anderson, 1991) and, although progress has been made, it should be recognised that evolutionary biology is more than population genetics and epidemiology is more than population dynamics.

The first use of evolutionary epidemiology we are able to trace was by Keymer *et al.* (1990) in a commentary on research describing the prevailing clonal structure in field populations of many parasitic protozoa, despite the commonness of sexual reproduction in the laboratory. Subsequent usage has been in a wide range of areas, including adaptiveness of the genetic basis of psychopathology (Wilson, 1992) and evolution of non-genetic cultural traits (Gatherer, 2002). In the medical and public health arena it has been claimed that models, which incorporate ideas from both evolutionary ecology and epidemiology, generate predictions that could not be made by either discipline alone (Galvani, 2003). With human influenza, the two-way interaction between epidemiological and evolutionary dynamics determines whether new epidemic strains invade a susceptible population (Boni *et al.*, 2004). In particular, the fast evolutionary rate of viruses opens possibilities for molecular epidemiological research and the tracking of virus strains through phylogenetic analysis. As noted by Moya *et al.* (2004) the rapidity of change in RNA viruses makes them useful experimental models for evolutionary epidemiology. Specific epidemiological components in host–parasite interactions can sometimes only be understood by integrating ecological and evolutionary ideas into more traditional epidemiology. For example, parasites may increase their transmission by manipulating the vector to greater fecundity or to bite more frequently, but if the latter the host may then in response kill the vector (Koella, 1999). Parasites do not always minimise damage to the vector or to the host.

A key priority for infectious disease research is to clarify how pathogen genetic variation, modulated by host response, transmission 'bottlenecks' and epidemic dynamics, determines pathogen phylogeny in individual hosts and host populations (Grenfell *et al.*, 2004). Grenfell and coworkers introduce a framework for analysis in which the diversity of epidemiological pathways (determining population size and spatial dynamics) and selection determines the shape of pathogen phylogenies and enables detection of specific mutations and estimation of population growth and migration rates (Fig. 1). The framework is applied to numerous examples for vertebrate RNA viruses.

II. Plant Virus Evolution

A. *General*

It is not the intention here to give a comprehensive account of plant virus evolution. Reviews for this purpose already exist, both for the mechanisms that govern evolution (Roossinck, 1997) and the genetic structure of virus populations that results (García-Arenal *et al.*, 2001). Similarly, there have been several reviews of virus evolution in relation to disease resistance. The probability of resistance-breaking virus variants appearing and spreading depends on their fitness in the absence of host resistance, the type of resistance encountered, and the number of resistance genes present (Harrison, 2002). Population genetic theory can be of considerable assistance in determining the evolutionary potential of plant pathogens (McDonald and Linde, 2002a,b)

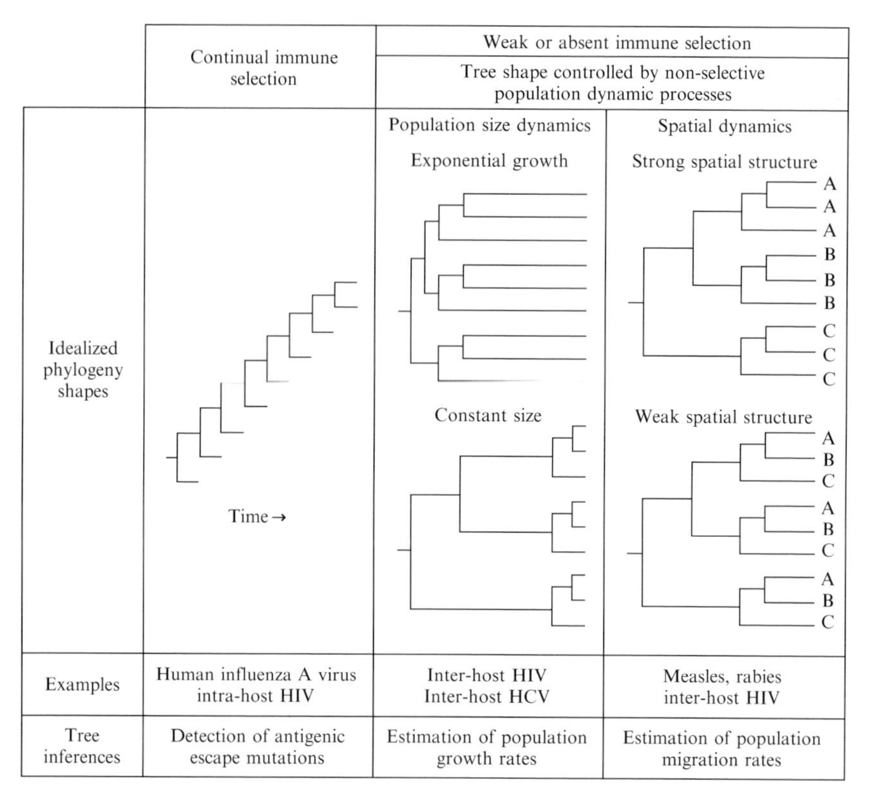

FIG 1. Idealised pathogen phylogeny in relation to epidemiological processes and selection. Reprintd with permission from Grenfell *et al.* (2004). Copyright 2004 AAAS. The shape phylogeny takes depends critically on temporal and spatial dynamics of population growth.

and can suggest rational approaches to breeding for durable resistance depending on the risk of resistance gene breakdown. McDonald and co-workers have suggested that high-risk pathogens are those with mixed sexual/asexual reproductive systems, a high potential for gene flow, large effective population sizes and high-mutation rates. Low-risk pathogens possess strictly asexual reproductive systems and the opposing other characteristics. This type of risk analysis was applied subsequently to 29 plant virus species and a compound risk index was devised (García-Arenal and McDonald, 2003) based on effective population size, the degree of recombination, and the amount of gene and genotype flow. Risk assessment has also been applied to the evolutionary and epidemiological risks from the use of virus-resistant transgenic plants (Hammond *et al.*, 1999; Tepfer, 2002).

Another criterion to take into account when considering plant virus evolution is the system used for classification, whether as a framework for diagnosis (Matthews, 1993), for finer separation using molecular techniques of groupings based on biological and physiochemical properties (Barnett, 1993), for predicting viral relationships based on genome organisation (Bradeen *et al.*, 1997), or for demarcation of species/genera within a family (Fauquet *et al.*, 2003). In the latter case, for example, the nine criteria given for demarcating geminivirus species can be seen as representing different aspects of the evolutionary process.

It is notable that in recent years, research in plant virus evolution has been viewed from a molecular rather than population standpoint (García-Arenal *et al.*, 2003). These authors consider that accumulated population-based data support the following interpretations: (1) high-mutation rates are not necessarily adaptive, (2) populations of plant viruses are not highly variable, (3) constraints on genetic variation in virus-encoded proteins are similar to those found in host plants and vectors and (4) genetic 'drift' may be important, as well as selection, in shaping evolution.

In this section sources of variation in plant viruses are considered, the various drivers of evolutionary change and the resulting molecular divergence and genetic structure in virus populations.

B. Sources of Variation

The sources of variation in plant viruses considered here are mutation, recombination, reassortment and migration.

1. Mutation

As noted in Section I, the rapidity of sequence change in RNA viruses makes them ideal experimental models for the study of evolution and mutation rates (Moya *et al.*, 2004). However, remarkably little is known about the mutational fitness effects associated with single-nucleotide substitution on RNA viral genomes (Sanjuan *et al.*, 2004a). A collection of single-site mutants of *Vesicular stomatitis virus* was created, where substitution mutations were already present in wild isolates, or were introduced randomly in the laboratory. Competition experiments were used to measure the relative fitness of the mutants (Fig. 2). With the randomly introduced mutations there were more deleterious (non-lethal) mutations giving higher reductions in fitness than with mutations already present in wild isolates. Similarly, for those mutations giving beneficial effects, the increase in fitness was greater for the previously described mutations than the random.

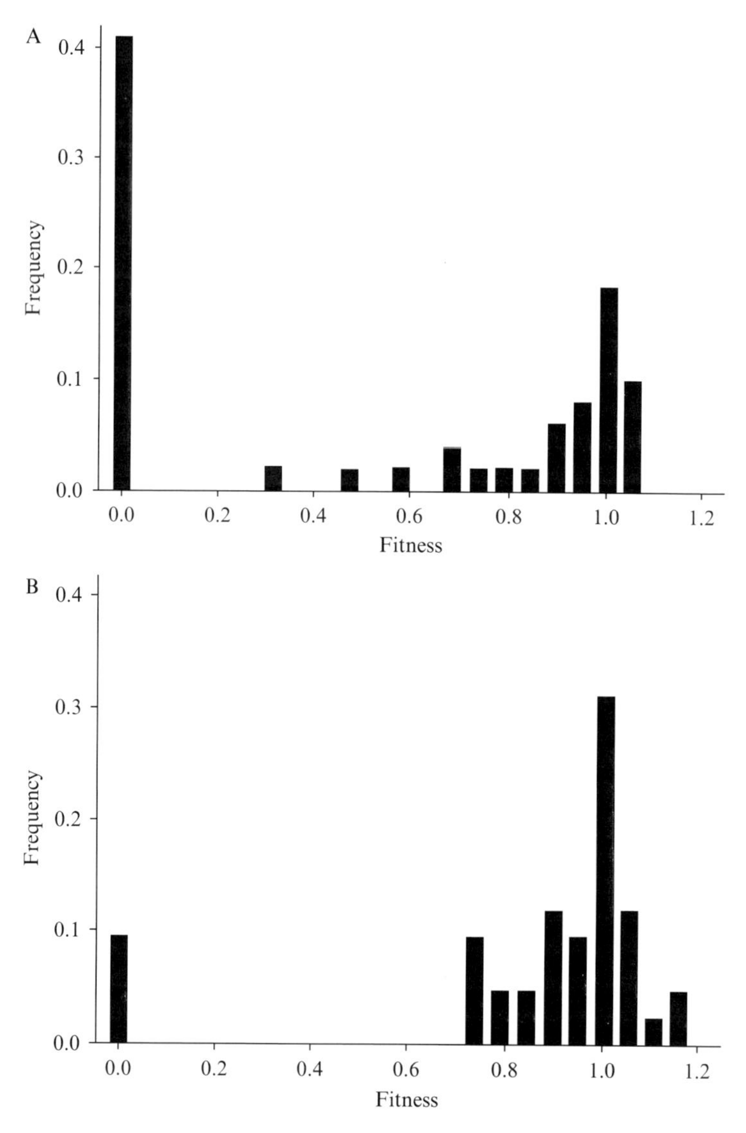

FIG 2. Frequency distribution of fitness effects with random and previously described mutations in *Vesicular stomatitis virus* (Sanjuan *et al.*, 2004a). Fitness was assessed relative to the wild type from the ratio of the intrinsic growth rates in titre of the mutant to the wild type clone. Copyright 2004 National Academy of Sciences, USA.

In other studies (Sanjuan *et al.*, 2004b) epistasis among pairs of mutations, an aspect rarely studied, had a considerable effect on the fitness of both deleterious and beneficial mutations.

Estimation of spontaneous mutation rates for RNA viruses can also be uncertain if the effects of selection are not considered. Malpica *et al.* (2002) provided estimates for the virus movement protein (MP) gene in *Tobacco mosaic virus* (TMV). They countered selection by providing MP function from a transgene. The estimated mutation rates were at the lower end of the range for animal RNA viruses and the proportion of base substitutions was lower than in most DNA-based organisms. The overall MP mutant frequency was 0.02–0.05 with some 35% of sequences having two or more mutations.

Extended passage in an alternative host can be seen as a major selective challenge to a plant virus. Kearney *et al.* (1999) took an initial population of TMV from tobacco and transferred this 11–12 times in seven plant host species over a period of 413–515 days. Some 14 unique mutations were detected from selected genome sequences and the overall mutation rate was calculated as 3.1×10^{-4} nucleotide substitutions/base/year (Table I). The evolutionary rates observed, albeit with less than 10% of the genome examined, were lower than for field isolates of other RNA viruses.

Longer-term studies of plant virus evolution are rare. The evolution of two tobamoviruses infecting the immigrant plant *Nicotiana glauca* in New South Wales, Australia was studied using isolates from herbarium and living specimens over the period 1899–1993 (Fraile *et al.*, 1997b). Before 1950, populations were infected with both TMV and

TABLE I
GENOME EVOLUTION OF *TOBACCO MOSAIC VIRUS* (TMV) DURING PASSAGE IN VARIOUS HOSTS (MODIFIED FROM KEARNEY *ET AL.*, 1999)

Host	Host response	No. of passages/ total days	Total numbers of mutations/base/year
Nightshade (*Solanum nigrum*)	Systemic mosaic	12/440	6×10^{-4}
Buckwheat (*Fagopyrum esculentum*)	Localised, no symptoms	11/514	2×10^{-4}
Plantain (*Plantago* sp.)	Localised, ringspots	12/413	3×10^{-4}

Tobacco mild green mosaic virus (TMGMV) (Table II) but only with TMGMV after that date. Half of the earlier infections were mixtures of the two viruses. There had been no increase in the estimated genetic diversity of TMGMV over the period. In experiments, plants were inoculated with the viruses recovered from the archive plant material and virus accumulation was measured when TMV was inoculated alone or doubly with all other TMGMV isolates, and similarly for TMGMV in single or double inoculations with TMV (Table II). The accumulation of TMV, but not of TMGMV, was much less in mixed than in single inoculations. Fraile *et al.* (1997b) interpreted these results as indicating that the population of TMV, an earlier or faster colonizer than TMGMV, decreased below a threshold at which deleterious mutations were eliminated, a phenomenon known as 'Muller's ratchet' or 'mutational meltdown'.

2. *Recombination*

High rates of spontaneous mutation, as reported earlier, are commonly held to be the main source of variation in virus evolution. However, major changes in viral genotypes are also possible through RNA or DNA recombination, for which the molecular mechanisms involved are well described for some species (Nagy *et al.*, 1998; White and Morris, 1995). In reality, both mutation and recombination will contribute together to the evolution of virus populations. Aranda *et al.* (1993) analysed the contribution of mutation and recombination to the evolution of the satellite RNA of *Cucumber mosaic virus* (CMV).

TABLE II

VIRAL ACCUMULATION OF *TOBACCO MOSAIC VIRUS* (TMV) AND *TOBACCO MILD GREEN MOSAIC VIRUS* (TMGMV) IN *NICOTIANA GLAUCA* LEAVES (MODIFIED FROM FRAILE *ET AL.*, 1997B)

		Mg viral RNA/g fresh weight			
Year collected	Viruses recovered	TMV alone	Double infection	TMGMV alone	Double infection
1899	TMV	2.69	0.21	–	–
1907	TMV/TMGMV	5.74	0.09	6.19	7.55
1909	TMV/TMGMV	0.25	0.16	5.11	8.55
1923	TMV/TMGMV	0.35	0.08	2.92	5.56
1938	TMGMV	–	–	2.52	5.40

Seventeen CMV-sat RNA isolates were sampled randomly from 62 field isolates collected over 3 years. Nucleotide sequences of all 17 isolates were unique with considerable genetic divergence. Most mutations did not disrupt base pairings and the need to maintain a functional structure was presumed to limit the extent of divergence. Phylogenetic analysis clustered the isolates into two sub-groups but three variants were shown to be recombinants of isolates from the two sub-groups, arising from at least two recombination events. Thus recombination made an important contribution to generating new variants in the population.

Sequencing of selected regions of the genome and phylogenetic analyses of the potyvirus *Yam mosaic virus* (YMV) structured the population into nine molecular groups, with most diversity and divergence apparent in isolates from Africa (Bousalem *et al.*, 2000). Recombination events with single and multiple crossover points largely contributed to the evolution of YMV. The authors suggested an African origin for YMV within *Dioscorea cayenensis–D. rotundata* followed by transfers to *D. alata* and *D. trifida* during virus evolution. As with estimation of mutation rates there can be major problems in estimating recombination rate in the absence of selection. Aaziz and Tepfer (1999) designed experiments minimizing selection pressure on two species of *Cucumovirus*, CMV and *Tomato aspermy virus* (TAV). They found that precise homologous recombination occurred in 3 of 82 doubly infected tobacco plants.

Recombination following multiple infections with different begomovirus species has been claimed as an explanation for the evolutionary divergence of the begomoviruses found in the Indian sub-continent and those causing similar diseases in other geographical regions (Fauquet *et al.*, 2005; Sanz *et al.*, 2000). Two distinct begomovirus species with different host ranges, *Tomato yellow leaf curl virus* (TYLCV) and *Tomato yellow leaf curl Sardinia virus* (TYLCSV) have caused a similar yellow leaf curl disease in Spain, with frequent doubly infected tomato plants. A recombinant strain is becoming more prevalent in the region where it was first detected (Monci *et al.*, 2002) (Table III). The novel phenotype exhibited might provide it with a selective advantage over the parental genotypes. Co-infection of nuclei, shown in both tomato and *Nicotiana benthamiana* (Table IV), may explain why recombination between the two begomoviruses occurs frequently (Morilla *et al.*, 2004). A recombinant of *African cassava mosaic virus* and *East African cassava mosaic virus* is associated with the severe form of cassava mosaic disease in Uganda (Legg *et al.*, this volume, pp. 355–418).

TABLE III

ANALYSIS OF *TOMATO YELLOW LEAF CURL VIRUS* (TYLCV) AND *TOMATO YELLOW LEAF CURL SARDINIA VIRUS* (TYLCSV) IN MIXED INFECTIONS OF TOMATO (MODIFIED FROM MONCI *ET AL.*, 2002)

		Proportion of samples infected with				
Sample type	*N*	TYLCV	TYLCSV	Both	Recombinant alone	Recombinant + TYLCV/TYLCSV
No symptoms	71	0.54	0	0.18	0.21	0.06
Symptoms	85	0.69	0.07	0.11	0.07	0.01
Total	156	0.62	0.04	0.14	0.14	0.03

TABLE IV

MIXED INFECTIONS OF *TOMATO YELLOW LEAF CURL VIRUS* (TYLCV) AND *TOMATO YELLOW LEAF CURL SARDINIA VIRUS* (TYLCSV) IN NUCLEI OF TOMATO CELLS (MODIFIED FROM MORILLA *ET AL.*, 2004)

			Percent of nuclei infected with			
					Both[a]	
	Number of nuclei analysed	Infected nuclei	TYLCSV	TYLCV	Observed	Calculated[a]
L. esculentum	15,000	200	37	42	22	25
N. benthamiana	4,200	272	48	34	18	25

[a] Calculated as the product of the frequencies for single viruses.

3. Reassortment

Reassortment in multipartite RNA or DNA viruses provides another opportunity for genetic exchange. For a multipartite virus to initiate a successful infection, each of the component particles must infect the same plant cell. An analysis of natural populations of the tripartite virus CMV in Spain was undertaken to estimate the frequencies of genetic exchange by recombination and reassortment (Fraile *et al.*, 1997a). Genetic exchange by segment reassortment occurred in 4% of field isolates, but in general such isolates were selected against and did not establish in the population. Recombinant isolates comprised 7% of the population and did not become established. These data did not support the hypothesis that multipartite viral genomes favour genetic exchange through reassortment. Contrasting data were presented by Lin *et al.* (2004) indicating that genetic exchange by reassortment contributed to the evolution of the CMV population. Further studies in Spain (Bonnet *et al.*, 2004) confirmed that reassortant and recombinant genotypes occurred in populations, with at least one recombinant as frequent as one parental genotype, but that reassortants and most recombinants were selected against. There appeared to be a higher-fitness cost of reassortment than recombination.

Viruses may also utilize segment reassortment to adapt rapidly to new host genotypes and this may result in suppression of a resistance reaction. In the presence of the *N*-gene for *Tomato spotted wilt virus* (TSWV), a mixed virus population formed a specific reassortment, whereas in its absence the virus population remained a heterogeneous mixture (Qiu and Moyer, 1999). The reassortment of the RNAs from two TSWV isolates under *N*-gene–derived resistance selection gave a specific combination, which was able to defeat the resistance.

More generally, reassortment among sympatric begomovirus species infecting cucurbits has been shown to occur in experimental studies and if generated in nature could result in begomoviruses with distinct biological properties (Brown *et al.*, 2002a). Both reassortment and recombination have been important factors in the emergence of novel damaging begomoviruses, including those causing cassava mosaic disease (Pita *et al.*, 2001; Seal *et al.*, 2006).

4. *Migration*

Migration of both vectors and viruses (with vectors or through movement of infected plant material) into or from other geographical areas is an important source of local genetic variation that is often ignored and can have major effects on evolutionary change as discussed in detail in Section III.

C. *Drivers of Evolutionary Change*

The different sources of genetic variation lead to the operation of various drivers of evolutionary change. We consider in turn the fitness of a population, selection and genetic drift, and the considerations that apply to a virus population where features such as within-host dynamics must be assessed.

1. *Fitness*

There are several controversies over the definition and use of the term *fitness*. For example, whether it refers to the contribution of an individual to the next generation within a population, or to the average fitness (the intrinsic rate of increase of a population). The controversies become more apparent when the question arises as to how natural selection (as part of the evolutionary process) is governed by individual or average fitness (Frank and Slatkin, 1992). These authors discuss Fisher's fundamental theorem of natural selection, which states that '*The rate of increase of fitness of any organism at any time is equal to its genetic variance in fitness at that time*' (Fisher, 1958); this is taken to refer to a particular environment, including the genomic environment, thus including the feedback caused by ecological changes due to natural selection. Furthermore, if fitness is broken down into n components then it can be shown that the expected average correlation between total fitness and any one component equals $\sqrt{1/n}$. If there is an extremely small negative correlation between components of fitness then there is no correlation between total fitness and these components, making them appear in effect neutral (Wallace, 1993). Aspects of the selectionist-neutral theories are discussed in detail by Dietrich (1994).

2. Selection

Given the differing views over fitness and the selectionist-neutral theories of molecular evolution, the question arises as to how much of the genetic diversity present in a population is adaptive, processed by natural selection and contributing to differences in fitness (Nevo, 1998). Comprehensive studies of plants and animals in Israel were designed to answer this key question. Genetic diversity assessed at the protein and DNA levels was found at all scales to be non-random, heavily structured, to display parallel trends in unrelated taxa, to be positively correlated with ecological and environmental diversity and negatively correlated with population size. The results were interpreted to be driven primarily by natural selection and were thus inconsistent with the neutral theory of molecular evolution. For plant viruses, selection pressure on their life history has been investigated for specific components. For example, Chare and Holmes (2004) considered selection in the capsid genes of a range of plant RNA viruses and the relationship with mode of transmission. They analysed patterns of non-synonymous (d_N) and synonymous (d_s) substitution in the capsid genes of 36 viruses and calculated d_N/d_s ratios as a basis for determining whether positive selection had occurred. They concluded that the capsid proteins of vector-borne viruses are subject to greater purifying selection than those transmitted by other routes, and that the virus–vector interactions mediated by cellular receptors in the arthropod vectors impose greater selective constraints than virus–plant host interactions. This conclusion was reached more generally for insect transmission by Power (2000). Chare and Holmes (2004) also made a comparison with vector-borne circulative animal viruses, but this was limited in scope as only two circulative plant viruses (*Tomato spotted wilt virus* and *Rice black streaked dwarf virus*) were included in the study.

Selection can act upon any individual variant in the population, including mutants in which substitutions have accumulated over time. However, some substitutions remain constant in frequency through host passaging (Hall *et al.*, 2001). This suggests that mutations arising by virus polymerase error are generated at a constant rate. However, as most newly generated mutations are sequestered in virions, they do not serve as replication templates. Thus relative fitness is not tested. Only those mutants that do serve as replication templates are subject to selection and genetic drift and thus may, or may not, become prominent as a consensus sequence.

3. *Genetic Drift*

Selection and genetic drift are not always easy to distinguish for disease in populations. As an illustration of this in a rather different context, both drivers occur and determine the frequency of genetic disease in human populations, but it has often proved difficult to decide to what extent each is responsible for the presence of a particular disease. The pattern of regionally specific mutations is best explained by selection for the haemoglobinopathies (Flint *et al.*, 1993). However, a selective driver is known (malaria) and investigations could be carried out in a population that is relatively homogeneous genetically, and with a known migratory history. For other genetic diseases, interpretation of gene frequencies on the basis of standard population genetics theory can still be difficult.

For some plant viruses, population 'bottlenecks' during the virus life cycle mean that the effective population size is much smaller than the numbers reported and genetic drift could be important in virus evolution. Sacristan *et al.* (2003) estimated the frequency of different genotypes of TMV in co-inoculated and systemically infected tobacco leaves as each leaf developed. A simple probability model was used to estimate the effective numbers of 'founders' for the populations in each leaf. Highly consistent estimates were obtained for different combinations of genotypes. For each leaf the number of founders was small resulting in a small effective population size.

Drift also operates at the population level. For example, influenza in humans is characterised by strong annual dynamics and antigenic evolution (Boni *et al.*, 2004). The amount of variation depends on the epidemic size and antigenic drift. Vaccination protects through classical 'herd immunity' and reduces the chance of new variants arising so that subsequent epidemics may be milder. With a high rate of antigenic variation the virus is always able to invade a susceptible population, but with less variation several introductions may be necessary. For the population-based influenza example, mathematical models proved useful in delineating the effects of drift. Models have also proved useful in analysing within-host dynamics and the role of drift. Frost *et al.* (2001) developed a metapopulation model of HIV replication. A population of infected cells consists of many small sub-populations each of which is established by a few founder viruses and undergoes turnover. The model demonstrated the action of founder effects and the genetic differentiation between sub-populations that results. Drift leads to an effective population size much lower than the actual size and is important in HIV evolution despite the often large number of infected cells.

French and Stenger (2003) explained the low level of sequence diversity in *Wheat streak mosaic virus* by considering the essentially linear intracellular replication of RNA viruses, that is, the 'stamping machine' model (Luria, 1951). RNA viruses may exhibit low variation despite very large population sizes and high-mutation rates, because the *effective* population size is very low. The variation that exists in field populations appears to be fitness neutral and thus the outcome of genetic drift in any single plant virus lineage is essentially unpredictable.

D. Molecular Divergence and Genetic Structure

As a consequence of the drivers of evolutionary change, molecular divergence occurs when change is amplified by population growth factors. Canine parvovirus (CPV) is an emerging single-stranded DNA virus, thought to have developed from feline panleukopenia parvovirus (FPLV) or a closely related virus (Shackelton *et al.*, 2005). The FPLV clade is characterised by a constant population size reflecting the endemic nature of the disease. Since 1970 CPV has caused an epidemic associated with a lineage of CPV with a broader host range and greater infectivity. Based on mutations in the capsid region the CPV population has effectively doubled in size every 2–5 years, depending on whether growth is characterised as exponential or logistic (Table V). Recombination played no role in the emergence of CPV, rather a high-rate of mutation and positive selection of mutations in the capsid gene was responsible. The rate of nucleotide substitution in this ssDNA virus is closer to RNA viruses than dsDNA viruses.

TABLE V

NUCLEOTIDE SUBSTITUTION RATES (PER SITE PER YEAR) AND POPULATION DYNAMICS OF FELINE PANLEUKOPENIA PARVOVIRUS (FPLV), CANINE PARVOVIRUS (CPV) AND A SUB-CLADE (CPV1A) (SELECTED DATA FROM SHACKELTON *ET AL.*, 2005)

Population model	FPLV clade constant	CPV clade logistic	CPV2a sub-clade exponential
Mean substitution rate	9.4×10^{-5}	1.7×10^{-4}	1.7×10^{-4}
Mean age of clade (years)	116	36	28
Mean growth rate (per year)	–	0.1	0.3
Doubling time (years)	–	5.0	2.3

The genetic structure of *Watermelon mosaic virus* (WMV) was studied in Spain, where Moreno *et al.* (2004) found a temporal replacement of genotypes, but no apparent spatial structure or population subdivision. Different genomic regions of the virus showed different evolutionary dynamics. Recombinant isolates accounted for at least 7% of the population, but there was selection against isolates with recombinant proteins. WMV isolates defined by mutation or recombination giving sequences, which fall outside acceptable regions of evolutionary space were eliminated from the population by purifying selection. A rather different situation was found with geminiviruses infecting wild plants in Japan (Ooi *et al.*, 1997). The molecular divergence of *Tobacco leaf curl virus* in *Eupatorium makinoi, E. glehni* and *Lonicera japonica* was compared. Wild populations of each plant species were found to have genetically diverse virus strains. Comparison with an RNA tobamovirus showed a higher-nucleotide substitution rate, more frequent migration among geographically isolated host populations and more frequent host changes. The rate of evolution in the ssDNA geminivirus appeared faster than in the RNA virus.

III. GEOGRAPHICAL SUB-DIVISION AND GENETIC VARIATION

Epperson (1993) wrote:

> Population genetic processes including natural selection exist in, and in some cases are inseparable from, the spatial context. Dispersal and migration connect temporal genetic processes at different locations. Spatial structure can be ignored a priori only if dispersal or migration reaches panmictic levels.

A wide range of spatial analysis techniques was considered in this review: the need to apply spatial statistics in comparing loci under selection or those that are selectively neutral, simulation of stochastic effects when genotypes interact through dispersal and reproduction, sampling design to test the patch structure, stochastic migration effects and generalising spatial statistics to space–time autoregressive processes.

More traditional population genetic models incorporate geographic sub-division with migration as well as mutation, recombination, selection (at a single locus) and genetic drift (at many linked loci). Geographic sub-division can substantially affect genetic variation provided that selection is sufficiently strong at a given locus (Kaplan *et al.*, 1991). Estimates of mutation rates and divergence times are very sensitive to population sub-division (Wakeley, 2003), although estimates of selection coefficients can be very robust. High levels of within-deme relatedness leads to low levels of intra-species polymorphism and increases the number of fixed differences between samples from diverging species. If sub-division is ignored mutation parameters are underestimated and species divergence times overestimated. Approximations for the frequency of a given allele in a population with many demes and migration can be obtained (Wakeley and Takahashi, 2004).

Geographical sub-division is one aspect of spatial heterogeneity at an appropriate scale. Selection may act differently in the different patches comprising the spatial heterogeneity. Gavrilets and Gibson (2002) modelled the situation in which a mutation may be advantageous in one patch but deleterious in another. In general, the perceived wisdom is that large populations are more responsive to selection than small populations. However, in the heterogeneous patch system modelled, it was found that under some conditions small populations respond faster to selection than do large populations and the effects of population size on the rate of evolution depend critically on how migration rates depend on population size (Fig. 3). Rapid speciation is a plausible outcome for populations divided into sub-populations (Gavrilets, 1999) and local adaptation is not necessary for rapid speciation.

There have been few explicit studies for plant viruses in which the effects of geographical sub-division are interpreted in terms of population genetics. For example, it had previously been suggested that genomic variation among begomoviruses is concerned primarily with geography rather than host range (Padidam *et al.*, 1995). Similarly, that isolates from different hosts in the same geographical area are more likely to be closely related than isolates from the same host in different geographic areas (Harrison and Robinson, 1999). An exception to this clustering was found by Hameed and Robinson (2004). They concluded that South Asian begomoviruses infecting legumes

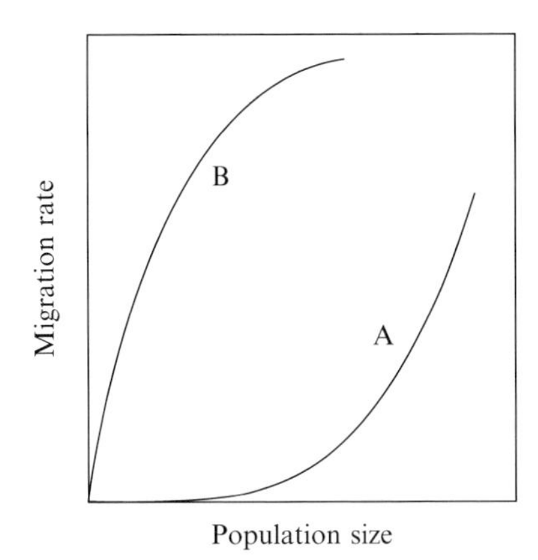

FIG 3. Dependence of migration rate on population size (Gavrilets and Gibson, 2002). In (A) migration rate remains low initially until the population size has increased substantially; in (B) the migration rate is high initially but levels off as the population size increases. With kind permission of Springer Science and Business Media.

represent a distinct lineage not closely related to other begomoviruses in the same geographical area.

Geographical sub-division can also be a factor in vector evolution as much as virus evolution. The genetic structure of field populations of begomoviruses and their whitefly vector *Bemisia tabaci* from different crop plants and weed species in Pakistan was analysed and compared to the distribution of cotton leaf curl disease (CLCuD) (Simón *et al.*, 2003). Analysis of the begomovirus coat protein gene sequences grouped isolates into three geographical clusters corresponding to the Punjab, Sindh and both provinces. Analysis of mitochondrial cytochrome oxidase I gene sequences showed that the *B. tabaci* population was structured into three genetic lineages corresponding to the previously described Indian, Mediterranean-African and Southeast Asian clades. It is notable that the Indian clade of *B. tabaci* was found only in the Punjab where CLCuD occurs, suggesting that the geographical distribution of virus and vector genotypes may be correlated.

IV. INTERACTIONS

Interactions are a key defining aspect of plant virus epidemiology. For most plant virus diseases there are two-way interactions between virus and vector, virus and host, and vector and host, to be considered

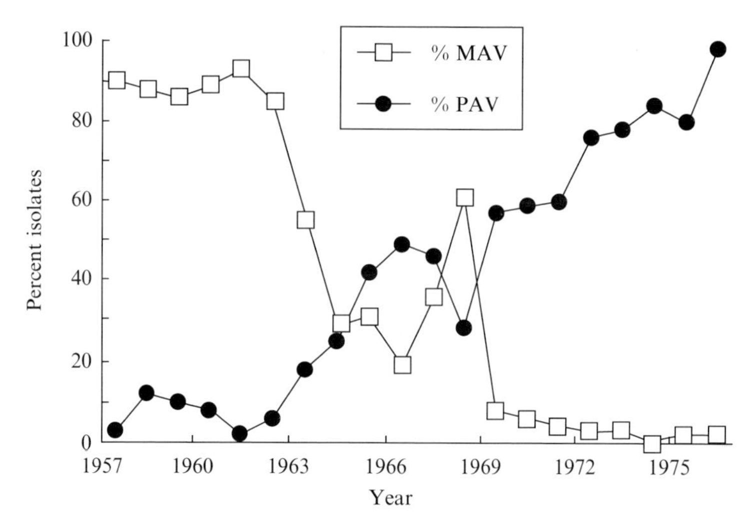

FIG 4. Dynamics of competition between MAV and PAV from 1957 to 1978 in barley yellow dwarf viruses (Power, 1996). With permission 1996, Ecological Society of America.

and also the three-way interaction between virus, vector and host. In many cases there are higher-order interactions with the biotic and abiotic environment. As stated by Power (1996), factors influencing the spread and success of plant viruses include direct competition between viruses in host plants, direct competition within and between vectors, differences in transmission rates, and virus influences on vector behaviour and population dynamics. For example, transmission rate plays an important role in determining the outcome of competition between different barley yellow dwarf viruses. Historical shifts in New York State have led to one virus strain PAV (transmitted by the aphids *Rhopalosiphum maidis* and *Sitobium avenae*) displacing another strain MAV (primarily transmitted by *S. avenae*) over an 18-year period 1957–1975 (Fig. 4). The interaction between transmission rate and vector behaviour appears important. PAV is the stronger competitor within the host, where double infections are more common than the occurrence of these viruses in the vector, and has a higher-overall transmission rate than MAV. There is also a greater production of winged morphs on PAV-infected plants that may lead to greater rates of virus spread for this strain. The various levels of interaction that occur are now considered.

A. Interactions Between Viruses

Interactions between plant viruses within hosts have practical implications as well as being of considerable theoretical interest, for example, in the use of cross protection using mild strains for plant virus disease control (Fulton, 1986; Lecoq, 1998) as practised for example against *Citrus tristeza virus* (Bar-Joseph *et al.*, 1989) and *Papaya ringspot virus* (Fuchs *et al.*, 1997; Yeh and Gonsalves, 1984). Opportunities for direct interaction within plant cells occur in many ways, between wild-type and mutant virus strains, between wild-type and defective interfering (DI) particles arising from deletion mutants, and with other viruses, where multiple particles are necessary for infection to occur ('co-viruses'). As mentioned earlier whether mutant strains of a virus compete with the wild-type depends on whether they serve as replication templates. DI particles cannot replicate unless they co-infect with the wild-type virus. With respect to human/animal RNA viruses, Frank (2000) developed a mathematical model for the population dynamics of wild-type virus and DI particles. Where there is a low rate of replacement of killed host cells, major epidemics occurred followed by a crash in viral abundance. As the rate of cell replacement increases there is an increased frequency of oscillation, but with decreasing amplitude, in DI and wild-type virus abundance. At the highest replacement rates almost all cells are infected by DI particles with a low-level persistence of the wild-type virus. Thus the long-term outcome of the co-infection modelled is highly dependent on the ability of the host to replace killed cells.

'Co-viruses' may have arisen from different deletions of the same self-sufficient virus or from two different viruses (satellite virus or RNA). *Pea enation mosaic virus* is notable in that it is neither a mixed infection of two unrelated viruses, nor is it a multicomponent virus, but rather it represents a seemingly unique form of symbiosis between two functionally defective viruses (de Zoeten and Skaf, 2001). Nee (2000) modelled the epidemiology and evolution of co-viruses using a simple metapopulation model and questioned whether co-viruses can co-exist with complete progenitor viruses. The model was applied to the example of *Tobacco rattle virus* (TRV) that has two particles—RNA1 encoding the replicase, and RNA2 encoding the coat protein. The model predicted that a complete autonomous progenitor of TRV could not co-exist with the RNA1 alone or both RNA1 and -2 together, unless transmission of the co-virus was seriously impaired by its divided nature, and for TRV was at least plausible in this respect.

Pepper huasteco virus (PHV) and *Pepper golden mosaic virus* (PepGMV) are both whitefly-borne begomoviruses with a bipartite genome, infecting a wide range of horticultural crops. Mixed infections occur commonly and were used as a model system to study virus–virus interactions (Méndez-Lozano *et al.*, 2003) in terms of symptom expression, gene expression, replication and systemic movement. The main findings were that: interaction for symptoms can be opposite in different hosts, with antagonism in pepper but synergism in tobacco and *Nicotiana benthamiana*; replication of both PHV and PepGMV was increased during mixed infections and there was an asymmetric complementation in movement function with PHV supporting systemic movement of PepGMV but not vice versa. The interaction of the two viral genomes with the plant genome during natural mixed infections might offer some advantages to both viruses and explain their widespread occurrence throughout the region.

In the previous example (Méndez-Lozano *et al.*, 2003) re-assortment between the two begomoviruses did not occur. Co-infection has been considered beneficial in virus populations because it permits genetic exchange between viruses and helps to purge mutational load. In contrast, by allowing virus complementation where 'inferior' genotypes benefit, it may be disadvantageous. Numerical simulations using a simple population genetic model predicted that mutational load would be removed faster in the presence of co-infection where there is alternating reassortment and selection, than in single infection where there is only selection (Froissart *et al.*, 2004). However, studies with the bacteriophage Ø6 and its mutants in *Pseudomonas syringae* showed the converse to hold, that is, deleterious mutants were purged faster in the absence of co-infection. This suggests that the disadvantages of complementation outweighed the benefits of reassortment. These findings also suggest that complementation may allow deleterious mutations to persist within the host sufficiently long for compensatory mutations to increase fitness, a process of some significance for plant viruses (Seal *et al.*, 2006).

A clear example of how mixed infection can affect the epidemiology of virus disease is seen with the epidemics of a particularly severe form of cassava mosaic disease in Uganda (Legg *et al.*, this volume, pp. 355–418). The etiology of the disease is complex involving *African cassava mosaic virus* (ACMV), and a recombinant of ACMV with *East African cassava mosaic virus* (EACMV) termed the Uganda variant (EACMV-[Ug]). The novel virus complex causing the most severe symptoms consists of the interspecific recombinant virus (EACMV-[Ug]) and one of its parents (ACMV). Mixed infection leads to higher

EACMV-[Ug] titres and more severe effects than EACMV-[Ug] alone (Harrison *et al.*, 1997). The severe cassava mosaic epidemic spread rapidly in East Africa and westwards through Central Africa and now poses a threat to the important cassava growing areas of West Africa (Legg *et al.*, this volume, pp. 355–418).

Finally, the question arises how best the within-host virus population can be modelled when there is more than one virus/virus strain present. A simple model based on the virus multiplication rate, parameters referring to cross-reactions and specific immune responses, and a virus-induced cell death rate, combined with the Simpson index for virus diversity has been proposed for HIV multiplication within CD4+ cells (Anderson, 1994). There are opportunities for such models to be developed for plant viruses but these have yet to be explored.

B. Interactions of Viruses with Vectors

Vector biology has a major impact on plant virus epidemics. It is unlikely that any single factor can be found to explain instances in which a major change in vector population dynamics or genetics has lead to an upsurge in virus disease (Henneberry and Castle, 2001), except through introductions (e.g. of the B-biotype of *B. tabaci*) to new areas. Virus vectors, such as aphids are capable of both short and long-range movement, and such movement has obvious implications for virus epidemiology and the spread of new genotypes (Loxdale *et al.*, 1993). Phenotypic variation exists even in clonal populations of aphids in various phases of flight behaviour (Reynolds *et al.*, this volume, pp. 453–517). These aspects are also present with *B. tabaci*. Laboratory observations have distinguished both migratory and trivial flying morphs and field observations confirm that a portion of the population are trivial flyers that do not engage in migration, whereas a portion respond to skylight rather than ground cues and fly for a period before alighting (Byrne *et al.*, 1996).

The transmission process is one obvious way in which viruses interact with vectors, and has led to the very useful classification into non-persistent, semi-persistent, and persistent-circulative and persistent-propagative that has been modelled successfully (Jeger *et al.*, 1998). Other aspects of vector transmission are much less studied. Horizontal transmission can be direct without the involvement of vectors, as in *Tobacco mosaic virus*, or indirect where transmission is through vectors. These can also be venereal transmission between vectors such as whiteflies (Czosnek *et al.*, 2001), and vertical, that is, transovarial, transmission again in whiteflies (Bosco *et al.*, 2004). The role of endosymbionts in the circulative transmission of begomoviruses (Morin *et al.*, 1999) is much less studied than for example aphid transmission of luteoviruses (van den Heuvel *et al.*, 1997).

We return again to the correlation and possible co-adaptation between plant viruses and vectors. Maruthi *et al.* (2002) compared Indian and African cassava mosaic begomoviruses for their transmissibility by one Indian and three African populations of *B. tabaci* collected in India and Africa. There was no relationship between transmission frequency and symptom severity. However, the African EAMCV was transmitted by all three African *B. tabaci* populations, whereas there was only limited transmission of this virus by the Indian *B. tabaci* (Table VI). There was a reciprocal pattern for the virus from India (ICMV). These findings were also apparent when the number of test adults/plant and the inoculation access period were increased. The authors considered this as support for virus–vector co-adaptation.

Insect transmission of plant viruses is clearly a constraint on virus variability (Power, 2000). Molecular studies have added to the understanding of transmission efficiency but evolutionary explanations for specificity are still required. Why, for example, should it be more difficult for a virus to increase its range of efficient vectors than its host range (Power, 2000)? To what extent does carrying the virus affect the fitness of the vector, especially with a persistent-propagative virus such as *Rice hoja blanca virus* (Ziegler and Morales, 1990)? Recent work has found that nanovirus-like DNA components associated with geminivirus diseases can adapt from whitefly to leaf hopper transmission when co-inoculated with a leaf hopper-transmitted virus (Saunders *et al.*, 2002). Such findings highlight the molecular complexities underlying virus epidemiology and evolution.

TABLE VI

COMPARATIVE TRANSMISSION OF CASSAVA MOSAIC GEMINIVIRUSES FROM AFRICA (*EAST AFRICAN CASSAVA MOSAIC VIRUS*, EACMV) AND INDIA (*INDIAN CASSAVA MOSAIC VIRUS*, ICMV) (MODIFIED FROM MARUTHI *ET AL.*, 2002) BY ADULT *B. TABACI*

	20 adults/plant (48 h)		50 adults/plant (5 days)	
Cassava mosaic geminivirus	African	Indian	African	Indian
EACMV	20/26[a]	0	13/14	3/29
ICMV	0	25/28	2/30	12/12

[a] Number of plants infected of number inoculated.

C. Interactions of Viruses with Hosts

Viruses interact with the host from the time of virion introduction to a plant cell, through the cellular processes underpinning viral replication and encapsidation, to cell-to-cell movement of virions. These interactions are discussed in Section V, dealing with resistance and virulence.

D. Interactions with Vectors and Hosts

Vectors themselves interact with the host plant and show preferences for particular host plants, irrespective of whether they are carrying virus or not. Also, pest resistance within a host species may affect the probability that transmission will occur both positively and negatively. Using simple mathematical models, McElhony *et al.* (1995) examined the probability of disease transmission where vectors have preference for diseased or healthy hosts. They found, that where spatial aspects of vector movement were not included in models with a high incidence of disease, spread was favoured by vectors preferring healthy plants; whereas with a low level of disease, spread was favoured by vectors preferring diseased plants. However, when spatial aspects were included an increase in patchiness can lead to a decrease in the rate of spread by a vector that moves only limited distances—a preference for diseased plants does not necessarily lead to an increase in disease spread. These results raise the possibility that the virus may manipulate adaptively either the preference behaviour of vectors or host attractiveness to increase fitness, although there have been few experimental studies of this possibility. Certainly host-plant dispersion and density can affect the abundance and behaviour of highly mobile virus vectors and hence rates of spread of disease (Power, 1992).

Many subtle interactions between virus, vector and host can occur. As one example with *Tomato mottle virus* (ToMoV) pathogenesis-related proteins were induced in tomato plants inoculated with ToMoV as much as with feeding by non-viruliferous *B. tabaci* (Mayer *et al.*, 2002). ToMoV infection significantly increased egg production by females. It seems possible that begomovirus infection and/or *B. tabaci* infestation isolate the plant for the selected reproduction of the virus and the whitefly. *B. tabaci*–begomovirus relationships are thus mutualistic, as argued elsewhere (Colvin *et al.*, 2004; this volume, pp. 419–452).

E. Higher-Order Interactions

The complexity of factors that could influence virus epidemiology and evolution becomes apparent when considering the numerous higher-order interactions that are possible. For example, the rate of herbivore (e.g. vector) adaptation to resistant host plants can be increased or decreased by natural enemies (Gould *et al.*, 1991). Similarly, the dynamics of host–parasitoid and host–pathogen interactions where pesticides are applied can affect the evolution of pesticide resistance (May and Hassell, 1988), an important aspect where natural enemies and/or insecticides are used for or contribute to vector control.

V. Host Resistance and Pathogen Virulence

Interactions between virus and host are almost entirely conditioned by the effect of the pathogen on the host and the host's response to pathogen challenge; in short by host resistance and pathogen virulence, although as will be seen these simple terms are easier to state than define. Non-host resistance or complete immunity to plant viruses are not considered in this context.

A. *Components of Resistance*

Resistance can be broken down into various components as listed in Table VII. All of these components or traits interact with different factors, including the physical environment, virus virulence and dose, and host genetic background, that influence the expression of those traits (Kegler and Schenk, 1990). Each of these cannot be discussed in detail here and only virus content is considered. Variation was found in the multiplication of *Beet necrotic yellow vein virus* (BNYVV) in relation to host resistance (Heijbroek *et al.*, 1999). The virus content in

TABLE VII
Components of Plant Resistance[a]

Components
Reduced infection rate
Prolonged incubation period
Lower virus concentration
Incomplete systemicity
Mild or no symptoms
Less or no growth inhibition
Less or no yield reduction

[a] All traits interact with different factors (environment, virus virulence and dose and host genetic background) influencing their expression (Kegler and Schenk, 1990).

taproots was always higher for one of the more damaging ('P') isolates, which was the one moving most rapidly in plants. The mean virus content, together with its frequency distribution across replicate plants, in relation to a threshold amount, gave a good measure of the level of resistance in a cultivar.

B. Resistance and Tolerance

The use of the terms *resistance* and *tolerance* in plant pathology has proved contentious, not least for virus diseases. Some plant breeders refer to tolerance as an ability to withstand infection by whatever means. Others regard it as a reduced yield loss for some cultivars showing the same level of disease as in non-tolerant cultivars. Moreover, some plant virologists refer to reduced symptoms in relation to identical virus contents; whereas others, almost colloquially, refer to the ability of a plant to tolerate the presence of virus. For this reason we deliberately go outside the mainstream of plant pathology and classify two distinct categories found useful in parasite biology and herbivory.

Conceptually we refer to *resistance* as reducing the success of an infection process or increasing the rate of parasite clearance, and *tolerance* as reducing the detrimental effects of the parasite (Restif and Koella, 2004). Thus resistance acts on the parasite directly, whereas tolerance acts on the effects of the parasite without altering its growth or development. It is entirely possible that resistance and tolerance are uncorrelated. Also the relative costs of resistance and tolerance in the absence of the parasite may be entirely different. If resistance and tolerance can vary independently what are the evolutionary pressures on these traits? What conditions favour resistance, tolerance or a combination of the two, taking into account the epidemiological dynamics of the systems. Restif and Koella (2004) predicted that resistance and tolerance are not necessarily mutually exclusive, but respond differently to changes in epidemiological parameters or the relative costs and benefits of the two traits, and that apparent associations among resistance, tolerance and host performance (in the absence of the parasite) can lead to false conclusions without some knowledge of the underlying mechanisms involved.

C. Virulence

There is also considerable ambiguity in the use of the term *virulence* in host–pathogen interactions. According to Nee (2000) the generalisations that abound are '*more an admission of ignorance than anything else*'. Broadly, the most established definition of virulence refers to the pathogen-induced death rate, either as a constant term or as a function of transmission and the population densities of susceptible and infected hosts. As with resistance there are components of virulence and similar arguments about correlations between total fitness and components of fitness can be made for virulence and its components (Ebert, 1998; Hochberg, 1998).

The evolution of virulence in a pathogen population has often been based on a game-theory approach to determine the fate of new mutant strains with perhaps different virulence characteristics. Problems with this approach have been pointed out by Day and Proulx (2004) who state that the approach does not allow the prediction of evolutionary dynamics, an experimental test of theory or the development of virulence management protocols, and it is only appropriate where epidemiological and evolutionary dynamics occur at comparable timescales. However, the evolutionary stable strategy approach has been used with some success to evaluate what plant breeding strategies, and selection of resistance traits, should best be followed to avoid increased virulence evolving in a plant virus population (Van den Bosch *et al.*, 2006). By contrast Day and Proulx (2004), constructed models similar to those used in quantitative genetics, including within-host mutation and the effects of infection between hosts. The models allow prediction of both short and long-term evolution of virulence and lead to clear interpretation of the selective forces involved.

D. Virulence and Transmission

The degree to which a pathogen evolves to be virulent depends in part on whether it is transmitted horizontally (either directly or indirectly through a vector) or vertically. Consequently, the virulence term used in epidemiological models may be expressed as a function of transmission, and the population densities of susceptible and infected hosts. There are some clear examples on the virulence-transmission relationships in areas other than plant virology. For honeybee diseases, pathogen fitness depends not only on infection and spread between individuals within a colony but also the ability to spread to new individuals in other colonies (Fries and Camazine, 2001). Vertical transmission may be the main transmission pathway for new colonies and consistent with the theory, most honeybee diseases including viruses exhibit low virulence, except where the viruses are transmitted horizontally (and indirectly) to new colonies by the *Varroa* mite.

The evolution of virulence in plant viruses has been studied comprehensively for CMV of tomato in Spain (Escriu *et al.*, 2000, 2003). Three types of isolates of CMV were described based on the presence or absence of satellite RNA. Necrogenic sat RNA isolates depressed CMV accumulation more than non-necrogenic isolates (Table VIII). Thus the efficacy of aphid transmissions was reduced where the former satellites occurred. The evolution of virulence was interpreted as a 'trade-off' between factors determining virulence and those affecting transmission. A single infection model was used to predict the long-term evolution of CMV to intermediate virulence. A co-infection model, with competition and an effect on transmission, explained the temporal sequence of invasion and the long-term evolutionary outcome. Parameters relating to virulence (the pathogen-induced *per capita* increase in host mortality) and transmissibility (the *per capita* rate of transmission) were estimated from these models (Table IX) and again supported the theory of an inverse relationship.

The remaining question to be posed in this section is whether infection with multiple strains affects the evolution of virulence, either by increasing or decreasing it. Classical models of virulence evolution conclude that increased competition will select for increased virulence, but in some circumstances [see arguments in Brown *et al.* (2002b)] multiple infections can select for reduced virulence. For cassava mosaic disease in Uganda, where mixed infection with ACMV and EACMV-[Ug] led to severe epidemics it is apparent that mild strains of EACMV-[Ug] are

TABLE VIII

Evolution of *Cucumber mosaic virus* (CMV) sat RNA to Deceased Virulence in Spain (Modified from Escriu *et al.*, 2000)

Years	Total plants tested	No. infected with CMV	No. with CMV sat RNA		
			Total	Necrotic	Non-necrotic
1989–1991	70	60	48	30	18
1992–1994	138	92	43	7	36
Accumulation of CMV particles in tomato (μ/g fresh weight)					
No sat RNA		1124			
Necrogenic		111			
Non-necrogenic		265			

TABLE IX

ESTIMATION OF VIRULENCE AND TRANSMISSION PARAMETERS FOR *CUCUMBER MOSAIC VIRUS* (CMV) WITH OR WITHOUT SAT RNA, BASED ON MAY AND ANDERSON MODEL (MODIFIED FROM ESCRIU *ET AL.*, 2003)

Status of CMV	Virulence[a] (α)	Transmission[b] (β)
No sat RNA	0.001	0.528
Necrogenic	0.011	0.199
Non-necrogenic	<0.001	0.288

[a] *Per capita* increase in mortality, calculated as the difference in reciprocal life spans of infected and non-infected plants.

[b] *Per capita* rate of transmission calculated from probability of aphid-mediated contact of virus with host $\times$ probability of transmission.

now common in post-epidemic areas (Owor *et al.*, 2004; Sseruwagi *et al.*, 2004).

VI. PLANT VIRUS EPIDEMIOLOGY

Plant virus epidemiology has matured as a science in recent years and mathematical models are being employed increasingly to represent the interacting dynamics of plant host, virus and vector, especially for arthropod vectors, often as a framework for evaluating the likely success of different control measures (Chan and Jeger, 1994; Jeger *et al.*, 1998, 2004; Madden *et al.*, 2000). Another trend can be seen in the use of molecular tools in epidemiology. Molecular techniques are being used increasingly to characterise virus populations, to elucidate virus–vector relationships and to study the interactions between virus strains so as to determine their origins and epidemiological effects. The information so obtained broadens the scope of classical epidemiology (Thresh *et al.*, 2003).

A. *Molecular Epidemiology*

Molecular analysis of virus isolates facilitates an understanding of virus epidemiology, as illustrated by studies on CMV. García-Arenal *et al.* (2000) characterised the genetic structure of CMV and its satellite RNA in Spain. Some 300 isolates were taken from 17 outbreaks representing sub-populations from different crops, regions and years and compared. All isolates conformed to one of three genetic types. The genetic structure of CMV appeared to vary randomly—a metapopulation structure with local extinction and random colonisation from virus reservoirs. The frequency of CMV isolates with a satellite RNA differed for each sub-population, but generally it was concluded that the sat RNA of CMV has spread epidemically in the existing virus population.

A 3-year analysis of the population dynamics of CMV in Spain has also been undertaken (Sacristan *et al.*, 2004). The dynamics differed in various weed habitats and CMV incidence was compared with the overall extent of vegetation cover. The population dynamics of CMV in melon crops was unrelated to that in weed habitats. No significant differences in the frequency of CMV genotypes were found for weeds or melon (Table X). It seems that movement between hosts and habitats has prevented observable population fragmentation.

B. *Epidemiology and Co-evolution of Pathogens and Hosts*

Most work in this area has dealt with host–parasite interactions. Models about evolutionary aspects of the relationship between parasite and host life history characteristics often ignore epidemiological feedback. For example, in theoretical studies Koella and Restif (2001) looked at the host's reproductive response as parasite virulence increases. Where the environment allows rapid host growth, a high-mortality rate of the host favours avirulent parasites and late reproduction of the host. When growth is slow, early reproduction and high virulence is favoured.

Most epidemiological models consider three main traits in the host–parasite interaction—transmission (infection of a susceptible host), virulence (detrimental effects on host fitness) and recovery (where this

TABLE X

FREQUENCY OF *CUCUMBER MOSAIC VIRUS* GENOTYPES IN MELON CROPS AND WEED HABITS IN SPAIN (MODIFIED FROM SACRISTAN, 2004)

		Haplotypes		
Host plant	No. of samples	1	2	3
Melon	33	0	19	14
Weeds	27	1	13	13
Total	60	1	32	27

occurs). Restif and Koella (2003) developed a co-evolutionary model in which traits depend on both parasite and host genotypes. These do not follow the classical gene-for-gene or matching allele systems proposed for plant pathogens. Instead they consider the questions—what is the optimal investment in host response and in a parasite's virulence if both organisms control the epidemiological trait? Evolution of this trait is constrained by trade-offs that account for costs of defence (resistance) and attack (virulence) as would be predicted from other approaches. Novel predictions were made however—the host evolves maximal investment in defence against parasites with intermediate multiplication, the evolution of the parasite depends strongly on the form of host defence, and pathogen virulence may decrease in relation to an increase in host (background) mortality.

Co-evolutionary models for plant–virus interactions and dynamics have yet to be developed in any detail.

C. *Evolution and Disease Management*

Epidemiology has long been considered the foundation on which robust disease management strategies can be developed (Jeger *et al.*, 2004). Its success in achieving this for plant virus diseases has been described by Jones (2001, 2004) mostly in the context of grain and pasture legumes. These approaches, however, do not always take into account evolutionary processes in virus populations. Nevertheless, pathogen evolution and variety development and deployment is seen as being at the heart of nearly all sustainable solutions in crop disease management (Jeger *et al.*, 2004). What approaches should be adopted by plant breeders, seed companies and farmers in response to the ongoing threat and periodic emergence of novel virulent viruses or virus strains? When a new pathogen emerges how can genetic resources best be deployed in time and space to combat the threats posed? Can so-called 'boom and bust' cycles of disease outbreaks (or cultivar usage) be avoided? What outcomes might be expected from plant breeding strategies involving virus resistance, vector resistance or a combination of both? To this question that is relevant for economic crops, might be added questions related to plants in natural or weed communities, both in terms of biological conservation and the role viruses play in their population dynamics, and their role as reservoirs of virus that can infect economic crops (Cooper and Jones, this volume, pp. 1–47). These questions are addressed in the final sections of this chapter.

VII. Evolution in Natural Plant Communities

There are increasing concerns for plants in relatively natural communities, where more and more ecosystems are being transformed and managed for economic crops, natural resources or for other human purposes. Paradoxically it is the human interest in remaining natural plant communities, whether from conservation or touristic motivation that has led to the increasing interest. Of particular concern is the dynamics of fragmented plant populations and the consequences of small population sizes (Ellstrand and Elam, 1993). Of course there are major considerations here both long-term with respect to global change, and more short-term such as the unpredictable occurrence of natural disasters, hurricanes, volcanic activity, earthquakes or floods. As well as these major factors, pests and diseases in natural plant communities have been shown to have a range of negative effects, including the potential to threaten whole populations, and positive effects in terms of maintaining diversity within populations and communities.

Some studies have been conducted with plant viruses including a geminivirus of *Eupatorium makinoi*, which is a short-lived perennial widely distributed in Japan. Local populations can become extinct following infection (Futayama-Noguchi, 2001; Futayama *et al.*, 2001). The virus causes impaired photosynthesis associated with the loss of the light-harvesting chlorophyll proteins associated with photosystem II. The lowered fitness of infected *E. makinoi* underlies the population dynamics observed. Virus infection was found to be an important biotic factor affecting ecological and evolutionary phenomena in natural plant communities.

Barley yellow dwarf virus (BYDV) strain SGV (transmitted by *Schizaphus graminum*) infects the perennial grass *Anthoxanthum odoratum* in which BYDV infection is asymptomatic. In experimental studies with transplanted plants in the field, BYDV-inoculated clones had significantly lower fitness than healthy controls (Kelley, 1994). BYDV infection appears to generate advantages for rare sexually produced genotypes (Table XI), presumably through frequency-dependent selection. Usually there was a complex interaction between vector, host, virus genetics and population structure, vector behaviour and host and vector dispersal patterns. Viral pathogens, often asymptomatic, play a significant evolutionary role in plant populations.

TABLE XI

PLANT PERFORMANCE[a] OF *A. ODORATUM* INFECTED OR UNINFECTED WITH *BARLEY YELLOW DWARF VIRUS* (BYDV) (MODIFIED FROM KELLEY, 1994)

Parent plants	Asexual progeny	Sexual progeny
Uninfected with BYDV		
Vegetative size	2.25	2.16
Fecundity	1.08	0.86
Infected with BYDV		
Vegetative size	1.71	1.77
Fecundity	0.14	0.51

[a] Values are mean numbers of vegetative and reproductive tillers, providing estimates of vegetative size and fecundity, respectively.

VIII. EVOLUTION AND EMERGING DISEASES

Wild plant hosts are known to harbour viruses that can cause disease in cultivated plants. *Tomato chlorotic mottle virus* occurs in tomato in some areas of Brazil, whereas in other regions an uncharacterised but distinct virus predominates with perhaps a further four uncharacterised viruses also present (Ambrozevicius *et al.*, 2002). The phylogenetic relationships observed suggest transfers of indigenous begomoviruses from wild hosts into tomato. The question arises—under what conditions can viruses from wild plants establish and persist (perhaps replacing existing viruses) in cultivated plants?

Approaches taken in human epidemiology may help in analysing such situations. New pathogens are believed to emerge from animal reservoirs (the wild host) when ecological changes increase the pathogen's opportunities to enter the human population (the cultivated host) and generate subsequent human-to-human (plant-to-plant) transmission (Antia *et al.*, 2003). The process is shown schematically in Fig. 5. Effective plant-to-plant transmission in the cultivated crop population requires that the virus's basic reproductive number R_0 should exceed one (where R_0 is the average number of secondary infections arising from one infected individual in a susceptible population). Initially the R_0 when the virus is transferred from the wild plant to a cultivated plant will be low and the disease will not establish or at best only transfer to a few other plants. However, if the virus evolves such that R_0 increases, then even though it is insufficient to generate an epidemic, the number of subsequently infected plants is increased, and so the probability of virus evolution to an $R_0 > 1$ and subsequent disease

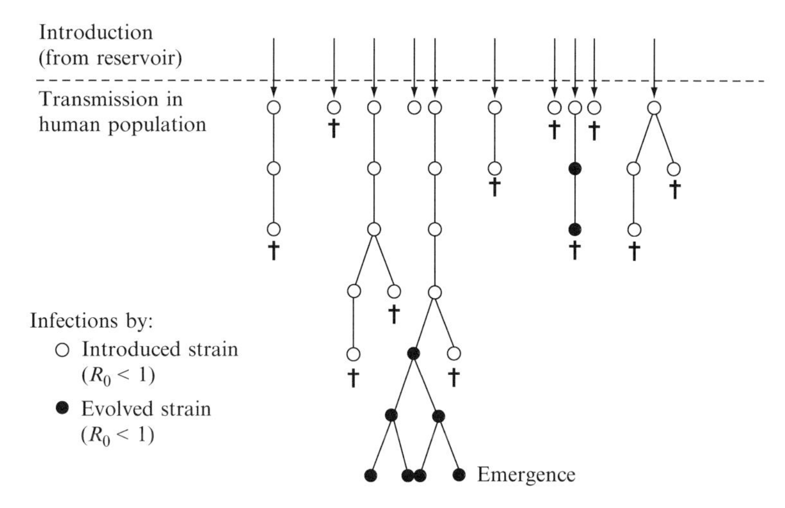

FIG 5. Emergence of a new evolved pathogen strain from a 'wild' reservoir (Antia *et al.*, 2003). Following introduction there is limited transmission in the challenged population until such time as the introduced strain has evolved to a basic reproductive number greater than 1. Reprinted by permission from MacMillan Publishers Ltd., Nature 2003.

emergence (with the potential for an epidemic to occur) can also increase markedly.

IX. CONCLUSIONS

The problem of managing plant virus disease can be approached from two differing time-scales. From an epidemiological and ecological perspective, an understanding of the population dynamics of host, vector and virus can contribute to decisions on how best to deploy control options involving host resistance, cropping practices, phytosanitation and chemical control of vectors; or if these options are not available, whether and in what form they should be developed. From an evolutionary perspective, considered over a longer time-scale, an understanding of the genetic variation present in host, vector and virus populations provides information on sources of variation, selective forces in operation, and the population structures that result. What has not been achieved to any significant extent is an integration of these two perspectives, and the formulation of questions and investigations that would not be apparent when approached from either perspective alone. Such an integrated approach is the subject matter of evolutionary epidemiology, whether applied to the retrospective analysis of historical epidemics and the causative factors involved, or to the much more difficult objective of predicting the consequences of current or proposed cropping or disease management practices for evolutionary change. A better appreciation of the potential for adverse evolutionary change, and its avoidance, will be critical for developing sustainable methods for managing future epidemics of plant virus disease.

Acknowledgments

We acknowledge the financial support from the UK Department for International Development Crop Protection Programme, managed by Natural Resources International Ltd., which supported in part research on which this review is based; and the organisers of the 12th International Plant Virus Epidemiology Symposium held in Lima, Peru, in March 2005 for the invitation to present an oral version of this review.

References

Aaziz, R., and Tepfer, M. (1999). Recombination between genomic RNAs of two cucumoviruses under conditions of minimal selection pressure. *Virology* **263:**282–289.

Ambrozevicius, L. P., Calegario, R. F., Fontes, E. P. B., de Carvalho, M. G., and Zerbini, F. M. (2002). Genetic diversity of a begomovirus infecting tomato and associated weeds in Southeastern Brazil. *Fitopatol. Bras.* **27:**372–377.

Anderson, R. M. (1991). Populations and infectious diseases: Ecology or epidemiology? *J. Anim. Ecol.* **60:**1–50.

Anderson, R. M. (1994). Populations, infectious disease and immunity: A very nonlinear world. *Phil. Trans. R. Soc. Lond. B.* **346:**457–505.

Antia, R., Regoes, R., Koella, J. C., and Bergstrom, C. T. (2003). The role of evolution in the emergence of infectious disease. *Nature* **426:**658–661.

Aranda, M. A., Fraile, A., and García Arenal, F. (1993). Genetic variability and evolution of the satellite RNA of cucumber mosaic-virus during natural epidemics. *J. Virol.* **67:**5896–5901.

Bar-Joseph, M., Marcus, R., and Lee, R. F. (1989). The continuous challenge of citrus tristeza virus control. *Annu. Rev. Phytopathol.* **27:**291–316.

Barnett, O. W. (1993). Modern technology improves plant virus taxonomy or melding the molecular and classical. *Phytopathology* **83:**33–34.

Boni, M. F., Gog, J. R., Andreason, V., and Christiansen, F. B. (2004). Influenza drift and epidemic size: The rate between generating and escaping immunity. *Theor. Pop. Biol.* **65:**179–191.

Bonnet, J., Fraile, A., Sacristan, S., Malpica, J. M., and García-Arenal, F. (2004). Role of recombination in the evolution of natural populations of *Cucumber mosaic virus*, a tripartite RNA plant virus. *Virology* **332:**359–368.

Bosco, D., Mason, G., and Accotto, G. P. (2004). TYLCSV DNA, but not infectivity, can be transovarially inherited by the progeny of the whitefly vector *Bemisia tabaci* (Gennadius). *Virology* **323:**276–283.

Bousalem, M., Douzery, E. J. P., and Fargette, D. (2000). High genetic diversity, distant phylogenetic relationships and intraspecies recombination events among natural populations of *Yam mosaic virus*: A contribution to understanding potyvirus evolution. *J. Gen. Virol.* **81:**243–255.
Bradeen, J. M., Timmermans, M. C. P., and Messing, J. (1997). Dynamic genome organization and gene evolution by positive selection in geminivirus (Geminiviridae). *Mol. Biol. Evol.* **14:**1114–1124.
Brown, J. K., Idris, A. M., Alteri, C., and Stenger, D. C. (2002a). Emergence of a new cucurbit-infecting begomovirus species capable of forming viable reassortment with related viruses in the squash leaf curl virus cluster. *Phytopathology* **92:**734–742.
Brown, S. P., Hochberg, M. E., and Grenfell, B. T. (2002b). Does multiple infection select for raised virulence? *Trends Microbiol.* **10:**401–405.
Byrne, D. N., Rathman, R. J., Orum, T. V., and Palumbo, J. C. (1996). Localised migration and dispersal by the sweet potato whitefly, *Bemisia tabaci*. *Oecologia* **105:**320–328.
Chan, Man-Suen, and Jeger, M. J. (1994). An analytical model of plant virus disease dynamics with roguing and replanting. *J. Appl. Ecol.* **31:**413–427.
Chare, E. R., and Holmes, E. C. (2004). Selection pressures in the capsid genes of plant RNA viruses reflect mode of transmission. *J. Gen. Virol.* **85:**3149–3157.
Colvin, J., Omongo, C. A., Maruthi, M. N., Otim-Nape, G. W., and Thresh, J. M. (2004). Dual begomovirus infections and high *Bemisia tabaci* populations: Two factors driving the spread of a cassava mosaic disease pandemic. *Plant Pathol.* **53:**577–584.
Czosnek, H., Morin, S., Rubinstein, G., Fridman, V., Zeidan, M., and Ghanim, M. (2001). Tomato yellow leaf curl virus: A disease sexually transmitted by Whiteflies. *In* "Virus-Insect-Plant-Interactions" (K. F. Harris, O. P. Smith, and J. E. Duffus, eds.), pp. 1–27. Academic Press, New York.
Day, T., and Proulx, S. R. (2004). A general theory for the evolutionary dynamics of virulence. *Am. Nat.* **163:**E41–E63.
de Zoeten, G. A., and Skaf, J. S. (2001). Pea enation mosaic and the vagaries of a plant virus. *Adv. Virus Res.* **57:**323–350.
Dietrich, M. R. (1994). The origins of the neutral theory of molecular evolution. *J. Hist. Biol.* **27:**21–59.
Ebert, D. (1998). Infectivity, multiple infections, and the genetic correlation between within-host growth and parasite virulence: A reply to Hochberg. *Evolution* **52:**1869–1871.
Ellstrand, N. C., and Elam, D. R. (1993). Population genetic consequences of small population size: Implications for plant conservation. *Annu. Rev. Ecol. Syst.* **24:**217–242.
Epperson, B. K. (1993). Recent advances in correlation studies of spatial patterns of genetic variation. *In* "Evolutionary Biology" (M. K. Hecht *et al.*, ed.), Vol. 27, pp. 95–155. Plenum Press, New York.
Escriu, F., Fraile, A., and García-Arenal, F. (2000). Evolution of virulence in natural populations of the satellite RNA of cucumber mosaic virus. *Phytopathology* **90:**480–485.
Escriu, F., Fraile, A., and García-Arenal, F. (2003). The evolution of virulence in a plant virus. *Evolution* **57:**755–765.
Fauquet, C. M., Bisaro, D. M., Briddon, R. W., Brown, J. K., Harrison, B. D., Rybicki, E. P., Stenger, D. C., and Stanley, J. (2003). Revision of taxonomic criteria for species demarcation in the family *Geminiviridae*, and an updated list of begomorivirus species. *Arch. Virol.* **148:**405–421.

Fauquet, C. M., Sawyer, S., Idris, A. M., and Brown, J. K. (2005). Sequence analysis and classification of apparent recombinant begomoviruses infecting tomato in the Nile and Mediterranean basins. *Phytopathology* **95:**549–555.

Fisher, R. A. (1958). *The genetical theory of natural selection*, 2nd revised edition. Dover, New York.

Flint, J., Harding, R. M., Clegg, J. B., and Boyce, A. J. (1993). Why are some genetic-diseases common: Distinguishing selection from other processes by molecular analysis of globin gene variants. *Hum. Genet.* **91:**91–117.

Fraile, A., Alonso Prados, J. L., Aranda, M. A., Bernal, J. J., Malpica, J. M., and García Arenal, F. (1997a). Genetic exchange by recombination or reassortment is infrequent in natural populations of a tripartite RNA plant virus. *J. Virol.* **71:**934–940.

Fraile, A., Escrui, F., Aranda, M. A., Malpica, J. M., Gibbs, A. J., and García-Arenal, F. (1997b). A century of tobamovirus evolution in an Australian population of *Nicotiana glauca*. *J. Virol.* **71:**8316–8320.

Frank, S. A. (2000). Within-host spatial dynamics of viruses and defective interfering particles. *J. Theor. Biol.* **206:**279–290.

Frank, S. A., and Slatkin, M. (1992). Fisher's fundamental theorem of natural selection. *Trends Ecol. Evol.* **7:**92–95.

French, R., and Stenger, D. C. (2003). Evolution of *Wheat streak mosaic virus*: Dynamics of population growth within plants may explain limited variation. *Annu. Rev. Phytopathol.* **41:**199–214.

Fries, I., and Camazine, S. (2001). Implications of horizontal and vertical pathogen transmission for honey bee epidemiology. *Apidologie* **32:**199–214.

Froissart, R., Wilke, C. O., Montville, R., Remold, S. K., Chao, L., and Turner, P. E. (2004). Co-infection weakens selection against epistatic mutations in RNA viruses. *Genetics* **168:**9–19.

Frost, S. D. W., Dumourier, K.-J., Wain-Hobson, S., and Leigh Brown, A. J. (2001). Genetic drift and within-host metapopulation dynamics of HIV-infection. *Proc. Natl. Acad. Sci.* **98:**6975–6980.

Fuchs, M., Ferreira, S., and Gonsalves, D. (1997). Management of virus-diseases by classical and engineered protection. *Mol. Plant Pathol. On-Line* [http://www.bspp.org.uk/mppol/]1997/0116fuchs.

Fulton, R. W. (1986). Practices and precautions in the use of cross protection for plant virus disease control. *Annu. Rev. Phytopathol.* **24:**67–81.

Futayama-Noguchi, S. (2001). Ecophysiology of virus-infected plants: A case study of *Eupatorium makinoi* infected by geminivirus. *Plant Biol.* **3:**251–262.

Futayama, S., Terashima, I., and Yahara, T. (2001). Effects of virus infection and light environment on populations of *Eupatorium makinoi* (Asteraceae). *Am. J. Bot.* **88:**616–622.

Galvani, A. P. (2003). Epidemiology meets evolutionary ecology. *Trends Ecol. Evol.* **18:**132–139.

García-Arenal, F., Escriu, F., Aranda, M. A., Alonson-Prados, J. L., Malpica, J. M., and Fraile, A. (2000). Molecular epidemiology of *Cucumber mosaic virus* and its satellite RNA. *Virus Res.* **71:**1–8.

García-Arenal, F., Fraile, A., and Malpica, J. M. (2001). Variability and genetic structure of plant virus populations. *Annu. Rev. Phytopathol.* **39:**157–186.

García-Arenal, F., and McDonald, B. A. (2003). An analysis of the durability of resistance to plant viruses. *Phytopathology* **93:**941–952.

García-Arenal, F., Fraile, A., and Malpica, J. M. (2003). Variation and evolution of plant virus populations. *Int. Microbiol.* **6:**225–232.

Gatherer, D. (2002). Identifying the cases of social contagion using memetic isolation: Comparison of the dynamics of a multisociety simulation with an ethnographic data set. *J. Artif. Soc. Social Simul.* **5:**105–125.

Gavrilets, S. (1999). A dynamical theory of speciation on holey adaptive landscapes. *Am. Nat.* **154:**1–22.

Gavrilets, S., and Gibson, N. (2002). Fixation probabilities in a spatially heterogeneous environment. *Popul. Ecol.* **44:**51–58.

Gould, F., Kennedy, G. G., and Johnson, M. T. (1991). Effects of natural enemies on the rate of herbivore adaptation to resistant host plants. *Entomol. Exp. Appl.* **58:**1–14.

Grenfell, B. T., Pybus, O. G., Gog, J. R., Wood, J. L. N., Daly, J. M., Mumford, J. A., and Holmes, E. C. (2004). Unifying the epidemiological and evolutionary dynamics of pathogens. *Science* **303:**327–332.

Hall, J. S., French, R., Morris, T. J., and Stenger, D. C. (2001). Structure and temporal dynamics of populations within wheat streak mosaic virus isolates. *J. Virol.* **75:**10231–10243.

Hameed, S., and Robinson, D. J. (2004). Begomoviruses from mung beans in Pakistan: Epitope profiles, DNA A sequences and phylogenetic relationships. *Arch. Virol.* **149:**809–819.

Hammond, J., Lecoq, H., and Raccah, B. (1999). Epidemiological risks for mixed virus infections and transgenic plants expressing viral genes. *Adv. Virus Res.* **54:**189–314.

Harrison, B. D. (2002). Virus variation in relation to resistance-breaking in plants. *Euphytica* **124:**181–192.

Harrison, B. D., and Robinson, D. J. (1999). Natural genomic and antigenic variation in whitefly-transmitted geminiviruses (begomoviruses). *Annu. Rev. Phytopathol.* **37:**369–398.

Harrison, B. D., Zhou, X., Otim-Nape, G. W., Liu, Y., and Robinson, D. J. (1997). Role of a novel type of double infection in the geminivirus-induced epidemic of sweet cassava mosaic in Uganda. *Ann. Appl. Biol.* **131:**437–448.

Heijbroek, W., Musters, P. M. S., and Schoone, A. H. L. (1999). Variation in pathogenicity and multiplication of beet necrotic yellow virus (BNYVV) in relation to the resistance of sugar-beet cultivars. *Eur. J. Plant Pathol.* **105:**397–405.

Henneberry, T. J., and Castle, S. J. (2001). *Bemisia*: Pest status, economics, biology and population dynamics. *In* "Virus-Insect-Plant Interactions," pp. 247–278. Academic Press, New York.

Hochberg, M. E. (1998). Establishing genetic correlations involving parasite virulence. *Evolution* **52:**1865–1868.

Jeger, M. J., Holt, J., vanden Bosch, F., and Madden, L. V. (2004). Epidemiology of insect-transmitted plant viruses: Modelling disease dynamics and control interventions. *Physiol. Entomol.* **29:**291–304.

Jeger, M. J., van Den Bosch, F., Madden, L. V., and Holt, J. (1998). A model for analysing plant-virus transmission characteristics and epidemic development. *IMA J. Math. Appl. Med. Biol.* **15:**1–18.

Jones, R. A. C. (2001). Developing integrated disease management strategies against non-persistently aphid-borne viruses: A model programme. *Int. Pest Manage. Rev.* **6:**15–46.

Jones, R. A. C. (2004). Using epidemiological information to develop effective integrated virus disease management strategies. *Virus Res.* **100:**5–30.

Kaplan, N., Hudson, R. R., and Iizuka, M. (1991). The coalescent process in models with selection, recombination and geographic subdivision. *Genet. Res.* **57:**83–91.

Kearney, C. M., Thomson, M. J., and Roland, K. E. (1999). Genome evolution of tobacco mosaic virus populations during long-term passaging in a diverse range of hosts. *Arch. Virol.* **144:**1513–1526.

Kegler, H., and Schenk, G. (1990). Connections and correlations of traits and influence factors in quantitative virus resistance of plants (Review). *Arch. Phytopathol. Pflanzenschutz, Berlin* **26:**427–439.

Kelley, S. E. (1994). Viral pathogens and the advantage of sex in the perennial grass. *Anthoxanthum odoratum. Phil. Trans. R. Soc. Lond. B* **346:**295–302.

Keymer, A. E., May, R. M., and Harvey, P. H. (1990). Evolutionary epidemiology: Parasite clones in the wild. *Nature* **346:**109–110.

Koella, J. C. (1999). Evolutionary ecology and epidemiology of interactions between Anopheles mosquitoes and malaria. *Schweiz. Med. Wochen.* **129:**1106–1110.

Koella, J. C., and Restif, O. (2001). Co-evolution of parasite virulence and host life history. *Ecol. Lett.* **4:**207–214.

Lecoq, H. (1998). Control of plant virus diseases by cross-protection. *In* "Plant Virus Disease Control" (A. Haddidi, R. K. Khetarpal, and H. Koganezawa, eds.), pp. 33–40. APS press, st. Paul, MN, USA.

Lin, H. X., Rubio, L., Smythe, A. B., and Falk, B. W. (2004). Molecular population genetics of *Cucumber mosaic virus* in California: Evidence for founder effects and reassortment. *J. Virol.* **78:**6666–6675.

Loxdale, H. D., Hardie, J., Halbert, S., Footit, R., Kidd, N. A. C., and Carter, C. I. (1993). The relative importance of short and long-range movement of flying aphids. *Biol. Rev.* **68:**291–311.

Luria, S. E. (1951). The frequency distribution of spontaneous bacteriophage mutants as evidence for the exponential rate of phage reproduction. *Cold Spring Harbour Symp. Quant. Biol.* **16:**463–470.

McDonald, B. A., and Linde, C. (2002a). The population genetics of plant pathogens and breeding strategies for durable resistance. *Euphytica* **124:**163–180.

McDonald, B. A., and Linde, C. (2002b). Pathogen population genetics, evolutionary potential, and durable resistance. *Ann. Rev. Phytopathol.* **40:**349–379.

McElhony, P., Real, L. A., and Power, A. G. (1995). Vector preference and disease dynamics: A study of barley yellow dwarf virus. *Ecology* **76:**444–457.

Madden, L. V., Jeger, M. J., and van den Bosch, F. (2000). A theoretical assessment of the effects of vector-virus transmission mechanism on plant virus disease epidemics. *Phytopathology* **90:**576–594.

Malpica, J. M., Fraile, A., Moreno, I., Obies, C. I., Drake, J. W., and García-Arenal, F. (2002). The rate and character of spontaneous mutation in an RNA virus. *Genetics* **162:**1505–1511.

Maruthi, M. N., Colvin, J., Seal, S., Gibson, G., and Cooper, J. (2002). Co-adaptation between cassava mosaic geminiviruses and their local vector populations. *Virus Res.* **86:**71–85.

Matthews, R. E. F. (1993). Overview. *In* "Diagnosis of Plant Virus Disease" (R. E. F. Matthews, ed.), pp. 1–14. CRC Press, Boca Raton.

May, R. M., and Hassell, M. P. (1988). Population dynamics and biological control. *Phil. Trans. R. Soc. London B* **318:**129–169.

Mayer, R. T., Inbar, M., McKenzie, C. L., Shatters, R., Borowicz, V., Albrecht, V., Powell, C. A., and Doostdar, H. (2002). Multitrophic interactions of the silverleaf whitefly, host plants, competing herbivores, and phytopathogens. *Arch. Ins. Biochem. Physiol.* **51:**151–169.

Méndez-Lozano, J., Torres-Pacheco, I., Fauquet, C. M., and Rivera-Bustamante (2003). Interactions between geminiviruses in a naturally occurring mixture: *Pepper huastecovirus* and *Pepper golden mosaic virus*. *Phytopathology* **93:**270–277.

Monci, F., Sánchez-Campos, S., Navas-Castillo, J., and Moriones, E. (2002). A natural recombinant between the geminiviruses *Tomato yellow leaf curl Sardinia virus* and *Tomato yellow leaf curl virus* exhibits a novel pathogenic phenotype and is becoming prevalent in Spanish populations. *Virology* **303:**317–326.

Moreno, I. M., Malpica, J. M., Diaz-Pendon, J. A., Moriones, E., Fraile, A., and García-Arenal, F. (2004). Variability and genetic structure of the population of watermelon mosaic virus infecting melon in Spain. *Virology* **318:**451–460.

Morilla, G., Krenz, B., Jesloe, H., Bejarano, E. R., and Wege, C. (2004). Tête à tête of *Tomato yellow leaf curl virus* and *Tomato yellow leaf curl Sardinia virus* in single nuclei. *J. Virol.* **78:**10715–10723.

Morin, S., Ghanim, M., Zeidan, M., Czosnek, H., Verbeek, M., and van den Heuvel, F. J. M. (1999). A GroEL homologue from endosymbiotic bacteria of the whitefly *Bemisia tabaci* is implicated in the circulative transmission of tomato yellow leaf curl virus. *Virology* **256:**75–84.

Moya, A., Holmes, E. C., and Gonzalez-Candelas, F. (2004). The population genetics and evolutionary epidemiology of RNA viruses. *Nature Rev. Microbiol.* **2:**279–288.

Nagy, P. D., Zhang, C. X., and Simon, A. E. (1998). Dissecting RNA recombination *in vitro*: Role of RNA sequences and the viral replicase. *EMBO J.* **17:**2392–2403.

Nee, S. (2000). Mutualism, parasitism and competition in the evolution of coviruses. *Phil. Trans. R. Soc. Lond. B* **355:**1607–1613.

Nevo, E. (1998). Molecular evolution and ecological stress at global, regional and local scales: The Israeli perspective. *J. Exp. Zool.* **282:**95–119.

Ooi, K., Ohshita, S., Ishii, I., and Yahara, T. (1997). Molecular phylogeny of geminivirus infecting wild plants in Japan. *J. Plant Res.* **110:**247–257.

Owor, B., Legg, J. P., Okao-Okuja, G., Obonyo, R., Kyamanywa, S., and Ogenga-Latigo, M. W. (2004). Field studies of cross protection with cassava mosaic geminiviruses in Uganda. *J. Phytopathol.* **152:**243–249.

Padidam, M., Beachy, R. N., and Fauquet, C. M. (1995). Classification and identification of geminiviruses using sequence comparisons. *J. Gen. Virol.* **76:**249–263.

Pita, J. S., Fondong, V. N., Sangare, A., Otim-Nape, G. W., Ogwal, S., and Fauquet, C. M. (2001). Recombination, pseudo-recombination and synergism of geminiviruses are determinant keys to the epidemic of the cassa mosaic disease in Uganda. *J. Gen. Virol.* **82:**655–661.

Power, A. G. (1992). Patterns of virulence and benevolence in insect-borne pathogens of plants. *Crit. Rev. Plant Sci.* **11:**351–372.

Power, A. (1996). Competition between viruses in a complex plant-pathogen system. *Ecology* **77:**1004–1010.

Power, A. (2000). Insect transmission of plant viruses: A constraint on virus variability. *Curr. Opin. Plant Biol.* **3:**336–340.

Qiu, W. P., and Moyer, J. W. (1999). Tomato spotted wilt tospovirus adapts to the TSWV N gene-derived resistance by genome. *Phytopathology* **89:**575–582.

Restif, O., and Koella, J. C. (2003). Shared control of epidemiological traits in a co-evolutionary model of host-parasite interactions. *Am. Nat.* **161:**827–836.

Restif, O., and Koella, J. C. (2004). Concurrent evolution of resistance and tolerance to pathogens. *Am. Nat.* **164:**E90–E102.

Roossinck, M. J. (1997). Mechanisms of plant virus evolution. *Ann. Rev. Phytopathol.* **35:**191–209.
Sacristan, S., Malpica, J. M., Fraile, A., and García-Arenal, F. (2003). Estimation of population bottlenecks during systemic movement of *Tobacco mosaic virus* in tobacco plants. *J. Virol.* **77:**9906–9911.
Sacristan, S., Fraile, A., and García-Arenal, F. (2004). Population dynamics of *Cucumber mosaic virus* in melon crops and in weeds in central Spain. *Phytopathology* **94:**992–998.
Sanjuan, R., Moya, A., and Elena, S. F. (2004a). The distribution of fitness effects caused by single-nucleotide substitutions in an RNA virus. *Proc. Natl. Acad. Sci.* **101:**8396–8401.
Sanjuan, R., Moya, A., and Elena, S. F. (2004b). The contribution of epistasis to the architects of fitness in an RNA virus. *Proc. Natl. Acad. Sci.* **101:**15376–15379.
Sanz, A. I., Fraile, A., García-Arenal, F., Zhou, X., Robinson, D. J., Khalid, S., Butt, T., and Harrison, B. D. (2000). Multiple infection, recombination and genome relationships among begomoviruses found in cotton and other plants in Pakistan. *J. Gen. Virol.* **81:**1839–1849.
Saunders, K., Bedford, I. D., and Stanley, J. (2002). Adaptation from whitefly to leaf hopper transmission of an autonomously replicating nanovirus-like DNA component associated with ageratum yellow vein disease. *J. Gen. Virol.* **83:**907–913.
Seal, S. E., Vandenbosch, F., and Jeger, M. J. (2006). Factors influencing begomovirus evolution and their increasing global significance: Implications for sustainable control. *Crit. Rev. Plant Sci.* **25:**23–46.
Shackelton, L. A., Parish, C. R., Truyen, V., and Holmes, E. C. (2005). High rate of viral evolution associated with the emergence of carnivore parvovirus. *Proc. Natl. Acad. Sci.* **102:**379–384.
Simón, B., Cenis, J. L., Beitia, F., Khalid, S., Moreno, I., Fraile, A., and García-Arenal, F. (2003). Genetic structure of field populations of begomoviruses and of their vector Bemisia tabaci in Pakistan. *Phytopathology* **93:**1422–1429.
Sseruwagi, P., Rey, M. E. C., Brown, J. K., and Legg, J. P. (2004). The cassava mosaic geminiviruses occurring in Uganda following the 1990s epidemic of severe cassava mosaic disease. *Ann. Appl. Biol.* **145:**113–121.
Tepfer, M. (2002). Risk assessment of virus-resistant transgenic plants. *Ann. Rev. Phytopathol.* **40:**467–491.
Thresh, J. M., Fargette, D., and Jeger, M. J. (2003). Epidemiology of tropical plant viruses. *In* "Virus and Virus-like Diseases of Major Crops in Developing Countries" (G. Lobenstein and G. Thottappilly, eds.), pp. 55–77. Kluwer Academic Publishers, The Netherlands.
Van den Bosch, F., Akudibilah, G., Seal, S. E., and Jeger, M. J. (2006). Host resistance and the evolutionary response of plant viruses. *J. Appl. Ecol.* (in press).
Van den Heuvel, J. F. J. M., Bruyere, A., Hogenhout, S. A., Ziegler-Graff, V., Brault, V., Verbeek, M., Vanderwilk, F., and Richards, K. (1997). The N-terminal region of the luteovirus read through domain determines virus binding to Buchnera Gro EL and is essential for virus persistence in the aphid. *J. Virol.* **71:**7258–7265.
Wakeley, J. (2003). Polymorphism and divergence for island-model species. *Genetics* **163:**411–420.
Wakeley, J., and Takahashi, T. (2004). The many-demes limit for selection and drift in a subdivided population. *Theor. Pop. Biol.* **66:**83–91.
Wallace, B. (1993). Towards a resolution of the neutralist-selectionist controversy. *Persp. Biol. Med.* **36:**450–459.

White, K. A., and Morris, T. J. (1995). RNA determinants of junction site selection in RNA virus recombinants and defective interfering RNAs. *RNA-A* **1:**1029–1040.
Wilson, D. R. (1992). Evolutionary epidemiology. *Acta Biotheor.* **40:**87–90.
Yeh, S. D., and Gonsalves, D. (1984). Evaluation of induced mutants of papaya ringspot virus for control by cross protection. *Phytopathology* **74:**1086–1091.
Ziegler, R. S., and Morales, F. J. (1990). Genetic determination of replication of rice hoja blanca virus within its planthopper vector, *Sogatodes oryzicola*. *Phytopathology* **80:**559–566.

ADVANCES IN VIRUS RESEARCH, VOL 67

CONTROL OF PLANT VIRUS DISEASES

Roger A. C. Jones*,†

*Agricultural Research Western Australia, Locked Bag No. 4
Bentley Delivery Centre, WA 6983, Australia
†School of Plant Biology, Faculty of Natural and Agricultural Sciences
University of Western Australia, 35 Stirling Highway, Crawley
WA 6009, Australia

The importance of establishing the effectiveness and reliability of control measures against plant virus epidemics, dissecting out the ways in which they operate and understanding how to use such information within integrated disease management (IDM) approaches is emphasised. An increasingly sophisticated range of control measures is becoming available to meet the challenge that virus epidemics pose to achieving satisfactory yields and quality of plant produce. Some are non-selective affecting a wide range of viral pathogens or their vectors, but others are selective targeting a particular virus or vector species. Control measures also differ in activity by targeting either the initial source of inoculum, which may be external or internal, or the rate of virus spread, which may be influenced at early or late stages of the epidemic. In general, cultural (phytosanitary and agronomic) and legislative measures are non-selective, whereas host resistance and most biological control measures are selective. Chemical control measures have low selectivity when a general pesticide is applied, but high selectivity when a selective pesticide is used. A drawback to selective measures is that they can cause inadvertent selection of variants of virus or vector that are more difficult to control. Host resistance, chemical and biological control measures decrease the rate of virus spread but

0065-3527/06 $35.00
DOI: 10.1016/S0065-3527(06)67006-1

are inactive against the initial virus source, whereas legislative and phytosanitary control measures and many agronomic measures address both source and spread. When combining control measures within IDM, success is optimized by including ones that are selective and others that are non-selective, that address both kinds of sources and phases of virus spread, and that operate in as many different ways as possible. Decisions over which combinations to include also depend on factors such as economic constraints, compatibility with standard agronomic practices and current measures against other threats, the likelihood of adoption, environmental issues and social concerns. The extent and magnitude of the economic losses resulting from virus epidemics are often underestimated, but even small virus-induced declines in yield or quality may damage profitability considerably. Therefore, deploying IDM in such circumstances can still provide a substantial boost to the revenue obtained. When 'interim' IDM strategies are devised from available epidemiological knowledge and 'generic' information on control measures to address short-term needs, subsequent validation is essential. Although much knowledge on how to control virus diseases has accumulated, success in addressing them is hindered by increasing difficulties and the overall situation continues to deteriorate. Fortunately, if employed intelligently, flexibly and innovatively, sophisticated and rapidly improving new technologies enable more effective use of available information, skills and experience when taking decisions over which combinations of measures to use.

I. Introduction

Virus epidemics occur in natural plant communities, including ones containing species that are ancestors of cultivated plants. However, spatial separation (isolation) between similar wild plant communities diminishes the inoculum sources needed to initiate epidemics, abundance of vectors is diminished by predators and parasites, and admixture with other wild plant species that are non-hosts decreases virus spread to susceptible hosts. Also, co-evolution of such host plants with viruses results in selection of natural host resistances and tolerances to viruses and vectors. Collectively, isolation, predators and parasites, admixture with non-hosts and host resistance/tolerance serve as natural control measures that help to prevent damaging virus epidemics. By contrast, domestication of plants and their widespread cultivation in monocultures has provided conditions suitable for frequent and damaging epidemics. Moreover, man's movement of cultivated plants away from their centres of origin disperses previously localised viruses and their vectors widely. It also exposes the introduced plants to 'new encounter' situations as they come into contact with viruses emerging from indigenous wild plants or already established in other cultivated plants. As such viruses have not co-evolved with wild ancestors of the introduced cultivated plants, damaging epidemics are likely to arise (Bos, 1992; Buddenhagen, 1977; Thresh, 1980, 1981, 1982).

Over the last 10,000–15,000 years since plants were first domesticated in their centres of origin in different parts of the world (Harlan, 1971, 1981; Lovisolo *et al.*, 2003), mankind has intervened to ameliorate the damage caused by virus epidemics in cultivated plants. This has been done by adopting practices that serve to ensure reliability of production of food for humans or domestic animals, fibre and raw materials used for fabrics, constructions or fuels, and plants used for ornamental and medicinal purposes. Since early plant domestication, practices that helped to control virus epidemics in subsistence agriculture included selection of propagules from the most vigorously growing plants for future planting, inter-planting of unrelated crops which are often non-hosts, use of cover crops, crop rotation and weeding. Selection of propagules also includes deliberately selecting 'land races' that produce more reliably than others in terms of yield and quality. Many of these 'land races' retain or concentrate virus and vector resistances from the wild ancestors of cultivated plants (e.g. Jones, 1981; Thresh, 1982). Until the 20th century, practices that helped to control virus epidemics were adopted without knowledge of the viruses involved or how they spread. Since then, however, knowledge has accumulated on what plant viruses are, how their epidemics develop, why they diminish the productivity of cultivated plants, and what factors limit their development and the damage they cause. This knowledge has fostered development of increasingly sophisticated and diverse types of control measures directed at removing virus sources or suppressing virus spread. Understanding the factors that restrict virus epidemics in natural plant communities and primitive subsistence cropping systems has helped in this process.

Virus control measures act by minimising the initial source(s) of virus infection (x_o), which delays the onset of virus spread, the rate of virus spread in the stand (r) (Zadoks and Schein, 1979) or both, thereby diminishing any impact on yield or quality of product. There are measures based on host resistance, cultural, chemical or biological means, and ones based on legislation. Here, the term 'cultural control' incorporates a diverse grouping of phytosanitary and agronomic measures that act to suppress virus epidemics in many different ways. A narrower definition of 'cultural control' that excludes phytosanitary measures is sometimes used by others (e.g. Thresh, 2003, this volume, pp. 245–295). Some virus control measures are generic, while others are so specific that they only apply to particular pathosystems in certain agro-ecological situations. This chapter does not provide a comprehensive account of the different types of control measures available, as this has been done previously (e.g. Broadbent, 1964; Jones, 2004a; Thresh, 1982, 1983, 2003; Zitter and Simons, 1980). Instead, it focuses on determining how diverse types of control measures operate and how to use such information about each measure when choosing which to combine together in IDM approaches. It also discusses the effectiveness and reliability of control measures, economic considerations in taking decisions about deploying them, current difficulties hindering success and opportunities to improve their effectiveness arising from new technology. Examples are drawn mainly from studies on virus control measures not used in previous reviews or by Thresh (this volume, pp. 245–295) in considering tropical virus diseases.

II. Effectiveness and Reliability of Control Measures

Control measures differ in their effectiveness in suppressing virus epidemics. At one extreme, although effective with one pathosystem or situation, a measure may be completely ineffective with another (Fig. 1). Inappropriate use of chemical control, apart from being environmentally undesirable, may actually accelerate an epidemic, for example, when pesticides without rapid knock-down properties stimulate migrating insect vectors of non-persistently transmitted viruses to move rapidly between plants and probe each of them in turn, or an insect vector population increases markedly following destruction of its natural enemies by pesticides (e.g. Perring *et al.*, 1999; Raccah, 1986). At the other extreme, a measure might be so effective that no infected plants develop and the epidemic is prevented or all plants infected previously are killed and no more appear, so the epidemic is eliminated. Unfortunately, this degree of effectiveness is rarely achieved in practice even when diverse control measures that act in different ways are combined and deployed simultaneously. For an epidemic to be stabilized, a control measure must prevent all further virus spread without killing and so removing previously infected plants. In most

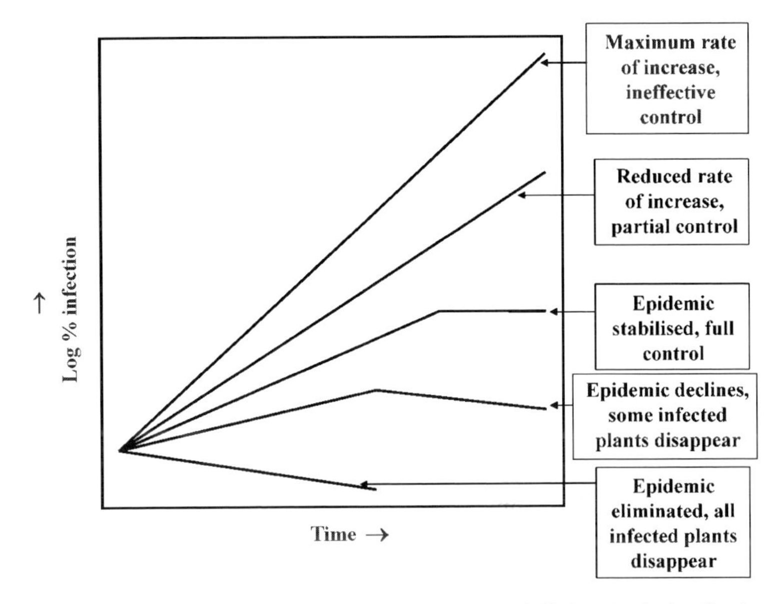

FIG 1. Virus epidemic suppression by control measures differing in their effectiveness.

instances, however, the farmer has to be satisfied with measures that suppress virus spread partially.

For various reasons, particular control measures sometimes fail to perform as well as expected when used alone. Thus, cultivars with single gene virus resistance are overcome when they encounter resistance-breaking strains of the virus. Moreover, chemical control can select vector populations with pesticide resistance or kill beneficial predators or parasites that help suppress vectors, and biological control of vectors by predators may not operate where the vector population is low. Also, a drawback to host resistance is that farmers generally prefer to sow better performing cultivars even if they are more susceptible than virus-resistant ones lacking other desirable traits such as high yield or improved quality (Jones, 2004a; Thresh, 2003). Moreover, a particular type of control measure may be unsuitable for environmental or socio-economic reasons. For example, chemical control may cause build up of toxic residues that are harmful to mankind, domestic animals and wildlife or there is unforeseen accumulation of damaging pests or other pathogens (Perring *et al.*, 1999; Raccah, 1986), and its use is prohibited entirely in true 'organic' production systems.

Obtaining information on the reliability and effectiveness of control measures used alone or in combination is best done with replicated field experiments. The control measures need to be deployed in the way that they would be used commercially, or as close to this as possible. However, unless the natural reservoir of virus infection is ample at the location concerned, it is usually necessary to introduce initial virus inoculum, for example, as infector plants transplanted in a uniform pattern that provides even exposure to ensure that sufficient virus spread occurs. Data on virus incidence are collected from experimental plots in the presence or absence of control measures and the data sets compared using appropriate statistical analyses, for example, when chemical control measures consisting of drenches applied to seedlings were used recently with *Tomato spotted wilt virus* (TSWV) in lettuce (Coutts and Jones, 2005b). In addition to data on virus incidence, wherever possible, data should also be collected on the effect of the control measures on yield and quality (e.g. Bwye *et al.*, 1994; Jones, 1993, 1994a; Jones and Nicholas, 1992, 1998; McKirdy and Jones, 1996, 1997a; McLean *et al.*, 1982). A drawback to field experiments on control measures is that they are expensive and need to be repeated in different years at representative sites in different agro-ecologies to ensure reliability. Where sufficient epidemiological information is available for the pathosystem concerned, a cheaper alternative is to use epidemiological models that identify which control strategies are likely to be effective for a particular pathosystem, as used with *Tomato leaf curl virus* (ToLCV) in India (Holt *et al.*, 1999), but subsequent validation is necessary. Preferably, this should involve replicated field experiments to evaluate the performance of the control measure or combination of measures identified by the model. When this is not feasible due to high cost or time constraints, collecting epidemiological information from naturally occurring virus epidemic situations or large, un-replicated plots is helpful (see Section IV).

III. Dissecting How Control Measures Operate

A. *Selectivity*

Different types of virus control measures differ widely in selectivity. Some are non-selective, acting broadly as 'blunt instruments' that affect a wide range of viral pathogens or their vectors. Others are selective, accurately targeting a particular virus or vector species, the former operating like a 'shot gun' and the latter with the accuracy of a snipers 'rifle'. In general, cultural and legislative measures are non-selective, whereas host resistance and most biological control measures are selective (Table I). Moreover, host resistance and biological control measures are sometimes so selective that they 'pinpoint' just one virus strain, as with strain-specific single gene host resistance (e.g. Fraser, 1986) or cross protection involving infecting plants with a mild stain to protect against a severe one, as with *Citrus tristeza virus* (CTV) in citrus (e.g. Bar-Joseph *et al.*, 1983). Chemical control measures have low selectivity when a general pesticide that kills several vector species is applied, but high selectivity when a more specific pesticide kills only one vector species. Although agronomic control measures are always non-selective, phytosanitary and legislative measures become selective if they are targeted at only one pathogen. For example, 'healthy' propagule or stock programmes that focus on just one virus, as with *Cucumber mosaic virus* (CMV) in lupin seed (Jones, 2000) or *Lettuce mosaic virus* in lettuce seed (Grogan, 1980), or when roguing concentrates only on removing plants with characteristic symptoms of a particular virus, as in removing *Cacao swollen shoot virus* (CSSV)-diseased trees in cocoa plantations (Thresh, 2003). In contrast, biological control with a pathogen, parasite or predator becomes non-selective if an agent that kills several vectors is used such that several viruses are controlled simultaneously.

When deployed alone, selective control measures are sometimes very effective in suppressing virus epidemics. However, there is a drawback to them that does not apply with non-selective cultural and legislative measures other than roguing. It is that deploying them widely can result in selection, albeit unwittingly, of variants of virus or vector which are more difficult to control. This is so particularly where, under heavy inoculum pressure, exclusive use of cultivars with single gene resistance to virus or vector selects resistance-breaking forms of either (e.g. Harrison, 2002), more damaging recombinants are generated through mixing of virus strains within infected plants during cross protection or overuse of biopesticides selects vector variants that are resistant to them. A well publicized example of this type of problem is the resistance-breaking strains of TSWV that develop readily when cultivars of tomato or pepper with single gene resistance are grown intensively and no other control measure is deployed (e.g. Latham and Jones, 1998; Moury *et al.*, 1997; Roggero *et al.*, 2002; Thomas-Carroll and Jones, 2003). Similarly, when stringent roguing removes symptomatic plants of one or more viruses, asymptomatic variants may be selected that cannot be controlled by this measure, as has often occurred with *Potato virus X* (PVX) in potato healthy stock schemes in different

TABLE I

EXAMPLES OF VIRUS CONTROL METHODS: THEIR SELECTIVITY AND ACTIVITY AGAINST THE INITIAL VIRUS SOURCE VERSUS VIRUS SPREAD

		Selectivity		Initial source (x_o)[a]		Rate of spread (r)	
Method	Measure	Low	High	External	Internal	Early	Late
Host resistance	Partial (partial resistance to infection, strain specific)	–	+	–	–	+	–
	Complete (comprehensive resistance)	–	+	–	–	+	+
Chemical	Specific, regular foliar application	–	+	–	–	+	+
	General, regular foliar application	+	–	–	–	+	+
	Specific, at, before or directly after planting	–	+	–	–	+	–
	General, at, before or directly after planting	+	–	–	–	+	–
	Oils and repellents, regular foliar application	+	–	–	–	+	+
Cultural–phytosanitary	Hygiene	+	–	+	+	+	–
	Roguing	+	+	–	+	+	–
	Healthy propagules	+	+	–	+	+	–
Cultural–agronomic	Isolation, safe planting distances	+	–	+	–	+	–
	Plant upwind, non-host barrier, large field size	+	–	+	–	+	–

	Mixture with non-host	+	−	+	+	+	+
	Manipulate sowing date	+	−	−	−	+	−
	Groundcover, reflective surfaces	+	−	−	−	+	−
	Early canopy cover, high plant density, narrow row spacing	+	−	−	+	−	+
	Manipulate grazing or mowing	+	−	−	+	−	+
	Early harvest, early maturing cultivar	+	−	−	−	−	+
	Crop and weed free period, crop rotation	+	−	+	+	+	−
Biological	Cross protection	−	+	−	−	+	+
	Specific predator, parasite or pathogen	−	+	−	−	−	+
	General predator, parasite or pathogen	+	−	−	−	−	+
	Biopesticide	−	+	−	−	−	+
Legislation	Quarantine	+	−	+	−	+	−
	Healthy stock	+	+	−	+	+	−

[a] Initial amount of inoculum.

countries (e.g. Wilson and Jones, 1990, 1995). Moreover, regardless of whether it operates selectively or non-selectively, chemical control has the drawback that applying particularly the older generation pesticides (pyrethroids, carbamates and organphosphates), selects for pesticide resistance readily in some virus vectors, including the whitefly *Bemisia tabaci*, thrips *Frankliniella occidentalis* and aphid *Myzus persicae* (e.g. Helyer and Brobyn, 1992; Perring *et al.*, 1999; Thackray *et al.*, 2000). Therefore, single gene host resistance, biological and chemical control measures are best not relied on alone. A safer approach is to deploy them as components of IDM strategies that also include cultural and legislative control measures that are non-selective, thereby not only restricting the selection processes acting on virus or vector populations but also optimizing the effectiveness of the virus epidemic suppression achieved (see Section IV).

B. Activity

In addition to differing in selectivity, the diverse types of control measures available act in different ways. As mentioned in Section I, they target either the initial source of virus infection (x_o), or the rate of virus spread in the stand (r) or both. In addition, the initial source they address may be external or internal, and spread can be influenced at early or late stages of the epidemic, or at both stages (Table I).

Host resistance, chemical and biological virus control measures decrease the rate of spread, but do not affect the initial source. They are most effective against spread that occurs early, when cultivars with partial host resistance to virus infection are used or systemic pesticides are applied to kill vectors at, before or directly after planting time. They are most effective against late spread when biological control agents consisting of predators, parasites and pathogens of vectors or biopesticides are deployed. Alternatively, they may be effective throughout the epidemic, as when cultivars with host resistance that is not strain-specific are used against viruses, foliar pesticides, oil sprays and vector repellents are applied at regular intervals to the stand, or cross-protection involving early infection with a mild virus strain is employed.

Legislative and phytosanitary virus control measures and many agronomic measures address both source and spread, in many instances the influence of such measures on spread being a direct consequence of targeting the initial source of virus inoculum. Here, the source addressed may be:

- External, as with quarantine measures and isolation, planting upwind, deploying a non-host barrier crop and using large fields with small perimeter to area ratios.
- Internal, as with roguing, planting healthy propagules, early canopy cover, high plant density, narrow row spacing and manipulation of defoliation pressure by grazing and mowing.
- Both external and internal, as with hygiene involving removal of infected weeds or volunteer plants within and outside a stand, mixture with a non-host, employing crop and weed free periods and crop rotation.

The spread affected by such measures is mostly early where the source addressed is external, as with phytosanitary and legislative measures and also the agronomic measures of isolation, planting upwind, large field size and planting a non-host barrier crop. However, where the source affected is internal, late spread is often influenced instead, as with early canopy cover, high plant density, narrow row spacing and manipulating defoliation dates within single species swards. Crop and weed-free periods and crop rotation help remove both external and internal sources, but are more effective in diminishing early than late spread. Depending how it is deployed, mixture with a non-host can act against both types of sources and phases of spread. Some agronomic control measures that only address spread are most active early, as with manipulating sowing date to avoid exposing vulnerable young plants to infection at peak vector population times and using groundcover or reflective mulch to repel incoming insect vectors before foliage growth covers the ground. Others are more active against late spread, as when the growing period is shortened by planting early maturing cultivars or harvesting early.

As mentioned previously in this section, partial host resistance mainly slows early virus spread. This is because once numbers of infected plants increase sufficiently to provide a substantial internal, secondary virus source, partial resistance becomes less effective, as with *Subterranean clover mottle virus* (SCMoV) in grazed pasture swards (Fig. 2; Ferris *et al.,* 1996). Similarly, although systemic pesticides applied to soil as granules or drenches at, before or directly after sowing or to seed as seed dressings are effective against virus spread by vectors early in the growth of a stand, unlike repeated foliar applications, they become less effective subsequently. This is because pesticide concentration gradually declines as plants grow larger, for example, with *Barley yellow dwarf virus* (BYDV) in cereals (McKirdy

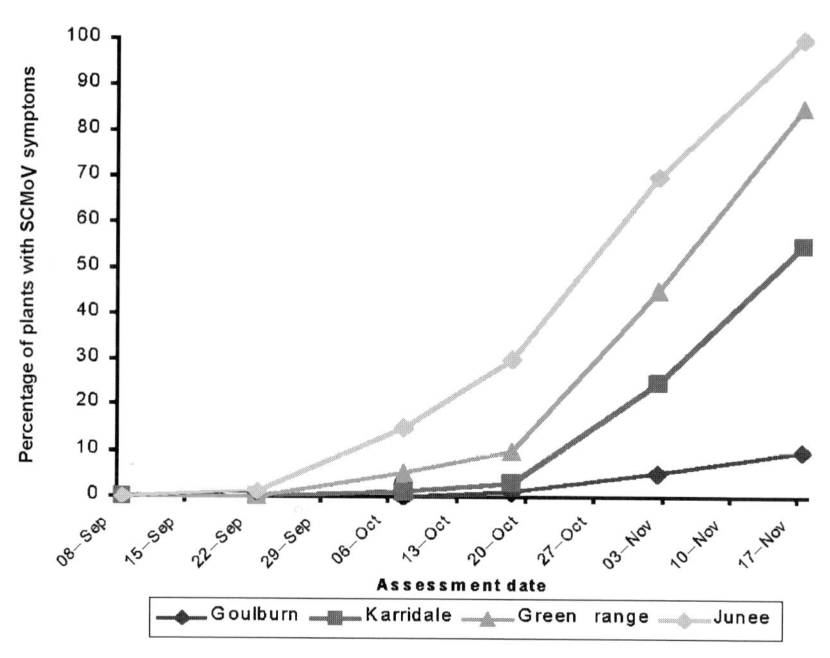

FIG 2. Disease progress curves for infection with the contact-transmitted virus SCMoV in subterranean clover cultivars with partial resistance. Virus vectored by grazing sheep. Cvs Goulburn, Karridale and Green range have different degrees of partial resistance; Junee is susceptible (Ferris *et al.,* 1996).

and Jones, 1996, 1997a; Thackray *et al.,* 2005). If, however, a crop is short-lived, they may last long enough to provide protection throughout the growing period, as when seedling drenches with insecticides were used to control TSWV in lettuce (Coutts and Jones, 2005b). Biological control agents and biopesticides generally require populations of the vectors they target to increase substantially before they become effective. This is because the agents responsible need to build up first in the vector population before sufficient are present to kill enough vectors to impact noticeably on virus spread and this takes time, as when predator *Orius* spp. are used to control *F. occidentalis* in protected crops (e.g. Cook *et al.,* 1996; Tavella *et al.,* 1996). In contrast, when cross protection is used in biological control, it is effective early as well as late when young plants are infected early with a mild strain to protect against later infection with a severe one, for example, with CTV in citrus plantations (Bar-Joseph *et al.,* 1983).

Quarantine measures that prohibit importation of virus-diseased plant materials to a country or region tackle the external virus source in its country or region of origin. The same applies if an introduced virus becomes established locally and cannot be eradicated. Here, legislative measures often concentrate on containing the incursion by banning movement of potentially diseased plant materials or by establishing healthy stock programmes to avoid distributing diseased planting stock.

Hygiene is a crucial phytosanitary control measure for many pathosystems. It helps diminish early virus spread by removing potential external and internal virus sources in many different ways, as with *Carrot virus Y* (CarVY) in carrots by destroying 'volunteer' carrot plants, removing cull piles, prompt removal of old crops, avoiding overlapping sowings and successive side-by-side plantings, and instigating carrot-free periods (Latham and Jones, 2004a). Where virus symptoms are conspicuous, roguing helps diminish early spread by removing internal virus sources. It is a practical proposition where stand-size is small and sufficient semi-skilled labour is available to identify and remove all the symptomatic plants before much virus spread occurs, as with daily roguing to remove infection with cucurbit potyviruses in zucchini and squash plantings in Northern Australia (Coutts and Jones, 2005a). When numerous primary infection foci are scattered at random, they constitute a potent virus source from which an epidemic can develop simultaneously throughout a stand. Using healthy vegetative propagules or seed minimises such internal sources, thereby, diminishing early virus spread in particular (e.g. Thresh, 1982, 2003). There are numerous examples from different parts of the world but one of the earliest and best known is healthy seed tuber production of potato (e.g. de Bokx, 1972). An example demonstrating the long-term benefits from planting healthy seed to avoid internal virus infection sources and, thereby, suppress early virus spread comes from a field experiment in which plots of annual burr medic (*Medicago polymorpha*) separated by non-host barriers (annual ryegrass) were sown with healthy or *Alfalfa mosaic virus* (AMV)-infected seed of two cultivars and the annually self-regenerating swards were grazed and monitored for infection. At the end of the brief annual growing season in the sixth year of the experiment, the incidence of AMV was still only 0.1–0.3% in plots originally sown with healthy seed but 25–47% in plots sown with infected seed. In the former, the growing period was too short for late spread from the latter to infect sufficient annual medic plants to cause any significant annual carry over via newly produced AMV-infected seeds (Jones and Nicholas, 1998).

The agronomic measures of planting upwind and deploying non-host barrier crops address external virus sources by diminishing viruliferous vector arrivals in plantings, thereby, slowing the early phases of virus epidemics. The effect of barriers is greatest if they are positioned at the windward edge closest to the source. With non-persistently aphid-borne viruses, the explanation is that when incoming migrant aphids probe non-host plants whilst in search of their preferred hosts, they lose a non-persistently aphid-borne virus from their mouthparts, thereby diminishing the amount of virus introduced once they arrive at the crop, as when non-host perimeter barriers protect lupin crops from *Bean yellow mosaic virus* (BYMV) spreading from nearby infected pasture (Jones, 1994a, 2001, 2005). Similarly, with persistently transmitted, non-propagative viruses, if the viruliferous insect vector stays and colonizes the non-host, by the time its progeny move on to the susceptible crop they are no longer viruliferous, as when the non-host cabbage barrier separated an external TSWV source from a lettuce planting (Coutts *et al.,* 2004). Thus, such non-host barriers act not just as 'physical barriers' but also as 'virus-cleansing barriers'. Planting large compact stands with low perimeter: area ratios rather than small ones in which the perimeter constitutes a substantial component of the stand, minimises the proportion of vulnerable peripheral areas, as with CSSV in cocoa plantations (Thresh *et al.*, 1988).

Agronomic measures active against internal virus sources operate by smothering over or crowding out within-crop infection foci. However, because it takes time for neighbouring healthy plants to grow sufficiently for this to occur, such measures mainly decrease late rather than early virus spread. Examples include when early canopy cover arising from planting early, sowing at high seeding rates or sowing at narrow row spacing is used to suppress CMV and BYMV spread in lupin stands (Bwye *et al.*, 1999; Jones, 1993, 2001), or when grazing or mowing is stopped to suppress BYMV or SCMoV spread in clover swards (Jones, 2004a). In addition to minimising the virus source, these measures may also suppress late spread of insect-transmitted viruses by influencing vector behaviour. This is because, once a canopy has formed, they decrease the landing rates of incoming insect vectors many of which, such as aphids and thrips species, are attracted to alight by open stands with bare earth visible between rows, but repelled by dense canopy cover (e.g. Jones, 2001; Thresh, 1982, 2003). Similarly, by influencing transmission, relaxing grazing or mowing pressure further diminishes late spread of contact-transmitted viruses within pasture swards, as with SCMoV and *White clover mosaic virus* (WCMV) in clover pasture (Coutts and Jones, 2002; Jones, 2004a). Some agronomic measures that are source-inactive but decrease early virus spread use groundcover to repel incoming viruliferous vectors. However, such measures are effective only in the period before canopy closure, as with stubble retention to suppress BYMV spread in lupin (Jones, 1994a). Where reflective surfaces are deployed against incoming vectors, the degree of repulsion is further magnified, as when reflective mulch suppresses spread of *Zucchini yellow mosaic virus* (ZYMV) by aphids in cucurbit crops (e.g. McLean *et al.*, 1982). Manipulation of sowing date can also decrease vector landings in the vulnerable early growth period of a crop. Here, the sowing dates selected are ones that avoid exposing vulnerable young plants at peak vector population times, as with CarVY in carrots (Latham and Jones, 2004a). Agronomic measures that shorten the growing period can be used to help minimise the late, exponential phase of virus spread, as with early harvesting or early foliage (haulm) destruction of seed potato crops to minimise virus spread to developing tubers (e.g. de Bokx, 1972). Planting early maturing cultivars helps because virus epidemics take time to build up within crops and physiologically old plants are generally more resistant to infection than young plants ('mature plant resistance'). Also, it is possible to time the sowing of short-duration cultivars so that they approach maturity and express 'mature plant resistance' to virus infection during the annual period of greatest vector activity.

Crop rotation is a generic agronomic control measure mainly used against weeds and 'volunteer' cultivated plants remaining from a previous crop. It helps limit both internal virus sources within fields and external sources in nearby fields. Using crop and weed-free periods that encompass neighbouring small farms or market gardens does both much more effectively, as with *Celery mosaic virus* (CeMV) in celery crops (e.g. Latham and Jones, 2003). By tackling both types of initial virus source, such measures help diminish early virus spread. However, once some spread occurs and virus infection foci are present within a stand, they have no effect on the spread that occurs from these foci.

Planting mixtures of virus-susceptible and non-host crops is often practised by subsistence farmers in developing countries even though the implications for virus spread are not understood (Thresh, 2003). Depending on how it is done, such mixtures have the potential to address both types of virus sources and both phases of virus spread. Thus, mixtures decrease virus spread from external virus sources when tall non-host crops are inter-planted with shorter susceptible crop plants, especially when used as cover crops. This is so because, as with non-host barriers, incoming vectors of non-persistently transmitted viruses that land on the non-host and probe it lose the virus before they move on to visit susceptible plants. Similarly, if vectors of persistently insect-transmitted viruses colonize the non-host, they may no longer be viruliferous when their progeny move on. Internal sources of virus infection are removed when less vigorous and competitive infected plants are smothered by profuse growth of nearby non-hosts. Diminishing external or internal virus sources like this results in decreased early virus spread within the susceptible crop. In addition, after canopy development by the non-host crop becomes substantial, late spread can be diminished by repelling viruliferous vectors migrating within the mixed species stand. This applies particularly when non-host cover crops are used. Other ways in which inter-planting with non-hosts can decrease virus spread include providing physical barriers and camouflage against vectors or a habitat for their natural enemies and increasing the separation between susceptible individuals within the mixture (Thresh, 1982, 2003). Examples of decreased virus spread in such situations include *Bean common mosaic virus* in common bean mixed with the taller non-host crop maize (van Rheenan *et al.*, 1981), cassava mosaic virus in cassava mixed with the non-host maize (Fondong *et al.*, 2002) and BYMV in lupin mixed with the non-host oats (Jones, 1993, 2001). However, there are disadvantages to using the mixed cropping approach in developed agricultural systems because sufficient inexpensive manual labour to thin out crop plants and remove weeds is often scarce or absent. Also, effective use of selective herbicides to remove weeds is difficult to achieve in practice without damaging one or other component of the mixture, as is obtaining a suitable balance between the different crop species so that neither dominates the other excessively (e.g. Jones, 1993, 2001).

Another situation where mixtures of host and non-host plants are commonly grown worldwide is in mixed species pasture swards. Here, in the absence of frequent grazing or mowing, healthy growth of the non-host shades out the less vigorous, more stunted, internal virus source plants, thereby, preventing virus acquisition from them and so slowing early virus spread. Also, loss of virus inoculum of non-persistently arthropod vectored or contact-transmitted viruses on non-host plants decreases early spread to host plants, as with AMV in annual medic mixed with capeweed (Jones and Nicholas, 1998) or annual ryegrass (Jones and Ferris, 2000), or AMV and WCMV in mixed swards of white clover and perennial ryegrass (Coutts and Jones, 2002).

IV. Combining Control Measures

The amount of virus control achieved is increased greatly by combining diverse measures that act in different ways. When this is done, their effects are always complementary and synergistic, resulting in more effective overall control. The knowledge that combinations of measures can greatly enhance effectiveness has led to development of IDM concepts for virus diseases that combine available host resistance, cultural, chemical, biological and legislative control measures, and can be applied together in farmers' plantings as one overall control 'package' (e.g. Cho *et al.,* 1989; Jones, 2001, 2004a; Makkouk *et al.,* 2005). Selecting the ideal mix of measures to use for each pathosystem and production situation requires knowledge of the epidemiology of the causal virus, and of the selectivity, mode of action, effectiveness and reliability of each individual control measure so that diverse responses can be devised that are tailored to meet the unique features of each of the different scenarios considered.

Along with knowing whether they are selective or non-selective (see Section III), understanding whether individual control measures address external or internal sources and early or late spread, or both is critical information when choosing which to include in an IDM strategy for a particular pathosystem and agro-ecology. Effectiveness of control is optimized by including measures that are selective and non-selective, address both kinds of sources and phases of virus spread, and that operate in as many different ways as possible. Decisions on what measures to include are also influenced by factors such as the extra cost involved, labour availability, the extent of disruption to standard agronomic practices and their compatibility with other control measures already in place for weeds, insects and other pathogens. They must also be robust, sustainable, environmentally friendly and socially acceptable, and be tailored to the needs of the end user concerned, who vary from resource-poor, subsistence farmers in developing countries and large-scale, low-input farmers in developed countries to high-input farmers producing high-value 'cash' crops in either type of country.

An example of an IDM approach being widely used for practical virus control in a developing country is one for *Faba bean necrotic yellows virus* (FBNYV) in faba bean (*Vicia faba*). This strategy combines planting late in the growing season, use of high seeding rate, application of one or two systemic insecticide sprays that are well-timed during the early stages of crop development and roguing infected plants early in the growing season. Its use has significantly improved the profitability of faba bean production by small-scale farmers in Egypt (Makkouk *et al.*, 2003). A similar example comes from subsistence farming of chickpea in northern Sudan, where a combination of partial virus resistance, delayed sowing and shorter intervals between irrigations diminished the incidence of *Chickpea chlorotic dwarf virus* and increased yield (Hamed and Makkouk, 2002).

An example for low input, large-scale farming in a developed country comes from lupin in southwest Australia. Here, lupin yields in high risk areas for BYMV and CMV are optimized by sowing virus-tested lupin seed stocks with minimal virus contents, sowing cultivars with inherently low seed transmission rates and isolation from neighbouring lupin crops (CMV only); using perimeter non-host oat barriers and avoiding fields with large perimeter:area ratios (BYMV only); promoting early canopy development, generating high plant densities, adjusting row spacing, direct drilling into retained stubble, sowing early maturing cultivars, maximising weed control and crop rotation (both viruses). The differences between the measures recommended for the different viruses reflect the internal seed-borne source with CMV compared with the external clover pasture source with BYMV (Jones, 2001). Another example from the same region is the strategy deployed to minimise infection with CMV in the lupin breeding programme. Here, protection of the small plots from infection is paramount rather than the cost of the control measures or environmental issues, so a very comprehensive range of individual measures can be used, including regular insecticide applications (Jones, 2001). Similarly, lettuce, peanut, pepper, tobacco and tomato 'cash' crops are sufficiently profitable in many countries for comprehensive packages of control measures to be included within IDM strategies against TSWV epidemics (Brown *et al.*, 1996; Cho *et al.*, 1989; Coutts *et al.*, 2004; Culbreath *et al.*, 1999; Jones, 2004a; Momol *et al.*, 2004; Riley and Pappu, 2000).

When information on the effectiveness of individual control measures is limited or lacking for a specific virus disease, crop and growing situation suffering from damaging virus epidemics, the farmers' need for advice on how to control the causal virus is still immediate. Obtaining information from field experiments on the effectiveness of individual control measures is expensive and may take several years, which is far too long. In these circumstances, an 'interim IDM strategy' is required that can be devised quickly using available epidemiological knowledge and 'generic' information on control measures known to work well with other similar pathosystems. An example of an 'interim IDM strategy' that is very comprehensive was devised by Jones *et al.* (2004) to control seed and aphid-borne CMV, AMV, BYMV and *Pea seed-borne mosaic virus* in chickpea, faba bean, field pea and lentil plots at breeding, selection and seed increase sites. An 'interim IDM strategy' that uses some of these measures and is affordable to farmers is available for use against the same four seed-borne viruses in commercial crops of these pulses (Jones, unpublished data). Other 'interim IDM' examples based entirely on knowledge of epidemiology of the pathosystem concerned and generic control measures include ones for CeMV in celery (for use where the extent of infection is insufficient to warrant employing a 'celery-free period'), CarVY in carrot and TSWV in potato (Jones *et al.*, 2004, 2005b; Latham and Jones, 2003, 2004a). However, to validate and refine them, such interim approaches should be followed up by field experimentation on the effectiveness and reliability of the control measures used (see Section II), or if this is too costly and slow, by collection of epidemiological information from field situations, and case history studies where the strategy is, or is not, being used.

How to undertake replicated field experiments designed specifically to determine the effectiveness of single control measures or, better still, combinations of measures in suppressing virus spread and improving yield and quality was described earlier. There are numerous examples in the literature for control measures applied alone (see Section II), but few for combinations of measures applied together (e.g. Latham and Jones, 2004b). Recent examples of how to collect epidemiological information for validation of control measures from naturally occurring epidemic situations or large un-replicated plots include: for non-host barriers, isolation, safe planting distances, planting upwind, prompt removal of virus sources and avoidance of side-by-side plantings when TSWV spread into lettuce (Coutts *et al.*, 2004); for isolation, safe planting distances, intervening fallow, planting upwind, prompt removal of virus sources, avoidance of side-by-side plantings and manipulation of planting date when CarVY spread into carrots (Jones *et al.*, 2005b); and for non-host barriers when BYMV spread into lupin (Jones, 2005). Although statistical analysis of the data collected from them may be impossible, case history studies cost less and still provide useful feedback on effectiveness, especially where combinations of measures are deployed in IDM. Such studies compare incidences of infection on farms where IDM is adopted with those on farms where it is not, as with CarVY in carrots where overall infection incidences of >65% in crops (both farms) changed to 0% (farm with IDM) and 45% (farm without IDM), 2 years later (Latham and Jones, 2004a).

Achieving adoption by farmers of IDM practices that address plant virus diseases is often difficult and this applies in both developing and developed countries. Thorough extension efforts are vital to ensure that IDM approaches are disseminated adequately and then adopted. However, the information required is often not disseminated well to farmers because of serious limitations in extension services, which, unfortunately, have often declined rather than improved in recent years in both developed and developing countries (see Section VI). Also, deploying a multi-faceted virus IDM strategy requires anticipatory action by the farmer from before planting time onwards and has to be instigated before there are any signs of virus epidemics. However, farmers tend not to worry about epidemics until they reach an advanced stage and are easily seen. They then may have unrealistic expectations that a 'silver bullet' control measure exists that will 'stop it in its tracks' and save them from financial losses. Advice that there is no 'quick fix' and that a multi-faceted IDM approach should have been adopted at or before planting time provides little or no relief from their immediate problem and so may not register as being worthwhile for the future.

There are numerous examples from different parts of the world of the difficulties often experienced in getting farmers to adopt IDM practices that address plant virus diseases. A typical example from developing countries involves the cassava mosaic disease pandemic in East Africa (Legg *et al.*, this volume, pp. 355–418; Thresh and Cooter, 2005). Despite warnings of the imminent threat of its spread from Uganda to neighbouring countries and the need to deploy measures involving hygiene and planting virus-resistant cassava cultivars, the response was limited until after the severe pandemic struck and major economic losses occurred. A typical example for a developed country involves deployment of an IDM package against TSWV by market gardeners growing lettuce, tomato and pepper crops in southern Australia (Jones, 2004a). Although informed about the package frequently via appropriate media outlets and local extension services, many ignored the simple phytosanitary and cultural procedures needed at and before planting time in the misplaced hope that, should a serious epidemic develop in their crops, repeated insecticide sprays to kill the thrips vector would suffice to control TSWV. In justifying their approach, farmers 'at risk' sometimes likened the recommendation that they deploy phytosanitary and agronomic measures at planting time to receiving unpalatable medical advice that 'taking a suitable medicine will not suffice to overcome their ailment, and they also need to go on a diet, eat different types of food, take more exercise and change there lifestyle'!

Persuading farmers to continue using IDM approaches against viruses once major economic losses abate can also be problematical.

There are many examples from different countries but one from southwest Australia and another from Uganda illustrate the point. Once the severe CMV epidemic in lupin crops in southwest Australia in the late 1980s abated following widespread adoption of the IDM strategy devised to tackle it, gradually many farmers stopped bothering to use it, laying the lupin industry open to future resurgence of significant losses (Jones, 2001). Similarly, once the cassava mosaic disease pandemic had abated in Uganda and it was no longer considered such a pressing problem, there was resurgence in cultivation of mosaic-sensitive cultivars of cassava, leaving the country vulnerable to substantial losses in the future.

V. Economic Considerations

Bos (1982) provided a detailed review of losses caused by plant virus infection and the economic consequences arising from these losses, so the subject is only briefly discussed here. The economic effects that result from virus epidemics in stands of cultivated plants include loss of revenue, increased production costs, shortages in supply and greater unreliability of production. They also include losses to consumers and society as a whole from wasted resources. An 'economic threshold' is used to decide if the expense and effort of deploying virus control measures is justified for a given pathosystem and situation. Calculating this 'threshold' requires information on the current market value of the product, the overall occurrence and economic importance of the virus in the locality, the yield and quality losses it induces in the cultivated plant concerned, the incidence of infection in the stand, the effectiveness of the control measures to be deployed and the cost of using them (Jones, 2004a).

The extent and magnitude of the economic damage that results from virus epidemics in stands of cultivated plants are often underestimated, for example, when symptoms are mild or confused with those arising from other factors, such as nutrient deficiencies, water-logging or drought, damage from application of chemicals, infestation with pests and infection with other types of pathogens. Also, where virus-induced yield losses are known to be relatively small and there is little obvious impairment to produce quality, they are often assumed to be unimportant. Such assumptions are often misguided as when the fixed costs in producing a commercial stand are deducted from final value of product, profits are often only a minor proportion of this overall value, so even small virus-induced declines in yield or quality may damage profitability considerably (Fig. 3). In such circumstances, after deducting the added cost involved in deploying them, virus control measures may still provide a substantial boost to the revenue obtained. Similarly, where larger declines in yield or greater impairment of quality render a virus-infected stand unprofitable, deploying control measures can ensure its return to profitability.

Obtaining information on the extent of virus-induced losses in a region requires large-scale surveys to establish virus incidence in commercial plantings. In herbaceous stands, such surveys are normally done by sampling mature plantings of widely grown cultivars that are representative of the crop, farm and overall region being surveyed, and using standard sampling patterns, such as Z or W patterns, to take 100 or more random samples by choosing one every 5 or 10 paces.

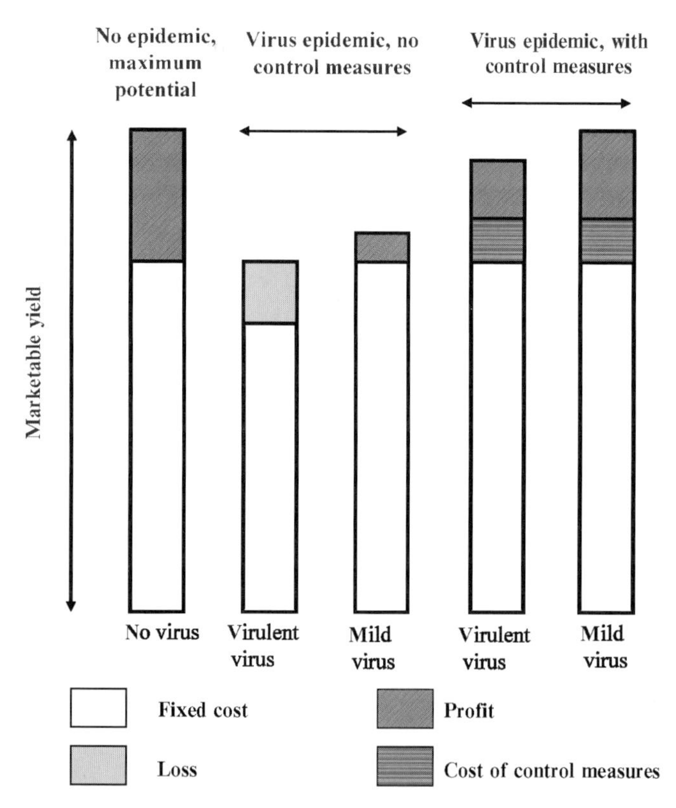

FIG 3. Losses in marketable yield due to infection with virulent or mild viruses with or without control measures, the cost of deploying control measures and impact on profits. (See Color Insert.)

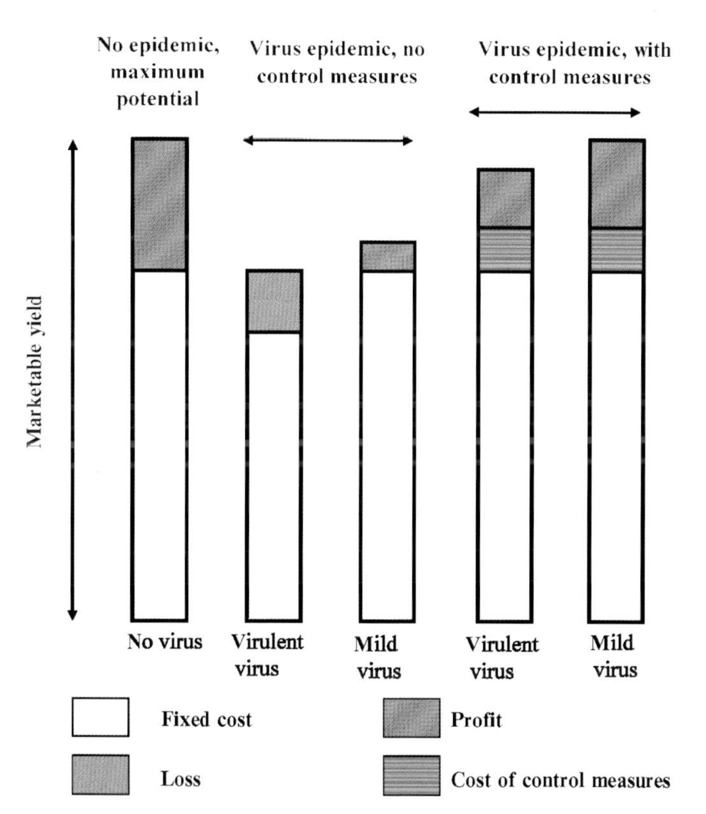

JONES, FIG 3. Losses in marketable yield due to infection with virulent or mild viruses with or without control measures, the cost of deploying control measures and impact on profits.

A small number of representative symptomatic plants are taken too and tested separately. To avoid collecting data that represent an atypical scenario for the pathosystem and situation concerned, the surveys need to be repeated over more than one growing season. Surveys of alternative hosts also need to include samples collected at random to assess virus incidence and identify any symptomlessly infected reservoir hosts. Enzyme-linked immunosorbent assay (ELISA) and tissue blot immunoassay are often the methods of choice for testing many plant samples. Recent large-scale surveys of crops, pastures and alternative hosts for virus incidence in southwest Australia provide examples (Cheng and Jones, 1999; Coutts and Jones, 1996, 2000, 2005a; Hawkes and Jones, 2005; Jones, 2004b; Latham and Jones, 1997, 2001, 2003; McKirdy and Jones, 1995, 1997b). Examples of large-scale surveys to establish the incidences of viruses infecting food legume crops in West Asia and North Africa are listed by Makkouk *et al.* (2005). Past large-scale surveys for CSSV in cocoa in Ghana (e.g. Thresh, 2003) provide an example from the tropics. A more recent tropical climate example is that of Sseruwagi *et al.* (2004), who described large-scale surveys for cassava mosaic disease in cassava involving 18 different countries in sub-Saharan Africa.

Obtaining information on the extent of virus-induced losses in a region requires field experiments to quantify the yield and quality impacts resulting from infection. Quantitative estimates of yield loss and quality impairment are best made in the field where infection spreads naturally rather than by comparing yields and quality effects from virus-infected and healthy plants grown singly in the field or in pots. This is because using single plants in such studies reflects an atypical 'worse case scenario' where there is no compensatory growth by nearby healthy plants (e.g. Jones, 1992; Latham and Jones, 2004a,b; Latham *et al.*, 2004a,b). Also, such data are often based on single plants infected simultaneously by artificial inoculation rather than on plants infected at different growth stages during natural spread of infection. However, the growth stage when plants first become infected and the proportion of plants infected within stands are critical factors in determining the extent of yield losses. These are generally greatest when plants become infected at vulnerable early growth stages, incidence approaches 100% and a sensitive cultivar is grown. When the incidence of infected plants is low and their distribution random, there may be little effect of infection on yield because compensatory growth of neighbouring healthy plants occupies the space left by the less vigorous infected ones. More often, the pattern of distribution of infected plants is clustered and compensatory growth of healthy plants then occurs only at the margins of the clusters, so the larger the clusters the smaller the amount of compensation and the greater the yield loss at lower virus incidences (Fig. 4; Hughes, 1988, 1996; Jones and Jones, 1964; Jones *et al.*, 1955; Thresh *et al.*, 1994). In situations when there are few, very large clusters of infected plants and a relatively tolerant cultivar is grown, there may be insufficient perimeter plants for compensatory growth of healthy plants to register, so the losses would then be directly proportional to virus incidence within the stand. Although there are apparently no recorded examples of such a relationship where growth of healthy plants is unaffected by other factors, such direct proportionality is found in virus-infected legume pasture swards when frequent grazing or mowing removes the new growth of foliage from healthy plants that would otherwise compete with neighbouring virus-infected ones around the edges of the clusters (Ferris and Jones, 1995; Jones, 1994b). Such a relationship is also found when plant density is so low that crop plants are too far apart for compensation to occur (e.g. Jones, 1993). Typical examples of quantitative estimates of yield losses from replicated field experiments involving stands where virus spread naturally include ones for BYDV

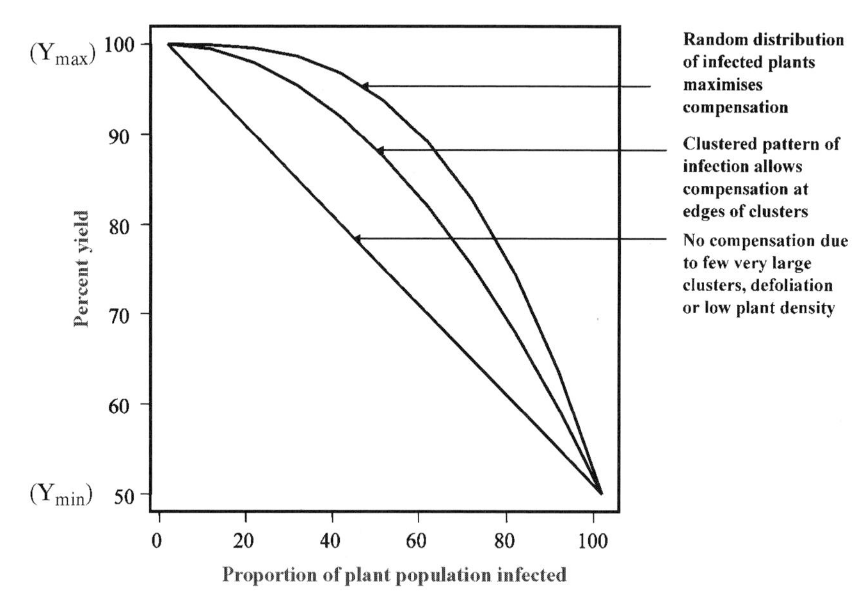

FIG 4. Effect of virus incidence and spatial distribution of infected plants on compensatory plant growth and yield losses. Maximum yield loss at 100% infection assumed to be 50%.

in wheat (McKirdy and Jones, 1996, 1997a; McKirdy *et al.*, 2002; Thackray *et al.*, 2005); CMV and BYMV in lupins (Bwye *et al.*, 1994; Jones, 1993, 1994a, 2001; Jones *et al.*, 2003); CMV, BYMV and SCoMV in clover swards (Ferris and Jones, 1995; Jones, 1991, 1994b; and AMV in annual medic swards (Jones and Nicholas, 1992, 1998). These studies sometimes also included assessments for quality impacts, such as seed size (with lupin and wheat), shrivelled seed or protein content (with wheat).

VI. Problems Hindering Success

Although much knowledge of how to control plant virus diseases has accumulated and increasingly powerful technologies are available to help, those seeking advice on how to control damaging virus epidemics in cultivated plants are faced with an increasingly difficult task. There are several 'big picture' reasons for this which, when combined together, increase the complexity and instability of the situation, so compounding the difficulties. Increasing concerns over sustainability, human health, the environment and social issues as well as the diminishing number of new pesticides being produced by chemical companies, all limit the control options available. Similarly, control options are limited by the ease of development of pesticide resistance in key arthropod vectors, such as *B. tabaci*, *F. occidentalis* and *M. persicae*, and of virus strains that break down single gene virus resistance. There is considerable opposition to the release of genetically modified plants, so their widespread use to help provide virus-resistant cultivars seems unlikely for many years to come. Also, the process of global warming is fuelling an increasingly unpredictable climate, providing conditions that favour more frequent occurrence of severe virus epidemics in many regions of the world. Moreover, to meet the needs of urbanization and the burgeoning human population, agriculture, horticulture and forestry are becoming more complex and diverse, with intensive production systems being used increasingly in different parts of the world, including in protected cropping systems. Traditional practices that have stood the test of time are being replaced by ones that have not been evaluated adequately with regard to their effects on virus epidemics. Also, such changes in practices increase the frequency of exposure of introduced cultivated plants to potentially damaging indigenous viruses emerging from local wild plants. Moreover, plant quarantine regulations and other restrictions on international trade in plants and plant products are being relaxed through international treaty and such trade is expanding rapidly. This trade deregulation and expansion combined with the increasing speed of modern transport systems, especially air freight, is accelerating the introduction of potentially damaging viruses and their vectors to regions where they were once absent, exposing cultivated plants to new encounters with introduced viruses they have not met previously (Bos, 1992; Buddenhagen, 1977; Thresh, 1980, 1981, 1982).

Two examples from tomato serve to illustrate the global problems arising from trade deregulation and expansion combined with the increasing speed of modern transport systems. Firstly, in 1974, *Pepino mosaic virus* (PepMV) was found infecting pepino (*Solanum muricatum*) in Peru and shown to have stable particles and be readily contact transmissible (Jones *et al.,* 1980). There were no further reports until 1999, when it appeared infecting tomato in the Netherlands (van der Vlugt *et al.*, 2000) and then rapidly spread to several other European countries, the USA, Canada and China. It damages the appearance of tomato fruits seriously and contaminates the surface of its seed. Its sudden appearance in other continents was attributed to the activities of international seed companies using South America to propagate seed crops and the increased speed and volume of international trade in tomato seeds and fruits (French *et al.,* 2001; Mumford and Jones, 2005; Verhoeven *et al.,* 2003). Secondly, *Tomato yellow leaf curl virus* (TYLCV) is an 'old world' *Begomovirus* first found in Israel. Its vector is *B. tabaci* and it causes serious losses to tomato production. Until the late 1980s, TYLCV was found only in eastern Mediterranean countries but since then it spread rapidly to western Mediterranean countries, Japan and the 'new world'. In the western hemisphere, it was reported first from the Dominican Republic in the Caribbean in 1994 (Polston *et al.,* 1994). This introduction is thought to have occurred through international trade in live tomato seedlings imported from Israel. Thereafter, TYLCV soon spread to additional countries in the region, including other Caribbean islands, Mexico and southeast USA. This spread apparently occurred through trade in tomato fruits and seedlings and local movement of viruliferous *B. tabaci* in wind currents (Pappu *et al.,* 2000; Polston and Anderson, 1997; Polston *et al.*, 1999).

Other factors causing the situation to deteriorate in developed countries include dwindling funding for epidemiological research, a decline in trained manpower, erosion of the skills base, loss of focus and a tendency to forget or ignore past lessons. In most developed countries, younger plant virologists are being trained in molecular and laboratory-based virus research but rarely in traditional plant virology involving epidemiology and field studies, so such skills are being lost as older plant virologists retire or move to other jobs. Consequently, there are increasing examples of non-virologists, for example, entomologists, fungal plant pathologists or plant breeders, who lack sufficient understanding of virus epidemiology, being called in to advise on virus control measures. Government-funded extension and advisory services, that in the past maintained disease surveillance programmes locally, including ones for virus diseases, and often provided sound advice over virus control measures to farmers, have been run down. Surveillance is rare or lacking now and the burden of giving virus control recommendations has fallen on commercial organizations. These include chemical or seed companies, which rarely give impartial advice, or the technical staff of monopoly supermarkets, who dictate to farmers exactly how the crops they purchase should be grown and have to take decisions based on short-term commercial considerations but whose knowledge of virus control strategies is rarely adequate. In developing countries also, there are serious limitations in extension services, which too, unfortunately, have often declined rather than improved in recent years. Complacency and a tendency to forget past lessons also contribute considerably to the deteriorating worldwide trend.

The run down of healthy stock schemes designed specifically to control viruses is particular cause for concern in many developed countries. For example, recent changes to the Australian national seed potato scheme that resulted from commercial pressures, complacency and a loss of focus. Basic virus control measures were relaxed, such as siting early seed potato propagation stages in regions with few aphids, ensuring adequate isolation between stands of different generations and crops sown with uncertified seed, labelling bags after harvest adequately so that their origins could be traced, cleaning of tuber-cutting implements after use and testing of randomly collected leaf samples to monitor for re-infection with viruses. Inevitably, in eastern Australia where these changes were made, such relaxation resulted in rapid re-invasion with the contact and aphid-transmitted viruses PVX, *Potato virus S* (PVS), *Potato virus Y* and *Potato leaf roll virus*, even into the first generation after mini-tubers, especially with PVX and PVS (Lambert *et al.*, 2005). Now that the extent of this re-contamination is realized, measures are being taken to improve the situation. By contrast, in southwest Australia, where measures more stringent than those used previously were introduced, the virus health of seed potatoes has improved considerably over the last 15 or so years (Holland and Jones, 2005; Wilson and Jones, 1990). Increasing distribution of virus-infected plant materials by nurseries is another cause for concern in different countries, for example, large-scale sale of lettuce seedlings infected with lettuce big-vein disease and its vector *Olpidium brassicae* (Latham *et al.*, 2004b). Such practices result from unwillingness to spend money on control measures and so avoid having to charge a higher price for the plants sold, and general complacency over the need to control viruses.

An example of the consequences of the move away from epidemiology and field studies and general lack of focus on practical plant virus issues in developed countries comes from *Wheat streak mosaic virus* (WSMV). This is a serious pathogen of wheat in the Great Plains region of the USA, where it has been studied for 50 or so years. Infected volunteer cereals and grasses surviving outside the growing season were always considered the sole reservoir for spread to wheat crops, although WSMV epidemics still occurred unpredictably even when such reservoirs were absent (e.g. Christian and Willis, 1993; French and Stenger, 2003; Thomas and Hein, 2003). However, routine quarantine checks elsewhere in the world where the virus is not known to occur recently revealed that WSMV is seed-borne in wheat, which explains the frequent occurrence of WSMV epidemics in the absence of this reservoir. It means that long accepted views over the epidemiology of the virus in wheat in the USA and elsewhere need to be revised and that control measures to limit the movement of WSMV by infected commercial seed stocks and germplasm should be considered (Jones *et al.*, 2005a).

VII. Opportunities from New Technologies

As mentioned in Section III, much information has accumulated over decades of plant virus research that provides an understanding of how epidemics develop and a diverse array of virus control measures which operate in different ways is available. Also, powerful, sophisticated and rapidly improving technologies can be used to improve the control of plant virus diseases. If employed intelligently, innovatively and flexibly, these enable much more effective use of available information, skills and experience when taking decisions over which control measures to deploy. Moreover, old technologies can now be used more effectively than previously. The main categories of new technology involved here are computing, new virus detection technologies and 'molecular epidemiology'.

Computers have greatly streamlined the tedious, time-consuming and labour-intensive procedures used previously in field data collection, collation and interpretation. They help enormously with processing copious field data on plant virus epidemics, extraction of the critical information needed to understand the pathosystem concerned and taking informed decisions over which combination of control measures to deploy. Computer programs are also used for cost–benefit analyses which, with appropriate assumptions, help to determine the 'economic threshold' above which using control measures becomes an economic proposition (see Section V). Where the population of an insect vector that transmits a virus persistently tends to be localized within a large crop, 'precision agriculture' employing Global Postioning Systems and Global Information Systems technology, potentially, is sufficiently accurate to identify which parts of the crop would benefit from insecticide application to diminish virus spread. Although currently mostly used to target fertilizer applications, such approaches to chemical control of virus spread are attractive because they are environmentally responsible and avoid unnecessary expenditure on chemicals. When spatial and temporal data from virus epidemics are analysed with suitable programs, they can be used to identify the principal means of virus spread, which in turn helps to indicate which combination of control measures is likely to be effective. For example, when Pethybridge *et al.* (2000) analysed such data from hop gardens, mowing to remove basal growth of hops was shown to be important in spreading *Hop mosaic virus* and *Hop latent virus* in addition to spread by aphids. They therefore suggested that desiccant herbicide be applied to remove basal growth instead of mowing. Moreover, computer programs, such as SADIE, which analyse spatial data are very useful for validating the different control measures recommended in 'established' or 'interim' IDM strategies, as discussed already for BYMV in lupin, CarVY in carrots and TSWV in lettuce (see Section IV). In plant quarantine, computer models are used widely to estimate the degree of threat posed by introduction of potentially damaging plant viruses ('risk assessment') and in intervention systems that recommend which control measures are most likely to provide effective virus control when used singly or in combination ('risk management') (e.g. Baker *et al.,* 2005).

Epidemiological models are playing an increasingly important role in evaluating the likelihood that different control tactics will be effective with a particular pathosystem and situation, and in deciding which combination of measures is likely to be useful and worth recommending. For example, Holt *et al.* (1999) used such a model for ToLCV in tomato in southern India. The model predicted that the best approach was to deploy protective netting around nursery beds to diminish immigration of *B. tabaci* vectors into the general vicinity and

ToLCV-resistant cultivars to minimise virus inoculation by vectors. The protective netting was treated with a persistent insecticide with a rapid knock down effect and coloured bright yellow to attract *B. tabaci*. Measures predicted to be less effective when adopted for this pathosystem and situation were isolation, regular insecticide applications and relying on natural predators. When adopted, the approach predicted proved effective. Epidemiological models are also often used to forecast the likelihood that damaging epidemics will develop for a particular pathosystem and situation. There are examples from different countries but validation with hard data is often lacking. Two examples of well validated models come from the Mediterranean-type climate of southwest Australia, where one forecasts epidemics of BYDV in cereals (Thackray and Jones, 2003) and the other epidemics of CMV in lupin (Thackray *et al.*, 2004). Crops are sown in late autumn and early winter in the region. Both models use rainfall during summer and early autumn to calculate an index of aphid build up in each district before the crop growing season starts. The index is used to forecast the timing and magnitude of aphid immigration into a crop, and the subsequent aphid build up and movement within the crop, virus spread and related yield losses. Cereal aphid vectors and BYDV persist in perennial grasses that survive the summer drought in isolated damp spots, for example, where dew deposited overnight on road surfaces runs off into roadside ditches that remain moist enough to support their growth (Hawkes and Jones, 2005). They then spread to germinating grass weeds and volunteer cereals in which they build up before the growing season starts and from which viruliferous aphids fly to cereal crops. Aphid vectors of CMV build up on weeds, crop volunteers and self-regenerating annual pastures before the growing season starts and fly from them to lupin crops, where they acquire and spread the virus from internal seed-borne inoculum sources. Both models determine the effects of different sowing dates and plant population densities on virus spread, and the CMV one also uses the percentage infection in the seed sown. For both, simulations were validated with field data from sites representing a range of pre-growing season rainfall scenarios, sowing dates, plant population densities and, for CMV, percentages of infection in the seed sown. Both models were incorporated into decision support systems that are accessible via the Internet for use by advisers, consultants or farmers in planning targeted insecticide sprays against the aphid vectors of BYDV or deciding which cultural control measures to deploy at planting time against CMV.

Thresh and Fargette (2003) described the potential practical benefits arising from 'molecular epidemiology'. In brief, molecular typing of isolates improves knowledge of virus epidemics by providing traceability, for example, in establishing the origins of outbreaks and sources of virus introductions and allowing virus population structure to be analysed. When different strains are present, molecular typing provides precise strain prevalence data for crops and alternative hosts and information on the geographical distribution of distinct strains. For example, when Tairo *et al.* (2005) analysed data from 46 isolates of *Sweet potato feathery mottle virus* from different parts of the world, they revealed that of the four distinct strains, strain EA was only found in East Africa, while the other three were widely distributed in different continents. This virus is a component of the damaging 'sweet potato virus disease' complex. Such detailed background knowledge for an economically important virus disease can help in choice of potential control measures, such as deployment of virus-resistant cultivars, especially where strain-specific virus resistance is involved, and phytosanitary or agronomic measures. It also assists with decisions over how to index planting materials effectively and the need to adjust quarantine measures associated with international trade and germplasm exchange in vegetatively propagated plants like sweet potato.

Rapid and reliable routine virus testing procedures are critical to the success of most healthy stock programmes, both with high-throughput sample testing in the laboratory and use in field test kits. They are also pivotal in quarantine testing, obtaining large-scale survey data, collecting spatial and temporal data from epidemics and determining the results of field experiments on control measures. Automated ELISA is widely used where samples are many and suitable antibodies available. Because they are labour-intensive, reverse-transcription polymerase chain reaction (RT-PCR) assays are only suitable for virus identifications with few samples, but 'real time' RT-PCR (TaqMan) assays are more applicable to large-scale, sensitive virus detection, as with TSWV (Dietzgen *et al.,* 2005). The latest innovations in medical research techniques are continually being applied in plant virus detection (e.g. Ward *et al.*, 2004). For example, microarray technology allows numerous samples to be tested simultaneously on the same microchip (Boonham *et al.,* 2002). Portable real-time PCR and lateral flow devices enable identification of quarantine viruses at port of entry, or common viruses from potentially symptomatic samples taken during visual inspection of crops during field certification. Baker *et al.* (2005) discuss several other innovations in testing procedures not yet applied in plant virology that offer considerable potential for improvement in the future.

VIII. Conclusions

Virus epidemics in cultivated plants pose a worldwide challenge to achieving satisfactory yields and quality of produce. An increasingly sophisticated and diverse range of host resistance, cultural (phytosanitary and agronomic), chemical, biological and legislative control measures are becoming available to meet this challenge. Knowledge of which factors restrict virus epidemics in natural plant communities and primitive subsistence cropping systems has helped in their development. When decisions are made over which control measures to deploy and whether to use a measure alone or together with others, the key ingredients are thorough epidemiological knowledge of the pathosystem concerned, and sound information on the selectivity, mode of action, effectiveness and reliability of each individual measure, and how to respond. There is an ever-increasing knowledge base and sophistication of technology to draw on to help. However, to be adopted control measures also need to be ecologically and socially sustainable, robust, affordable and compatible with standard agricultural practices.

This chapter emphasises the importance of dissecting the different ways in which diverse types of virus control measures operate before taking decisions on which to deploy. Some control measures are non-selective affecting a wide range of viral pathogens or their vectors, but others are selective targeting a particular virus or vector species and may not be transferable between pathosystems. Control measures also differ in activity by targeting either the initial source of inoculum, which may be external or internal, or the rate of virus spread in the stand, which may be influenced at early or late stages of the epidemic. In general, phytosanitary, agronomic and legislative measures are non-selective, whereas host resistance and most biological control measures are selective. Chemical control measures have low selectivity when a general pesticide that kills several vector species is applied, but high selectivity when a specific pesticide kills only one vector species. However, a drawback to selective measures, including the phytosanitary measure of roguing, is that deploying them widely may result in inadvertent selection of variants of virus or vector that are more difficult to control. Host resistance, chemical and biological control measures decrease the rate of virus spread but are inactive against the initial virus source, whereas legislative and phytosanitary control measures, and many agronomic measures address both source and spread.

When combining control measures within IDM strategies, success is optimized by including measures that are selective and others that are non-selective, that address both kinds of sources and phases of virus spread, and that operate in as many different ways as possible. This also applies when devising 'interim IDM strategies' to address short-term control needs. Such strategies are necessary when information on the effectiveness of control measures for the pathosystem concerned is insufficient, but the farmers' need to take remedial action is immediate. All available 'generic' information on measures from related pathosystems and situations is used to help devise them, but each should be validated subsequently for the pathosystem concerned and the tactics included modified as required. When recommended, both 'established' and 'interim' IDM strategies must address the limitations of the situation concerned. The level of sophistication suitable for resource-poor subsistence farms in developing countries and large-scale, low-input farms in developed countries differs greatly from what is feasible to deploy on high-input farms producing valuable 'cash' crops in either type of country (Jones, 2004a; Thresh, 2003). Nevertheless, depending on their individual circumstances, the farmer still has the option of deciding whether to employ all measures within a recommended IDM package, or to leave out one or more of them that are incompatible with individual circumstances.

This chapter also addresses how to determine the effectiveness and reliability of control measures used alone or in combination, economic issues associated with deciding whether one or more measures should be deployed, the 'big picture' problems that increasingly hinder success in tackling virus diseases and the exciting opportunities provided by new technologies. We are faced with increasing difficulties, not least from emerging virus diseases that result from the rapidly expanding nature of world trade in plants and plant products, rapidity of modern transport systems and greater intensity of cultivation to feed the growing human population (e.g. Anderson *et al.*, 2004). Fortunately, powerful new technologies offer many opportunities to optimize control tactics and delivery, especially in developed countries. The way forward depends on intelligent, innovative and flexible use of available experience, information and technology. The challenge for the practical plant virologist is, within an environment of skills erosion, lack of focus and diminishing resources for field studies, to apply new technologies to greatest effect while still ensuring adequate epidemiological research and field validation of control tactics.

References

Anderson, P. K., Cunningham, A. A., Patel, N. G., Morales, F. J., Epstein, P. R., and Daszak, P. (2004). Emerging infectious diseases of plants: Pathogen pollution, climate change and agrotechnology drivers. *Trends Ecol. Evol.* **19:**535–544.

Baker, R., Cannon, R., Bartlett, P., and Barker, I. (2005). Novel strategies for assessing and managing the risks posed by invasive alien species to global crop production and biodiversity. *Ann. Appl. Biol.* **146:**177–191.

Bar-Joseph, M., Roistacher, C. N., and Garnsey, S. M. (1983). The epidemiology and control of citrus tristeza disease. *In* "Plant Virus Epidemiology–The Spread and Control of Insect-Borne Pathogens" (R. T. Plumb and J. M. Thresh, eds.), pp. 61–72. Blackwell Scientific Publishers, Oxford, UK.

Boonham, N., Walsh, K., Smith, P., Madagan, K., Graham, I., and Barker, I. (2002). Detection of potato viruses using microarray technology: Towards a generic method for plant viral disease diagnosis. *J. Virol. Methods* **108:**181–187.

Bos, L. (1982). Crop losses caused by viruses. *Crop Prot.* **1:**263–282.

Bos, L. (1992). New plant virus problems in developing countries: A corollary of agricultural modernisation. *Adv. Virus Res.* **38:**349–407.

Broadbent, L. (1964). Control of plant virus diseases. *In* "Plant Virology" (M. K. Corbett and H. D. Sisler, eds.), pp. 330–364. University of Florida Press, Gainsville, USA.

Brown, S. L., Todd, J. W., and Culbreath, A. K. (1996). Effect of selected cultural practices on incidence of tomato spotted wilt virus and populations of thrips vectors in peanuts. *Acta Hortic.* **431:**491–498.

Buddenhagen, I. W. (1977). Resistance the vulnerability of tropical crops in relation to their evolution and breeding. *Ann. New York Acad. Sci.* **287:**309–326.

Bwye, A. M., Jones, R. A. C., and Proudlove, W. (1994). Effects of sowing seed with different levels of infection, plant density and the growth stage at which plants first develop symptoms on cucumber mosaic virus infection of narrow-leafed lupins (*Lupinus angustifolius*). *Aust. J. Agric. Res.* **45:**1395–1412.

Bwye, A. M., Jones, R. A. C., and Proudlove, W. (1999). Effects of different cultural practices on spread of cucumber mosaic virus in narrow-leafed lupins (*Lupinus angustifolius*). *Aust. J. Agric. Res.* **50:**985–996.

Cheng, Y., and Jones, R. A. C. (1999). Distribution and incidence of the necrotic and non-necrotic strains of bean yellow mosaic virus in wild and crop lupins. *Aust. J. Agric. Res.* **50:**589–599.

Cho, J. J., Mau, R. F. L., German, T. L., Hartman, R. W., Yudin, L. S., Gonsalves, D., and Provvidenti, R. (1989). A multidisciplinary approach to management of tomato spotted wilt virus in Hawaii. *Plant Dis.* **73:**375–383.

Christian, M. L., and Willis, W. G. (1993). Survival of wheat streak mosaic virus in grass hosts in Kansas from wheat harvest to fall wheat emergence. *Plant Dis.* **77:**239–242.

Cook, D. F., Houlding, B. J., Steiner, E. C., Hardie, D. C., and Postle, A. C. (1996). The native anthocorid bug (*Orius armatus*) as a field predator of *Frankliniella occidentalis* in Western Australia. *Acta Hort.* **431:**507–512.

Coutts, B. A., and Jones, R. A. C. (1996). Alfalfa mosaic and cucumber mosaic virus infection in chickpea and lentil: Incidence and seed transmission. *Ann. Appl. Biol.* **129:**491–506.

Coutts, B. A., and Jones, R. A. C. (2000). Viruses infecting canola (*Brassica napus*) in south-west Australia: Incidence, distribution, spread and infection reservoir in wild radish (*Raphanus raphinistrum*). *Aust. J. Agric. Res.* **51:**925–936.

Coutts, B. A., and Jones, R. A. C. (2002). Temporal dynamics of spread of four viruses within mixed species perennial pastures. *Ann. Appl. Biol.* **140:**37–52.

Coutts, B. A., and Jones, R. A. C. (2005a). Incidence and distribution of viruses infecting cucurbit crops in the Northern Territory and Western Australia. *Aust. J. Agric. Res.* **56:**847–858.

Coutts, B. A., and Jones, R. A. C. (2005b). Suppressing spread of *Tomato spotted wilt virus* by drenching infected source or healthy recipient plants with neonicotinoid insecticides to control thrips vectors. *Ann. Appl. Biol.* **146:**95–103.

Coutts, B. A., Thomas-Carroll, M. L., and Jones, R. A. C. (2004). Patterns of spread of *Tomato spotted wilt virus* in field crops of lettuce and pepper: Spatial dynamics and validation of control measures. *Ann. Appl. Biol.* **145:**231–245.

Culbreath, A. K., Todd, J. W., Brown, S. L., Baldwin, J. A., and Pappu, H. R. (1999). A genetic and cultural "package" approach for management of tomato spotted wilt virus in peanut. *Biolog. Cult. Tests* **14:**1–8.

de Bokx, J. A. (ed.) (1972). *"Viruses of Potatoes and Seed-Potato Production."* Centre for Agricultural Publishing and Documentation, Wageningen, The Netherlands.

Dietzgen, R. G., Twin, J., Talty, I., Selladurai, S., Carroll, M. L., Coutts, B. A., Berryman, D. I., and Jones, R. A. C. (2005). Genetic variability of *Tomato spotted wilt virus* in Australia and validation of real time RT-PCR for its detection in single and bulked leaf samples. *Ann. Appl. Biol.* **146:**517–530.

Ferris, D. G., and Jones, R. A. C. (1995). Losses in herbage and seed yields caused by subterranean clover mottle sobemovirus in grazed subterranean clover swards. *Aust. J. Agric. Res.* **46:**775–791.

Ferris, D. G., Jones, R. A. C., and Wroth, J. M. (1996). Determining the effectiveness of resistance to subterranean clover mottle sobemovirus in different genotypes of subterranean clover in the field using the grazing animal as virus vector. *Ann. Appl. Biol.* **128:**303–315.

Fondong, V. N., Thresh, J. M., and Zok, S. (2002). Spatial and temporal spread of cassava mosaic virus in cassava grown alone and when intercropped with maize and/or cowpea. *J. Phytoplathol.* **150:**1–10.

Fraser, R. S. S. (1986). Genes for resistance to plant viruses. *CRC Crit. Rev. Plant Sci.* **3:**257–294.

French, C. J., Bouthillier, M., Bernardy, M., Sabourin, M., Johnson, R. C., Masters, C., Godkin, S., and Mumford, R. (2001). First report of *Pepino mosaic virus* in Canada and the United States. *Plant Dis.* **85:**1121.

French, R., and Stenger, D. C. (2003). Evolution of *Wheat streak mosaic virus*: Dynamics of population growth within plants may explain limited variation. *Annu. Rev. Phytopathol.* **41:**199–214.

Grogan, R. G. (1980). Control of lettuce mosaic virus with virus-free seed. *Plant Dis.* **64:**446–449.

Hamed, A. A., and Makkouk, K. M. (2002). Occurrence and management of *Chickpea chlorotic dwarf virus* in chickpea fields in northern Sudan. *Phytopathol. Medit.* **41:**193–198.

Harlan, J. R. (1971). Agricultural origins: Centers and non centers. *Science* **174:**468–474.

Harlan, J. R. (1981). Ecological settings for the emergence of agriculture. *In* "Pests, Pathogens and Vegetation" (J. M. Thresh, ed.), pp. 3–22. Pitman Press, London, UK.

Harrison, B. D. (2002). Virus variation in relation to resistance-breaking in plants. *Euphytica* **124:**181–189.

Hawkes, J. R., and Jones, R. A. C. (2005). Incidence and distribution of *Barley yellow dwarf virus* and *Cereal yellow dwarf virus* in over-summering grasses in a Mediterranean-type environment. *Aust. J. Agric. Res.* **56:**257–270.

Helyer, N. L., and Brobyn, P. J. (1992). Chemical control of western flower thrips (*Frankliniella occidentalis* Pergande). *Ann. Appl. Biol.* **121:**219–231.

Holland, M. B., and Jones, R. A. C. (2005). Benefits of virus testing in seed schemes. *In* "Proceedings of 'Potato 2005'—Australian National Potato Conference" (A. J. Pitt and C. Donald, eds.), pp. 81–87. Cowes, Victoria, Australia.

Holt, J., Colvin, J., and Muniyappa, V. (1999). Identifying control strategies for tomato leaf curl virus disease using an epidemiological model. *J. Appl. Ecol.* **36:**625–633.

Hughes, G. (1988). Modelling the effect of spatially heterogenous pest injury on crop yields. *Crop Res.* **28:**137–144.

Hughes, G. (1996). Incorporating spatial pattern of harmful organisms into crop loss models. *Crop Prot.* **15:**407–421.

Jones, F. G. W., and Jones, M. G. (1964). Pests of field crops, 1st Ed. Edward Arnold Publishers, London, UK.

Jones, F. G. W., Dunning, R. A., and Humphries, K. P. (1955). The effect of defoliation and loss of stand upon yield of sugar beet. *Ann. Appl. Biol.* **43:**63–70.

Jones, R. A. C. (1981). The ecology of viruses infecting wild and cultivated potatoes in the Andean Region of South America. *In* "Pests, Pathogens and Vegetation" (J. M. Thresh, ed.), pp. 89–107. Pitman, London.

Jones, R. A. C. (1991). Losses in productivity of subterranean clover swards caused by sowing cucumber mosaic virus infected seed. *Ann. Appl. Biol.* **119:**273–288.

Jones, R. A. C. (1992). Further studies on losses in productivity caused by infection of annual pasture legumes with three viruses. *Aust. J. Agric. Res.* **43:**1229–1241.

Jones, R. A. C. (1993). Effects of cereal borders, admixture with cereals and plant density on the spread of bean yellow mosaic potyvirus into narrow-leafed lupins (*Lupinus angustifolius*). *Ann. Appl. Biol.* **122:**501–518.

Jones, R. A. C. (1994a). Effects of mulching with cereal straw and row spacing on spread of bean yellow mosaic potyvirus into narrow-leafed lupins (*Lupinus angustifolius*). *Ann. Appl. Biol.* **124:**45–58.

Jones, R. A. C. (1994b). Infection of subterranean clover swards with bean yellow mosaic potyvirus: Losses in herbage and seed yields and patterns of virus spread. *Aust. J. Agric. Res.* **45:**1427–1444.

Jones, R. A. C. (2000). Determining "threshold" levels for seed-borne virus infection in seed stocks. *Virus Res.* **71:**171–183.

Jones, R. A. C. (2001). Developing integrated disease management strategies against non-persistently aphid-borne viruses: A model program. *Int. Pest Man. Rev.* **6:**15–46.

Jones, R. A. C. (2004a). Using epidemiological information to develop effective integrated virus disease management strategies. *Virus Res.* **100:**5–30.

Jones, R. A. C. (2004b). Occurrence of virus infection in seed stocks and 3-year-old pastures of lucerne (*Medicago sativa*). *Aust. J. Agric. Res.* **55:**757–764.

Jones, R. A. C. (2005). Patterns of spread of two non-persistently aphid-borne viruses in lupin stands under four different infection scenarios. *Ann. Appl. Biol.* **146:**337–350.

Jones, R. A. C., and Ferris, D. G. (2000). Suppressing spread of alfalfa mosaic virus in grazed legume pasture swards using insecticides and admixture with grass, and effects of insecticides on numbers of aphids and three other pasture pests. *Ann. Appl. Biol.* **137:**259–271.

Jones, R. A. C., and Nicholas, D. A. (1992). Studies on alfalfa mosaic virus infection of burr medic (*Medicago polymorpha*) swards: Seed-borne infection, persistence, spread and effects on productivity. *Aust. J. Agric. Res.* **43:**697–715.

Jones, R. A. C., and Nicholas, D. A. (1998). Impact of an insidious virus disease in the legume component on the balance between species within self-regenerating annual pasture. *J. Agric. Sci. Cambridge* **131:**55–170.

Jones, R. A. C., Koenig, R., and Lesseman, D. E. (1980). Pepino mosaic virus, a new potexvirus from pepino (*Solanum muricatum*). *Ann. Appl. Biol.* **94:**61–68.

Jones, R. A. C., Coutts, B. A., and Cheng, Y. (2003). Yield limiting potential of necrotic and non-necrotic strains of *Bean yellow mosaic virus* in narrow-leafed lupin (*Lupinus angustifolius*). *Aust. J. Agric. Res.* **54:**849–859.

Jones, R. A. C., Latham, L. J., and Coutts, B. A. (2004). Devising integrated disease management tactics against plant viruses from "generic" information on control measures. *Agric. Sci. Aust.* **17:**10–18.

Jones, R. A. C., Coutts, B. A., Mackie, A. E., and Dwyer, G. I. (2005a). Seed transmission of *Wheat streak mosaic virus* shown unequivocally in wheat. *Plant Dis.* **89:**1048–1050.

Jones, R. A. C., Smith, L. J., Gajda, B. E., and Latham, L. J. (2005b). Patterns of spread of *Carrot virus Y* in carrot plantings and validation of control measures. *Ann. Appl. Biol.* **147:**57–67.

Lambert S., Hay, F., Pethybridge, S., and Wilson, C. (2005). Spread of potato carlavirus S (PVS) and potato potexvirus X (PVX) in seed potato crops in Tasmania, Australia. *In* "Proceedings of the IXth International Plant Virus Epidemiology Symposium–Applying Epidemiological Research to Improve Virus Disease Management," p. 93 (Abstr.). Lima, Peru, April 4–7, 2005.

Latham, L. J., and Jones, R. A. C. (1997). Occurrence of tomato spotted wilt tospovirus in native flora, weeds and horticultural crops. *Aust. J. Agric. Res.* **48:**359–369.

Latham, L. J., and Jones, R. A. C. (1998). Selection of resistance breaking strains of tomato spotted wilt tospovirus. *Ann. Appl. Biol.* **133:**385–402.

Latham, L. J., and Jones, R. A. C. (2001). Incidence of virus infection in experimental plots, commercial crops, and seed stocks of cool season crop legumes. *Aust. J. Agric. Res.* **52:**397–413.

Latham, L. J., and Jones, R. A. C. (2003). Incidence of *Celery mosaic virus* in celery crops in south-west Australia and its management using a "celery-free period." *Australasian Plant Pathol.* **32:**527–531.

Latham, L. J., and Jones, R. A. C. (2004a). Carrot virus Y: Symptoms, losses, incidence, epidemiology and control. *Virus Res.* **100:**89–99.

Latham, L. J., and Jones, R. A. C. (2004b). Deploying partially resistant genotypes and plastic mulch on the soil surface to suppress spread of lettuce big-vein disease in lettuce. *Aust. J. Agric. Res.* **55:**131–138.

Latham, L. J., Jones, R. A. C., and Coutts, B. A. (2004a). Yield losses caused by virus infection in four combinations of non-persistently aphid-borne virus and cool-season crop legume. *Aust. J. Exp. Agric.* **44:**57–63.

Latham, L. J., Jones, R. A. C., and McKirdy, S. J. (2004b). Lettuce big-vein disease: Sources, patterns of spread, and losses. *Aust. J. Agric. Res.* **55:**125–130.

Lovisolo, O., Hull, R., and Rosler, O. (2003). Coevolution of viruses with hosts and vectors and possible paleontology. *Adv. Virus Res.* **62:**325–379.

Makkouk, K. M., Kumari, S. G., Hughes, J.d'A., Muniyappa, V., and Kulkarni, N. K. (2003). Other legumes: Faba bean, chickpea, lentil, pigeonpea, mungbean, blackgram, lima bean, horsegram, bambara groundnut and winged bean. *In* "Virus and Virus-like Diseases of Major Crops in Developing Countries" (G. Loebenstein and G. Thottappilly, eds.), pp. 447–476. Kluwer Academic Publishers, Dordrecht, The Netherlands.

Makkouk, K. M., Jones, R. A. C., Morales, F., and Kumari, S. G. (2005). Management of virus diseases in food legumes. *In* "Proceedings of Xth International Food Legumes Conference." New Delhi, India (in press).

McKirdy, S. J., and Jones, R. A. C. (1995). Occurrence of alfalfa mosaic and subterranean clover red leaf viruses in legume pastures in Western Australia. *Aust. J. Agric. Res.* **46:**763–774.

McKirdy, S. J., and Jones, R. A. C. (1996). Use of imidacloprid and newer generation synthetic pyrethroids to control spread of barley yellow dwarf luteovirus in cereals. *Plant Dis.* **80:**895–901.

McKirdy, S. J., and Jones, R. A. C. (1997a). Effect of sowing time on barley yellow dwarf virus infection in wheat: Virus incidence and grain yield losses. *Aust. J. Agric. Res.* **48:**99–206.

McKirdy, S. J., and Jones, R. A. C. (1997b). Further studies on the incidence of virus infection in white clover pastures. *Aust. J. Agric. Res.* **48:**31–37.

McKirdy, S. J., Jones, R. A. C., and Nutter, F. (2002). Quantification of yield losses caused by barley yellow dwarf virus in cereals. *Plant Dis.* **86:**769–773.

McLean, G. D., Burt, J. R., Thomas, D. W., and Sproul, A. N. (1982). The use of reflective mulch to reduce the incidence of watermelon mosaic virus in Western Australia. *Crop Prot.* **1:**491–496.

Momol, M. T., Olson, S. M., Funderburk, J. E., Stravisky, J., and Marois, J. J. (2004). Integrated management of tomato spotted wilt on field-gown tomatoes. *Plant Dis.* **88:**882–890.

Moury, B., Palloix, A., Gebre Selassie, K., and Marchoux, G. (1997). Hypersensitive resistance to tomato spotted wilt virus in three *Capsicum chinense* accessions is controlled by a single gene and is overcome by virulent strains. *Euphytica* **94:**45–52.

Mumford, R., and Jones, R. A. C. (2005). *Pepino mosaic virus*. AAB Descriptions of Plant Viruses, No. 411, 11 pp.

Pappu, S. S., Pappu, H. R., Langston, D. B., Jr., Flanders, J. T., Riley, D. G., and Diaz-Perez, J. C. (2000). Outbreak of *Tomato yellow leaf curl virus* (Family *Geminiviridae*) in Georgia. *Plant Dis.* **84:**370.

Perring, T. M., Gruenhagen, N. M., and Farrar, C. A. (1999). Management of plant viral diseases through chemical control of insect vectors. *Annu. Rev. Entomol.* **44:**457–481.

Pethybridge, S. J., Wilson, C. R., Ferrandino, F. J., and Leggett, G. W. (2000). Spatial analyses of viral epidemics in Australian hop gardens: Implications for mechanisms of spread. *Plant Dis.* **84:**513–515.

Polston, J. E., and Anderson, P. K. (1997). The emergence of whitefly-transmitted geminiviruses in tomato in the western hemisphere. *Plant Dis.* **81:**1358–1369.

Polston, J. E., Bois, D., Serra, C. A., and Concepción, S. (1994). First report of a tomato yellow leaf curl-like geminivirus in the western hemisphere. *Plant Dis.* **78:**831.

Polston, J. E., McGovern, R. J., and Brown, L. G. (1999). Introduction of tomato yellow leaf curl virus in Florida and implications for the spread of this and other geminiviruses of tomato. *Plant Dis.* **83:**984–988.

Raccah, B. (1986). Nonpersistent viruses: Epidemiology and control. *Adv. Virus Res.* **31:**350–429.

Riley, D. G., and Pappu, H. R. (2000). Evaluation of tactics for management of thrips-vectored tomato spotted wilt virus in tomato. *Plant Dis.* **84:**847–852.

Roggero, P., Masenga, V., and Tavella, L. (2002). Field isolates of *Tomato spotted wilt virus* overcoming resistance in pepper and their spread to other hosts in Italy. *Plant Dis.* **86:**950–954.

Sseruwagi, P., Sserubombwe, W. S., Legg, J. P., Ndunguru, J., and Thresh, J. M. (2004). Methods of surveying for the incidence and severity of cassava mosaic disease and whitefly vector populations on cassava in Africa: A review. *Virus Res.* **100:**129–142.

Tairo, F., Musaka, S. B., Jones, R. A. C., Kullaya, A., Rubaihayo, P. R., and Valkonen, J. P. T. (2005). Unravelling the genetic diversity of the three main viruses involved in sweet potato virus disease (SPVD), and its practical implications. *Mol. Plant Pathol.* **6:**199–211.

Tavella, L., Alma, A., Conti, A., and Arzone, A. (1996). Evaluation of the effectiveness of *Orius* spp. in controlling *Frankliniella occidentalis*. *Acta Hort.* **431:**499–506.

Thackray, D. J., and Jones, R. A. C. (2003). Forecasting aphid outbreaks and epidemics of *Barley yellow dwarf virus*—a decision support system for a Mediterranean-type climate. *Australasian Plant Pathol.* **32:**438(Abstr.).

Thackray, D. J., Jones, R. A. C., Bwye, A. M., and Coutts, B. A. (2000). Further studies on the effects of insecticides on aphid vector numbers and spread of cucumber mosaic virus in narrow-leafed lupins (*Lupinus angustifolius*). *Crop Prot.* **19:**121–139.

Thackray, D. J., Diggle, A. J., Berlandier, F. A., and Jones, R. A. C. (2004). Forecasting aphid outbreaks and epidemics of *Cucumber mosaic virus* in lupin crops in a Mediterranean-type environment. *Virus Res.* **100:**67–82.

Thackray, D. J., Ward, L. T., Thomas-Carroll, M. L., and Jones, R. A. C. (2005). Role of winter-active aphids spreading *Barley yellow dwarf virus* in decreasing wheat yields in a Mediterranean-type environment. *Aust. J. Agric. Res.* **56:**1089–1099.

Thomas, J. A., and Hein, G. L. (2003). Influence of volunteer wheat plant condition on movement of the wheat curl mite, *Aceria tosichella*, in winter wheat. *Exp. Appl. Acarol.* **31:**253–268.

Thomas-Carroll, M. L., and Jones, R. A. C. (2003). Selection, biological properties and fitness of resistance-breaking strains of *Tomato spotted wilt virus* in pepper. *Ann. Appl. Biol.* **142:**235–243.

Thresh, J. M. (1980). The origins and epidemiology of some important plant virus diseases. *Appl. Biol.* **5:**1–65.

Thresh, J. M. (ed.) (1981). "Pests Pathogens and Vegetation." Pitman, London, UK.

Thresh, J. M. (1982). Cropping practices and virus spread. *Annu. Rev. Phytopathol.* **20:**193–218.

Thresh, J. M. (1983). Plant virus epidemiology and control: Current trends and future prospects. *In* "Plant Virus Epidemiology—the Spread and Control of Insect-Borne Pathogens" (R. T. Plumb and J. M. Thresh, eds.), pp. 349–360. Blackwell, Oxford, UK.

Thresh, J. M. (2003). Control of plant virus diseases in sub-Saharan Africa: The possibility and feasibility of an integrated approach. *African Crop Sci. J.* **11:**199–223.

Thresh, J. M., and Cooter, R. J. (2005). Strategies for controlling cassava mosaic virus disease in Africa. *Plant Pathol.* **54:**587–614.

Thresh, J. M., and Fargette, D. (2003). The epidemiology of African plant viruses: Basic principles and concepts. *In* "Plant Virology in Sub-Saharan Africa" (J. d'A. Hughes and B. O. Odu, eds.), pp. 61–111. International Institute for Tropical Agriculture, Ibadan, Nigeria.

Thresh, J. M., Owusu, G. L. K., and Ollennu, L. A. A. (1988). Cocoa swollen shoot: An archetypal crowd disease. *J. Plant Dis. Prot.* **95:**428–446.

Thresh, J. M., Fargette, D., and Otim-Nape, G. W. (1994). Effect of African cassava mosaic geminivirus on the yield of cassava. *Trop. Sci.* **34:**26–42.

van der Vlugt, R. A. A., Stijger, C. C. M. M., Verhoeven, J. Th. J., and Lesemann, D. E. (2000). First report of *Pepino mosaic virus* on tomato. *Plant Dis.* **84:**103.

van Rheenan, H. A., Hasselbach, O. E., and Muigai, S. G. S. (1981). The effect of growing beans together with maize on the incidence of bean diseases and pests. *Neth. J. Plant Pathol.* **87:**193–199.

Verhoeven, J. Th. J., van der Vlught, R. A. A., and Roenhorst, J. W. (2003). High similarity between isolates of *Pepino mosaic virus* suggests a common origin. *Eur. J. Plant Pathol.* **109:**419–425.

Ward, E., Foster, S. J., Fraaije, B. A., and McCartney, H. A. (2004). Plant pathogen diagnostics: Immunological and nucleic acid-based approaches. *Ann. Appl. Biol.* **145:**1–16.

Wilson, C. R., and Jones, R. A. C. (1990). Virus content of seed potato stocks produced in a unique seed potato production scheme. *Ann. Appl. Biol.* **116:**103–109.

Wilson, C. R., and Jones, R. A. C. (1995). Occurrence of potato virus X strain group 1 in seed stocks of potato cultivars lacking resistance genes. *Ann. Appl. Biol.* **127:**479–487.

Zadoks, J. C., and Schein, R. D. (1979). "Epidemiology and Plant Disease Management." Oxford University Press, Oxford, UK.

Zitter, T. A., and Simons, J. N. (1980). Management of viruses by alteration of vector efficiency and cultural control practices. *Annu. Rev. Phytopathol.* **18:**289–310.

ADVANCES IN VIRUS RESEARCH, VOL 67

CONTROL OF TROPICAL PLANT VIRUS DISEASES

J. M. Thresh

Natural Resources Institute, University of Greenwich, Chatham Maritime
Kent ME4 4TB, United Kingdom

Plant viruses and virus diseases have been studied for more than 100 years and much attention has been given to their control. However, this has been difficult to achieve because of the lack of any effective means of curing virus-infected plants. Chemotherapy, thermotherapy and meristem-tip culture can be successful in eliminating viruses from plant tissue, but they cannot be used on a large scale. Consequently, the main approach has been to prevent or delay virus infection or to ameliorate its effects. Various means have been used to achieve these objectives, including phytosanitation (involving quarantine measures, crop hygiene, virus-free planting material and eradication), changes in cropping practices, the use of pesticides to control vectors, mild strain protection and the deployment of resistant or tolerant varieties. These measures can be used singly or in combination so as to exploit synergistic interactions. This chapter considers the advantages of an integrated approach to the control of tropical plant virus diseases and provides selected examples from experience with cacao swollen shoot, cassava mosaic, groundnut rosette and rice tungro. It is emphasised that much detailed research and a thorough understanding of agricultural and horticultural practices is required before effective

0065-3527/06 $35.00
DOI: 10.1016/S0065-3527(06)67007-3

integrated disease management programmes can be developed and promoted. There are also formidable problems in ensuring adoption because of the generally limited education and resources of farmers in many tropical areas and the severe constraints imposed by the patterns of land use and cropping practices adopted. Nevertheless, integrated control measures have evident benefits and should be fostered and promoted as a means of enhancing crop productivity to meet the increasing demands of a burgeoning human population and to do so in a sustainable manner and without damaging the environment.

I. Introduction

Higher plants provide habitats for a wide range of pathogens of which viruses are some of the most prevalent. They affect virtually all crop species, including many that are of great importance in agriculture and horticulture. The effects of viruses are seldom benign and they usually decrease crop growth and yield and may cause serious losses. This has long been recognized and provided an early incentive to study viruses of crop plants. One of the main objectives is to develop effective control measures that can be used on a large scale to increase crop productivity and to make the most effective use of the land, labour and other resources being utilized.

The main approaches to control are considered here, with the emphasis on virus diseases of tropical crops. These are of particular importance because of the crucial role played by agriculture in the tropics where there is a greater dependence on crop production for subsistence, employment and export earnings than in other more developed regions of the world (Thresh, 2003). The constraints to the adoption of control measures are assessed and the scope for developing and utilizing an integrated approach. Four particularly important tropical diseases are discussed in detail as specific 'case histories'.

II. Control Measures

Some viruses can be eliminated from infected plants by heat or meristem-tip therapy, or by using chemicals (Faccioli and Marani, 1998; Mink *et al.*, 1998). These methods are used widely to develop virus-free seed or vegetative propagules for further multiplication and release to growers. However, therapy cannot be used on a large scale and the lack of any feasible means of curing infected plants is an important constraint to control. Consequently, other approaches have been adopted. These are to:

- prevent plants from becoming infected;
- delay infection to such a late stage of crop growth that yields are not seriously impaired;
- decrease the deleterious effects of infection.

These objectives can be achieved in different ways as discussed in the following sections under five main headings.

A. Phytosanitation

This term is applied to various approaches to control achieved by decreasing the number of foci of infection from which further virus spread can occur. There are five main ways of doing this:

- quarantine measures to avoid introducing viruses and their vectors to areas where they are not already established;
- sanitation practices involving the removal of all surviving plants, debris and self-sown 'volunteer' seedlings of previous crops;
- removal from within and around crops of any weed or wild plants known to be alternative hosts of viruses;
- use of virus-free stocks of seed or vegetative propagules for all new plantings;
- removal ('roguing') of diseased plants from within plantings, especially those found during the early most vulnerable stages of crop growth.

1. Quarantine

The information available on the geographic distribution of viruses and their vectors is inadequate, especially in the tropics where there is often a lack of adequate facilities and trained personnel to carry out the necessary surveys and virus identifications. This creates difficulties in interpreting reports of a seemingly 'new' disease, or of the apparent 'spread' of a known disease into a new area. Nevertheless, it is apparent that some viruses and vectors are restricted to certain regions and absent from others. Some of the most important of the potential threats to crops in other tropical regions are:

- *Rice hoja blanca virus* (currently restricted to South and Central America);
- rice tungro viruses (South and South-East Asia);

- *Rice yellow mottle virus* (Africa);
- *Indian cassava mosaic virus* (India/Sri Lanka);
- *Sri Lankan cassava mosaic virus* (Sri Lanka);
- African cassava mosaic viruses (Africa);
- *Cassava brown streak virus* (Africa);
- *Cassava common mosaic virus* (South America);
- *Cassava vein mosaic virus* (South America);
- *Maize rough dwarf virus* (Mediterranean);
- *Maize rayado fino virus* (South/Central America);
- *Groundnut bud necrosis virus* (Asia);
- groundnut rosette viruses (Africa).

There are obvious advantages in adopting quarantine and other measures in attempts to maintain the situation and to avoid introducing viruses or their vectors to other areas where they could become established and cause problems (Foster and Hadidi, 1998). This has long been recognized and there has been considerable expenditure on establishing and operating quarantine facilities at several centres in the tropics or subtropics. In Africa, they include Muguga in Kenya, Ibadan in Nigeria and Pretoria in Republic of South Africa. There are also special quarantine facilities for banana and plantain at the University of Louven, Belgium, and for cacao at the University of Reading, United Kingdom (Hadley *et al.*, 1989). Regulations and procedures are collated and co-ordinated by the United Nations Food and Agriculture Organization (FAO), which has published a series of guidelines for the safe movement of germplasm (Frison and Putter, 1989). However, there are many difficulties in implementation and enforcement due to the continually increasing scale of international travel and trade and in the movement of plant material for breeding and crop improvement programmes. There are also particular problems in controlling the movement across land borders and difficulties associated with the inevitable disruptions caused by natural disasters, insecurity and civil unrest.

These problems have led to the suggestion that quarantine controls are of limited value because pests and pathogens will eventually become established in all the areas where agroecological conditions are suitable (Bennett, 1952; Kahn, 1979). However, this 'inevitability concept' is unduly pessimistic and there are cogent arguments for enforcing quarantine controls to delay the introduction of exotic pests and pathogens for as long as possible and to provide the opportunity to introduce resistant varieties and make other contingency arrangements for use should the need arise (Ebbels, 2003; Hewitt and

Chiarappa, 1977; Kahn, 1989). This emphasises the importance of maintaining and improving quarantine procedures and the need to exploit the advances being made in virus diagnostics to develop new and more sensitive techniques that can be used routinely to increase the throughput of samples and overcome previously intractable problems in virus detection. One of these is posed by *Banana streak virus*, which can be integrated within the host-plant genome and activated to cause symptoms in response to environmental stress factors that are as yet ill defined (Ndowora *et al.*, 1999). This has complicated quarantine procedures and led to constraints on the movement of *Musa* germplasm to avoid the risk of introducing the virus to new areas. The effect has been to greatly impede banana improvement and breeding programmes.

2. Crop Sanitation

The epidemiological literature contains numerous examples of the hazards posed by the debris of previous crops and by regrowth from the tubers, roots, stems or other infected plant material left in or on the ground at harvest. There are also problems due to the growth of 'self-sown' seedling 'volunteers' of crops such as cereals, rice, pigeon pea and groundnut. This facilitates the survival and perennation of viruses and their vectors and can provide a 'green bridge' between successive growing seasons. An African example is provided by groundnut rosette disease and the main aphid vector (*Aphis craccivora*). These are most prevalent in areas and in seasons when there is abundant growth of volunteer groundnut seedlings from previous crops that act as 'bridging' hosts (Evans, 1954; Storey and Bottomley, 1928).

Rice provides another example because of the problems posed by self-sown seedlings and regenerating stubbles of previous crops, especially in situations in which growth is facilitated by the availability of adequate soil moisture throughout all or much of the year. Perennation in this way is a feature of the epidemiology of all three of the main virus diseases of tropical rice: tungro in Asia, yellow mottle in Africa and hoja blanca in South/Central America. This explains why adequate land preparation to eliminate stubbles and volunteer seedlings before planting begins is an important feature of control recommendations against these and other rice diseases (Hibino, 1996). For example, strict measures are enforced in the Muda and other large-scale irrigation schemes for commercial rice production in Malaysia. All crop residues must be destroyed and ploughing must be completed throughout the entire area before water is released for new plantings (Nozaki *et al.*, 1984). Such measures are seldom adopted in the many other areas where rice is grown mainly by individual smallholders, although there are undoubtedly benefits to be gained by doing so.

The advantages of adopting husbandry practices that decrease the amount of crop debris and impede survival were appreciated at an early stage in studies on cotton leafcurl disease in the Sudan Gezira irrigation scheme. Special implements were devised to facilitate the removal and destruction of the cotton stumps remaining after harvest. These would otherwise have survived and regenerated to become initial foci of infection in subsequent plantings (Tarr, 1951). Similar measures are also adopted to avoid the carry-over of inoculum in sugarcane, tobacco and other commercial crops and the recommendations have at times been enforced by legislation to ensure the removal of all crop residues before new planting begins, as with tobacco in what is now Zimbabwe (Hopkins, 1932; Shaw, 1976).

The Dominican Republic provides an example of the use of legislation to enforce crop sanitation. The losses caused by whitefly-borne viruses of melon, tomato and other export crops in the late 1980s became so great that a crop-free period was introduced. There was a ban on bean, tomato, melon, pepper and eggplant cultivation during a 4-month period of each year to break the otherwise continuous cycle of whitefly reproduction and virus spread (Alvarez and Abu-Antún, 1995; Salati *et al.*, 2002). Compliance was intended to be compulsory and the measure was regarded as successful (Morales and Anderson, 2001). However, government enforcement was uneven and the main practitioners were the large commercial growers with the necessary financial resources to adopt a full range of control measures. These were seldom used by the many small-scale producers (Polston and Anderson, 1997).

3. *Removal of Weed or Wild Hosts*

Many viruses have weed or wild hosts that act as foci of infection from which there is spread into or within crops (Bos, 1981; Thresh, 1981; Cooper and Jones, this volume, pp. 1–47). For example, the initial distribution of maize dwarf mosaic disease is often associated with patches of the perennial grass weed *Sorghum halepense* that occur commonly within and around crop stands (Damsteegt, 1976). Similarly, the common composite weed *Tridax procumbens* is the host of the aphid-borne virus or viruses that cause ringspot and leafcurl diseases of sunflower in eastern and southern Africa (Theuri *et al.*, 1987). These examples emphasise the importance of adopting effective weed control and other measures to avoid competition and also to eliminate sources of inoculum. The advantages to be gained are apparent from experience with *Cacao swollen shoot virus* in the western region of Ghana, where many of the outbreaks in cacao are associated with the understorey forest tree *Cola chlamydantha* (Attafuah, 1965; Todd, 1951). These trees are indigenous hosts of the virus from which spread occurs to nearby cacao and the wild hosts are removed when outbreaks are treated during the official eradication campaign that is being operated (pp. 267–271).

Elsewhere, there have been restrictions on the cultivation of okra and malvaceous ornamentals in and around the Gezira irrigation scheme in the Sudan, as these plants are known to be hosts of *Cotton leafcurl virus* and the whitefly vector *Bemisia tabaci* (Tarr, 1951). Such measures are likely to be more widely applicable, but they are difficult to enforce in the tropics where little attempt is made to control virus diseases by removing weed or alternative hosts.

4. Virus-Free Propagules

The use of virus-free propagules for all new plantings is a basic approach to control that is beneficial for several reasons:

- virus-free plants usually establish more readily and are more productive than infected ones;
- if virus-free material is adopted, there are no initial foci of infection within crops at the outset, during the early and most vulnerable stages of crop growth. This delays and curtails the period over which any subsequent spread can occur;
- plants not infected until a late stage of crop growth are usually affected less severely than those infected early;
- infected propagules are particularly dangerous sources of inoculum because they tend to be distributed randomly within crops. This facilitates virus spread from infected to neighbouring healthy plants, whether this is by contact or by vectors.

For these reasons, much attention has been given in technologically advanced countries to producing virus-free stocks of seed and of the tubers, bulbs, cuttings or other propagules of crops that are propagated vegetatively (Hollings, 1965). There are not usually any major technical problems in obtaining stocks that are free or largely free of infection by careful selection from those already available, or by using some form of therapy, as discussed previously (pp. 267–271). When such stocks are obtained, they are multiplied at carefully selected sites where there is spatial or temporal isolation from sources of infection and likely to be little or no contamination. The 'elite' stocks are then released to seed merchants, wholesale distributors, specialist nurserymen or direct to farmers. Official inspection and certification procedures are used to designate and maintain the health status of the stocks being distributed and they are widely used for temperate fruit, potato, ornamentals and several other crops in North America, Europe, Australasia and elsewhere (Ebbels, 1979; Waterworth, 1998).

A similar approach would undoubtedly bring considerable benefits with many tropical crops and facilitate the control of virus diseases including those of cassava, potato, sweet potato, yams, sugarcane, citrus and other vegetatively propagated crops. There are also advantages to be gained from producing stocks that are free of seed-borne viruses, including *Bean common mosaic virus, Cowpea mosaic virus, Soybean mosaic virus, Tomato mosaic virus* and others that are causing serious losses. However, only limited attempts have been made to produce and deploy virus-free stocks, except to meet international quarantine requirements and so facilitate the movement of germplasm, commercial stocks and other plant material within and between regions. An exception to this generalization is in the Republic of South Africa where seed and stock certification schemes are available for potato, sweet potato, citrus and a range of other crops. Banana is another crop of which selected stocks are available and virus-free micropropagated material is being used widely in Taiwan and also in Kenya, Uganda and elsewhere. There has also been considerable use of virus-free stocks of potato in India, Pakistan, North Africa, Ethiopia, Sudan, Kenya and several South American countries; 'seed' tubers are produced locally from stocks selected in the country or imported from Europe. In addition, there have been imports to Africa and South/Central America of certified stocks of rose, chrysanthemum, carnation and other ornamentals, and seed of horticultural crops including tomato, lettuce, common bean (*Phaseolus vulgaris*) and cucurbits. These are multiplied for local use and also re-exported to Europe or North America in commercial consignments.

These examples are few, but they indicate the scope for a generally increased use of virus-free material of a much wider range of tropical crops. However, this will not be achieved quickly or easily because of the need to develop appropriate seed production and marketing systems and to educate farmers and encourage a transition from traditional subsistence farming to more productive commercial practices utilizing certified stocks of improved quality and health status. Meanwhile, commercially oriented seed production and distribution systems in many tropical countries and especially in sub-Saharan Africa (SSA) are poorly developed. The usual practice of farmers in the tropics is to retain seed or other propagules from their previous plantings, or from neighbours or relatives. Only limited attention is given to health status, which is frequently unsatisfactory (DeVries and Toenniessen, 2001).

5. Roguing

The removal of diseased plants from within crop stands is usually termed roguing. This is a well-known means of virus disease control that is achieved by eliminating initial sources of infection from which further spread can occur. Roguing is widely applicable and it has been used in attempts to control or at least contain diseases of diverse crops in both temperate and tropical regions (Thresh, 1988). As with other phytosanitation measures, the approach is most effective against viruses that do not spread quickly or far in any considerable amount (Putter, 1980). However, roguing is generally unpopular with farmers, who are seldom prepared to allocate the time and effort required to inspect crops with the thoroughness and frequency required to identify and remove diseased plants at an early stage of infection. There is also a reluctance to rogue any diseased plants that may contribute at least some yield and an even greater unwillingness to remove neighbouring symptomless plants, although this may be necessary to eliminate latent infections (p. 268).

For these reasons, roguing has been most successful when adopted by government or commercial organizations, which take on the responsibility of employing and training staff to carry out or supervise the regular inspections and treatments required. This approach has been adopted in Australia against banana bunchy top and sugarcane Fiji diseases and in parts of Europe and North America against citrus tristeza, plum pox and other diseases of fruit crops (Thresh, 1988). Roguing has also been used widely to maintain the health status of 'seed' potato stocks. The inspection and treatment procedures adopted in operating such schemes should be based on epidemiological information; where infection is prevalent it can be advantageous to remove whole plantings as a means of safeguarding less severely affected plantings nearby.

The eradication campaign against cacao swollen shoot disease in Ghana has been the biggest and most ambitious ever undertaken in Africa or elsewhere (pp. 267–271). There has, at times, been considerable opposition to the campaign from farmers and such measures are seldom appropriate or feasible in the tropics. This explains why so little use has been made of roguing as a means of controlling cassava mosaic or other diseases that could be amenable to this approach. However, roguing has been used to improve the health status of planting material of cassava or other crops being propagated by researchers or extentionists for distribution to farmers. There is also scope for roguing stands of sugarcane, banana, citrus, papaya, pineapple or other plantation crops and especially on large commercial farms. For example, banana plantations in the Philippines that produce crops for export are inspected regularly to detect banana bunchy top and other systemic diseases. The aim is to remove all infected plants before they have become important sources of inoculum (Smith *et al.*, 1998). However, these circumstances are exceptional, individual small-scale farmers make little use of roguing and the situation is unlikely to change in the immediate future.

B. Changing Cropping Practices

There is abundant evidence from a wide range of crops of the importance of cropping practices in determining the prevalence of virus diseases and the losses they cause (Thresh, 1982). A further detailed consideration of the many husbandry practices that are known to influence virus spread and control is beyond the scope of this chapter. Nevertheless, some of the most important are listed in Table I in which it is convenient to distinguish between practices adopted before or at planting and those adopted later.

Pre-planting factors that influence disease prevalence include the choice of site, as based on previous cropping history, site suitability and availability and the selection of planting material. Factors at planting include the date(s) chosen, the spacing adopted, the size and shape of the area planted and the choice of variety or varieties and of any intercrop(s) grown. Post-planting factors include thinning and harvesting dates, the method, timing, frequency and effectiveness of any weed control measures and use of irrigation, fertilizers, herbicides or pesticides.

Virus spread is facilitated by some cropping practices and impeded by others. This provides an opportunity for farmers and plantation managers to adopt measures that are beneficial or at least 'neutral' in their effects and to avoid those that are detrimental. Farmers may do so empirically from the experience gained by themselves and their predecessors over many years indicating that some practices usually lead to satisfactory and reliable yields whereas others do not. However, it is more advantageous for decisions to be based on, or at least influenced by, a detailed knowledge of the epidemiology of the virus or viruses of concern. Such information is not easily obtained and is seldom available for diseases of tropical crops. Moreover, even if the information is available it is not easily disseminated to farmers,

TABLE I

CROPPING PRACTICES THAT INFLUENCE VIRUS SPREAD

- Pre-planting
  - Site selection:
    - Cropping history/isolation
    - Field size/shape/orientation/aspect
  - Crop/cultivar selection:
    - Single/multiple crops
    - Single/multiple cultivars
    - Seed/vegetative propagules
    - Source of propagules
- Planting
  - Sowing/planting:
  - Direct planted/transplanted
  - Planting/transplanting dates
- Crop spacing/arrangement:
  - Plant population
  - In-row/between row spacing
- Pesticide/herbicide/fertilizer application:
  - At or before planting
  - Amount/type
- Post-planting
  - Weed control/tillage:
    - Method/frequency/effectiveness
  - Fertilizer:
    - Amount/type/timing/method of application
  - Thinning/pruning:
    - Crop growth stage/extent/method
  - Roguing:
    - Intensity/timing/frequency/extent
  - Irrigation:
    - Amount/mode/frequency
  - Harvesting:
    - Method/date

especially those who are largely illiterate and with little or no awareness of even the most rudimentary principles of plant pathology.

A further problem is that cropping practices known to be beneficial in decreasing disease incidences are not always feasible or appropriate in relation to other agricultural or horticultural requirements, or on socio-economic grounds. This is evident from experience with maize streak disease in the Democratic Republic of Congo. It is well known from experiences in Congo and elsewhere in Africa that the disease is more prevalent in late than in early plantings. However, the area of land that can be cleared and planted early often depends on the amount of manual labour available within the family or local community, which can be so inadequate as to become a serious constraint (Vogel *et al.*, 1993). Consequently, late plantings are not made or they are seriously affected by streak. Other examples of the importance of cropping practices are provided in the case histories of specific diseases (pp. 267–282).

The benefits of close spacing in decreasing the proportion of infected plants within a stand have been demonstrated in experiments on groundnut rosette (A'Brook, 1964; Davies, 1976, Farrell, 1976a), groundnut bud necrosis (Reddy *et al.*, 1983) and cassava mosaic diseases (Egabu *et al.*, 2002; Fargette *et al.*, 1990). It has also been shown that the incidence of *Tomato spotted wilt virus* in tobacco is decreased if selective thinning to the final spacing required is delayed until after the main influx of thrips vectors has occurred (Vanderplank and Anderssen, 1944a,b). This finding has been exploited to provide what was referred to as 'a mathematical solution to a problem of disease'.

Several other cropping practices merit further discussion in relation to tropical virus diseases. One relates to the sites used for new plantings. It may be advantageous to adopt fresh sites or suitable rotations to avoid any carry-over of inoculum from previous crops. For example, crop rotation is a means of decreasing persistent infestations of Johnson grass (*Sorghum halepense*), which is an important perennial host of *Maize dwarf mosaic virus* (Eberwine and Hagood, 1995). There may also be advantages in planting crops in large compact blocks of uniform age to decrease the proportion of plants in the vulnerable peripheral areas. This is because incoming insect vectors tend to alight preferentially and lead to a high incidence of infection around the margins. This has been reported with cassava mosaic disease and the whitefly vector (Colvin *et al.*, 1998; Fargette *et al.*, 1985) and with groundnut rosette disease and its aphid vector (Hayes, 1932). Moreover, it is advantageous to provide at least some degree of isolation from known sources of inoculum and to consider the direction of the prevailing wind. Plantings should be upwind rather than downwind and as far as possible from major foci of infection (Thresh, 1976).

These principles are utilized widely in temperate agriculture and especially in selecting sites for producing virus-free stocks of seed or vegetative propagules. Comparable benefits could be gained in the tropics as apparent from studies on several diseases in which the incidence of infection decreased with increasing distance from the source. Such gradients tend to be curvilinear and of concave shape, which means that the decrease is usually most rapid over quite short distances near the sources, as reported with cacao swollen shoot (Thresh and Lister, 1960), cassava mosaic (Fargette *et al.*, 1990), cotton leafcurl (Giha and Nour, 1969), maize streak (Gorter, 1953; Rose, 1973), okra mosaic (Atiri and Varma, 1991), okra leaf curl (Fargette *et al.*, 1993) and rice tungro diseases (Satapathy *et al.*, 1997). The advantages of isolation and siting in relation to wind direction and known sources of infection have been exploited to decrease the incidence of maize streak disease in South Africa (Gorter, 1953) and also with sugarcane and other tropical plantation crops. However, this is seldom feasible in subsistence agriculture, especially in the many areas where land is scarce, individual holdings are small and there is little or no separation between them and it is difficult to ensure synchrony in planting date. This facilitates spread into and between plantings of a wide range of tropical crops, as exemplified in studies on cacao swollen shoot disease in Ghana (Thresh *et al.*, 1988a) and from experience with maize, rice and other crops that are commonly grown in overlapping sequence throughout all or much of the year.

Genetic diversity is a marked feature of many tropical crops due to the cultivation of numerous farmer-selected local landraces. Examples include common bean, cassava, cowpea, maize, sorghum, potato, sweet potato and yams. Commonly, two or more landraces are grown in each field and the mixtures are sometimes complex. This is exemplified by data for cassava in Uganda (Bua *et al.*, 2005) and in northern areas of Mozambique, where many landraces were recorded in each locality that was surveyed and up to five occurred, even in individual small plantings (Hillocks *et al.*, 2002). Moreover, interplanting two or more different crops is a common practice in many tropical areas and one that is facilitated by the common practice of using hand labour for planting and weeding.

In recent decades, both varietal mixtures and interplanting with other crops have received considerable attention from agronomists, plant pathologists and other researchers. It is now recognized that both types of crop diversity are important in providing farmers with at least some degree of resilience and stability that contributes to reliable yields and decreases the risk of total crop failure (Smithson and Lenné, 1996). Such an outcome is achieved if one or more constituents of the mixture can withstand adverse environmental conditions, or the damaging effects of pests and diseases, even though others may succumb. Beneficial effects of this type are likely to occur with virus diseases. This is because of a decrease in vector populations, or in their movement within crop stands, or by providing physical barriers or camouflage and also by increasing the mean separation distance between susceptible individuals within the mixture. However, little attention has been given to decreasing virus spread in this way and there is need for much additional research. This is apparent from the few published results which show a decrease in the spread of cassava mosaic disease to a susceptible variety of cassava when grown together with mosaic-resistant varieties (Sserubombwe *et al.*, 2001) and also when interplanted with maize or other crops (Ahohuendo and Sarkar, 1995; Fondong *et al.*, 2002). There are also reports of a decrease in the spread of aphid and beetle transmitted viruses of bean when intercropped with maize (van Rheenan *et al.*, 1981), or with cassava and plantain (Gámez and Moreno, 1983). However, the situation can be complex and inconsistent results have been reported in other intercropping trials with cassava (Fargette and Fauquet, 1988) and on the incidence of *Maize streak virus* when maize was intercropped with millet in comparison with sole stands of maize in Uganda (Page *et al.*, 1999). There is also a report of the adverse effect of *Phaseolus* beans on the growth and yield of a groundnut intercrop in Malawi, even though the incidence of rosette disease was decreased (Farrell, 1976b). Clearly, many additional studies of this type and detailed cost/benefit analyses are required to elucidate the role of biodiversity and to assess the full implications of the trend away from the use of crop and varietal mixtures. It may eventually be possible to devise cropping systems that combine the advantages of modern varieties and methods of cultivation with those of traditional practices.

C. Host-Plant Resistance

At least some degree of genetic diversity is a feature of all crop species. It is one of great importance that has been exploited by farmers for millennia, and by agriculturalists and horticulturalists to increase crop productivity and as a means of avoiding the most damaging effects of pests and pathogens. This is achieved by selecting and adopting genotypes that yield satisfactorily and avoid or in some way withstand biotic and abiotic constraints.

In considering host-plant resistance to virus diseases it is helpful to distinguish between *positive* and *negative* selection. *Negative selection* occurs when particularly vulnerable crop genotypes are discarded by farmers or researchers, either because they are recognized as being severely diseased, or simply because they do not grow or yield satisfactorily for reasons that may be unclear. *Positive selection* is the outcome of deliberately selecting particularly resistant genotypes when heterogeneous populations are exposed to infection. Consequently, positive selection requires considerable scientific input and expertise, whereas negative selection is practised even within the most rudimentary of traditional cropping systems. Both approaches to selection have been widely used and intentionally or unintentionally host-plant resistance has made a big contribution to virus disease control by decreasing the incidence and/or consequences of infection.

The use of disease-resistant genotypes has obvious advantages because it provides a convenient means of control that can be used alone or to supplement or complement other measures. Moreover, there are no harmful side effects and no additional costs once resistant varieties are made available that are acceptable to farmers and comparable or superior in performance to the susceptible ones being grown. This explains why breeding for host-plant resistance has long featured so prominently in research on many tropical crops and especially since the establishment of the various International Agricultural Research Centres in the tropics (Bos, 1992; Thurston, 1977). Breeding for resistance to viruses of maize (CIMMYT, IITA), rice (IRRI, WARDA), wheat (CIMMYT), cassava (CIAT, IITA), potato and sweet potato (CIP), common bean (CIAT), groundnut, pigeon pea and sorghum (ICRISAT), chickpea (ICARDA), cowpea (IITA) and vegetables (AVRDC) has featured in research at one or more of the Institutes or regional substations based in Africa, Asia or South/Central America.

The scope for exploiting host-plant resistance is apparent from experience with many tropical crops and diseases. For example, it has long been known in SSA that cassava varieties differ considerably in their response to cassava mosaic disease. Some of the many varieties being grown are more severely damaged by the disease than others and farmers can alleviate the problem of mosaic by retaining the most productive varieties and discarding those that are particularly vulnerable. This occurred during the early epidemics reported in what is now Ghana and elsewhere in West Africa (Dade, 1930) and later in Madagascar (Cours, 1951) and more recently during the severe pandemic in Uganda (Bua *et al.*, 2005; Otim-Nape *et al.*, 2000, 2001; Thresh *et al.*, 1994b; Legg *et al.*, this volume, pp. 355–418; Thresh, this volume, pp. 89–125). The 1990s pandemic in Uganda was most damaging in districts where few varieties were being grown; farmers in other districts were able to select types that were somewhat resistant or tolerant from the much wider range of varieties available. One of the reasons why farmers were able to adapt so quickly to the problem of mosaic is that cassava is propagated vegetatively from stem cuttings collected from previous crops at harvest. Severely diseased plants produce few cuttings that are suitable for further propagation and so tend to be under-represented in subsequent plantings. This occurs even if only some farmers discriminate in favour of vigorous unaffected plants, or those that are only slightly affected.

Similar considerations apply with the many other vegetatively propagated crops that are affected by virus diseases, including banana, yams, potato, sugarcane and sweet potato. This provides farmers with an effective 'coping strategy' and explains why these crops can usually be grown and yield satisfactorily despite the often high incidence of virus infection. Infected plants tend to express inconspicuous symptoms that are not associated with obvious deleterious effects on growth or yield. Such symptoms tend to be ignored, even though growth may be impaired and this explains why such limited attention is given to the use of virus-free planting material.

The value of resistance breeding has long been evident from experience with sugarcane mosaic wherever this disease has been a problem (Summers *et al.*, 1948). Resistant varieties have also been sought in resistance breeding programmes on cassava mosaic, cacao swollen shoot disease, groundnut rosette and rice tungro diseases (see case histories in Section III). Moreover, attention has been given to many other diseases, including bean common mosaic, maize streak, pigeon pea sterility, rice hoja blanca and rice yellow mottle. Considerable success has been achieved and various types of host-plant resistance have been exploited. They include hypersensitivity or other forms of resistance to virus infection, tolerance of infection and resistance to the arthropod vector. These traits have been obtained by selection from amongst existing or introduced varieties, or by crosses to related species.

Virus-resistant varieties are used widely in agriculture and horticulture and they could make an even greater contribution to disease control but for several constraints:

- A considerable research effort is needed to develop effective resistance breeding programmes. These must also take account of other biotic and abiotic constraints and meet the often exacting and sometimes conflicting requirements of farmers, consumers and processors. The necessary funds, personnel and resources are not always available for a sufficiently long period, or they may be inadequate. This is more

PLATE 1. Dead and dying cacao trees in Ghana affected by cacao swollen shoot disease (Thresh, this volume, pp. 97–99, 267–271).

PLATE 2. Rice planting in Philippines affected by rice tungro disease (Thresh, this volume, pp. 103–105, 279–282).

PLATE 3. Groundnut planting affected by groundnut rosette disease (Thresh, this volume, pp. 277–279).

likely to occur if the disease against which resistance is required occurs sporadically and unpredictably than with diseases that consistently cause serious and widespread problems.

• There have been instances of resistant varieties being released without adequate on-farm testing to ensure that the new varieties are suitable for general adoption and that they meet all requirements. Some of the criticism of this so-called 'top down' approach and of the alleged lack of interaction of researchers with subsistence farmers is excessive and unjustified. Nevertheless, there is undoubtedly scope for greater involvement of farmers and consumers in the evaluation and selection of new varieties and a participatory approach is now a requirement of many donors funding crop improvement programmes in the tropics.

• Even if resistant varieties are developed they may not be available for use in quantity because of the lack of an effective multiplication and distribution system, or because farmers are unaware of the benefits to be gained from their adoption.

• The resistance may be associated with undesirable traits or resistant varieties may lack some of the desirable attributes of the susceptible varieties being grown. For example, the first varieties of groundnut that were selected for resistance to rosette disease required a long growing season and were not suitable for general use (pp. 277–279).

• The need to adopt resistant varieties is not necessarily compelling, especially if the disease occurs sporadically and attracts less attention than other factors decreasing yield.

• Resistance may be overcome due to the emergence or increased prevalence of virus strains that damage previously unaffected varieties (García-Arenal and McDonald, 2003; Harrison, 2002). This has been experienced in breeding for resistance to *Sugarcane mosaic virus* (Summers *et al.*, 1948) and to *Rice grassy stunt virus* (Hibino *et al.*, 1985) and *Cotton leafcurl virus* (Ali, 1997). Moreover, varieties that are resistant in some areas may be susceptible in others, as reported with *Rice grassy stunt virus* in India and the Philippines (Ghosh *et al.*, 1979) and with sweet potato virus disease in East and West Africa (Mwanga *et al.*, 1991). For these reasons it may be difficult to develop and exploit broad-based resistance that is also durable, in the sense of long lasting.

Collectively, these constraints are of great importance. Nevertheless, virus-resistant varieties of many tropical crops are already available through conventional plant breeding and usually the resistance has been durable (García-Arenal and McDonald, 2003). This approach could make an even greater contribution to virus disease control if extended to other crops and diseases, and if improvements are made in the means by which resistant varieties are evaluated, selected, multiplied and made available to farmers. Moreover, progress will be facilitated by using modern technologies to improve conventional breeding programmes. Such approaches are being used already to ensure viable crosses between hitherto recalcitrant parents or species. Moreover, resistance genes can be located and so permit marker-assisted selection as now possible in breeding varieties resistant to cassava mosaic disease (Akano *et al.*, 2002).

There is also an opportunity to exploit the novel forms of resistance that have become possible since the seminal studies in the 1980s on pathogen-derived resistance to *Tobacco mosaic virus* (Powell-Abel *et al.*, 1986). Several viruses of tropical crops have featured in subsequent studies and resistance to papaya ringspot disease achieved by incorporating the coat protein of the causal virus into the genome of papaya is being used successfully to control the disease in Hawaii and elsewhere (Ferreira *et al.*, 2002; Gonsalves, 1998; Gonsalves, this volume, pp. 317–354). A similar biotechnological approach is being considered with several other viruses of tropical crops, including *Rice yellow mottle virus,* cassava mosaic viruses, *Banana streak virus,* groundnut rosette viruses and *Sweet potato feathery mottle virus*. Transgenic plants with putative resistance to these viruses are being tested in laboratory containment facilities in Europe or North America and there have been field trials with a genetically transformed sweet potato cultivar in Kenya. An advantage of the transgenic approach is that resistance can be introduced to improve widely grown varieties that are popular because they have many of the attributes required by farmers and consumers, even though susceptible to virus infection. However, there is considerable opposition to the release of genetically modified organisms (GMOs) and formidable regulatory issues and public misconceptions will have to be overcome before GMOs can be released for general use (Gonsalves, this volume, pp. 317–354). As for conventional resistance, effective methods of multiplication and distribution will be required if GMOs are to be widely adopted.

D. Chemical Control

Many viruses have animal or fungal vectors that spread into, between and within crops. This has led to the use of pesticides or other chemicals to prevent such spread by decreasing vector populations or by impeding transmission (Perring *et al.*, 1999; Satapathy, 1998). Insecticides, acaricides, nematicides and fungicides have all been used successfully to prevent or at least decrease virus spread. Mineral oils and repellents have also been used to impede virus transmission by vectors (Simons and Zitter, 1980). However, several difficulties have become apparent and the use of chemicals is not always applicable or appropriate.

One of the problems with the use of chemicals is that the relationships between rates of virus spread and vector populations are complex and a decrease in vector population density does not necessarily achieve a commensurate decrease in virus incidence. This is apparent from the limited effectiveness of insecticides used in attempts to control the aphid vectors of non-persistent viruses (Raccah, 1986). These are acquired and transmitted most effectively in brief probes and often by aphid species that visit but do not colonize the crop being sprayed. Moreover, there are problems due to the difficulty and cost of treating crops repeatedly throughout the entire vulnerable period of growth. There may also be damage to the natural enemies of vectors and risks to the environment and to human health. Consequently, pesticides play only a limited role in plant virus control, even in developed countries of temperate regions. There they are mainly used against the vectors of persistently transmitted viruses of sugar beet, cereals, potato and a wide range of horticultural/vegetable crops. Mineral oils are also used to protect horticultural crops and ornamentals that are of sufficiently high value to justify the cost of repeated applications.

Despite the problems and limitations, insecticides have been used extensively on rice, cotton, tomato and other vegetable crops in many tropical regions of Asia and South/Central America and to a much lesser extent in SSA. Insecticides are sometimes used somewhat indiscriminately to control a diverse array of direct pests, including virus vectors. However, they are also used more specifically in response to the damage caused by a particular virus diseases, as discussed in more detail in the section on rice tungro disease in South and South-East Asia (pp. 279–282). Another example is rice hoja blanca and its planthopper vector in South/Central America. Others include diseases of cotton, tomato and other vegetable crops caused by whitefly-borne viruses in South/Central America and in many Asian countries.

The use of insecticides has brought benefits in achieving at least some degree of virus control, but there have also been problems due to the emergence of pesticide-resistant vector populations. These have been reported with rice leafhoppers and rice planthoppers and also with many aphids and the whitefly *B. tabaci*. There may also be a resurgence of vector populations following the destruction of natural enemies by use of insecticides, as recorded with the rice brown planthopper (*Nilaparvata lugens*) (Kenmore *et al.*, 1984) and with *B. tabaci* on cotton and other crops (Dittrich *et al.*, 1985). Furthermore, there has been increasing concern with the threat to wildlife and to the environment and with the risk to the health of spray operators and to those consuming the produce of sprayed crops (Loevinsohn, 1987).

The magnitude and severity of the problems caused by the use and misuse of insecticides on rice, cotton and other crops grown in the tropics has had an importance influence on attitudes towards pest and disease control and in moulding scientific approaches and public opinion. One outcome has been to question the validity, suitability, effectiveness and sustainability of the entire range of approaches being used to enhance agricultural productivity by adopting modern innovations and more intensive methods of crop production. Another has been a realization of the need to decrease the use of pesticides and to use these more selectively than previously and only when necessary. Moreover, there has been an appreciation of the value of an holistic integrated approach to pest management (IPM) that is based on ecological principles and utilizes natural enemies, host-plant resistance and cultural methods of control (Kenmore *et al.*, 1987).

The FAO-sponsored rice IPM project that operated in several of the most populous countries of South-East Asia has been particularly influential in developing and promoting such practices (Matteson, 2000). Innovative methods were used to train farmers in the principles of IPM and in seeking government support, including the introduction of legislation to ban the use of some of the most harmful pesticides being used on rice. Another international project with similar objectives has been mounted in Asia to promote the use of IPM in vegetable crop production. The devastating losses caused by whitefly-borne diseases in the New and Old Worlds led to the Global Whitefly Project and the promotion of IPM (Anderson and Morales, 2005; Oliveira *et al.*, 2001).

Compared with South-East Asia and South/Central America, there is relatively little use of chemicals to control viruses in SSA, where the main use of pesticides is to control direct pests of export crops including coffee, cacao, tea, cotton, pineapple, sugarcane, tobacco and vegetable/horticultural crops. Otherwise, except on the relatively few large-scale commercial farms, only a small proportion of farmers have access to pesticides or means of application and the value of the crops being grown is seldom sufficient to justify or cover the cost of treatment. Moreover, it is inappropriate to promote the use of pesticides on subsistence crops, many of which are consumed fresh or after minimal processing.

The scope for using insecticides to control viruses in SSA is indicated by trials with maize streak (van Rensberg *et al.*, 1991) and with cacao swollen shoot and groundnut rosette as discussed in the detailed case histories (pp. 267–271 and 277–279). However, such treatments are seldom cost-effective and they have not been recommended for general use. An exception is the considerable use of insecticides to control the beetle vectors of *Rice yellow mottle virus* in Madagascar (Reckhaus and Andriamasintseheno, 1997) and some other African rice-growing countries. The main use of insecticides in SSA is in attempts to control viruses of tomato and vegetable crops. Moreover, insecticides are likely to become increasingly important in horticulture as vector populations continue to become more prevalent on such crops. It is notable that the damaging B biotype of *B. tabaci* has been reported already in some parts of Africa (Bedford *et al.*, 1993) and it is likely to become more prevalent and lead to problems of the type encountered already in Asia and the Americas (Brown, 1994; Morales, this volume, pp. 127–162).

E. Mild Strain Protection

A possible means of alleviating the effects of virus infection on crop growth and yield is by prior inoculation with a mild strain of the same virus. The ability of mild strains to protect plants from the damaging effects of closely related virulent ones has long been recognized and the phenomenon has been used extensively by virologists to assess the relationships between viruses and virus strains (Bawden, 1950). However, there were several reasons for an initial reluctance to consider this approach to controlling virus diseases. There was concern that mild strains might mutate to more virulent forms, or that they would spread and cause damage to other more virus-sensitive crops. It was also feared that mild strains might have synergistic effects in combination with other unrelated viruses. Moreover, the deliberate dissemination of mild strains was incompatible with control strategies based on phytosanitation.

Despite these reservations, mild strains have been disseminated widely and successfully to control *Tomato mosaic virus* and several other viruses that are so prevalent and difficult to control by other means that the deliberate introduction and dissemination of mild strains was considered to pose no additional hazards (Fulton, 1986). This is the situation in tropical areas of Asia that are severely affected by *Papaya ringspot virus* (Yeh *et al.*, 1988). Moreover, mild strain protection has been used against *Citrus tristeza virus* in the Republic of South Africa (Garnsey *et al.*, 1998; van Vuuren *et al.*, 1993) and even more widely in Brazil (Costa and Muller, 1980). There have also been trials on the scope for using mild strain protection against *Cacao swollen shoot virus* (p. 269) in Ghana and against cassava mosaic viruses in Uganda (pp. 276–277). With these viruses the approach was promising, but it was concluded that additional research was required before mild strain protection could be recommended for general use.

F. Integration and Uptake of Control Measures

In considering the effectiveness of control measures and their integration, it is convenient to follow previous approaches (Thresh, 1983; Zadoks and Schein, 1979) and distinguish between measures that decrease the initial sources of inoculum from which spread occurs (Xo) and those that decrease the rate of disease progress (*r*). Clearly, phytosanitation decreases values of Xo, whereas resistant varieties, pesticides, intercrops and several other measures decrease *r*.

There are obvious advantages in combining two or more different control measure to decrease both Xo and *r*. Examples are, the use of virus-free seed together with intercropping to achieve partial control of *Bean common mosaic virus* and the use of resistant varieties to facilitate the control of *Cacao swollen shoot virus* by sanitation. Failure to use virus-free propagules or other means of decreasing Xo can undermine the otherwise beneficial effects of isolation and other measures. This is evident from experience with cassava mosaic disease (pp. 271–277), which also indicates the benefits of adopting a selected combination of measures. The problem is to devise an integrated approach of this type that is effective and also one that farmers are able and prepared to adopt. Herein lie the difficulties as there are serious obstacles to developing and adopting an appropriate and effective combination of control recommendations. This is due to the inadequate information on the range of options available, or to the failure of farmers to utilize the measures proposed, either because they are inappropriate or they have not been promoted adequately. For example:

- Virus-free propagules are not generally available and other phytosanitary measures tend to be unpopular and seldom adopted.
- There are limitations on the extent to which site selection and other cropping practices can be adapted to avoid or alleviate virus disease problems. Moreover, the use of sole cropping, a restricted range of varieties and some of the other changes associated with modern practices are likely to undermine the effectiveness and resilience of traditional cropping systems.

- Resistant varieties have been identified for only some of the many diseases encountered and they are not always readily available. Moreover, even if available they are not always durable and may have undesirable features, which limit their adoption by farmers or acceptance by processors.
- Chemicals have been used to control vectors or to impede virus transmission, especially in South-East Asia and South/Central America. However, their use may be prohibitively expensive or inappropriate and there have been serious problems because of the risks to human health, natural enemies and the environment.
- There has been only limited research on the scope for utilizing mild strain protection and there are likely to be few situations in which this approach to control is appropriate and effective.
- There are continuing delays in the development and use of transgenic forms of resistance, in part due to the lack of regulatory procedures and public concern over the use of GMOs.

Collectively, these are powerful constraints and explain why such limited progress has been made in controlling many tropical virus diseases and why they continue to cause serious losses. These comments and conclusions apply generally and they are illustrated by the following case histories. Other examples in which an integrated approach has been developed to control are with *Tomato spotted wilt virus* in Hawaii (Cho *et al.*, 1989) and *Chickpea chlorotic dwarf virus* in Sudan (Hamed and Makkouk, 2002).

III. Control of Four Important Tropical Diseases

Four tropical diseases of particular importance are discussed here to illustrate how the general principles presented in the foregoing sections have been used in developing control measures.

A. *Cacao Swollen Shoot Disease*

CSSD was first reported in what is now Ghana in 1936, although there is anecdotal evidence that the disease occurred earlier in the 1920s, when large patches of dead and dying trees were seen in one of the main cacao-growing areas (Posnette, 1947). Swollen shoot was soon shown to be caused by a mealybug-transmitted pathogen that was assumed to be a virus, but it was not possible to confirm this or to implement control measures on a large scale until after the Second World War (1939–1945). The cacao-growing areas were then mapped and surveyed and an official eradication campaign was introduced and enforced initially by staff of the Department of Agriculture. This was despite considerable opposition from farmers to the removal of established trees, especially those still bearing pods. The eradication procedure adopted was based on field studies, which established that outbreaks could be brought under control by regular inspections and treatments of cacao farms so as to find and remove all trees with the stem or leaf symptoms of CSSD (Posnette, 1943). This approach was shown to be most effective in treating small outbreaks and when all neighbouring symptomless 'contact' trees were also removed in attempts to eliminate latently infected trees that could be foci of infection (Thresh and Owusu, 1986).

Based on these findings eradication measures have been used on a large scale in Ghana for more than 50 years and they still continue. However, a serious limitation of the approach adopted is that for long periods it was possible to remove only trees with symptoms. This was because of opposition from farmers to more drastic measures. Inevitably, latently infected trees remained as foci of infection within treated farms until they were found with symptoms and removed during subsequent retreatment operations, or farms were totally denuded and had to be replanted.

There has been no recent assessment of the eradication campaign but by 1988 a total of 187 million cacao trees had been removed. This is equivalent to 187,000 ha at conventional spacings, or *c*. 9% of the total area of cacao recorded in the 1970s (Ollennu *et al*., 1989; Thresh, 1988). The number of trees removed had increased to more than 192.5 million by 1996 (Ampofo, 1997) and continues to rise. Millions of other trees have been killed by the disease before they could be removed and there have, at times, been millions more awaiting removal because of delays in implementation. Moreover, there have been periods of little or no activity due to opposition from the farmers, logistical problems, or to a lack of funding. A major problem has been the vast scale and expense of the undertaking, which has necessitated a huge allocation of national resources. This was because of the need to ensure that all cacao farms were located, mapped and inspected regularly and to carry out the treatments and retreatments required. Such a big allocation of manpower and resources was justified by the importance of cacao as the main Ghanaian export crop and source of 'hard' currency. Inevitably, there has been a major diversion of funds that otherwise could have been spent on other aspects of cacao production or on agricultural improvement projects.

Despite the massive allocation of budget and resources in Ghana it has seldom been possible to inspect and treat the worst-affected *'Areas of Mass Infection'*. Consequently, the main emphasis has been on treating the immediately adjoining areas to form a *'Cordon Sanitaire'* and on the other less affected *'Areas of Scattered Outbreaks'*. An eradication policy was also adopted in Nigeria where the CSSD problem was less than in Ghana and farmers were less hostile to the measures adopted. Accordingly, it was possible to adopt more drastic cutting out procedures and initially all trees with symptoms were removed and also all symptomless trees within a distance of *c.* 30 m around each outbreak (Lister and Thresh, 1957). The measures were revised later on the basis of field studies and outbreaks were then treated less drastically according to the size of the outbreak found (Thresh and Lister, 1960).

The extensive *Areas of Mass Infection* in Ghana have been left largely untreated and virulent strains of virus have become prevalent. In these circumstances, the deliberate dissemination of mild strains of virus to protect plants from the damaging effects of virulent strains poses no additional hazards. The scope for using mild strains in this way was suggested from the results of field studies (Posnette and Todd, 1955). However, these were not pursued because the intention at the time was to implement the eradication campaign in all areas and the contrasting approaches of phytosanitation and mild strain protection were incompatible. It was later realized that eradication was not feasible in the worst affected areas and research on mild strain protection was resumed (Hughes and Ollennu, 1994). Promising results were obtained, but the approach has not been adopted.

Various attempts have been made to devise means of improving the effectiveness of the eradication campaign in Ghana. One approach is to apply an insecticide to the symptomless trees around the outbreaks being treated so as to restrict further spread by the mealybug vectors. A decrease in vector populations and in rate of virus spread were apparent in a trial with a systemic organophosphorus insecticide applied to the soil or injected into the main trunks of the trees (Hanna and Heatherington, 1957). However, these benefits were only achieved by repeated applications, which were expensive and caused an undesirable taint of the cacao beans harvested from the treated trees.

Different problems were encountered when the chlorinated hydrocarbon insecticides endrin and dieldrin were used to decrease vector populations by controlling the crematogasterine ants that protect and tend the mealybug colonies on cacao (Entwistle, 1972). This led to undesirable side effects and to the emergence of secondary insect pests of cacao that had hitherto caused little or no damage to the shoots or pods. It has also been difficult to decrease vector populations by introducing or augmenting natural enemies. For these reasons, vector control currently plays no part in the routine measures adopted against CSSD in Ghana or elsewhere in West Africa.

There is obvious scope for using virus-resistant cacao varieties to help overcome the CSSD problem by decreasing the rate of spread and so facilitating control by eradication. This was appreciated at an early stage of the investigations, but the range of varieties available in West Africa was limited at the outset and initial attempts to identify resistant genotypes from those available locally were unsuccessful (Posnette and Todd, 1951). Accordingly, seed was introduced to Ghana from the trees being grown in a collection of cacao in Trinidad that originated in the Amazon region of South America. Genotypes from Peru were identified that are infected less readily and develop less severe symptoms than the Amelonado cacao that predominated at the time in Ghana and elsewhere in West Africa. Seed or seedlings of selected Upper Amazon types have been widely distributed to farmers and for official replanting schemes to replace trees removed during cutting out operations. Moreover, selected Upper Amazon trees have been intercrossed or crossed with Amelonado trees to produce vigorous hybrids that yield well and begin to bear pods earlier than Amelonado trees. Some of the hybrids are also resistant to virus infection and seed gardens have been established to provide progeny for use in the worst affected areas (Thresh *et al.*, 1988b).

Inter-Amazon hybrids are now widely grown in Ghana and elsewhere in West Africa and they make a substantial contribution to cacao production, which has continued despite the prevalence of CSSD. The contribution of the resistant varieties would have been even greater and reinfection would have been lessened if all new plantings had been made in large compact blocks separated from other cacao by clear alleys, or stands of a non-host barrier crop. The scope for adopting such practices has long been recognized (Cornwell, 1958; Ollennu *et al.*, 1989; Thresh, 1958). However, there has been little attempt to demonstrate or exploit the benefits, mainly because of the limitations imposed by the complex patterns of land use and land tenure.

Cacao production in Ghana and Nigeria is largely by peasant farmers who establish mainly small plots of irregular shape. There is seldom any physical separation between plantings of different age or under different ownership and collectively these may form large contiguous almost uninterrupted stands (Thresh *et al.*, 1988a). This greatly facilitates the spread of CSSD. Moreover, the work of the inspection and treatment teams is hampered by the need to identify individual farm boundaries and ownership so that compensation can be paid for trees removed and for replanting to begin. Cutting out and replanting may have to be delayed until all the owners are traced, or the operations are discontinued and untreated farms are left and remain as sources of inoculum within partially treated areas. Consequently, there is little opportunity to establish the large isolated blocks of compact shape that would decrease the proportion of trees in the vulnerable peripheral areas and so decrease the risk of infection, especially if subdivided into smaller separate units (Ollenu *et al.*, 1989).

CSSD is exceptional in several respects. For example, for many years it received far more research attention than any other virus disease of an African crop (Thresh, 1991). Moreover, the control of the disease has been largely the responsibility of government or quasi-government organizations and farmers play only a very limited role. They do not carry out the routine inspections or cutting out treatments. At times the subsequent replacement of trees removed, or complete replanting of treated farms, has also been done on behalf of and not by the farmers concerned. This largely circumvented the need to educate and train farmers in disease control and methods of replanting. It also facilitated the adoption of resistant varieties and enabled the authorities to enforce measures that would otherwise have been unacceptable. However, there were big financial implications and for years the high cost of CSSD control measures dominated the budget of the Ghana Department of Agriculture. Responsibility for the eradication campaign later passed to the Ghana Cacao Marketing Board and recently part of the cost of cutting out operations has been borne by the European Community Aid Programme as a means of boosting Ghanaian export earnings. Nevertheless, there has been little evident scrutiny of the effectiveness of the measures adopted and no cost/benefit analysis of the operation. This emphasizes the exceptional nature of the swollen shoot disease problem and the approach used is not one that can be applied readily to other tropical virus diseases.

B. Cassava Mosaic Disease

The disease now known as cassava mosaic (CMD) was first reported in 1894 in what is now Tanzania. It has since been recorded in India and Sri Lanka and in all the cassava-growing areas of Africa including the offshore islands of Cape Verde to the west and Zanzibar,

Madagascar and Seychelles to the east (Calvert and Thresh, 2002). It has long been known that the disease is prevalent in some areas of SSA and relatively unimportant in others and that the incidence of infection can change markedly within only a few years. This makes it difficult to assess the importance of mosaic and the benefits to be gained from adopting control measures. These are likely to be substantial, as apparent from surveys of the prevalence of the disease in 18 important cassava-growing countries (Sseruwagi *et al.*, 2004b) and from studies in several countries on the effects of the disease on the yield of representative varieties (Thresh *et al.*, 1994a). Based on these findings, it was estimated that total production in Africa would be increased by 15–24% but for the effects of mosaic, which was estimated to cause annual losses of 12–23 million tonnes, worth US$ 1200–2300 million (Thresh *et al.*, 1997). The estimate of losses was later increased to 19–27 million tonnes to take account of the findings of later disease surveys (Legg and Thresh, 2003). These figures and the estimate by Legg *et al.* (this volume, pp. 355–418) emphasise the scope for adopting control measures to enhance production, or to increase productivity and so release land for other purposes, or to permit longer periods of fallow and so restore soil fertility.

Although the prevalence and importance of CMD has long been recognized in SSA and more recently in India, there have been only limited attempts to control the disease except at times and places when/where losses have been so severe as to threaten food security and rural livelihoods. This is partly because cassava has been somewhat neglected and it has received inadequate attention from researchers and extensionists and from those who were promoting commercial crops for export or more nutritious foods. Moreover, cassava in SSA is grown mainly by small-scale farmers and originally the productivity per unit of land was largely unimportant. This was because much of the crop was grown for local consumption and for food security, to be used in times of drought. The status of cassava is now changing as it becomes increasingly important because of its ability to grow and yield even in poor soils and under adverse conditions. Cassava has also become a substantial source of income through sales of roots to urban areas and for processing into chips, flour, alcohol, starch, animal feed or other products (Nweke *et al.*, 2002). Furthermore, there is a need to increase crop productivity to meet the demands of a continually increasing human population and despite decreasing soil fertility and increased pressure on the land available. The pandemic of a particularly severe form of CMD in East and

Central Africa has been a further incentive and led to the recent surge of research on the disease in several African countries (Otim-Nape and Thresh, 2006; Legg *et al.*, this volume, pp. 355–418).

In considering the different approaches to controlling CMD, it is helpful to distinguish between the measures used to some extent inadvertently by farmers, and even by researchers, from the specific control recommendations of pathologists. The important role played by farmers in overcoming the mosaic problem is apparent from experience in West Africa soon after the disease was first reported there. It was then realized that some of the many varieties being used in Ghana were growing much more satisfactorily than others and the least productive varieties were discarded (Dade, 1930). Whether this was done because of the severity of the mosaic symptoms expressed or because of the poor growth and yield of the affected plants is unclear. Nevertheless, the outcome would have been similar in leading to the selection of varieties that were to some extent resistant to or tolerant of infection.

There is much additional evidence of the ability of farmers to cope in this way and to continue producing cassava satisfactorily by adopting suitable locally selected landrace varieties. For example, the 1990s epidemic in Uganda led to big changes in the main varieties being grown (Otim-Nape *et al.*, 2000, 2001; Thresh, this volume, pp. 89–125). At the outset, the locally selected variety Ebwanateraka predominated in several of the most important cassava-growing districts. It proved to be extremely vulnerable to infection with the severe form of CMD and was soon largely displaced by other varieties, only some of which had been introduced and distributed officially because of their resistance to infection. Local varieties that are somewhat tolerant of infection have undoubtedly made a substantial contribution to the recovery in production that has occurred since the peak of the epidemic in the mid-1990s (Bua *et al.*, 2005; Otim-Nape and Thresh, 2006; Otim-Nape *et al.*, 2000).

In much of Uganda, in many other African countries and also in India, CMD undoubtedly decreases yields but not to an unacceptably low level and farmers may be unaware or unconcerned that the disease is present, or they regard it as less important than other factors influencing productivity. This is consistent with the observation that it is not uncommon for mosaic-infected plants of some popular varieties to outyield healthy plants of others. Moreover, the ability of farmers to 'live with' mosaic also explains why the disease has attracted so little attention in India or in much of Ghana, Nigeria and other West African countries where infection is endemic and prevalent in many of the varieties being grown (Calvert and Thresh, 2002). There is much use of infected cuttings as planting material and seldom any attempt to select healthy stocks. The failure of farmers and even researchers to practise selection is not readily explained, especially in areas where circumstances are propitious. That is, sufficient healthy plants are available to provide the cuttings required for new plantings, the symptoms of CMD are conspicuous and there is an obvious difference in the growth and yield of diseased and healthy plants. This emphasises the extent of 'the information gap' between farmers and researchers that must be bridged if progress is to be made in utilizing the knowledge that is already available to improve the health status and productivity of cassava and other crops.

There were periods in the twentieth century when CMD was so damaging in some areas of SSA that there was an urgent need for varieties that were more resistant and better able to withstand infection than those being grown. This occurred in Madagascar soon after the first reports of CMD on the island in the 1930s (Cours, 1951; Cours *et al.*, 1997) and in what is now Tanzania in the 1940s. Severe epidemics were also reported later in Cape Verdé (Anonymous, 1992), Akwa Ibom State of Nigeria (Anonymous, 1993) and in Uganda during the 1950s (Jameson, 1964) and in the 1990s (Otim-Nape *et al.*, 2000; Legg *et al.*, this volume, pp. 355–418). The severe losses encountered and the threat to food security led to government-funded virus resistance breeding programmes in Madagascar, Tanzania and later in Nigeria, Uganda and elsewhere. There were also efforts to introduce exotic varieties for local evaluation so that the most resistant could be multiplied and distributed to farmers.

Initially, there was little success using varieties introduced from Asia or South America, or by intercrossing African varieties. However, working independently the programmes in Madagascar (Cours, 1951) and Tanzania (Nichols, 1947) produced highly resistant varieties from hybrids between cassava (*Manihot esculenta*) and ceara rubber (*M. glaziovii*). These were then backcrossed to cassava to produce progeny that gave satisfactory yields of tuberous roots of acceptable quality (Jennings, 1994). Some of the varieties so produced were used successfully in Madagascar and parts of mainland East Africa. Seed from the Tanzania breeding programme was also introduced to Nigeria and selections made there were used as parents in the national cassava breeding programme and later at the International Institute of Tropical Agriculture (IITA), Ibadan, to produce the Tropical *Manihot* Series (TMS) of varieties (Beck, 1982).

The selection of TMS varieties was based on their overall performance, including at least some degree of resistance to CMD that enables them to yield satisfactorily, even in conditions of high-virus inoculum pressure. Resistant varieties are not readily infectible and when infected they develop generally mild symptoms that usually become even less conspicuous as the plants mature. Moreover, infection does not become fully systemic and some of the cuttings collected from infected plants are not infected and grow into healthy plants (Fargette *et al.*, 1996; Thresh *et al.*, 1998a). These features explain why the selection of healthy planting material has not been regarded as necessary and it is seldom practised, even in some official multiplication and distribution schemes. Consequently, there is a generally high incidence of CMD in some TMS varieties, although not in the most resistant ones.

The implication on yield of adopting or not adopting phytosanitation with TMS varieties is not known because of the limited amount of information available on the effects of CMD on the growth and yield of resistant varieties. Consequently, there is continuing uncertainty on the most appropriate method of utilizing these varieties. Some argue that sanitation and host-plant resistance are complementary and should be used together. Others regard sanitation as unnecessary if the varieties grown are sufficiently resistant. Thresh *et al.* (1998b) and Legg *et al.* (this volume, pp. 355–418) discuss the main issues involved in the continuing debate that has not yet been resolved.

Despite the divergent views and uncertainties, TMS varieties and others derived from them have been adopted widely in Nigeria and elsewhere. They have also been used as parents in several National Breeding Programmes in SSA to provide varieties that are suitably adapted to local conditions. The methods used for multiplying and distributing TMS varieties to farmers in Nigeria and Uganda are particularly well documented. In Nigeria, the uptake of virus-resistant material by farmers was facilitated by the demand for improved varieties to meet the requirements of processors who were supplying the rapidly expanding urban markets (Nweke *et al.*, 1996). By comparison, the main incentive in Uganda was for varieties that could withstand the severe form of CMD that became prevalent during the epidemic (Otim-Nape *et al.*, 2000; Thresh *et al.*, 1994b). This created an enormous demand for planting material of resistant varieties to replace the vulnerable genotypes being grown. It was supplied with considerable financial assistance from the UK-based Gatsby Charitable Foundation and many other governmental or non-governmental organizations (Otim-Nape *et al.*, 1994, 2000).

CMD-resistant varieties predominated in *c.* 25% of all plantings recorded in surveys between 1997 and 2001 in representative districts of Uganda and in *c.* 32% of those assessed in 2003 (Bua *et al.*, 2005). These varieties have made a disproportionate contribution to yield, because of their superior performance compared with the mainly infected local landraces being grown previously (Sserubombwe *et al*, 2005). However, some of the introduced varieties do not have all the flavour or other desirable attributes required by farmers. Consequently, it is likely that some of the landraces will be retained, at least until other resistant varieties are introduced which have superior taste and other attributes compared with those released originally.

This accounts for the current interest in the tropical *Manihot esculenta* ('TME') sources of resistance. It was identified in some of the landraces collected in Nigeria and elsewhere in West Africa and later attributed to a dominant major gene designated *CMD-2* (Akano *et al.*, 2002; Mignouna and Dixon, 1997). Some of the landraces and their progeny have the upright growth habit and other characteristics sought by many farmers. Accordingly, they are being intercrossed with TMS varieties to produce hybrids that are both high-yielding and very highly resistant to CMD and desirable in other respects. TME clones and hybrids have been assessed in on-station and on-farm trials in Uganda, Kenya, Madagascar, Rwanda and elsewhere and they could make a big contribution in further decreasing the losses due to mosaic. However, the vulnerability of some TME genotypes to cassava brown streak disease has become apparent recently and may restrict their use (Legg *et al.*, this volume, pp. 355–418).

Another outcome of the damaging pandemic in Uganda and elsewhere in East Africa, and of the recognition of the merits of the widely grown landraces, is that there has been increased research on other approaches to controlling CMD. This could lead to means of deploying resistant varieties more effectively and also avoid undue reliance on such varieties. For example, it may be possible to develop methods of maintaining the health status of landraces through the use of varietal mixtures, intercropping or by an appropriate deployment of plantings to provide some degree of isolation and so decrease the risk of serious infection (Thresh and Cooter, 2005). Moreover, it may eventually be possible to utilize mild strain protection or the transgenic forms of resistance now being developed (Legg *et al.*, this volume, pp. 355–418; Owor *et al.*, 2004).

Early studies on mild strain protection with CMD in Tanzania were not encouraging (Storey and Nichols, 1938). However, additional research was initiated when seemingly mild strains of virus were encountered and became increasingly prevalent following the 1990s epidemic in Uganda (Pita *et al.*, 2001; Sseruwagi *et al.*, 2004a). Plants grown from mild strain-infected cuttings continued to grow satisfactorily when exposed to infection with virulent strains. Moreover, the mild strain-infected plants outyielded the initially healthy controls that became severely affected during the course of the trials (Owor *et al.*, 2004). These results justify additional research on the effectiveness and sustainability of this approach when cuttings are collected for further cycles of propagation. There is also a need for trials in different locations in Uganda and elsewhere in Africa to determine whether there could be problems due to the diversity of cassava mosaic viruses encountered.

Genetically engineered resistance to CMD was one of the objectives of the original 'Cassava Trans' project at the Scripps Institute in California (Fauquet and Beachy, undated). Studies have continued elsewhere in the United States and also in Europe and some success has been achieved in engineering resistance to cassava mosaic geminiviruses in herbaceous host plants and in cassava (Fregene and Puonti-Kaerlas, 2002; Legg *et al.*, this volume, pp. 355–418). However, this approach has yet to be evaluated in the field and there are formidable problems to be overcome before it is acceptable and can be advocated for use on a large scale. The difficulties likely to be encountered are apparent from experience in developing transgenic varieties of virus-resistant papaya in Hawaii and elsewhere (Gonsalves, this volume, pp. 317–354).

C. Groundnut Rosette Disease

Groundnut rosette disease was first described in 1907 and it has since been reported in many other countries of eastern, western and southern Africa (Naidu *et al.*, 1999; Taliansky *et al.*, 2000). The disease has received considerable attention from researchers because of its importance in decreasing the productivity of crops grown for local consumption or for export. Very damaging epidemics and almost total crop failure have been reported and farmers are well aware of the need for control measures, especially in areas that are particularly prone to infection.

Groundnuts in Africa are usually produced under rain-fed conditions in areas subject to a prolonged dry season and it has long been known that rosette disease tends to be most prevalent in crops that are sown late in the growing season, some weeks after the beginning of the main rains (Hayes, 1932). By that time, there has been a considerable build-up of the principal aphid vector (*Aphis craccivora*) on crops sown early and also on self-sown 'volunteers', weeds and wild host plants. It was also recognized by farmers and researchers that rosette tends to be particularly severe in groundnut stands that are sown at wide spacings, or in which there are many gaps in the stand due to poor establishment and weak growth. Other observations were that rosette is particularly prevalent at the margins of plantings and in crops that have been weeded early to leave open stands having a discontinuous canopy of foliage (Hayes, 1932). These observations led to the view that vector colonization and virus spread are facilitated where the crop is grown at wide spacing and the canopy is open, and not at close spacings, or where a cereal or other non-host intercrop is grown. Based on these observations and the supporting evidence of field trials in Nigeria, Malawi, Tanzania and Uganda, farmers were advised to sow early and at close spacing to avoid serious losses due to rosette (A'Brook, 1964; Davies, 1976; Farrell, 1976a). However, they were not advised to delay weeding because it was considered that this 'might encourage indolence' (Hayes, 1932). Moreover, intercropping did not feature in the recommendations. This practice is considered to be beneficial in decreasing virus spread, but there is little evidence from field trials on the most suitable crop combination to adopt. In one trial, a bean intercrop decreased the incidence of rosette, but impeded the growth of groundnut and decreased yields substantially (Farrell, 1976b).

The recommendations on groundnut planting date and spacing have been difficult to implement in many areas because of constraints on the availability of labour for land preparation and planting. Farmers tend to give priority to sowing maize or other cereals and often groundnut is not planted until later. Moreover, groundnut seed is often scarce and expensive, especially after a previous season of serious disease and low yields. This explains the general reluctance to adopt close spacings. There are also constraints on the use of large compact fields to decrease edge effects. Such sites are seldom available and socio-economic considerations also restrict the use of insecticides, although these have been shown to be effective in decreasing the spread of rosette into and within field trials (Davies, 1975; Evans, 1954).

In these circumstances, there are obvious advantages to be gained from adopting rosette-resistant varieties. These were first obtained from landraces of the Virginia type being grown in West Africa (Sauger and Catherinet, 1954) and they were used later in breeding programmes in Nigeria and Malawi. Some of these varieties have been adopted by farmers, but a serious obstacle to their use more widely is that they have an inherently long growth period. Consequently, they are not suitable for use in the many areas of SSA where short-duration varieties are required to decrease the risk of losses due to drought, especially when these are likely to occur during the late stages of crop growth.

Repeated attempts were made to transfer the resistance from long-duration to short-duration types, but initially these were unsuccessful. However, short-duration rosette-resistant varieties have been obtained from other groundnut accessions introduced from Asia (van der Merwe and Subrahmanyam, 1997). These varieties seldom express rosette symptoms as they are highly resistant to *Groundnut rosette virus* and the associated RNA satellite that is linked with symptom expression. However, they are not resistant to *Groundnut rosette assistor virus* that is required for transmission by the aphid vectors. The assistor virus alone does not cause obvious symptoms, but it may become prevalent and decrease growth under conditions of high-inoculum pressure (Subrahmanyam *et al.*, 2002).

Groundnut breeding programmes have also utilized accessions that are resistant to the aphid vector (Padgham *et al.*, 1990). These are susceptible to rosette, but largely escape infection in the field. Short-duration varieties with resistance to rosette and/or *A. craccivora* have been developed by the International Centre for Research in the Semi-arid Tropics sub-station at Chitedze, Malawi. These have been released to governmental and non-governmental organizations in Malawi, Mozambique and elsewhere in the region (Subrahmanyam *et al.*, 2002). The varieties have also been introduced in Uganda, where they have been evaluated in comparisons with the susceptible local varieties being used (Chancellor, 2002). This was done in on-station and on-farm trials in close collaboration with farmers. The most successful varieties are resistant to rosette and crop at least as well as the locals, even in seasons when there is little or no rosette in the locality. Rosette-resistant material is now being multiplied in quantity for release to farmers. However, a general experience with groundnut is that rates of seed multiplication are low and various approaches are being adopted in attempts to increase supplies. The demand for the new varieties is high and they are likely to have a big impact in both enhancing and stabilizing production. They will be most valuable in areas that are particularly prone to rosette attack and where farmers are unable or unwilling to sow early or to adopt the other cropping practices recommended to control the disease.

D. Rice Tungro Disease

Rice tungro disease was first described in 1965 following outbreaks in the Philippines and successful transmission experiments with the rice green leafhopper *Nephotettix 'impicticeps'* (now known as *N. virescens*) (Rivera and Ou, 1965). What is likely to have been tungro disease had been reported much earlier in the Philippines, Malaysia, Thailand and Indonesia under different local names. The earliest reports were in 1859, although no transmissible pathogen was implicated at the time or until much later studies and the symptoms were for long attributed to unfavourable soil conditions (Ou, 1984).

Tungro is now known to be widespread and sometimes prevalent in all the main rice-growing areas of South-East Asia, where it appears to be indigenous. Severe epidemics occurred at the time of the 'green revolution' in tropical rice production in the late 1960s in India, Indonesia, Malaysia, Philippines, Sri Lanka and Thailand soon after the release of the first International Rice Research Institute (IRRI) varieties (Sõgawa, 1976). There have since been more epidemics in these countries and elsewhere in Asia (Azzam and Chancellor, 2002; Chancellor and Thresh, 1997). An unexplained feature of tungro is that the disease can suddenly become prevalent and cause serious losses throughout whole regions. Consequently, it has attracted considerable notoriety and politicians and administrators as well as agriculturalists are well aware of the serious consequences of devastating epidemics. These cause famine and severe hardship and change substantial areas from being net exporters to net importers of rice. Such considerations explain the allocation of large budgets for research and extension activities in attempts to combat tungro disease. Moreover, disease and vector monitoring and survey operations have been mounted by several of the National Agricultural Research Programmes in the region in efforts to determine damage thresholds and reliable methods of forecasting.

There has also been widespread and in some instances misguided and irresponsible use of insecticides to control the leafhopper vectors of tungro. This created hazards to human health, domestic animals and wildlife (Loevinsohn, 1987). Another undesirable consequence of the use of insecticides was to decrease populations of the natural enemies of the rice brown planthopper (*Nilaparvata lugens*), which became increasingly important in many parts of South-East Asia as a direct pest and as a virus vector (Kenmore *et al.*, 1984). Pest management programmes were introduced to decrease the use of insecticides and restore natural enemy populations. However, these efforts were thwarted when farmers resorted to the use of insecticide as a prophylactic measure to avoid tungro epidemics, or to treat areas already affected. In some areas, the use of insecticides was promoted by government campaigns and subsidies to enable farmers to purchase sprayers and insecticides. This was done in attempts to maintain or enhance rice production for local consumption or for export. The use of insecticides also became an essential requirement of some loan agreements arranged by farmers to facilitate the purchase of improved rice seed, fertilizer and other inputs.

There is considerable information on the two viruses involved in tungro disease and on their transmission by the leafhopper vectors (Hibino and Cabunagan, 1986). Until recent years, much less was known about the epidemiology of the disease, on the patterns, sequence and distance of spread into and within plantings and on the role of weed hosts (Chancellor *et al.*, 1996). This was a serious limitation of attempts to achieve control and the initial emphasis was on the use of resistant varieties and insecticides. Little attempt was made to isolate or protect seed beds, or to plant away from known sources of infection and the possible benefits of such measures were not assessed until later (Holt *et al.*, 1996). However, farmers were encouraged to plant synchronously within the locality, to allow a break between successive plantings and to avoid very late planting in areas where there was the highest risk of infection. In Sulewezi, Indonesia, they were also advised to adopt recommended sowing dates. This was so that young plants at the most vulnerable stage of growth were not exposed to infection at times when vectors were likely to be particularly abundant (Manwan *et al.*, 1985; Sama *et al.*, 1991).

The need for varieties that are resistant to tungro and its main vector *Ne. virescens* became apparent at an early stage of the IRRI breeding programme and routine screening tests have been made since the late 1960s (Khush, 1977). With one exception, all the IRRI varieties released since 1966 were rated as having at least some degree of vector resistance in the Philippines at the time they were introduced. Such varieties tend to escape infection with tungro (Heinrichs and Rapusas, 1983) and they have played a major role in decreasing the losses that are otherwise likely to have occurred due to the disease. It seems that tungro does not become a problem where vector-resistant varieties predominate, even though they are unable to withstand high-inoculum pressure. This suggests that epidemics occur only where susceptibles are widely grown, or when vector populations have become adapted to previously resistant varieties.

Problems due to the emergence and build-up of vector populations able to thrive on previously resistant varieties have been less acute than those encountered with the rice brown planthopper. Nevertheless, there have been various reports of tungro epidemics associated with marked shifts in the behaviour of varieties that were originally regarded as virus and vector resistant (Dahal, 1988). Thus, previous views on the apparent durability of resistance to *Ne. virescens* were changed (Khush, 1984) and it was recognized that populations are likely to adapt to vector-resistant varieties within a few years of their general release (Ruangsook and Khush, 1987). This led to detailed studies on mechanisms of resistance, gene deployment and breeding strategies to provide varieties with more effective and durable forms of resistance, not only to the vector but also to both tungro viruses (Hibino *et al.*, 1988).

Screening for resistance to each of the viruses began once they had been characterised and specific antisera became available to facilitate their detection (Bajet *et al.*, 1985; Sta. Cruz *et al.*, 1999). Sources of resistance have not been found to *Rice tungro bacilliform virus* (RTBV), which causes the usual symptoms of tungro disease (Hibino *et al.*, 1990). However, some breeding lines are tolerant and develop only inconspicuous symptoms on infection. Other accessions, including the IRRI varieties IR 20 and IR 26, were found to be resistant to *Rice tungro spherical virus* (RTSV), which is responsible for transmission by vectors (Hibino *et al.*, 1988). These lines are susceptible to RTBV and express symptoms, but they do not become sources of infection from which further spread can occur. This explains their effectiveness in preventing secondary spread of RTBV within plantings, even if there is primary infection from outside sources (Cabunagan *et al.*, 1989).

These developments have made it possible to develop varieties that are resistant to RTSV and *Ne. virescens* and also tolerant of RTBV. Promising selections were grown at a trial site in the Philippines and at two sites in Indonesia (Azzam and Chancellor, 2002). The most successful were released for use by farmers in Bali and were unaffected by tungro disease. This suggests that the latest virus and vector-resistant varieties will be adopted more widely and that they will also become an important feature of control measures elsewhere.

IV. Discussion

It is apparent from the foregoing that there are many different approaches to controlling plant virus diseases and many of those causing serious losses in the tropics could be controlled through the application of existing knowledge. There are also likely to be important contributions from the new techniques and approaches to control being developed by biotechnologists. The challenge to researchers and extensionists is to ensure that this information is utilized in developing and promoting control measures on a suitably large scale. Moreover, these should be cost-effective and appropriate for use by the farmers concerned. The measures should also avoid harmful effects on human health or on the environment and should complement and be fully compatible with those being used against other pathogens and pests and also with sound cropping practices.

There are many reasons why these are formidable and exacting requirements that are seldom met in tropical areas. An obvious technical problem is that it is difficult to evaluate the different options for control when used singly or in different combinations. Such studies are very demanding in terms of the personnel and resources required for field experiments. Furthermore, these should be done on a suitably large scale, in different agroecologies and over a sufficiently long period to provide a reliable indication of the feasibility and cost-effectiveness of the measures when adopted widely by farmers. This is evident from experience in developed countries with several diseases of temperate crops, including cereal yellow dwarf, potato leafroll, sugar beet yellows and plum pox. These and other diseases continue to cause problems and farmers incur considerable expense in implementing control measures, even though there has been much detailed research in many countries and over a prolonged period.

Cacao swollen shoot, maize streak, rice tungro, groundnut rosette and cassava mosaic are prime examples of some of the few tropical virus diseases that have been studied intensively and yet it is apparent from the four case histories (pp. 267–282) that many uncertainties remain concerning their control and they continue to cause serious losses. Much less information is available on many other virus diseases of tropical crops, some of which have been largely ignored by researchers. This is an unsatisfactory situation and a serious impediment to increased crop production and to the development of effective control measures. However, there is little prospect of any immediate improvement because of the limited allocation of personnel and resources to plant virology in many tropical countries. Morales (this volume, pp. 127–162) discusses some of the problems arising in South and Central America due to current budget constraints and policy considerations. There are also particular problems in SSA, as discussed at a conference at the IITA, Ibadan, in 2001. This led to recommendations on possible ways of improving the unsatisfactory situation. However, it was agreed that this will not be achieved quickly or easily, or without considerable expenditure (Hughes and Odu, 2003).

A further even more intractable problem is that even if researchers are successful in developing effective control measures it will be difficult to ensure their adoption on a sufficiently large scale. This is because of the serious limitations of the extension services in many parts of the tropics and the limited ability of farmers to access and adopt the technical advice and recommendations made available. Extension services lack personnel, resources and funding, and the various attempts made to improve the situation in different countries have been largely unsatisfactory and not sustainable. One outcome has been for researchers to become directly involved in technology transfer and this is a current trend at the International Agricultural Research Centres, even though they have a limited expertise or capacity to undertake this role. National researchers also became involved in technology transfer in Uganda in the 1990s following the onset of the severe pandemic of cassava mosaic disease (Otim-Nape *et al.*, 2000). The Ugandan researchers who identified and evaluated mosaic-resistant varieties of cassava in on-station and on-farm trials assumed the responsibility for disseminating the new varieties to farmers. This was done in collaboration with extension staff, local authorities and non-governmental organizations (Otim-Nape *et al.*, 1994).

In developing and deploying control measures in the tropics, it is essential to consider the requirements and capacity of the farmers involved and their ability to access and utilize research findings. It is difficult to generalize because of the many different agroecologies that are exploited and the wide range of crops and cropping systems adopted. At one extreme there are large commercial farms or plantations, many of which are under strong central management. In such circumstances, there is ready access to the inputs and new technology that are available to enhance productivity and profitability. Examples of export crops so produced include banana, citrus, coconut, cotton, oil palm, papaya, pineapple and tobacco plantations, and horticultural enterprises in several regions of Africa, Asia and South/Central America. There are also large commercial farms growing food crops including rice, maize, wheat, barley, cassava and groundnut. However, such cropping systems are exceptional and typical landholdings in the tropics are usually small, farmers have had little or no formal education, and they lack resources and ready access to new varieties, fertilizers and pesticides and other technological innovations.

These are very powerful constraints to the adoption of control measures, as farmers are likely to be aware of only the most obvious and damaging of the pests and diseases present. This explains why the history of plant pathology in the tropics is dominated by damaging epidemics and periodic crises in production that necessitate action by researchers, extensionists and those allocating funds for use in attempts to alleviate the problem. Farmers and policymakers are largely unaware of the benefits to be gained from controlling less important diseases, or the options available for them to do so. This creates a particular difficulty because much of the funding available for agricultural research is allocated by donors in response to the perceived needs of farmers. These are seldom able to articulate their requirements and so inevitably tend to be neglected.

In such circumstances, it is not surprising that many of the poorest and least educated farmers make little use of specific control measures against virus diseases, although they do adopt traditional practices that are believed to avoid or alleviate disease problems. This was apparent in Uganda during the pandemic of cassava mosaic disease when farmers removed the shoot tips of newly infected plants as a well intentioned but naïve form of phytosanitation and applied wood ash and urine in attempts to control the whitefly vector. There is no evidence that this was beneficial, but similar 'indigenous' practices are also adopted against other diseases and merit further study to determine their effectiveness (Bentley and Thiele, 1999; Thurston, 1990). Meanwhile, it is apparent that peasant farmers who have little or no knowledge of crop science or plant pathology can avoid serious losses due to disease, even though they do so unwittingly by adopting varieties and cropping practices that provide yields that are satisfactory, consistent and sustainable.

The ability of farmers to produce yields that are usually adequate despite the occurrence of viruses and other pathogens is an important feature of traditional agriculture that is associated with biodiversity (Smithson and Lenné, 1996). It is one that should be retained in any attempt to enhance yields by introducing higher-yielding varieties, sole cropping and more intensive methods of crop production. These innovations include mechanization, irrigation, use of fertilizers and the adoption of specially bred varieties having a short growing season. Such developments are required despite any decrease in soil fertility and increased pressure on the land available for crop production (Lal, 2001). It is necessary to increase exports and also to feed the burgeoning human population in the tropics, of which an increasing proportion will be in urban areas and not directly involved in agriculture. This will decrease the availability of farm labour and a further constraint will be the increasing impact of the HIV/AIDS pandemic in decreasing the numbers and productivity of the rural work force (Cockcroft, 2003).

Fifty years ago the eminent virologist F. C. Bawden (1955) considered some of the ways in which modern agricultural practices facilitate the spread of pests and diseases. He concluded that the full benefits of improved technologies would be attained only by developing more effective means of controlling pests and diseases. This remains a crucial challenge to researchers, extensionists and farmers in the twenty-first century, as they must develop methods that are effective and sustainable and also benign in their effects on human health and the environment. Improved methods of virus control have an important role to play in enhancing productivity and it will be possible to utilize the experience gained already in developed countries and also the new biotechnological approaches being introduced. It will also be important to integrate virus disease control measures with those being developed against other pathogens and pests and with improved crop management practices. To date, only limited attempts have been made to develop or promote such an holistic ecological approach. Major research efforts that are well funded and sustained will be required in the whole range of different agroecologies if success is to be achieved. However, the need is great and the economic and social benefits to be gained provide a powerful incentive to such undertakings.

References

A'Brook, J. (1964). The effects of planting date and spacing on the incidence of groundnut rosette disease and of the vector, *Aphis craccivora* Koch, at Mokwa, northern Nigeria. *Ann. Appl. Biol.* **54:**199–208.

Ahohuendo, B. C., and Sarkar, S. (1995). Partial control of the spread of African cassava mosaic virus in Benin by intercropping. *Z. Pflanzenkr. Pflanzensch.* **102:**249–256.

Akano, A. O., Dixon, A. G. O., Mba, C., Barrera, E., and Fregene, M. (2002). Genetic mapping of a dominant gene conferring resistance to cassava mosaic disease. *Theor. Appl. Genet.* **105:**521–525.

Ali, M. (1997). Breeding cotton varieties for resistance to the cotton leaf curl virus. *Pak. J. Phytopathol.* **9:**1–7.

Alvarez, P. A., and Abud-Antún, A. J. (1995). Report de República Dominica. *CEIBA* **36:**39–47.

Ampofo, S. T. (1997). The current cocoa swollen shoot virus disease situation in Ghana. *In* "Proceedings 1st International Cocoa Pests and Diseases Seminar. Accra, Ghana, 1995," pp. 175–178.

Anderson, P. K., and Morales, F. J. (eds.) (2005). "Whitefly and Whitefly-Borne Viruses in the Tropics: Building a Knowledge Base for Global Action," pp. 286. CIAT, Colombia.

Anonymous (1992). Quarantine Implications: Cassava Program 1987–1991. Working document No. 116, CIAT, Colombia.

Anonymous (1993). How Akwa Ibom overcame a crisis in cassava production. *Cassava Newsl.* **17**(2)**:**9–10.

Atiri, G. I., and Varma, A. (1991). Some climatic considerations in the epidemiology and control of two okra viruses in southern Nigeria. *In* "Proceedings CTA/IFS Seminar on the Influence of Climate on the Production of Tropical Crops," 23–28 September 1991, Burkina Faso, pp. 328–335.

Attafuah, A. (1965). Occurrence of cacao viruses in wild plant species in Ghana. *Ghana J. Sci.* **5:**97–101.

Azzam, O., and Chancellor, T. C. B. (2002). The biology, epidemiology and management of rice tungro disease in Asia. *Plant Dis.* **86:**88–100.

Bajet, N. B., Daquioag, R. D., and Hibino, H. (1985). Enzyme-linked immunosorbent assay to diagnose tungro. *J. Plant Prot. Trop.* **2:**125–129.
Bawden, F. C. (1950). "Plant Viruses and Virus Diseases," pp. 335. Chronica Botanica Company, Waltham, MA, USA.
Bawden, F. C. (1955). The spread and control of plant virus diseases. *Ann. Appl. Biol.* **42:**140–147.
Beck, B. D. A. (1982). Historical perspectives of cassava breeding in Africa. *In* "Root Crops in Eastern Africa" (S. K. Hahn and A. D. R. Ker, eds.). Proceedings of a Workshop held in Kigali, Rwanda, 1980, pp. 13–18. IDRC, Ottawa, Canada.
Bedford, I. D., Briddon, R. W., Markham, P. G., Brown, J. K., and Rossell, R. C. (1993). A new species of *Bemisia* or biotype of *Bemisia tabaci* (Genn.) as a future pest of European agriculture. *In* "Plant Health and the European Single Market," BCPC Monograph No. 54, pp. 381–386.
Bennett, C. W. (1952). Origin and distribution of new or little-known virus diseases. *Plant Dis. Rep. Suppl.* **211:**43–46.
Bentley, J. M., and Thiele, G. (1999). Farmer knowledge and management of crop disease. *Agr. Human Values* **16:**75–81.
Bos, L. (1981). Wild plants in the ecology of virus diseases. *In* "Plant Diseases and Vectors: Ecology and Epidemiology" (K. Maramorosch and K. F. Harris, eds.), pp. 1–33. Academic Press, New York.
Bos, L. (1992). New plant virus problems in developing countries: A corollary of agricultural modernization. *Adv. Virus Res.* **38:**349–407.
Brown, J. K. (1994). The status of *Bemisia tabaci* Genn. as a plant pest and virus vector in agro-ecosystems worldwide. *FAO Plant Prot. Bull.* **42:**3–32.
Bua, A., Sserubombwe, W. D., Alicai, T., Baguma, Y. K., Omongo, C. A., Akullo, D., Tumwesigye, S., Apok, A., and Thresh, J. M. (2005). The incidence and severity of cassava mosaic virus disease and the varieties of cassava grown in Uganda: 2003. *Roots* (in press).
Cabunagan, R., Flores, Z. M., Hibino, H., Muis, A., Talanca, H., Sudjak, S. M., and Bastian, A. (1989). Sporadic occurrence of tungro (RTV) in rice resistant to rice tungro spherical virus (RTSV) in rice germplasm. *Int. Rice Research Newsletter* **18:**287.
Calvert, L. A., and Thresh, J. M. (2002). The viruses and virus diseases of cassava. *In* "Cassava: Biology, Production and Utilization" (R. J. Hillocks, J. M. Thresh, and A. C. Bellotti, eds.), pp. 237–260. CABI, Wallingford, UK, pp. 332.
Chancellor, T. C. B. (2002). Groundnut rosette disease management. Final Technical Report on Project R7445 (ZA0317) to Crop Protection Programme, Department for International Development, London.
Chancellor, T. C. B., and Thresh, J. M. (eds.) (1997). "Epidemiology and Management of Rice Tungro Disease," pp. 108. Natural Resources Institute. Chatham, UK.
Chancellor, T. C. B., Teng, P. S., and Heong, K. L. (1996). Rice Tungro Disease Epidemiology and Vector Ecology. *IRRI Discussion Paper Series* No. 19, IRRI, Manila.
Cho, J. J., Mau, R. F. L., German, T. L., Hartmann, R. W., Yudin, L. S., Gonsalves, D., and Provvidenti, R. (1989). A multidisciplinary approach to management of tomato spotted wilt virus in Hawaii. *Plant Dis.* **73:**375–383.
Cockcroft, L. (2003). Current and projected trends in African agriculture: Implications for research strategy. *In* "Plant Virology in sub-Saharan Africa" (J.d'A. Hughes and B. O. Odu, eds.), pp. 172–188. IITA, Ibadan, Nigeria.
Colvin, J., Fishpool, L. D. C., Fargette, D., Sherington, J., and Fauquet, C. (1998). *Bemisia tabaci* (Hemiptera: Aleyrodidae) trap catches in a cassava field in Côte

d'Ivoire in relation to environmental factors and the distribution of African cassava mosaic disease. *Bull. Entomol. Res.* **88:**369–378.

Cornwell, P. B. (1958). Movements of the vectors of virus diseases of cacao in Ghana I. Canopy movement in and between trees. *Bull. Entomol. Res.* **49:**613–630.

Costa, A. S., and Muller, G. W. (1980). Tristeza control by cross protection: A U.S.-Brazil cooperative success. *Plant Dis.* **64:**538–541.

Cours, G. (1951). Le manioc à Madagascar. *Mèmoires de l'Institut Scientifique de Madagascar, Série B, Biologie Végétale* **3:**203–400.

Cours, G., Fargette, D., Otim-Nape, G. W., and Thresh, J. M. (1997). The epidemic of cassava mosaic virus disease in Madagascar in the 1930s–1940s; lessons for the current situation in Uganda. *Trop. Sci.* **37:**238–248.

Dade, H. A. (1930). Cassava Mosaic. Paper 28. Year Book, Department of Agriculture, Gold Coast, Bulletin **23:**245–247.

Dahal, G. (1988). Transmission of tungro-associated viruses by field and selected colonies of *Nephotettix virescens* distant and their mode of feeding on selected rice cultivars. Ph. D. thesis, University of Philippines, Lõs Banos.

Damsteegt, V. D. (1976). A naturally occurring corn virus epiphytotic. *Pl. Dis. Reptr.* **60:**858–861.

Davies, J. C. (1975). Use of menazon insecticide to for control of rosette disease of groundnut in Uganda. *Trop. Agric. Trin.* **52:**359–367.

Davies, J. C. (1976). The incidence of rosette disease in groundnut in relation to plant density and its effect on yield. *Ann. Appl. Biol.* **82:**489–501.

DeVries, J., and Toenniessen, G. (2001). "Securing the Harvest: Biotechnology, Breeding and Seed Systems for African Crops," pp. 208. CABI Publishing, Oxon, UK.

Dittrich, V., Hassan, S. O., and Ernst, G. H. (1985). Sudanese cotton and the whitefly: A case study of the emergence of a new primary pest. *Crop. Prot.* **4:**161–176.

Ebbels, D. L. (1979). A historical review of certification schemes for vegetatively-propagated crops in England and Wales. *ADAS Quarterly Review* **32:**21–58.

Ebbels, D. L. (2003). "Principles of Plant Health and Quarantine." CAB Int., UK.

Eberwine, J. W., and Hagood, E. S. (1995). Effects of johnson grass (*Sorghum halepense*) control on the severity of virus diseases of corn *(Zea mays)*. *Weed Technol.* **9:**73–79.

Egabu, J., Osiru, D. S. O., Adipala, E., and Thresh, J. M. (2002). The influence of plant population on cassava mosaic disease epidemiology in central Uganda. *Afr. Crop Sci. J.* **5:**439–443.

Entwistle, P. F. (1972). "Pests of Cacao," pp. 779. Longman, London.

Evans, A. C. (1954). Groundnut rosette disease in Tanganyika. I. Field studies. *Ann. Appl. Biol.* **41:**189–206.

Faccioli, G., and Marani, F. (1998). Virus elimination by meristem tip culture and tip micrografting. Chapter 27, pp. 346–380. *In* Hadidi *et al.* (1998).

Fargette, D., and Fauquet, C. (1988). A preliminary study on the influence of intercropping maize and cassava on the spread of African cassava mosaic virus by whiteflies. *Asp. Appl. Biol.* **17:**195–202.

Fargette, D., Fauquet, C., and Thouvenel, J.-C. (1985). Field studies on the spread of African cassava mosaic. *Ann. Appl. Biol.* **106:**285–294.

Fargette, D., Fauquet, C., Grenier, E., and Thresh, J. M. (1990). The spread of African cassava mosaic virus into and within cassava fields. *J. Phytopathol.* **130:**289–302.

Fargette, D., Muniyappa, V., Fauquet, C. M., N'Guessan, P., and Thouvenel, J.-C. (1993). Comparative epidemiology of three tropical whitefly-transmitted geminiviruses. *Biochimie* **75:**547–554.

Fargette, D., Colon, L. T., Bouveau, R., and Fauquet, C. (1996). Components of resistance of cassava to African cassava mosaic virus. *Eur. J. Plant Pathol.* **102:**645–654.
Farrell, J. A. K. (1976a). Effects of groundnut sowing date and plant spacing on rosette virus disease in Malawi. *Bull. Entomol. Res.* **66:**159–171.
Farrell, J. A. K. (1976b). Effects of intersowing with beans on the spread of groundnut rosette virus by *Aphis craccivora* Koch (Hemiptera, Aphididae) in Malawi. *Bull. Entomol. Res.* **66:**331–333.
Fauquet, C., and Beachy, R. N. (undated). Cassava viruses and genetic engineering. International Cassava-Trans Project. CTA, Wageningen. p. 30.
Ferreira, S. A., Pitz, K. Y., Manshardt, R., Zee, F., Fitch, M., and Gonsalves, D. (2002). Virus coat protein transgenic papaya provides practical control of *Papaya ringspot virus* in Hawaii. *Plant Dis.* **86:**101–105.
Fondong, V. N., Thresh, J. M., and Zok, S. (2002). Spatial and temporal spread of cassava mosaic virus disease in cassava grown alone and when intercropped with maize and/or cowpea. *J. Phytopathol.* **150:**365–374.
Foster, J. A., and Hadidi, A. (1998). Exclusion of plant viruses. Chapter 16, pp. 208–229. *In* Hadidi *et al.*, 1998.
Fregene, M., and Puonti-Kaerlas, J. (2002). Cassava biotechnology. *In* "Cassava Biology, Production and Utilization" (R. J. Hillocks, J. M. Thresh, and A. C. Bellotti, eds.), pp. 179–207. CABI, Wallingford, UK.
Frison, E. A., and Putter, C. A. J. (1989). FAO/IBPGR Technical Guidelines for Safe Movement of Germplasm. Food and Agriculture Organization of the United Nations/ National Board for Plant Genetic Resources, Rome.
Fulton, R. (1986). Practices and precautions in the use of cross protection for plant virus disease control. *Annu. Rev. Phytopathol.* **24:**67–81.
Gámez, R., and Moreno, R. A. (1983). Epidemiology of beetle-borne viruses of grain legumes in Central America. *In* "Plant Virus Epidemiology" (R. T. Plumb and J. M. Thresh, eds.), pp. 103–113. Blackwell Scientific, Oxford, UK.
García-Arenal, F., and McDonald, B. A. (2003). An analysis of the durability of resistance to plant viruses. *Phytopathology* **93:**941–952.
Garnsey, S. M., Gottwald, T. R., and Yokomi, R. K. (1998). Control strategies for citrus tristeza virus. Chapter 48, pp. 639–658. *In* Hadidi *et al.*, 1998.
Ghosh, A., John, V. T., and Rao, J. R. K. (1979). Studies on grassy stunt disease of rice in India. *Pl. Dis. Reptr.* **63:**523–525.
Giha, O. H., and Nour, M. A. (1969). Epidemiology of cotton leafcurl virus in the Sudan. *Cotton Growers Rev.* **46:**105–118.
Gonsalves, D. (1998). Control of papaya ringspot virus in papaya: A case study. *Annu. Rev. Phytopathol.* **36:**415–437.
Gorter, G. J. M. A. (1953). "Studies on the Spread and Control of the Streak Disease of Maize," pp. 20, Science Bulletin 341, Department of Agriculture and Forestry, Union of South Africa.
Hadidi, A., Khetarpal, R. K., and Koganezawa, H. (1998). "Plant Virus Disease Control," pp. 684. American Phytopathological Society, St. Paul, Minnesota.
Hadley, P., Lee, T., and Thresh, J. M. (1989). The University of Reading cocoa quarantine project. *Cocoa Growers' Bull.* **42:**5–11.
Hamed, A. A., and Makkouk, K. M. (2002). Occurrence and management of *Chickpea chlorotic dwarf virus* in chickpea fields in northern Sudan. *Phytopathol. Mediterr.* **41:**193–198.

Hanna, A. D., and Heatherington, W. (1957). Arrest of the swollen-shoot virus disease of cacao in the Gold Coast by controlling the mealybug vectors with the systemic insecticide Dimefox. *Ann. Appl. Biol.* **45:**473–480.
Harrison, B. D. (2002). Virus variation in relation to resistance-breaking in plants. *Euphytica* **124:**181–192.
Hayes, T. R. (1932). Groundnut rosette disease in the Gambia. *Trop. Agric. Trin.* **19:**211–217.
Heinrichs, E. A., and Rapusas, H. (1983). Correlation of resistance to the green leafhopper *Nephotettix virescens* (Homoptera: Cicadellidae) with tungro virus infection in rice varieties having different genes for resistance. *Environ. Entomol.* **12:**201–205.
Hewitt, W. B., and Chiarappa, L. (1977). "Plant Health and Quarantine in International Transfer of Genetic Resources." CRC Press, Cleveland, Ohio.
Hibino, H. (1996). Biology and epidemiology of rice viruses. *Annu. Rev. Phytopathol.* **34:**249–274.
Hibino, H., and Cabauatan, R. C. (1986). Rice tungro associated viruses and their relation to host plants and vector leafhoppers. *Tropical Agriculture Research Series*, No. 19, pp. 173–181, Tropical Agriculture Research Centre, Japan.
Hibino, H., Cabuatan, P. Q., Omura, T., and Tsuchizaki, T. (1985). Rice grassy-stunt virus strain causing tungro-like symptoms in the Philippines. *Plant Dis.* **69:**538–541.
Hibino, H., Daquioag, R. D., Cabauatan, P. Q., and Dahal, G. (1988). Resistance to rice tungro spherical virus in rice. *Plant Dis.* **72:**843–847.
Hibino, H., Daquioag, R. D., Mesina, E. M., and Aguiero, V. M. (1990). Resistance in rice to tungro-associated viruses. *Plant Dis.* **74:**923–926.
Hillocks, R. J., Thresh, J. M., Tomas, J., Botao, M., Macia, R., and Zavier, R. (2002). Cassava brown streak disease in northern Mozambique. *Int. J. Pest Manag.* **48:**179–182.
Hollings, M. (1965). Disease control through virus-free stock. *Annu. Rev. Phytopathol.* **3:**367–396.
Holt, J., Chancellor, T. C. B., Reynolds, D. R., and Tiongco, E. R. (1996). Risk assessment for rice planthopper and tungro disease outbreaks. *Crop Prot.* **15:**359–368.
Hopkins, J. C. F. (1932). Leafcurl of tobacco in Southern Rhodesia. *Rhodesia Agric. J.* **29:**680–686.
Hughes, J.d'A., and Odu, B. O. (eds.) (2003). "Plant Virology in Sub-Saharan Africa," pp. 589. IITA, Nigeria.
Hughes, J.d'A., and Ollennu, L. A. A. (1994). Mild strain protection of cacao in Ghana against cacao swollen shoot virus—a review. *Plant Pathol.* **43:**442–457.
Jameson, J. D. (1964). Cassava mosaic disease in Uganda. *East Afric. Agric. For. J.* **29:**208–213.
Jennings, D. L. (1994). Breeding for resistance to African cassava mosaic geminivirus in East Africa. *Trop. Sci.* **34:**110–122.
Kahn, R. P. (1979). A concept of pest risk analysis. *EPPO Bull.* **9:**119–130.
Kahn, R. P. (1989). "Plant Protection and Quarantine." Volumes 1–3, pp. 226, 240, 265. CRC Press, Boca Raton, Florida.
Kenmore, P. E., Cariño, F. O., Perez, C. A., Dych, V. A., and Gutierrez, A. P. (1984). Population regulation of the rice brown planthopper (*Nilaparvata lugens* Stål) within rice fields in the Philippines. *J. Plant Prot. Trop.* **1:**19–37.
Kenmore, P. E., Litsinger, J. A., Bandong, J. P., Santiago, A. C., and Salac, M. M. (1987). Philippine rice farmers and insecticides: Thirty years of growing dependency and new options for change. *In* "Management of Pests and Pesticides: Farmers' Perceptions and Practices." (J. Tait and B. Napompeth, eds.), pp. 98–108. Westview Press, Boulder and London.
Khush, G. S. (1977). Breeding for resistance in rice. *Ann. NY Acad. Sci.* **287:**296–308.

Khush, G. S. (1984). Breeding rice for resistance to insects. *Prot. Ecol.* **7**:147–165.
Lal, R. (2001). Managing world soils for food security and environmental quality. *Adv. Agron.* **74**:155–192.
Legg, J. P., and Thresh, J. M. (2003). Cassava virus diseases in Africa. *In* "Plant Virology in Sub-Saharan Africa" (J.d'A. Hughes and B. O. Odu, eds.), pp. 517–552. IITA, Ibadan.
Lister, R. M., and Thresh, J. M. (1957). The history and control of cacao swollen shoot disease in Nigeria. *In* "Proceedings of Cacao Conference, London," pp. 132–142. Cacao, Chocolate and Confectionery Alliance.
Loevinsohn, M. E. (1987). Insecticide use and increased mortality in rural Central Luzon, Philippines. *Lancet*, June 13, pp. 1359–1362.
Manwan, I., Sama, S., and Rizvi, S. A. (1985). Use of varietal rotation in the management of tungro disease in Indonesia. *Indones. Agric. Res. Dev. J.* **7**:43–48.
Matteson, P. C. (2000). Insect pest management in tropical Asian irrigated rice. *Annu. Rev. Entomol.* **45**:549–574.
Mignouna, H. D., and Dixon, A. G. O. (1997). Genetic relationships among cassava clones with varying levels of resistance to African cassava mosaic disease using RAPD markers. *Afr. J. Root Tuber Crops* **2**:28–32.
Mink, G. I., Wample, R., and Howell, W. E. (1998). Heat treatment of perennial plants to eliminate phytoplasmas, viruses and viroids while maintaining plant survival. Chapter 26, pp. 332–345. *In* Hadidi *et al.* (1998).
Morales, F. J., and Anderson, P. K. (2001). The emergence and dissemination of whitefly-transmitted geminiviruses in Latin America. *Arch. Virol.* **146**:415–441.
Mwanga, R. O. M., Obwoya, C. N. O., Otim-Nape, G. W., and Odongo, B. (1991). Sweet potato improvement in Uganda. *In* "Proceedings: 4th Eastern and Southern Africa Regional Root Crops Workshop" (M. N. Alvarez and R. Asiedu, eds.), pp. 59–67. IITA, Ibadan, Nigeria.
Naidu, R. A., Kimmins, F. M., Deom, C. M., Subrahmanyam, P., Chiyembekeza, A. J., and van der Merwe, P. J. A. (1999). Groundnut rosette: A virus disease affecting groundnut production in sub-Saharan Africa. *Plant Dis.* **83**:700–709.
Ndowora, T., Dahal, G., LaFleur, D., Harper, G., Hull, R., Olszewski, N. E., and Lockhart, B. (1999). Evidence that badnavirus infection in *Musa* can originate from integrated pararetroviral sequences. *Virology* **255**:214–220.
Nichols, R. F. W. (1947). Breeding cassava for virus resistance. *East Afr. Agric. J.* **12**:184–194.
Nozaki, M., Wong, H. S., and Ho, N. K. (1984). A new double-cropping system proposed to overcome instability of rice production in the Muda irrigation area of Malaysia. *Jpn. Agric. Res. Q.* **18**:60–68.
Nweke, F. I., Ugwu, B. O., and Dixon, A. G. O. (1996). "Spread and Performance of Improved Cassava Varieties in Nigeria," pp. 34. Working Paper No. 15. Collaborative Study of Cassava in Africa, IITA, Ibadan, Nigeria.
Nweke, F. I., Spencer, D. S. C., and Lynam, J. K. (2002). "The Cassava Transformation," pp. 273, Michigan State University Press, East Lansing.
Oliveira, M. R. V., Henneberry, T. J., and Anderson, P. (2001). History, current status and collaborative projects for *Bemisia tabaci*. *Crop Prot.* **20**:709–723.
Ollennu, L. A. A., Owusu, G. K., and Thresh, J. M. (1989). The control of cocoa swollen shoot disease in Ghana. *Cocoa Growers' Bull.* **42**:25–35.
Otim-Nape, G. W., and Thresh, J. M. (2006). The recent epidemic of cassava mosaic virus diseases in Uganda. *In* "The Epidemiology of Plant Diseases" (M. B. Cooke, D. Gareth Jones, and B. Kaye, eds.), 2nd Ed., pp. 521–549. Springer, Dordrecht, Netherlands.

Otim-Nape, G. W., Bua, A., and Baguma, Y. (1994). Accelerating the transfer of improved production technologies: Controlling African cassava mosaic virus disease epidemics in Uganda. *Afr. Crop Sci. J.* **2:**479–495.

Otim-Nape, G. W., Bua, A., Thresh, J. M., Baguma, Y., Ogwal, S., Ssemakula, G. N., Acola, G., Byabakama, B., Colvin, J., Cooter, R. J., and Martin, A. (2000). "The Current Pandemic of Cassava Mosaic Virus Disease in East Africa and its Control," pp. 100. NARO, NRI, DFID, Natural Resources Institute, Chatham, UK.

Otim-Nape, G. W., Alicai, T., and Thresh, J. M. (2001). Changes in the incidence and severity of cassava mosaic virus disease, varietal diversity and cassava production in Uganda. *Ann. Appl. Biol.* **138:**313–327.

Ou, S. H. (1984). Exploring tropical rice diseases: A reminiscence. *Annu. Rev. Phytopathol.* **22:**1–10.

Owor, B., Legg, J. P., Okao-Okuja, G., Obonyo, R., Kyamanywa, S., and Ogenga-Latigo, M. W. (2004). Field studies of cross protection with cassava mosaic geminiviruses in Uganda. *J. Phytopathol.* **152:**243–249.

Padgham, D. E., Kimmins, F. M., and Ranga Rao, G. V. (1990). Resistance in groundnut (*Arachis hypogaea* L.) to *Aphis craccivora* (Koch). *Ann. Appl. Biol.* **117:**285–294.

Page, W. W., Smith, M. C., Holt, J., and Kyetere, D. (1999). Intercrops, *Cicadulina* spp., and maize streak virus disease. *Ann. Appl. Biol.* **135:**385–393.

Perring, T. M., Gruenhagen, N. M., and Farrar, C. A. (1999). Management of plant viral diseases through chemical control of insect vectors. *Annu. Rev. Entomol.* **44:**457–481.

Pita, J. S., Fondong, V. N., Sangaré, A., Otim-Nape, G. W., Ogwal, S., and Fauquet, C. M. (2001). Recombination, pseudorecombination and synergism of geminiviruses are determinant keys to the epidemic of severe cassava mosaic disease in Uganda. *J. Gen. Virol.* **82:**655–665.

Polston, J. E., and Anderson, P. K. (1997). The emergence of whitefly-transmitted geminiviruses in tomato in the western hemisphere. *Plant Dis.* **81:**1358–1369.

Posnette, A. F. (1943). Control measures against swollen shoot virus disease of cacao. *Trop. Agric. Trin.* **20:**116–123.

Posnette, A. F. (1947). Virus diseases of cacao in West Africa. I. Cacao viruses 1A, 1B, 1C and 1D. *Ann. Appl. Biol.* **34:**388–402.

Posnette, A. F., and Todd, J. McA. (1951). Virus diseases of cacao in West Africa. VIII. The search for virus-resistant cacao. *Ann. Appl. Biol.* **38:**785–800.

Posnette, A. F., and Todd, J. McA. (1955). Virus diseases of cacao in West Africa IX. Strain variation and interference in virus 1A. *Ann. Appl. Biol.* **43:**433–453.

Powell-Abel, P., Nelson, R. S., De, B., Hoffmann, N., Rogers, S. G., Fraley, R. T., and Beachy, R. N. (1986). Delay of disease development in transgenic plants that express the tobacco mosaic virus coat protein gene. *Science* **232:**738–743.

Putter, C. A. J. (1980). The management of epidemic levels of endemic diseases under tropical subsistence farming conditions. *In* "Comparative Epidemiology: A Tool for Better Disease Management" (J. Palti and J. Kranz, eds.), pp. 93–103. Pudoc, Wageningen, Netherlands. 122 pp.

Raccah, B. (1986). Non-persistent viruses: Epidemiology and control. *Adv. Virus Res.* **31:**387–429.

Reckhaus, P. M., and Andriamasintseheno, H. F. (1997). Rice yellow mottle virus in Madagascar and its epidemiology in the northwest of the island. *J. Plant Dis. Prot.* **104:**289–295.

Reddy, D. V. R., Amin, P. W., McDonald, D., and Ghanekar, A. M. (1983). Epidemiology and control of groundnut bud necrosis and other diseases of legume crops in India

caused by tomato spotted wilt virus. *In* "Plant Virus Epidemiology" (R. T. Plumb and J. M. Thresh, eds.), pp. 93–102. Blackwell Scientific, Oxford, UK.
Rivera, C. T., and Ou, S. H. (1965). Leafhopper transmission of "tungro" disease of rice. *Plant Dis. Rep.* **49:**127–131.
Rose, D. J. W. (1973). Distances flown by *Cicadulina* spp. (Hem., Cicadellidae) in relation to distribution of maize streak disease in Rhodesia. *Bull. Entmol. Res.* **62:**497–505.
Ruangsook, B., and Khush, G. S. (1987). Genetic analysis of resistance to green leafhopper *Nephotettix virescens* (Distant) in some selected rice varieties. *Crop Prot.* **6:**244–249.
Salati, R., Nahkla, M. K., Rojas, M. R., Guzman, P., Jaquez, J., Maxwell, D. P., and Gilbertson, R. L. (2002). *Tomato yellow leaf curl virus* in the Dominican Republic: Characterization of an infectious clone, virus monitoring in whiteflies and identification of reservoir hosts. *Phytopathology* **92:**457–496.
Sama, S., Hasanuddin, A., Manwan, I., Cabunagan, R. C., and Hibino, H. (1991). Integrated management of rice tungro disease in South Sulawesi, Indonesia. *Crop Prot.* **10:**34–40.
Satapathy, M. K. (1998). Chemical control of insect and nematode vectors of plant viruses. Chapter 14, pp. 188–195. *In* Hadidi *et al.* (1998).
Satapathy, M. K., Chancellor, T. C. B., Teng, P. S., Tiongco, E. R., and Thresh, J. M. (1997). Effect of introduced sources of inoculum on tungro disease spread in different rice varieties. *In* "Epidemiology and Management of Rice Tungro Disease" (T. C. B. Chancellor and J. M. Thresh, eds.), pp. 11–21. Natural Resources Institute, Chatham, UK.
Sauger, L., and Catherinet, M. (1954). La rosette chlorotique de l'arachide et les lignées selectionnées. *Agron. Trop.* **9:**28–36.
Shaw, M. J. P. (1976). Insect-borne diseases of tobacco in Rhodesia and the role of the tobacco-free period. *Rhodesia Agric. J.* **76:**87–90.
Simons, J. N., and Zitter, T. A. (1980). Use of oils to control aphid-borne viruses. *Plant Dis.* **64:**542–546.
Smith, M. C., Holt, J., Kenyon, L., and Foot, C. (1998). Quantitative epidemiology of banana bunch top virus disease and its control. *Plant Pathol.* **47:**177–187.
Smithson, J. B., and Lenné, J. M. (1996). Varietal mixtures: A viable strategy for sustainable productivity in subsistence agriculture. *Ann. Appl. Biol.* **128:**127–158.
Sõgawa, K. (1976). Rice tungro and its vectors in tropical Asia. *Rev. Plant Prot. Res.* **9:**21–46.
Sserubombwe, W. S., Thresh, J. M., Otim-Nape, G. W., and Osiru, D. O. S. (2001). Progress of cassava mosaic virus disease and whitefly vector populations in single and mixed stands of four cassava varieties grown under epidemic conditions in Uganda. *Ann. Appl. Biol.* **138:**161–170.
Sserubombwe, W. S., Bua, A., Baguma, Y. K., Alicai, T., Omongo, C. A., Akullo, D., Tumwesigye, S., Apok, A., and Thresh, J. M. (2005). The relative productivity of local and improved cassava mosaic disease-resistant varieties of cassava in Uganda in 1999 and 2003. *Roots* **9**(2)**:**15–20.
Sseruwagi, P., Rey, M. E. C., Brown, J. K., and Legg, J. P. (2004a). The cassava mosaic geminiviruses occurring in Uganda following the 1990s epidemic of severe cassava mosaic disease. *Ann. Appl. Biol.* **145:**113–121.
Sseruwagi, P., Sserubombwe, W. S., Legg, J. P., Ndunguru, J., and Thresh, J. M. (2004b). Methods of surveying the incidence and severity of cassava mosaic disease and whitefly vector populations on cassava in Africa: A review. *Virus Res.* **100:**129–142.

Sta. Cruz, F. C., Boulton, M. I., Hull, R., and Azzam, O. (1999). Agroinoculation allows the screening of rice for resistance to rice tungro bacilliform virus. *J. Phytopathol.* **147:**653–659.

Storey, H. H., and Bottomley, A. M. (1928). The rosette disease of peanuts (*Arachis hypogaea* L.). *Ann. Appl. Biol.* **15:**26–45.

Storey, H. H., and Nichols, R. F. W. (1938). Studies of the mosaic diseases of cassava. *Ann. Appl. Biol.* **25:**790–806.

Subrahmanyam, P., van der Merwe, P. J. A., Chiyembekeza, A. J., and Chandra, S. (2002). Integrated management of groundnut rosette disease. *Afr. Crop Sci. J.* **10:**99–110.

Summers, E. M., Brandes, E. W., and Rands, R. D. (1948). "Mosaic of Sugarcane in the United States, with Special Reference to Strains of the Virus," pp. 104, Technical Bulletin 955, USDA, Washington.

Taliansky, M. E., Robinson, D. J., and Murant, A. F. (2000). Groundnut rosette disease virus complex: Biology and molecular biology. *Adv. Virus Res.* **55:**357–400.

Tarr, S. A. J. (1951). "Leaf Curl Disease of Cotton." Commonwealth Mycological Institute, Kew.

Theuri, J. M., Bock, K. R., and Woods, R. D. (1987). Distribution, host range and some properties of a virus disease of sunflower. *Trop. Pest Manag.* **33:**202–206.

Thresh, J. M. (1958). "The Spread of Virus Disease in Cacao," pp. 36, Technical Bulletin No. 5, West African Cacao Research Institute, Tafo, Ghana.

Thresh, J. M. (1976). Gradients of plant virus diseases. *Ann. Appl. Biol.* **82:**381–406.

Thresh, J. M. (1981). The role of weeds and wild plants in the epidemiology of plant virus diseases. *In* "Pests, Pathogens and Vegetation" (J. M. Thresh, ed.), pp. 53–70. Pitman, London. 517 pp.

Thresh, J. M. (1982). Cropping practices and virus spread. *Annu. Rev. Phytopathol.* **20:**193–218.

Thresh, J. M. (1983). Progress curves of plant virus disease. *Adv. Appl. Biol.* **8:**1–85.

Thresh, J. M. (1988). Eradication as a virus disease control measure. *In* "Control of Plant Disease: Costs and Benefits" (B. C. Clifford and E. Lester, eds.), pp. 155–194. Blackwell, Oxford.

Thresh, J. M. (1991). The ecology of tropical plant viruses. *Plant Pathol.* **40:**324–339.

Thresh, J. M. (2003). The impact of plant virus diseases in developing countries. *In* "Virus and Virus-like Diseases of Major Crops in Developing Countries" (G. Loebenstein and G. Thottappilly, eds.), pp. 1–30. Kluwer Academic, London.

Thresh, J. M., and Cooter, R. J. (2005). Strategies for controlling cassava mosaic virus disease in Africa. *Plant Pathol.* **54:**587–614.

Thresh, J. M., and Lister, R. M. (1960). Coppicing experiments on the spread and control of cacao swollen shoot-disease in Nigeria. *Ann. Appl. Biol.* **48:**65–74.

Thresh, J. M., and Owusu, G. K. (1986). The control of cocoa swollen shoot disease in Ghana: An evaluation of eradication procedures. *Crop Prot.* **5:**41–52.

Thresh, J. M., Owusu, G. K., and Ollennu, L. A. A. (1988a). Cocoa swollen shoot: An archetypal crowd disease. *J. Plant Dis. Prot.* **95:**428–446.

Thresh, J. M., Owusu, G. K., Boamah, A., and Lockwood, G. (1988b). Ghanaian cocoa varieties and swollen shoot virus disease. *Crop Prot.* **7:**219–231.

Thresh, J. M., Fargette, D., and Otim-Nape, G. W. (1994a). Effects of African cassava mosaic geminivirus on the yield of cassava. *Trop. Sci.* **34:**26–42.

Thresh, J. M., Otim-Nape, G. W., and Jennings, D. L. (1994b). Exploiting resistance to African cassava mosaic virus. *Asp. Appl. Biol.* **39:**51–60.

Thresh, J. M., Otim-Nape, G. W., Legg, J. P., and Fargette, D. (1997). African cassava mosaic virus disease: The magnitude of the problem. *Afr. J. Root Tuber Crops* **2**:13–19.

Thresh, J. M., Otim-Nape, G. W., and Fargette, D. (1998a). The components and deployment of resistance to cassava mosaic virus disease. *Int. Pest Manag. Rev.* **3**:209–224.

Thresh, J. M., Otim-Nape, G. W., and Fargette, D. (1998b). The control of African cassava mosaic virus disease: Phytosanitation and/or resistance? Chapter 50, pp. 670–677. *In* Hadidi *et al.* (1998).

Thurston, H. D. (1977). International crop development centers: A pathologist's perspective. *Annu. Rev. Phytopathol.* **15**:223–247.

Thurston, H. D. (1990). Plant disease management practices of traditional farmers. *Plant Dis.* **74**:96–102.

Todd, J. McA. (1951). An indigenous source of swollen shoot disease of cacao. *Nature (London)* **167**:952–953.

van der Merwe, P. J. A., and Subrahmanyam, P. (1997). Screening of rosette-resistant short-duration groundnut breeding lines for yield and other characteristics. *Int. Arachis Newsl.* **17**:14–15.

Vanderplank, J. E., and Anderssen, E. E. (1944a). Kromnek disease of tobacco: A promising method of control. *Farming S. Afr.* **19**:391–394.

Vanderplank, J. E., and Anderssen, E. E. (1944b). "Kromnek Disease of Tobacco; A Mathematical Solution to a Problem of Diseases," pp. 6, Science Bulletin No. 240, Department of Agriculture and Forestry, South Africa.

van Rensberg, G. D. J., van Rensberg, J. B. J., and Giliomee, J. H. (1991). Towards cost effective insecticidal control of the maize leafhopper *Cicadulina mbila*, and the stalk borers *Busseola fusca* and *Chilo partellus*. *Phytophylactica* **23**:137–140.

van Rheenan, H. A., Hasselbach, O. E., and Muigai, S. G. S. (1981). The effect of growing beans together with maize on the incidence of bean diseases and pests. *Neth. J. Plant Pathol.* **87**:193–199.

van Vuuren, S. P., Collins, R. P., and Da Graça, J. V. (1993). Evaluation of citrus tristeza virus isolates for cross protection of grapefruit in South Africa. *Plant Dis.* **77**:24–28.

Vogel, W. O., Hennessey, R. D., Berhe, T., and Matungulu, K. M. (1993). Yield losses to maize streak disease and *Busseola fusca* (Lepidoptera: Noctuidae) and economic benefits of streak-resistant maize to small farmers in Zaire. *Int. J. Pest Manag.* **39**:229–238.

Waterworth, H. E. (1998). Certification for plant viruses—an overview, pp. 325–331. *In* Hadidi *et al.* (1998).

Yeh, S.-D., Gonsalves, D., Wang, H.-L., Namba, R., and Chiu, R.-J. (1988). Control of papaya ringspot virus by cross protection. *Plant Dis.* **72**:375–380.

Zadoks, J. C., and Schein, R. D. (1979). "Epidemiology and Plant Disease Management," pp. 427. Oxford University Press, New York.

ADVANCES IN VIRUS RESEARCH, VOL 67

BEGOMOVIRUS EVOLUTION AND DISEASE MANAGEMENT

S. E. Seal,* M. J. Jeger,† and F. Van den Bosch‡

*University of Greenwich at Medway, Chatham Maritime
Kent ME4 4TB, United Kingdom
†Division of Biology, Imperial College London, Wye Campus, Wye
Ashford, Kent TN25 5AH, United Kingdom
‡Biomathematics and Bioinformatics Division, Rothamsted Research, Harpenden
Herts AL5 2JQ, United Kingdom

Begomoviruses (Family *Geminiviridae*, Genus *Begomovirus*) are whitefly-transmitted and include some of the most destructive plant viruses in tropical and sub-tropical regions. The increased prevalence of these diseases in recent decades appears to be closely related to human activities, including agricultural intensification, increased movement of plants and plant material, and changes in cropping practices, such as the introduction of cultivars that support high populations of *Bemisia tabaci*, the only known vector species. This chapter discusses factors implicated in the global emergence of these whitefly-transmitted diseases, and key research questions that must be addressed in order to develop more sustainable management strategies.

I. Introduction

Harrison (1985) in a prescient review emphasised the need for concerted research efforts to facilitate the control of diseases caused by geminiviruses. Nevertheless, in recent decades, there has been a marked

0065-3527/06 $35.00
DOI: 10.1016/S0065-3527(06)67008-5

increase in the emergence of whitefly-transmitted geminiviruses (now ascribed to the Family *Geminiviridae*, Genus *Begomovirus*) and of populations of their vector *Bemisia tabaci* (Hemiptera: Aleyrodidae) (Morales, this volume, pp. 127–162; Morales and Anderson, 2001; Polston and Anderson, 1997; Seal *et al.*, 2006; Varma and Malathi, 2003). Research in recent decades has been intense and many new begomovirus species have been designated and the polyphyletic nature of *B. tabaci* has been recognised (Brown *et al.*, 1995; De Barro *et al.*, 2005; Fauquet *et al.*, 2003). Considerable progress has been made in understanding geminivirus replication (Gutierrez, 1999; Gutierrez *et al.*, 2004), gene function and silencing (Chellappan *et al.*, 2004; Hanley-Bowdoin *et al.*, 1999; Moissiard and Voinnet, 2004), sources of genetic variation (Mansoor *et al.*, 2003b) and the role of DNA satellite molecules (Briddon *et al.*, 2003; Stanley, 2004). Similarly, much progress has been made in understanding the phylogenetic relationships between different *B. tabaci* populations (De Barro *et al.*, 2005). The above references are generally those of recent reviews, where additional details are presented.

Research is now needed that integrates biological and molecular knowledge of viruses, vectors and host plants within an epidemiological framework, and to relate this to new and changing cropping practices in general. Increased knowledge in these areas should lead to a better understanding of the factors that promote the selection of more virulent recombinant or other novel strains, and more fecund or increased vector populations with extended host ranges. This may assist the development of control measures that limit the exposure of the viruses or their vector to selection pressures that are undesirable from the human perspective. In this chapter, we consider first knowledge of the roles of vector, viruses and plant hosts in the emergence of begomovirus epidemics and outline areas in which further research is needed. We then look at the vector–virus–host interaction from an epidemiological and evolutionary perspective and consider how the understanding gained can be used to modify cropping practices (in the widest sense) to achieve sustainable disease management. Finally, consideration is given to the implications of international traffic in plants for the continuing spread of begomovirus epidemics.

II. Roles of Vector, Viruses and Host Plants in the Emergence of Begomovirus Epidemics

A. The Vector

Increased populations of the vector *B. tabaci* are a general characteristic of begomovirus disease epidemics, and facilitate the spread of begomoviruses into and within crops, and their transmission to and from weed hosts. Increased populations of *Bemisia tabaci* are associated with a range of factors including conducive climatic conditions (Morales and Jones, 2004), the spread of the more fecund B-biotype (Perring, 2001; Polston and Anderson, 1997), the cultivation of particular crops or varieties (Costa, 1975; Morales and Anderson, 2001; Varma and Malathi, 2003), and virus infection of the host (Colvin *et al.*, 2004, this volume, pp. 419–452; Mayer *et al.*, 2002). The development of insecticide resistance in *B. tabaci* has also resulted in insecticide usage sometimes being counterproductive in reducing *B. tabaci* parasitoid populations, or favouring the selection of the more fecund insecticide-resistant B-biotype over other biotypes (Bos, 1992; Seal *et al.*, 2006).

For many epidemics, the increased vector populations are merely characterised as being '*B. tabaci*', with no information obtained, or published, on the host-plant feeding preferences and virus transmission properties of such populations. These characteristics differ greatly amongst distinct but morphologically indistinguishable *B. tabaci* populations (Bird, 1957; Brown *et al.*, 1995). To date, such distinct populations have been termed 'biotypes' A to T (Demichelis *et al.*, 2005; Diehl and Bush, 1984; Perring, 2001), but there remains uncertainty over their degree of reproductive isolation (De Barro *et al.*, 2005). There are molecular differences between morphologically indistinguishable *B. tabaci* populations, and ribosomal and mitochondrial gene sequences of *B. tabaci* populations from around the world cluster into at least six phylogenetic groups, predominantly linked to their geographic region (Brown, 2000; De Barro *et al.*, 2005; Frohlich *et al.*, 1999).

Distinct vector biotypes and genotypes have been associated with epidemics such as the many that followed the introduction of the B-biotype to the Western Hemisphere (Morales, this volume, pp. 127–162; Morales and Anderson, 2001; Polston and Anderson, 1997). A specific vector population genotype might be linked to cotton leaf curl disease in Pakistan (Simon *et al.*, 2003), but further studies are needed to clarify this. The association of a specific *B. tabaci* genotype with the recent devastating cassava mosaic disease epidemic in Uganda (Legg *et al.*, 2002) was questioned by Colvin *et al.* (2004), as no differences were found in vector fecundity, virus transmission abilities, or mating barriers from pre-epidemic and epidemic vector populations (Colvin *et al.*, 2004; Maruthi *et al.*, 2001, 2002).

Vector–virus co-adaptation has been demonstrated for some cassava mosaic begomoviruses (Maruthi *et al.*, 2002), and research is now required to determine the extent to which *B. tabaci* is a driving force for the diversity of begomoviruses that has characterised outbreaks in different regions in the last 20 years. Power (2000) considered that begomoviruses have a much stronger specificity for their vector than their hosts, and congruent phylogenetic clusters for some vector and virus genes support the view of co-adaptation (Brown, 2000). Research should now target more detailed studies of the diversity of whitefly population(s) and begomoviruses isolated from the same individual host plants. Detailed vector diversity studies may also demonstrate the natural feeding preferences of particular whitefly genotypes. This could facilitate an understanding of which crops favour vector movement, the emergence of mixed virus infections and hence the potential for more aggressive virus variants to arise through recombination.

Studies in southern India in the early 1990s showed that, at the time, tomato was not favoured as a host by indigenous *B. tabaci* populations and as a result there was significant whitefly movement into and out of tomato fields (Ramappa *et al.*, 1998). Mixed begomovirus infections are prevalent in tomato in this region, and tomato begomovirus diversity in India is very high with recombination shown to have contributed significantly to their evolution (Chowda Reddy *et al.*, 2005; Kirthi *et al.*, 2002). It remains to be seen what changes in begomovirus diversity will result from the displacement of the local Indian biotypes by the relatively recently introduced B-biotype (Banks *et al.*, 2001), for which tomato is a favoured host.

Further research is required on the life history traits in whitefly populations that lead to the emergence of biotypes with increased fecundity and/or host range. An improved understanding of the most important traits of the B-biotype that have enabled it to be such a successful invader may assist in selecting more effective control measures against the rapid spread of this particularly destructive biotype. De Barro *et al.* (2006) describe a complex interaction of factors affecting the displacement of the indigenous Australian (AN) biotype by the B-biotype, with the establishment of the invader biotype being dependent on it having sufficient hosts that preferentially support its development over that of the AN biotype. On hosts mutually acceptable to the AN and B-biotype, no displacement of the indigenous AN population occurs, provided it maintains at least a 20 times greater population size than the B-biotype. In contrast, a host that is strongly preferred by the B-biotype will allow its establishment even when the initial AN:B population size ratio is 50:1. The broader host range of the B-biotype thus reduces the impact of mating-interference (De Barro and Hart, 2000) by the indigenous biotype(s).

The significance of transovarial transmission in *B. tabaci* requires clarification. Some strains of begomovirus *Tomato yellow leaf curl virus* (TYLCV) have been reported to be transmitted transovarially ('vertically') and sexually in *B. tabaci* (Ghanim and Czosnek, 2000), whereas others have not (Bosco *et al.*, 2004). DNA of *Tomato yellow leaf curl Sardinia virus* was reported to be transmitted transovarially through eggs and nymphs to first generation adult progeny of *B. tabaci*. However, these progeny did not infect the tomato plants on which they fed (Bosco *et al.*, 2004). It thus appears that transovarial transmission is not playing an important epidemiological role, but the importance of this possible avenue for transmission should be elucidated and the need for further studies identified. The significance of sexual (venereal) transmission will depend on the male:female ratio in whitefly populations and the haplo-diploid nature of whitefly reproduction. This form of transmission has not been modelled, and was not considered in previous reviews of vector transmission and epidemic development (Jeger *et al.,* 1998; Madden *et al.,* 2000).

B. *Viruses*

Begomoviruses have genomes consisting either of one (monopartite) or two (bipartite) circular ssDNA components of *c.* 2.6–2.8-kb size, encapsidated in geminate quasi-isometric 20–30-nm virion particles (Harrison, 1985). The replication of the DNA-A and DNA-B genome components depends on host-plant enzymes and occurs in plant cell nuclei by rolling-circle (Hanley-Bowdoin *et al.*, 1999; Saunders *et al.*, 1991) and recombination-dependent replication mechanisms (Jeske *et al.*, 2001; Preiss and Jeske, 2003). The wide availability of polymerase chain reaction (PCR) methodology (Mullis and Faloona, 1987) has greatly facilitated the sequencing of begomovirus genes and genomes (Briddon *et al.*, 1993). As a result immense progress has been made in recent years in knowledge not only of gene functions, but also of the enormous diversity of the genomes of begomoviruses and their evolution through mutation, recombination and the exchange of genomic components termed pseudorecombination (Fauquet *et al.*, 2003, 2005; Kirthi *et al.*, 2002; Padidam *et al.*, 1999; Pita *et al.*, 2001; Saunders *et al.*, 2002; Seal *et al.*, 2006; Zhou *et al.*, 1997). Mutation frequencies reported for geminiviruses are often equivalent to that of RNA viruses and the primary role of recombination might be to repair ssDNA defects that have arisen through mutation (Preiss and Jeske, 2003). Recombinants that have a selective advantage would become prevalent in field populations, potentially leading to disease epidemics. The best-known example is the devastating cassava mosaic disease pandemic associated with the recombinant cassava mosaic begomovirus *East African cassava mosaic virus*-Uganda (EACMV-[UG]) (Zhou *et al.*, 1997).

It is not clear how different cropping systems, or even host crops, affect begomovirus diversity and the relative contribution they make to the various factors driving evolution. Research concentrating on whether there are selection pressures that operate and drive the evolution of begomoviruses towards increased virulence and an extended host range has been rather limited. The most that has been achieved is to show the extreme genetic plasticity of viruses that is possible. The inference is that this enables begomoviruses to evolve very rapidly in response to changing cropping systems. As such this provides little more than an initial hypothesis on which to base detailed epidemiological studies in relation to evolutionary change.

Research on the molecular diversity of plant virus populations needs to target the population rather than 'molecular' level (García-Arenal *et al.,* 2003), as simply determining the number of different molecular sequences present in a host plant, crop or region, is inadequate to track evolutionary change and determine the influence of factors such as the introduction of host-plant resistance, or changes in cropping system. Additionally, there is still a lack of information on the true frequency of virus variants and inevitably there will be biases in the current knowledge of virus diversity. Diagnostic techniques, such as PCR, are selective even when degenerate begomovirus PCR primers are used. Furthermore, many studies of gene or genome function have dealt only with the properties of infectious clones of one sequence. In the field, the biological function of the virus may depend on the interaction between a 'swarm' of variant sequences upon which selection acts (Roossinck, 1997).

One crucial area that needs clarification is the role and mode of interaction of the relatively recently discovered circular ssDNA satellites with each other, their helper viruses and their role in begomovirus epidemiology. These DNA satellites share no significant sequence homology with begomovirus sequences and are of various types. The first to be discovered was 682 nucleotides in size and associated with *Tomato leaf curl virus* in Australia. The satellite depended on this helper virus for replication, movement and encapsidation but had no apparent effect on disease symptoms (Dry *et al.*, 1997). Subsequently, similar but larger satellites (~1350 nucleotides) were found associated with other monopartite begomoviruses (*Ageratum yellow vein virus, Cotton leaf curl Multan virus*) and these DNA-β satellites were found to be essential for sufficient accumulation of DNA-A and symptom development (Briddon *et al.*, 2001; Saunders *et al.*, 2000). DNA-β molecules are widespread in the Old World (Briddon *et al.*, 2003; Bull *et al.*, 2004; Zhou *et al.*, 2003) and plants infected with these molecules usually contain another group of ssDNA satellites of similar size termed DNA1s (Briddon *et al.*, 2004; Mansoor *et al.*, 1999; Saunders and Stanley, 1999; Stanley, 2004). DNA1s are, however, not required for disease symptom induction and reduce viral accumulation (Saunders *et al.*, 2001; Wu and Zhou, 2005).

The epidemiological role of DNA-β satellite molecules appears to be in extending the host range of begomoviruses. For example, at least five diverse begomovirus species, including *Papaya leaf curl virus*, can cause cotton leaf curl disease in Pakistan but only when associated with a particular DNA-β molecule (Mansoor *et al.*, 2003a). The epidemiological role of DNA1 satellites is less clear. However, their presence in nearly all monopartite viruses that contain DNA-β components suggests that the ability to reduce viral accumulation presents a selective advantage for such monopartite helper viruses associated with DNA-β satellites. A similar function appears to be provided by an independently evolved mechanism in bipartite begomoviruses, where the DNA-B component is usually essential for systemic symptom development. Defective DNA-B components occur commonly in some bipartite begomoviruses and are considered to compete during replication with full-size functional components (Patil and Dasgupta, 2006; Stanley *et al.*, 1990). Down-regulation of viral infection may be advantageous by maintaining host metabolism suitable for sustained viral accumulation (Moissiard and Voinnet, 2004; Patil and Dasgupta, 2006). Further molecular and epidemiological modeling-based research is needed on the mechanisms of interaction between different satellites and of these molecules with their helper virus. An improved understanding of these molecules could assist in developing novel control strategies (Patil and Dasgupta, 2006).

C. *Host Plants*

Begomoviruses infect only dicotyledonous plants, and the plant hosts and varieties grown will influence virus diversity through selecting for particular viruses as well as vector populations. The fecundity of different *B. tabaci* populations differs greatly on different hosts (Colvin *et al.*, this volume, pp. 419–452; Gerling *et al.*, 1986) and changes in the crops cultivated may result in marked changes in vector abundance. For example, the increased cultivation of cotton, soybean and horticultural crops in Latin America in the 1970s led to greater *B. tabaci* populations and ensuing begomovirus disease epidemics (Costa, 1975; Morales and Anderson, 2001).

Begomoviruses transmitted by vectors will need to avoid or overcome host-plant defence mechanisms, such as post-transcriptional gene silencing (PTGS) (Moissiard and Voinnet, 2004; Voinnet *et al.*, 1999) to enable successful infection. This performs a critical function in determining the outcome of begomovirus–plant host interactions (Chellappan *et al.*, 2004; Vanitharani *et al.*, 2004). PTGS may be the as yet unknown mechanism for the cross-protection of cassava by mild strains of EACMV-[UG] against severe strains (Owor *et al.*, 2004). Further research to understand the mechanisms of PTGS is essential and could lead to improved control strategies. Many questions remain on the mechanism, such as why does PTGS target different genes (and DNA components) for different cassava mosaic begomoviruses (Chellappan *et al.*, 2004; Vanitharani *et al.*, 2004)? Such research may also clarify the role of PTGS in 'driving' the evolution of begomoviruses.

The deployment of host-plant resistance to begomoviruses has been the most desirable disease control option, particularly for the many farmers with such limited resources that it restricts the use of other measures (Lapidot and Friedman, 2002; Thresh, this volume, pp. 245–295; Thresh and Cooter, 2005). The widespread use of resistant cultivars has been shown to alter virus diversity. For example, highly virulent EACMV-[UG] strains were more common in resistant than more susceptible cassava varieties in Uganda (Alicai, 2003). This could possibly be due to such strains being largely self-eliminating in the most susceptible varieties. In Spain, the use of TYLCV-resistant cultivars appears to select for TYLCV-Israel over TYLCV-Sardinia (Hernandez-Gallardo *et al.*, 2003).

Many of the 'resistant' plant cultivars grown are in fact 'tolerant', in the sense that they express less conspicuous symptoms but have the same virus content as sensitive cultivars. Tolerant host-plant varieties should be deployed with caution as there is the potential for high-yield losses if tolerance breaks down. Tolerant host plants also act as an important virus reservoir. Epidemiological models suggest that tolerance, in the sense used here, may lead to the selection of virus strains with higher-virus accumulation in the host (van den Bosch *et al.*, 2006). In contrast, mechanisms reducing virus inoculation or acquisition rates are selectively neutral in this respect. Further research is needed to clarify the selective forces that host-plant resistance and tolerance place on the evolution of begomoviruses.

Research should also target the potential dangers of using begomovirus-resistant or tolerant cultivars that support high-vector populations. These include some of the cassava mosaic disease-resistant cultivars grown in Uganda (Colvin *et al.*, 2004; Legg *et al.*, this volume, pp. 355–418; Thresh and Cooter, 2005). These appear to present a real danger to cassava production in the future, as well as causing direct feeding damage. Should a resistance-breaking virus strain evolve, its management will be much more difficult given the accompanying high-*B. tabaci* populations. Moreover, the increased vector populations are likely to have facilitated the appearance and spread in Uganda of *Cassava brown streak virus* (CBSV, Genus *Ipomovirus*; Family *Potyviridae*), which is also transmitted by *B. tabaci* (Maruthi *et al.*, 2005). CBSV has a geographic distribution, which overlaps partly with CMD and it now poses a serious threat to other areas of East and Central Africa (Legg *et al.*, this volume, pp. 355–418). Resistance in cassava that reduces vector populations directly would be desirable. The approaches towards achieving this to African *B. tabaci* are outlined in a review on CMD control strategies (Thresh and Cooter, 2005). A reduction in vector population size together with the phytosanitation measures of roguing and the use of healthy cuttings as planting material, as advocated by Legg *et al.* (this volume, pp. 355–418), offer much promise for control of whitefly-borne virus diseases of cassava. These factors are postulated to select for less virulent viruses that persist through further cycles of vegetative propagation (van den Bosch *et al.*, 2006).

The use of resistant varieties for control of begomoviruses thus should be managed carefully and research needs to target features of cropping practices that minimise the evolution of resistance-breaking virus variants and more fecund or polyphagous vector populations. McDonald and Linde (2002) highlight that resistance will be more durable where there is spatial or temporal patterning of the selective force. Resistance gene rotations and use of crop or varietal mixtures need to be evaluated to determine the role they might play in providing disruptive selection in crops prone to begomoviruses. There is little information on the implications of these cropping practices, although the prevalence of CMD is reduced in susceptible cassava cultivars when grown with resistant ones or when intercropped (Sserubombwe *et al.*, 2001; Thresh and Cooter, 2005).

III. Host Plant–Virus–Vector Interactions and the Evolution of Begomoviruses

Plant virus research of the type described under the previous headings is often highly focused, concentrating on aspects of the host, virus or vector separately, or at most in pair wise interactions. However, any attempt to understand the dynamics of a virus disease epidemic must consider the full range of plant–virus–vector interactions that can occur (Bos, 1992; Thresh, 1980). Complex mutualistic interactions between virus infection, vector fecundity and host attributes have been reported for begomoviruses (Colvin *et al.*, 1999, this volume, pp. 419–452), mastreviruses (Bosque-Pérez, 2000) and luteoviruses (Jiménez-Martínez *et al.*, 2004; Mayer *et al.*, 2002). The key driving force for the devastating cassava mosaic pandemic that has spread rapidly in East Africa since the late 1980s (Legg, 1999, this volume, pp. 355–418) appears to be an interaction between particular virus strains, vector populations and host genotypes rather than a single factor (Colvin *et al.*, 2004, this volume, pp. 419–452). The fecundity of *B. tabaci* increases significantly on cassava plants infected with the recombinant EACMV-[UG], leading to a much higher-population density of the whiteflies on the limited green areas of severely affected leaves and an increased emigration rate of infective adults.

The above studies suggest that additional virus-induced driving forces for begomovirus and other virus epidemics could be the alteration of plant biochemistry so that infected plants increase vector fecundity, emit volatiles as vector attractants (Jiménez-Martínez *et al.*, 2004) and alter feeding behaviour so enabling increased virus acquisition (Bosque-Pérez, 2000). These interactions could have the effect of increasing the fitness of the virus population. Biological and genetic studies to elucidate such interactions are a high priority for future research. Moreover, experimental research should be combined with modelling studies, which offer opportunities for dissecting and integrating the various layers of interaction and exploring the implications for virus epidemiology (Jeger *et al.*, 1998, 2004; Madden *et al.*, 2000).

IV. Cropping Practices and Sustainable Disease Management

The adoption of particular cropping practices will assist controlling the spread of begomovirus diseases, and appearance of diseases caused by variant begomoviruses, if the practices reduce vector transmission rates and virus population densities or both. However, little is known about the effects of cropping practices on the evolution of vector and begomovirus populations. Few studies have considered the genetic stability of virus and vector under field conditions and the effects imposed by changed cropping systems. There have been comprehensive studies for viruses causing tomato yellow leaf curl diseases and *B. tabaci* biotypes in Spain (Sanchez-Campos *et al.*, 1999a,b). It would be of considerable interest to carry out similar studies in tropical agro-ecosystems, where cultivation is year-round and there is a greater abundance of alternative crop and weed hosts.

Pathogens that maintain a high population size throughout the year can respond faster to changing selection pressures than those in which high populations only occur for a few months of the year (Bos, 1992; Seal *et al.*, 2006). This is an important consideration that makes the control of begomovirus diseases more difficult in tropical than temperate regions of the world, where crop growth is limited to fewer months of the year (Bos, 1992; Polston and Anderson, 1997). Irrigation has exacerbated the situation, particularly in tropical production systems, by leading to a continuous and abundant sequence of crop and weed hosts enabling *B. tabaci* and viruses to sustain relatively high-population levels year-round. Irrigation can also encourage the carry-over of viruses and vectors to subsequent crops by facilitating weed growth and enhancing the regeneration of plant remains left after harvesting (Thresh, 1982). Some problems associated with irrigation can be alleviated where large-scale control over planting times, crop rotations and irrigation schedules is possible, but this is not achievable in many tropical regions due to numerous independent smallholders (Thresh, 1982). In more temperate regions, the introduction of winter cultivation of ornamentals and vegetables in plastic tunnels or greenhouses has also increased the prevalence and severity of begomovirus diseases by enabling *B. tabaci* to overwinter on these plants (Bos, 1992; Thresh, 1982).

Consequently, successful control of begomoviruses has been achieved only rarely and in general only in discontinuous cropping systems, where production is synchronised and growers can maintain a host-free period. For example, in Florida *Tomato mottle virus* (ToMoV) has been controlled using this approach (Polston and Anderson, 1997). In contrast, TYLCV-Is has persisted following its introduction in the 1990s, despite great efforts to eradicate it (Polston *et al.*, 1999). TYLCV-Is has a wider host range than ToMoV, which has no significant weed hosts, making it easier to generate and maintain a host-free period and for applications of the insecticide imidacloprid to have a much greater impact on ToMoV than TYLCV-Is (Polston, J., personal communication). Although it will be difficult and seldom feasible for smallholders in the tropics to generate similar host-free periods, research does need to identify appropriate cropping practices that minimise infection rates and create a disruptive selection force. Reduced yield losses due to begomoviruses can be achieved by such measures in tropical regions, as illustrated by a 4-month ban on the cultivation of bean, tomato, pepper, eggplant and melon cultivation in the Dominican Republic, which enabled melon production to be resumed in some regions (Morales and Anderson, 2001).

Knowledge of sources of infection, including the importance of wild hosts needs to be explored further, as does the effectiveness of physical and crop barriers, varietal mixtures, intercropping, restricted planting dates and crop spacing. Mathematical models could have a major impact in this area by providing an initial screen to assess possible cropping system scenarios for their impact on disease management. The model of Holt *et al.* (1999) used to identify control strategies for tomato leaf curl virus disease in southern India predicted that a very low rate of vector immigration was sufficient to cause near total infection of a tomato crop. It suggested that the use of protective netting combined with resistant varieties would be much more effective than insecticide use. Subsequent research has shown brinjal and cucumber to be particularly effective as border crops to reduce the migration of *B. tabaci* into tomato crops, as were nylon nets placed at different heights (see www.tomatoleafcurlandwhitefly.org).

The traditional practice of intercropping two or more different species is one means of reducing the spread of virus diseases of crops that are usually grown singly (Thresh, 1982). However, agricultural intensification has led to crops being grown singly to facilitate cultivation, uniformity of produce and weed control and this trend will be difficult to reverse (Bos, 1992; Thresh, 1982). Although little progress has been made in restoring mixed cropping systems to reduce the incidence of begomovirus or other diseases, intercropping has been shown to be effective in decreasing whitefly numbers and the spread of tomato yellow leaf curl disease (Picó *et al.*, 1996) and cassava mosaic disease (Fondong *et al.*, 2002). However, the effects of intercropping on the genetic diversity of virus populations and on vector behaviour are not known. More research is required on which crop mixtures (either inter or intra-specific) minimize the build up of vector and virus populations, vector movement, the prevalence of mixed infections, and thus presumably the frequency of virus recombination. Equally, large-scale sequencing of begomovirus genomes and *B. tabaci* sequences from crop and weed host plants in small geographical areas would facilitate an understanding of the relative effects of prevailing host and weed genotypes and vector populations on virus diversity.

V. International Traffic and Spread of Begomovirus Diseases

Strict quarantine measures offer an important control measure by minimising the introduction and spread of *B. tabaci* and begomovirus genotypes to new areas. However, such measures are difficult to enforce as small numbers of *B. tabaci* eggs may not be visible on the exposed plant parts and begomovirus infections can be asymptomatic. Moreover, with the increased intercontinental movement of germplasm and commercial assignments in the past few decades, it is not feasible to hold material for long periods in quarantine, or to test all material for asymptomatic virus infection. Consequently, it is not surprising that international trade in plants has spread the more polyphagous *B. tabaci* B-biotype across continents and movement of infected plants is also known to have spread TYLCV to many new countries across the world (Polston and Anderson, 1997; Seal *et al.*, 2006; Thresh, this volume, pp. 89–125). Symptomless tomato fruit has recently been reported as a source of TYLCV for *B. tabaci* under experimental conditions (Delatte *et al.*, 2003) and there is a need to determine if this mode of transmission is of epidemiological or quarantine significance. It should also be realised that plants or plant produce may present a route for symptom-modulating DNA satellite complexes to reach new geographical regions. Such complexes could have a serious impact, even in the Western Hemisphere where bipartite begomoviruses predominate, as DNA-β molecules may be able to increase the disease severity of bipartite begomoviruses (Rouhibakhsh and Malathi, 2005). Ways to improve within-country and international quarantine measures are required urgently, despite the many problems involved in their introduction and enforcement. The development of sensitive new diagnostic techniques, such as specific PCR amplification of begomovirus DNA sequences, often enables suitably resourced laboratories to detect very low-titer infections, or for quarantine samples to be tested in bulked lots to minimise symptomless infected plants being missed due to insufficient sampling.

Risk models, often based on climate, can be used to predict countries that should pay particular attention to the import of viruses or vectors in vegetative planting material or plant produce. For example, Morales and Jones (2004) highlight Chile as a country that needs to enforce highly effective quarantine controls to prevent the introduction of the B-biotype, as the Central Valley region of the country has an ideal climate for *B. tabaci*. Such pest risk analysis models have application and can be extended worldwide to assess potential countries at risk and the impacts that global warming may have.

VI. Conclusions

The emergence of begomovirus epidemics can be seen as a direct result of human activities related to agricultural intensification to try to improve crop yields (Bos, 1992; Morales and Anderson, 2001; Seal *et al.*, 2006; Thresh, 1982; Varma and Malathi, 2003). The trend towards extensive monocultures coupled with irrigation, protective cropping, fertilizers and pesticide applications have led to increases in *B. tabaci* populations. These have been accompanied by increases in yield losses caused by direct feeding damage and begomovirus diseases. An increase in international movement of vegetative planting material has led to the highly fecund and polyphagous *B. tabaci* B-biotype being distributed around the world, as well as begomoviruses with wide host ranges that are difficult to eradicate. In some instances attempts to manage begomovirus diseases may have contributed to the severity of the epidemics reported. For example, the use of insecticides has in a number of locations led to upsurges in vector populations, due to the development of insecticide resistances in *B. tabaci* (Denholm *et al.*, 1996). Moreover, insecticides sometimes have a greater effect on *B. tabaci* predators and parasitoids than the target vector population (Bos, 1992; Dittrich *et al.*, 1985; Eveleens, 1983). Furthermore, insecticides will have little impact on reducing incoming viruliferous whiteflies and hence on begomovirus diseases for which the main factor controlling disease incidence is spread into, rather than within the crop (Fargette *et al.*, 1990; Holt *et al.*, 1999; Thresh and Cooter, 2005).

Knowledge of vector and virus diversity and evolution, and the complex interactions between plant host, vector and virus, leads inevitably to the conclusion that no single or sustainable solution will be found to control begomovirus disease epidemics. The emergence of future epidemics will only be reduced by limiting some of the selection pressures that have arisen in the past: including some of the negative aspects of agricultural intensification, the introduction of highly susceptible virus or vector hosts, or conversely the wide-scale cultivation of some types of plant host resistance that are vulnerable to the evolution of new virulent strains. Currently it is not clear whether more virulent virus variants and more fecund or polyphagous whitefly biotypes will inevitably be bound together as a co-evolutionary consequence of their interactions with each other. The research suggested here will assist in addressing such questions through a combination of epidemiological and ecological studies, together with data on molecular evolution and divergence of the viruses. Accompanying these experimental studies, mathematical models can be used to predict the effectiveness and robustness of disease management strategies for reducing begomovirus disease problems worldwide.

References

Alicai, T. (2003). Cassava mosaic-resistant cassava: Selective effects on cassava mosaic begomoviruses and reversion and recovery from cassava mosaic disease, p. 244. PhD Thesis. University of Greenwich, UK.

Banks, G. K., Colvin, J., Chowda Reddy, R. V., Maruthi, M. N., Muniyappa, V., Venkatesh, H. M., Kiran Kumar, M., Padmaja, A. S., Beitia, F. J., and Seal, S. E. (2001). First report of the *Bemisia tabaci* B biotype in India and an associated tomato leaf curl virus disease epidemic. *Plant Dis.* **85:**231.

Bird, J. (1957). A whitefly transmitted mosaic of *Jatropha gossypifolia.* Technical Paper. *Univ. Puerto Rico Agric. Exp. Stn.* **22:** 1–35.

Bos, L. (1992). New plant virus problems in developing countries: A corollary of agricultural modernization. *Adv. Virus Res.* **38:**349–407.

Bosco, D., Mason, G., and Accotto, G. P. (2004). TYLCSV DNA, but not infectivity, can be transovarially inherited by the progeny of the whitefly vector *Bemisia tabaci* (Gennadius). *Virology* **323:**276–283.

Bosque-Pérez, N. A. (2000). Eight decades of maize streak virus research. *Virus Res.* **71:**107–121.

Briddon, R. W., Prescott, A. G., Lunness, P., Chamberlin, L. C. L., and Markham, P. G. (1993). Rapid production of full-length infectious geminivirus clones by abutting primer PCR (AbP-PCR). *J. Virol. Methods* **43:**7–20.

Briddon, R. W., Mansoor, S., Bedford, I. D., Pinner, M. S., Saunders, K., Stanley, J., Zafar, Y., Malik, K., and Markham, P. G. (2001). Identification of DNA components required for induction of cotton leaf curl disease. *Virology* **285:**234–243.

Briddon, R. W., Bull, S. E., Amin, I., Idris, A. M., Mansoor, S., Bedford, I. D., Dhawan, P., Rishi, N., Siwatch, S. S., Abdel-Salam, A. M., Brown, J. K., and Zafar, Y. (2003). Diversity of DNA-β, a satellite molecule associated with some monopartite begomoviruses. *Virology* **312:**106–121.

Briddon, R. W., Bull, S. E., Amin, I., Mansoor, S., Bedford, I. D., Rishi, N., Siwatch, S. S., Zafar, Y., Abdel-Salam, A. M., and Markham, P. G. (2004). Diversity of DNA 1: A satellite-like molecule associated with monopartite-DNA-β complexes. *Virology* **324:**462–474.

Brown, J. K. (2000). Molecular markers for the identification and global tracking of whitefly vector-*Begomovirus* complexes. *Virus Res.* **71:**233–260.

Brown, J. K., Frohlich, D. R., and Rosell, R. C. (1995). The sweet potato or silver leaf whiteflies: Biotypes of *Bemisia tabaci* or a species complex? *Annu. Rev. Entomol.* **40:**511–534.

Bull, S. E., Tsai, W.-S., Briddon, R. W., Markham, P. G., Stanley, J., and Green, S. K. (2004). Diversity of begomovirus DNA-β satellites of non-malvaceous plants in East and South East Asia. *Arch. Virol.* **149:**1193–1200.

Chellappan, P., Vanitharani, R., and Fauquet, C. M. (2004). Short interfering RNA accumulation correlates with host recovery in DNA virus-infected hosts, and gene silencing targets specific viral sequences. *J. Virol.* **78:**7465–7477.

Chowda Reddy, R. V., Colvin, J., Muniyappa, V., and Seal, S. (2005). Diversity and distribution of begomoviruses infecting tomato in India. *Arch. Virol.* **150:**845–867.

Colvin, J., Omongo, C. A., Maruthi, M. N., Otim-Nape, G. W., and Thresh, J. M. (2004). Dual begomovirus infections and high-*Bemisia tabaci* populations: Two factors driving the spread of a cassava mosaic disease pandemic. *Plant Pathol.* **53:**577–584.

Colvin, J., Otim-Nape, G. W., Holt, J., Omongo, C., Seal, S., Stevenson, P., Gibson, G., Cooter, R. J., and Thresh, J. M. (1999). Factors driving the current epidemic of severe cassava mosaic disease in East Africa. *In* "VIIth International Plant Virus Epidemiology Symposium–Plant Virus Epidemiology: Current Status and Future Prospects," pp. 76–77. International Society of Plant Pathology, Aquadulce (Almeria), Spain.

Costa, A. S. (1975). Increase in the population density of *Bemisia tabaci*, a threat of widespread virus infection of legume crops in Brazil. *In* "Tropical Disease of Legumes" (J. Bird and K. Maramorosch, eds.), pp. 27–49. Academic Press, New York.

De Barro, P. J., and Hart, P. J. (2000). Mating interactions between two biotypes of the whitefly, *Bemisia tabaci* (Hemiptera: Aleyrodidae) in Australia. *Bull. Entomol. Res.* **90:**103–112.

De Barro, P. J., Trueman, J. W. H., and Frohlich, D. R. (2005). *Bemisia argentifolii* is a race of *B. tabaci* (Hemiptera: Aleyrodidae): The molecular differentiation of *B. tabaci* populations around the world. *Bull. Entomol. Res.* **95:**1–11.

De Barro, P. J., Bourne, A., Khan, S. A., and Brancatini, V. A. L. (2006). Host plant and biotype density interactions: Their role in the establishment of the invasive B biotype of *Bemisia tabaci. Biol. Invasions* **8:**287–294.

Delatte, H., Dalmon, A., Rist, D., Soustrade, I., Wuster, G., Lett, J. M., Goldbach, R. W., Peterschmitt, M., and Reynaud, B. (2003). *Tomato yellow leaf curl virus* can be acquired and transmitted by *Bemisia tabaci* (Gennadius) from tomato fruit. *Plant Dis.* **87:**1297–1300.

Denholm, I., Cahill, M., Byrne, F. J., and Devonshire, A. L. (1996). Progress with documenting and combating insecticide resistance in *Bemisia*. *In* "*Bemisia* 1995, Taxonomy, Biology, Damage, Control and Management" (D. Gerling and R. T. Mayer, eds.), pp. 577–603. Intercept, Andover, Hants.

Demichelis, S., Arno, C., Bosco, D., Marian, D., and Caciagli, P. (2005). Characterization of Biotype T of *Bemisia tabaci* associated with *Euphorbia characias* in Sicily. *Phytoparasitica* **33:**196–208.

Diehl, S. R., and Bush, G. L. (1984). An evolutionary and applied perspective of insect biotype. *Annu. Rev. Entomol.* **29:**471–504.

Dittrich, V., Hassan, S. O., and Ernst, G. H. (1985). Sudanese cotton and the whitefly: A case study of the emergence of a new primary pest. *Crop Prot.* **4:**161–176.

Dry, I. B., Krake, L. R., Rigden, J. E., and Rezaian, M. A. (1997). A novel subviral agent associated with a geminivirus: The first report of a DNA satellite. *Proc. Natl. Acad. Sci. USA* **94:**7088–7093.

Eveleens, K. G. (1983). Cotton-insect control in the Sudan Gezira: Analysis of a crisis. *Crop Prot.* **2:**273–287.

Fargette, D., Fauquet, C., Grenier, E., and Thresh, J. M. (1990). The spread of African cassava mosaic virus into and within cassava fields. *J. Phytopathol.* **130:**289–302.

Fauquet, C. M., Bisaro, D. M., Briddon, R. W., Brown, J. K., Harrison, B. D., Rybicki, E. P., Stenger, D. C., and Stanley, J. (2003). Revision of taxonomic criteria for species demarcation in the family Geminiviridae, and an updated list of begomovirus species. *Arch. Virol.* **148:**405–421.

Fauquet, C. M., Sawyer, S., Idris, A. M., and Brown, J. K. (2005). Sequence analysis and classification of apparent recombinant begomoviruses infecting tomato in the Nile and Mediterranean basins. *Phytopathology* **95:**549–555.

Fondong, V. N., Thresh, J. M., and Zok, S. (2002). Spatial and temporal spread of cassava mosaic virus disease in cassava grown alone and when intercropped with maize and/or cowpea. *J. Phytopathol.* **150:**365–374.

Frohlich, D. R., Torres-Jerez, I., Bedford, I. D., Markham, P. G., and Brown, J. K. (1999). A phylogeographical analysis of *Bemisia tabaci* species complex based on – mitochondrial DNA markers. *Mol. Ecol.* **8:**1683–1691.

García-Arenal, F., Fraile, A., and Malpica, J. M. (2003). Variation and evolution of plant virus populations. *Int. Microbiol.* **6:**225–232.

Ghanim, M., and Czosnek, H. (2000). Tomato yellow leaf curl geminivirus (TYLCV-Is) is transmitted among whiteflies (*Bemisia tabaci*) in a sex-related manner. *J. Virol.* **74:**4738–4745.

Gutierrez, C. (1999). Geminivirus DNA replication. *Cell. Mol. Life Sci.* **56:**313–329.

Gutierrez, C., Ramirez-Parra, E., Castellano, M. M., Sanz-Burgos, A. P., Luque, A., and Missich, R. (2004). Geminivirus DNA replication and cell cycle interactions. *Vet. Microbiol.* **98:**111–119.

Hanley-Bowdoin, L., Settlage, S. B., Orozoo, B. M., Nagar, S., and Robertson, D. (1999). Geminiviruses: Models for plant DNA replication, transcription, and cell cycle regulation. *Crit. Rev. Plant Sci.* **18:**71–106.

Harrison, B. D. (1985). Advances in geminivirus research. *Annu. Rev. Phytopathol.* **23:**55–82.

Hernandez-Gallardo, M. D., Guerrero, M. M., Barcelo, N., Cenis, J. L., Lacasa, A., and Martinez, M. A. (2003). Effects of the use of TYLCV-resistant cultivars on the proportion of TYLCV-Is/TYLCV-Sar in tomato crops of southeast Spain. *In* "Proceedings of the 3rd International Bemisia Workshop," 17–20, March 2003, Barcelona, Spain.

Holt, J., Colvin, J., and Muniyappa, V. (1999). Identifying control strategies for tomato leaf curl virus disease using an epidemiological model. *J. Appl. Ecol.* **36:**625–633.

Jeger, M. J., Holt, J., van den Bosch, F., and Madden, L. V. (2004). Epidemiology of insect-transmitted plant viruses: Modelling disease dynamics and control interventions. *Physiol. Entomol.* **29:**291–304.

Jeger, M. J., van den Bosch, F., Madden, L. V., and Holt, J. (1998). A model for analysing plant virus transmission characteristics and epidemic development. *IMA J. Math. Appl. Med. Biol.* **14:**1–18.

Jeske, H., Lutgemeier, M., and Preiss, W. (2001). Distinct DNA forms indicate rolling circle and recombination-dependent replication of Abutilon mosaic geminivirus. *EMBO J.* **20:**6158–6167.

Jiménez-Martínez, E. S., Bosque-Pérez, N. A., Berger, P. H., Zemetra, R. S., Ding, H., and Eigenbrode, S. D. (2004). Volatile cues influence the response of *Rhopalosiphum padi* (Homoptera: Aphididae) to Barley yellow dwarf virus-infected transgenic and untransformed wheat. *Environ. Entomol.* **33:**1207–1216.

Kirthi, N., Maiya, S. P., Murthy, M. R. N., and Savitri, H. S. (2002). Evidence of recombination among the tomato leaf curl virus strains/species from Bangalore, India. *Arch. Virol.* **147:**255–272.

Lapidot, M., and Friedman, M. (2002). Breeding for resistance to whitefly-transmitted geminiviruses. *Ann. Appl. Biol.* **140:**109–127.

Legg, J. P. (1999). Emergence, spread and strategies for controlling the pandemic of cassava mosaic virus disease in East and Central Africa. *Crop Prot.* **18:**627–637.

Legg, J. P., French, R., Rogan, D., Okao-Okuja, G., and Brown, J. K. (2002). A distinct *Bemisia tabaci* (Genn.) genotype cluster is associated with the epidemic of severe cassava mosaic virus disease in Uganda. *Mol. Ecol.* **11:**1219–1229.

Madden, L. V., Jeger, M. J., and van den Bosch, F. (2000). A theoretical assessment of the effects of vector-virus transmission mechanism on plant virus disease epidemics. *Phytopathology* **90:**576–594.

Mansoor, S., Briddon, R. W., Bull, S. E., Bedford, I. D., Bashir, A., Hussain, M., Saeed, M., Zafar, Y., Malik, K. A., Fauquet, C., and Markham, P. G. (2003a). Cotton leaf curl disease is associated with multiple monopartite begomoviruses supported by single DNA-β. *Arch. Virol.* **148:**1969–1986.

Mansoor, S., Briddon, R. W., Zafar, Y., and Stanley, J. (2003b). Geminivirus disease complexes: An emerging threat. *Trends Plant Sci.* **8:**128–134.

Mansoor, S., Khan, S. H., Bashir, A., Saeed, M., Zafar, Y., Malik, K. A., Briddon, R., Stanley, J., and Markham, P. G. (1999). Identification of a novel circular single stranded DNA associated with cotton leaf curl disease in Pakistan. *Virology* **259:**190–199.

Maruthi, M. N., Colvin, J., and Seal, S. E. (2001). Mating compatibility, life history traits and RAPD-PCR variation in *Bemisia tabaci* associated with the cassava mosaic disease pandemic in East Africa. *Entomol. Exp. Appl.* **29:**13–23.

Maruthi, M. N., Colvin, J., Seal, S. E., Gibson, G., and Cooper, J. (2002). Co-adaptation between cassava mosaic geminivirus and their local vector population. *Virus Res.* **86:**71–85.

Maruthi, M. N., Hillocks, R. J., Mtunda, K., Raya, M. D., Muhanna, M., Kiozia, H., Rekha, A. R., Colvin, J., and Thresh, J. M. (2005). Transmission of Cassava brown streak virus by *Bemisia tabaci* (Gennadius). *J. Phytopathol.* **153:**307–312.

Mayer, R. T., Inbar, M., McKenzie, C. L., Shatters, R., Borowicz, V., Albrecht, U., Powell, C. A., and Doostdar, H. (2002). Multitrophic interactions of the silverleaf whitefly, host plants, competing herbivores, and phytopathogens. *Arch. Insect Biochem. Physiol.* **51:**151–169.

McDonald, B. A., and Linde, C. (2002). Pathogen population genetics, evolutionary potential, and durable resistance. *Ann. Rev. Phytopathol.* **40:**349–379.

Moissiard, G., and Voinnet, O. (2004). Viral suppression of RNA silencing in plants. *Mol. Plant Pathol.* **5:**71–82.

Morales, F. J., and Anderson, P. K. (2001). The emergence and dissemination of whitefly-transmitted geminiviruses in Latin America. *Arch. Virol.* **146:**415–441.

Morales, F. J., and Jones, P. G. (2004). The ecology and epidemiology of whitefly-transmitted viruses in Latin America. *Virus Res.* **100:**57–65.

Mullis, K. B., and Faloona, F. A. (1987). Specific synthesis of DNA *in vitro* via a polymerase-catalyzed chain reaction. *Methods Enzymol.* **155:**335–351.

Owor, B., Legg, J. P., Okao-Okuja, G., Obonyo, R., Kyamanywa, S., and Ogenga-Latigo, M. W. (2004). Field studies of cross protection with cassava mosaic geminiviruses in Uganda. *J. Phytopathol.* **152:**243–249.

Padidam, M., Sawyer, S., and Fauquet, C. M. (1999). Possible emergence of new geminiviruses by frequent recombination. *Virology* **265:**218–224.

Patil, B. L., and Dasgupta, I. (2006). Defective interfering DNAs of plant viruses. *Crit. Rev. Plant Sci.* **25:**47–64.

Perring, T. M. (2001). The *Bemisia tabaci* species complex. *Crop Protect.* **20:**725–737.

Picó, B., Diez, M. J., and Nuez, F. (1996). Viral disease causing the greatest economic losses to the tomato crop. II. The *Tomato yellow leaf curl virus*: A review. *Sci. Hortic.* **67:**151–196.

Pita, J. S., Fondong, V. N., Sangare, A., Otim-Nape, G. W., Ogwal, S., and Fauquet, C. M. (2001). Recombination, pseudo-recombination and synergism of geminiviruses are determinant keys to the epidemic of the cassava mosaic disease in Uganda. *J. Gen. Virol.* **82:**655–661.

Polston, J. E., and Anderson, P. L. (1997). The emergence of whitefly transmitted geminiviruses in tomato in the Western Hemisphere. *Plant Dis.* **81:**1358–1369.

Polston, J. E., McGovern, R. J., and Brown, L. G. (1999). Introduction of *Tomato yellow leaf curl virus* in Florida, and implications for the spread of this and other geminiviruses of tomato. *Plant Dis.* **83:**984–988.

Power, A. G. (2000). Insect transmission of plant viruses, a constraint on virus variability. *Curr. Opin. Plant Biol.* **3:**336–340.

Preiss, W., and Jeske, H. (2003). Multitasking in replication is common among geminiviruses. *J. Virol.* **77:**2972–2980.

Ramappa, H. K., Muniyappa, V., and Colvin, J. (1998). The contribution of tomato and alternative host plants to tomato leaf curl virus inoculum pressure in different areas of South India. *Ann. Appl. Biol.* **133:**187–198.

Roossinck, M. J. (1997). Mechanisms of plant virus evolution. *Ann. Rev. Phytopathol.* **35:**191–209.

Rouhibakhsh, A., and Malathi, V. G. (2005). Severe leaf curl disease of cowpea: A new disease of cowpea in northern India caused by *Mungbean yellow mosaic India virus* and a satellite DNA-β. *Plant Pathol.* **54:**259.

Sanchez-Campos, S., Diaz, J. A., Monci, F., Bejarano, E. R., Reina, J., Navas-Castillo, J., Aranda, M. A., and Moriones, E. (1999a). High genetic stability of the begomovirus *Tomato yellow leaf curl Sardinia virus* in southern Spain over an 8-year period. *Phytopathology* **92:**842–849.

Sanchez-Campos, S., Navas-Castillo, J., Camero, R., Soria, C., Diaz, J. A., and Moriones, E. (1999b). Displacement of tomato yellow leaf curl virus (TYLCV)-Sr by TYLCV-Is in tomato epidemics in Spain. *Phytopathology* **89:**1038–1043.

Saunders, K., and Stanley, J. (1999). A nanovirus-like DNA component associated with yellow vein disease of *Ageratum conyzoides*: Evidence for interfamilial recombination between plant DNA viruses. *Virology* **264:**142–152.

Saunders, K., Bedford, I. D., Briddon, R. W., Markham, P. G., Wong, S. M., and Stanley, J. (2000). A unique virus complex causes *Ageratum* yellow vein disease. *Proc. Natl. Acad. Sci. USA* **97:**6890–6895.

Saunders, K., Bedford, I. D., and Stanley, J. (2001). Pathogenicity of a natural recombinant associated with ageratum yellow vein disease: Implications for begomovirus evolution and disease aetiology. *Virology* **282:**38–47.

Saunders, K., Lucy, A., and Stanley, J. (1991). DNA forms of the geminivirus African cassava mosaic virus consistent with a rolling-circle mechanism of replication. *Nucl. Acids Res.* **19:**2325–2330.

Saunders, K., Salim, N., Mali, V. R., Malathi, V. G., Briddon, R., Markham, P. G., and Stanley, J. (2002). Characterization of *Sri Lankan cassava mosaic virus* and *Indian cassava mosaic virus*: Evidence for acquisition of a DNA B component by a monopartite begomovirus. *Virology* **293:**63–74.

Seal, S. E., van den Bosch, F., and Jeger, M. J. (2006). Factors influencing begomovirus evolution and their increasing global significance: Implications for sustainable control. *Crit. Rev. Plant Sci.* **25:**23–46.

Simon, B., Cenis, J. L., Beitia, F., Khalid, S., Moreno, I. M., Fraile, A., and García-Arenal, F. (2003). Genetic structure of field populations of begomoviruses and of their vector *Bemisia tabaci* in Pakistan. *Phytopathology* **93:**1422–1429.

Sserubombwe, W. S., Thresh, J. M., Otim-Nape, G. W., and Osiru, D. O. S. (2001). Progress of cassava mosaic virus disease and whitefly vector populations in single and mixed stands of four cassava varieties grown under epidemic conditions in Uganda. *Ann. Appl. Biol.* **138:**161–170.

Stanley, J. (2004). Subviral DNAs associated with geminivirus disease complexes. *Vet. Microbiol.* **98:**121–129.

Stanley, J., Frischmuth, T., and Ellwood, S. (1990). Defective viral DNA ameliorates symptoms of geminivirus infection in transgenic plants. *Proc. Natl. Acad. Sci. USA* **87:**6291–6295.

Thresh, J. M. (1980). The origin and epidemiology of some important plant virus diseases. *In* "Applied Biology" (T. H. Coaker, ed.), Vol. V, pp. 1–65. Academic Press, London.

Thresh, J. M. (1982). Cropping practices and virus spread. *Annu. Rev. Phytopathol.* **20:**193–218.

Thresh, J. M., and Cooter, R. J. (2005). Strategies for controlling cassava mosaic virus disease in Africa. *Plant Pathol.* **54:**587–614.

van den Bosch, F., Akudibilah, G., Seal, S., and Jeger, M. J. (2006). Host resistance and the evolutionary response of plant viruses. *J. Appl. Ecol.* (in press).

Vanitharani, R., Chellappan, P., Pita, J. S., and Fauquet, C. M. (2004). Differential roles of AC2 and AC4 of cassava geminiviruses in mediating synergism and suppression of posttranscriptional gene silencing. *J. Virol.* **78:**9487–9498.

Varma, A., and Malathi, V. G. (2003). Emerging geminivirus problems. A serious threat to crop production. *Ann. Appl. Biol.* **142:**145–164.

Voinnet, O., Pinto, Y. M., and Baulcombe, D. C. (1999). Suppression of gene silencing: A general strategy used by diverse DNA and RNA viruses of plants. *Proc. Natl. Acad. Sci. USA* **96:**14147–14152.

Wu, P.-J., and Zhou, X.-P. (2005). Interaction between a nanovirus-like component and the Tobacco curly shoot virus/satellite complex. *Acta Biochim. Biophys. Sin.* **37:**25–31.

Zhou, X., Liu, Y., Clavert, L., Munoz, C., Otim-Nape, G. W., Robinson, D. J., and Harrison, B. D. (1997). Evidence that DNA-A of a geminivirus associated with severe mosaic disease in Uganda has arisen by interspecific recombination. *J. Gen. Virol.* **78:**2101–2111.

Zhou, X., Xie, Y., Tao, X., Zhang, Z., Li, Z., and Fauquet, C. M. (2003). Characterisation of DNA-β associated with begomoviruses in China and evidence for co-evolution with their cognate viral DNA-A. *J. Gen. Virol.* **84:**237–247.

ADVANCES IN VIRUS RESEARCH, VOL 67

TRANSGENIC PAPAYA: DEVELOPMENT, RELEASE, IMPACT AND CHALLENGES

Dennis Gonsalves

USDA Pacific Basin Agricultural Research Center
99 Aupuni St., Suite 204, Hilo, Hawaii 96720

0065-3527/06 $35.00
DOI: 10.1016/S0065-3527(06)67009-7

Although the technology for developing virus-resistant transgenic plants by using the coat protein of a virus was first reported 20 years ago, it is surprising to note that only a three such virus-resistant plants (squash, potato and papaya) have been commercialized in the United States. The transgenic papaya cultivars Rainbow and SunUp were released commercially in Hawaii in 1998 and virtually saved Hawaii's papaya industry from the devastation being caused by *Papaya ringspot virus* (PRSV). This chapter focuses on the development, release and impact of the PRSV-resistant transgenics. Primary consideration is given to the factors that affected the timely development and deployment, rather then specifics on the technical aspects of the transgenic papaya. Our efforts to transfer the technology to Jamaica, Venezuela and Thailand are described to point out the factors that have influenced the practical deployment of the transgenic papaya to countries outside the United States. Finally, challenges such as durability of resistance, growing of non-transgenic papaya in Hawaii for the Japanese market, deregulation of the transgenic papaya in Japan and the co-existence of transgenic and non-transgenic papaya in Hawaii are discussed.

I. Introduction

Squash and papaya were the first virus-resistant transgenic crops that were commercialized in the United States and the papaya was released in 1998. Given the numerous reports on the effectiveness of transgenic crop resistance against many viruses, it is rather surprising that the transgenic squash and papaya are still the only virus-resistant plants that have been commercialized, other than virus-resistant transgenic potato, which is not grown commercially in the United States. This chapter considers the development, testing, release and impact of the virus-resistant transgenic papaya in Hawaii. The efforts to transfer the technology to Jamaica, Venezuela and Thailand are also covered because they reflect the range of factors that affect deployment of transgenic crops in various countries. The focus is on the factors that contributed to the timely development and commercialization of the transgenic papaya and those that are affecting its impact. This chapter does not deal with specifics on the molecular aspects of the transgenic papaya or the mechanisms of transgenic resistance. A number of other laboratories have developed transgenic papaya (Tripathi *et al.*, 2005), but this review deals only with transgenic papaya developed in my laboratory. A number of reviews (Fermin and Gonsalves, 2004; Fermin *et al.*, 2004b; Gonsalves, 1998; Gonsalves and Fermin, 2004; Gonsalves *et al.*, 1998, 2004a,b, 2006; Tripathi *et al.*, 2005) have been written on transgenic papaya. Extensive use is made of my previous reviews in this chapter.

II. Papaya and *Papaya ringspot virus* (PRSV)

A. *Papaya*

Much of the information in Section II is taken from a previous review (Gonsalves, 1998). Papaya *(Carica papaya)* is an important fruit crop grown widely in tropical and subtropical lowland regions. It is largely consumed as a fresh dessert fruit and the green papaya fruit is often used as salad. Papain is also recovered from the latex of green fruit. The tree is widely planted in home gardens because it is relatively easy to grow from seed; the first mature fruits can be harvested 9 months after sowing seeds and fruit is produced continuously year-round. Papaya is a herbaceous plant with a single stem bearing a crown of large palmately shaped leaves. Flowers are produced at the axils of the leaf petiole. The plant is polygamous and the plants are male, female or hermaphrodite. In the wild, dioecious plants predominate, whereas cultivated plants are both dioecious and hermaphrodite. In breeding cultivars, the latter can be inbred, which results in stable characteristics from generation to generation. Papaya trees are fast growing (they can be 4-m tall in a year), and they produce mature fruit within 9–12 months after seeds are planted. Commercially, when trees are grown at a density of 1500–2500 per ha, annual production can range from 57,000 to 136,000 kg per ha. Fruit are harvested for 1–2 years, after which the trees are usually too tall for efficient harvesting.

Papaya is the second most important fruit crop grown in Hawaii, next to pineapple. The commercial papaya grown in Hawaii is broadly known as the Hawaiian Solo type. These are gynodioecious cultivars (i.e. produce female and hemaphrodite plants) that are rather small, a fruit being *c.* 454–1135 g. Before 1998, the yellow-fleshed 'Kapoho' was the dominant papaya grown in Hawaii, distantly followed by the red-fleshed Sunrise. In fact, Kapoho made up 95% of Hawaii's papaya production in 1992.

B. Papaya ringspot virus

PRSV is by far the most widespread and damaging virus that infects papaya. The name of the disease it causes, papaya ringspot, is taken from the ringed spots on fruit of infected trees. Trees infected with PRSV develop a range of symptoms—mosaic and chlorosis of leaf lamina, water-soaked oily streaks on the petiole and upper part of the trunk, and distortion of young leaves that resembles mite damage. Infected plants lose vigour and become stunted. When infected at the seedling stage or within 2 months after planting, trees do not normally produce mature fruit. Production of fruit by trees infected at progressively later stages is severely reduced and of poor quality, owing to the presence of ringspots and generally lower sugar concentrations. PRSV is transmitted by numerous species of aphids in a non-persistent manner to a limited host range of cucurbits and papaya. PSRV also produces local lesions in *Chenopodium quinoa* and *C. amaranticolor.* Evidence largely suggests that PRSV is not seed transmitted, although there has been a report of seed transmission. PRSV is grouped into two types: type P (PRSV-p) infects cucurbits and papaya, whereas type W (PRSV-w) infects cucurbits but not papaya. The latter type was previously referred to as *Watermelon mosaic virus-1* (WMV-1). Although both types are serologically closely related, observations suggest that papaya is the major primary and secondary source for the spread of PRSV-p in large plantations and small orchards alike.

Much progress has been made in the molecular characterization of PRSV. Strains of PRSV-p from Hawaii and Taiwan have been completely sequenced. The genomic RNA consists of 10,326 nucleotides and has the typical array of genes found in potyviruses. The genome is monocistronic and is expressed via a large polypeptide that is subsequently cleaved to functional proteins. There are two possible cleavage sites, 20 amino acids apart, for the N terminus of the coat protein. These two sites may be functional; the upstream site produces a functional Nib protein, and the other, produces an aphid-transmissible coat protein (CP). It is impossible to segregate PRSV-p and PRSV-w types by their CP sequences. Within the P types, however, the coat protein sequences can diverge by as much as 12%.

III. Rationale for PRSV Work in Hawaii

A. *Proactive Approach to Control PRSV in Hawaii by Cross-Protection*

Hawaii's papaya industry started in the 1940s on the island of Oahu, on *c.* 200 ha (Ferreira and Mau, 1994). By the 1950s, production on Oahu was affected and the industry subsequently moved to the Puna area of Hawaii Island, which previously had no commercial production. The area increased to 650 ha by 1960 and by the 1970s it was producing *c.* 95% of Hawaii's papaya. Although Puna was still free of PRSV-p in the late 1970s, PRSV was only *c.* 30 km away from the town of Hilo. Thus the potential for severe damage to Hawaii's papaya industry was immense if PRSV became established in the papaya fields of Puna. In 1979, we began to investigate the possibility of using cross-protection as a control measure. Cross-protection is the phenomenon whereby plants that are systemically infected with a mild strain of a virus are protected, against the effects of infection by a more virulent related strain (Yeh and Gonsalves, 1984, 1994). Our efforts on using cross-protection have been documented elsewhere (Yeh and Gonsalves, 1994) and are not covered here. Cross-protection was used for a short time to control PRSV on Oahu Island, but was never fully adopted. However, several important developments came out of the efforts: (1) it started the attempts to develop control methods in advance of the potential problem in Puna, (2) information and technology were developed on detection and characterization of PRSV, (3) it provided the impetus for research on the molecular biology of PRSV, which helped to prepare us for later transgenic research and (4) it provided the 'mild' strain from which the CP gene was cloned and used later in developing transgenics. In retrospect, this proactive approach was the key to the timely control of PRSV in Hawaii.

B. *Transgenic Papaya Team*

Much of the success on the timely development of the transgenic papaya was due to synergy among the researchers comprising the team. This included Jerry Slightom (molecular biologist), Richard Manshardt (horticulturist), Maureen Fitch (plant physiology–tissue culture), and Steve Ferreira (plant pathologist). John Sanford was involved in the early phase of biolistic transformation with the gene gun. Two important features of the team were the balance of expertise, and the focused attitude and goal of developing a practical control method for PRSV in Hawaii.

IV. Development of Transgenic Papaya

As is normal in science, the approaches adopted were largely influenced by recent breakthroughs that held much promise and for which there was a good body of evidence for its validity. Our approach for developing PRSV-resistant transgenic papaya was based on the 1986 report of Beachy and colleagues (Powell-Abel *et al.*, 1986), which showed that tobacco expressing the *CP* gene of *Tobacco mosaic virus* (TMV) exhibited significant delay of symptoms following inoculation by TMV. A feature of this finding was that delay or resistance was genetically inherited, and that logically it could likely be applied to other plants and viruses. The implication of this achievement is hugely significant because previously the only available avenue for obtaining resistance to viruses was by transferring resistant genes from one germplasm to another through conventional breeding. Since resistant germplasm is not always available for specific viruses and can be difficult to introgress into acceptable cultivars, the potential of obtaining resistance through insertion of the *CP* gene of the virus into the plant genome was enormous. About the same time, the general concept of parasite-derived resistance was presented by Sanford and Johnston (1985). Parasite-derived resistance can be defined as a phenomenon whereby transgenic plants containing genes or sequences of a parasite (in this case, the *CP* gene of a virus) are protected against detrimental effects of the same or related pathogens. Ours was one of numerous laboratories that started research to utilize the approach of CP-mediated protection. Naturally, this approach required the adoption of several new technologies, mainly cloning and engineering of the target genes and regulatory elements, transforming cells with these target genes, and subsequent regeneration of transformed cells into plants that could be tested. In other words, 'molecular biology' and 'tissue culture' were key elements. An additional aspect was that many of the elements that were needed in the approach were protected by intellectual property rights, which was rather new to agricultural scientists. The following account of the transgenic papaya is taken largely from a previous review (Gonsalves, 1998).

The specific *CP* gene that was used for the project was from PRSV HA 5-1, a mild mutant of PRSV HA from Hawaii that we had used for cross-protection, and had been recently cloned and sequenced. Because of various technical difficulties and the belief that the gene had to be expressed as a protein, the 'coat protein' gene was engineered as a chimeric protein containing 17 amino acids of *Cucumber mosaic virus* at the N terminus of the full-length coat protein gene of PRSV HA 5-1. Whether this would enhance or decrease the chances of obtaining resistant plants was unclear at the time.

The ability to transform a plant is almost totally dependent on the availability or development of protocols for successful regeneration of transformed cells into plants. While engineering genes and vectors are fairly straightforward molecular procedures, tissue culture and regeneration of cells can be daunting and in many cases remains the limiting factor to successful mobilization of transgenes into plants. In 1987, when efforts to transform papaya began, protocols for transformation and regeneration of papaya were not yet known and in fact very few laboratories had contemplated transforming papaya. Our target cultivars were the dominant yellow flesh Hawaiian solo cultivar Kapoho which accounted for 95% of Hawaii papaya at that time, and to a lesser extent the red-fleshed Sunrise and Sunset (a sib selection of Sunrise). Numerous efforts by Fitch to develop a papaya regeneration system via organogenesis failed. The research moved rapidly once efforts shifted to transforming embryogenic tissue, resulting in the development of a procedure to produce highly embryogenic tissue starting from immature zygotic embryos.

In 1988–1989, embryogenic papaya tissues were bombarded with tungsten particles coated with DNA of the PRSV HA 5-1 *CP* gene using the gene gun in Sanford's laboratory. Transgenic plants were obtained and were growing in the greenhouse 15 months later. Overall, however, the process was relatively inefficient and only nine transgenic lines were obtained that were ready for screening in the greenhouse—three of Kapoho and six of Sunset. To speed up the process of testing, clones of the R_0 plants were made and screened for resistance under greenhouse conditions using PRSV HA. This was the severe parent of the mild mutant that was used as a source of the *CP* gene for the transgenic papaya. The conventional practice was to first raise R_0 plants to seeds and subsequently test the R_1 plants. Taking this approach would have resulted in a delay of at least 14 months before plants could be screened. Furthermore, screening for resistance was the main objective and thus the transgenic lines were not characterized fully before obtaining results on their resistance to PRSV.

With so few transgenic papaya lines to test, it was fortunate that one R_0 line (designated 55-1) of Sunset showed resistance to inoculations using PRSV HA. Three other lines showed varying degrees of delay in the onset of symptoms, while the other lines developed symptoms at the same time as control plants. Were we lucky? Yes, but at the same time, we enhanced our fortunes by not being so 'academic' and waiting at least a year to screen R_1 plants. This decision became pivotal, as will be seen later. Line 55-1 was female and thus seeds could not be obtained directly from the R_0 plants, as for a hermaphrodite. A dual approach was instituted to move the research ahead aggressively and to determine whether line 55-1 would be resistant to PRSV under field conditions and have suitable horticultural characteristics. First, a decision was made to conduct a field trial using R_0 plants instead of waiting a year to obtain R_1 plants. Second, the original R_0 plants that were screened and found to be resistant were raised for seed under greenhouse conditions at Cornell University and University of Hawaii. Doing the virus inoculation work at Cornell allowed us to determine the resistance of R_1 seedlings against PRSV strains collected from the two different countries. It was unwise to do these types of tests in Hawaii since papaya is a major crop and thus one would not want to risk the escape of foreign PRSV strains in Hawaii.

V. Field Tests Coincide with PRSV Invasion of Puna

A. R_0 Field Test, April 1992

Much of information for this section was taken from a review by Gonsalves (1998). The resistance of R_0 plants of line 55-1 under greenhouse conditions had been well established by April 1991. To move rapidly to the field, an application for a small field trial of R_0 plants at the University of Hawaii's experimental farm at Waimanalo, on Oahu Island, was submitted to Animal Plant Health Inspection Service (APHIS) and a permit was received in 1991.

Plants were set in the field by the end of June 1992. The field trial was designed to determine the resistance of R_0 plants to mechanical and aphid inoculations of PRSV. Non-transgenic plants in border rows of the plot were inoculated with a PRSV isolate from Oahu Island to create high-virus pressure in the field plot. Data were taken on total soluble solid levels of fruit, growth characteristics and virus symptoms. The transgenic papaya showed excellent resistance throughout the 2-year trial. Nearly all (95%) of the non-transgenic plants and those of a transgenic line that lacked the *CP* gene showed PRSV symptoms by 77 days after the start of the field trial, whereas none of the line 55-1 plants showed symptoms. Virus was not recovered from line 55-1 plants except for two plants, which showed virus symptoms on side shoots but none on the leaves of the main canopy. Plants grew normally, and fruit appearance and total soluble solids of *c.* 13% were within the expected range. By the end of 1992, the trial had provided convincing evidence that line 55-1 would be useful for controlling PRSV in Hawaii, or at least on Oahu Island.

B. PRSV Detected in Puna, May 1992

The move to establish a field trial rapidly became even more critical when PRSV was discovered in Puna in the first week of May 1992, only 1 month before the R_0 plants were established in the field. The inevitable entry of PRSV into the Puna district on Hawaii Island was discovered in a papaya field in Pahoa, within 5 km of the major papaya growing areas in Puna. Apparently, infection had been established in this area for several months, as judged by symptoms on the fruits and the fact that many plants were infected in one location. Surveys of the immediate area revealed PRSV in abandoned orchards, as well as in young orchards that were not yet producing fruit. PRSV was poised to invade the major papaya growing areas of Puna, which included Kapoho, Opihikau, Kahuawai and Kalapana.

The Hawaii Department of Agriculture (HDOA) immediately launched an eradication and containment programme that delayed the onslaught of the virus. However, the hope of containment was short-lived. The incidence of PRSV increased dramatically in Kapoho, which was closest to Pahoa, as the programme of voluntary cutting of trees was not strictly followed. Moreover, farmers experiencing high-infection rates abandoned their fields, creating huge reservoirs of virus for aphids to acquire and spread the virus. By late 1994, nearly all the papaya of Kapoho was infected. In October 1994, the HDOA declared that PRSV was uncontrollable and stopped the practice of marking trees for roguing. In less than 3 years, a third of the Puna papaya area was infected. By 1997, Pohoiki and Kahuawai were completely infected. Kalapana was the last place to become heavily infected. Five years after the onset of the virus in Pahoa, the entire Puna area was severely affected.

C. SunUp and Rainbow Varieties

The establishment of the early R_0 field trial in 1992 played another crucial role by serving as the source of germplasm to create the SunUp and Rainbow varieties. Line 55-1 is a transgenic Sunset, which is a commercial red-fleshed cultivar. The dominant cultivar growing in Puna, however, was the yellow-fleshed Kapoho. Thus transgenic Sunset (line 55-1) was not a suitable yellow-flesh substitute for Kapoho. To obtain a yellow-fleshed cultivar, Manshardt first crossed the female line 55-1 with non-transgenic Sunset to get seeds, and subsequently did selection and screening to obtain a line 55-1 that was homozygous for resistance. Southern blots had shown that line 55-1 had only a single insert of the *CP* gene. This homozygous transgenic Sunset was named SunUp. To create a yellow-flesh transgenic line, Manshardt crossed the SunUp with non-transgenic Kapoho. The resulting progenies produced fruit that was yellow fleshed since yellow is dominant over red. Importantly, this F_1 hybrid was resistant to PRSV since all F_1 plants had the *CP* gene in a hemizygous state. The F_1 hybrid was named Rainbow. The creation of SunUp and Rainbow was done in a timely manner by quickly establishing an R_0 field trial of line 55-1. It should be noted, however, that these cultivars had not yet been evaluated fully under field conditions.

D. Establishment of Transgenic Field Trial in Kapoho

By 1994, the severity of PRSV in Puna, the complete devastation caused by PRSV in the Kapoho area, the results of the R_0 field trial in Oahu and the progress in developing PRSV-resistant cultivars set the stage to establish a field trial in Kapoho to determine if the transgenic papaya could be used to rescue the papaya industry. Arguments could be marshalled for and against establishing such a field trial within a devastated commercial growing region of papaya. Line 55-1 had performed very well in field trials on Oahu Island and the line was resistant to PRSV-Panaewa, which is a greenhouse isolate from Hawaii island; the industry needed drastic actions to survive; the plan for moving the industry might not succeed and PRSV might not be eradicated from Puna; the risk of PRSV-resistant papaya becoming a weed was not relevant because papaya is not a weed in areas that do not have PRSV-p; wild relatives of *C. papaya* are not grown in Hawaii; and the potential benefits of transgenic papaya far outweighed the risks. Arguments against the field trial could be that pollen from the transgenic papaya might contaminate commercial plantings, resulting in the potential sale of commercial fruit containing a non-deregulated transgene. Moreover, it would be difficult to prevent pilferage in a trial installed in a farmer's field, and thus there might be serious consequences if stolen fruit reached commercial markets.

By late 1994, an application was submitted to APHIS for a field trial. Our previous experience in applying for such a permit, combined with the helpful cooperation of APHIS, facilitated the review process. The field trial was allowed with the stipulation that: (1) the field must be sufficiently isolated from commercial orchards to minimize the chance of transgenic pollen escaping to non-transgenic material outside the field test, (2) all abandoned papaya trees in the area must be monitored for the introgression of the transgene into fruits of these trees and (3) all fruits had to be buried on site.

Approval was obtained in early 1995 and the field trial was set up in Kapoho in October 1995 (Ferreira *et al.*, 2002). The trial was on the property of a farmer who had ceased growing papaya because of PRSV. One part of the trial consisted of replicated blocks to compare the virus-resistance performances of SunUp and Rainbow, of Kapoho cross-protected with PRSV HA 5-1, and of PRSV-tolerant lines that were being developed by Ferreira and Francis Zee (of USDA Germplasm Repository, Hilo) and other tolerant lines developed in Thailand. Another part of the trial was established to simulate commercial conditions. A square 0.4-ha solid block of Rainbow was planted adjacent to the replicated blocks. Several rows of non-transgenic Sunrise were planted on the perimeter of the replicated and solid blocks. An abandoned papaya field alongside the field plot was used as a primary source of the virus; the field had to be destroyed before the plants flowered.

The results of the field trial clearly demonstrated the potential value of the transgenic papaya for restoring papaya production in Puna (Ferreira *et al.*, 2002; Gonsalves, 1998). Except for three plants that showed infection at the beginning of the trial, none of the transgenic plants became infected. In contrast, 50% of the non-transgenic control plants within the experiment and in the border rows were infected within 4.5 months after transplanting and all were infected by 7 months. The growth differences between the transgenic and non-transgenic trees were remarkable; transgenic plants grew vigorously, with dark green leaves and full fruit columns, whereas non-transgenic plants were stunted, with yellow and mosaic-affected leaves and very sparse fruit columns. In the solid block, yield averaged about 113,000 kg of marketable fruit per ha/year, whereas non-transgenic plants averaged about 5500 kg per ha/year. Although these data were from only one trial, observations suggested that Rainbow out-yielded Kapoho. Also, the transgenic papaya performed far better than the PRSV-tolerant lines and the cross-protected plants.

Despite their excellent resistance, would farmers accept these cultivars as an acceptable substitute for Kapoho, which had been a mainstay of the Hawaiian papaya industry for several decades? The performance of Rainbow was especially critical because it was targeted as the alternative to Kapoho. Taste, production, colour, size and packing and shipping qualities of Rainbow were analysed. In addition to tests by research personnel, field days were held to allow farmers, packers, politicians and University personnel to observe the field and fruit at the test site. The consensus was that Rainbow is a more than adequate substitute for Kapoho. The fruit is larger than Kapoho, but commercial packers did not see this as a major impediment.

E. Narrow Resistance of Rainbow

It is appropriate to provide a brief description of some data that influenced our approach in deploying the transgenic papaya. As noted in an earlier section, as the 1992 field test was being conducted, steps were taken to obtain seeds from the R_0 cloned plants of line 55-1. With line 55-1 being a female and having only a single insert of the *CP* gene, crossing line 55-1 with a hermaphrodite Sunset or its Sunrise sib would yield an R_1 population in which 50% of the population are hemizygous for the *CP* gene. These R_1 plants were tested for resistance to a number of strains of PRSV originating outside Hawaii. In 1994, it was reported (Tennant *et al.*, 1994) that R_1 plants were highly resistant to Hawaiian isolates, but showed variable levels of resistance (largely susceptible) to other isolates. For example, plants inoculated with an isolate from Thailand developed severe symptoms with no delay in symptom appearance. In contrast, isolates from Jamaica infected line 55-1, but symptoms were delayed and attenuated. Isolates from Florida and Mexico infected only a percentage of the plants and symptoms were milder. Nevertheless, our data showed clearly that hemizygous line 55-1 and Rainbow would not be resistant outside Hawaii. Later studies confirmed that Rainbow also had narrow resistance (Tennant *et al.*, 2001). It also showed that SunUp, which is homozygous for the *CP* gene, is resistant to many strains of PRSV. The broader resistance of SunUp as compared to Rainbow is because SunUp is homozygous for the *CP* gene. The looming question was—Were there isolates in Hawaii that would overcome the resistance of Rainbow and SunUp? Fortunately, Rainbow was resistant to all the isolates of PRSV that were collected from Oahu and Hawaii Island.

VI. Deregulation and Commercialization of Transgenic Papaya

A. Deregulation

Various information in Section VI was taken from reviews by the author and a colleague (Gonsalves, 1998; Gonsalves and Fermin, 2004). Deregulation of the transgenic papaya required approval from APHIS and Environmental Protection Agency (EPA) and full consultation with Food and Drug Administration (FDA). APHIS was largely concerned with the potential risk of transgenic papaya to the environment. Two main risks were hetero-encapsidation of the incoming virus with CP produced by the transgenic papaya and of recombination of the transgene with incoming viruses. The former might allow non-vectored viruses to become vector transmissible, whereas the latter might result in the creation of novel viruses. A third concern, that escape of the transgenic genes to wild relatives might make the relatives more weedy, or even make papaya more weedy because of resistance to PRSV, was of no consequence since there are no papaya relatives in the wild in Hawaii. Moreover, papaya is not considered to be a weed there, even in areas were there is no PRSV. In November 1996, transgenic line 55-1 and its derivatives were deregulated by APHIS. This action greatly increased the efficiency of the ongoing field trial in Kapoho because fruit no longer had to be buried at the test site, which allowed us to sample and send fruit to various laboratories and to the packing house without undue constraints.

The EPA regards the CP transgenes as a pesticide because they confer resistance to plant viruses. A pesticide is subjected to tolerance levels in the plant. In the permit application, we petitioned for an exemption from this stipulation. We contended that CP was already present in many of the papaya fruits consumed by the public, since many of those produced in the tropics are from PRSV-infected plants. Moreover, we had earlier used cross-protection to control PRSV and fruit from these trees was sold to consumers. Furthermore, there is no evidence to date that the CP of PRSV or other plant viruses is allergenic or in any way detrimental to human health. Finally, the amounts of CP in transgenic plants were much lower than those of infected plants. In response to these submissions an exemption from tolerance to lines 55-1 was granted in August 1997.

The FDA is concerned with the food safety of transgenic products. This agency follows a consultative process whereby the investigators submit an application with data and statements corroborating that the product is not harmful to human health. Several aspects of the transgenic papaya were considered—the concentration range of some important vitamins, including vitamin C; the presence of *GUS* and *nptll* genes; and whether transgenic papaya had abnormally high concentrations of benzyl isothiocyanate. This latter compound has been reported in papaya. FDA approval for the use of transgenics was granted in September 1997.

B. Commercialization

In the United States, a transgenic product cannot legally be commercialized unless it is fully deregulated and until licences are obtained for the use of the intellectual property rights for processes or components that are part of the product or that have been used to develop the product. The processes in question were the gene gun and parasite-derived resistance, particularly CP-mediated protection. The components were translational enhancement leader sequences and genes (*nptll*, *GUS* and *CP*). This crucial hurdle involved legal and financial considerations beyond our means and expertise. These tasks were undertaken by the industry's Papaya Administrative Committee (PAC), which was formed through a USDA marketing order that was established to obtain funds through a small levy on sales of papaya.

Even before license agreements had been obtained, the PAC had commissioned the Hawaii Agricultural Research Center to produce Rainbow seeds so that they would be available when needed. The seeds were produced on Kauai, where PRSV is not present. Seeds sufficient for *c.* 400 ha were scheduled for distribution in May 1998. The seeds were distributed free to growers under the stipulations that people would register to receive seeds, attend an educational session or watch a video on the transgenic papaya and sign a sublicence which mandated that seeds would be used to plant papaya only in Hawaii. Information brochures were available in English and in the Ilocano Filipino language. It is important to note that due to an equitable distribution plan, all of the farmers did not receive seeds at the same time, but over a number of months. The allocation was based on three levels of priority favouring those who were 'currently and historically most affected' by PRSV, and on four prescribed distributions. For example, in the first distribution, top priority farmers received two allotments each of 57 g. Each 57-g packet contained *c.* 4000 seeds that sufficed to plant a 0.2 ha parcel (Gonsalves *et al.*, 2004a).

VII. Early Adoption Rate of Transgenic Papaya

The rather small size of the papaya industry, the confined area where the transgenic papaya would be released, the proximity of the research team to the papaya growers and the PAC, and the practical orientation of the papaya provided a unique opportunity to assess the adoption of the transgenic papaya before and soon after its release. The effort was taken up in a volunteer research project by Carol Gonsalves

(Gonsalves, 2001; Gonsalves *et al.*, 2004a). The time period started several months before the release of the transgenic seeds in May 1998 and ended in September 1999.

In 1998, it was estimated that Hawaii had *c.* 300 papaya farmers. A more definitive list was obtained by finding the number of farmers who had petitioned the PAC for possible use of transgenic seed. This list yielded 256 farmers, 171 of whom farmed in Puna. Diligent efforts were made to contact all Puna farmers, and 91 (54%) of these farmers were interviewed. Some interesting statistics on the background of the farmers were: (1) 90–91% were of Filipino descent and lived in Puna, (2) many of the farmers (46%) held off-farm jobs and (3) 38% of the families of the farmers derived greater than half their income from raising papaya. These statistics show that the Hawaii papaya industry mainly consisted of small growers who were family orientated.

Of the 92 farmers who qualified to receive seeds, 90% obtained the seeds, 76% planted the seeds and 19% were actually harvesting fruit within a year of seed distribution. Those who planted seeds did so quite soon after receiving them—of the 71 farmers surveyed, 38% planted less than a month after receiving seeds, 42% planted after the first but before the third month, and 20% planted between the fourth and ninth month. The reason for adopting the transgenic papaya seeds was overwhelmingly (96%) for their resistance to PRSV. In summary, growers anxiously waited for the seeds and planted the seeds soon after receiving them. This provides a reflection of the severity of the disease problem at the time when seeds were released.

VIII. Impact of Transgenic Papaya

A. Papaya Industry in 1998

The impact of the transgenic needs to be measured based on the situation of the papaya industry in 1998. As noted earlier, the efforts of HDOA to contain the virus were discontinued in late 1994, people were subsequently abandoning fields and thus creating more sources of virus. Moreover, all of Puna district was infected and only Kalapana district had less than total infection. In Kapoho, where one-third of Puna's papaya had been growing before PRSV invaded in 1992, plantings were completely infected and many were abandoned. This made it impossible to economically establish new papaya fields in Kapoho and many of the papaya fruits that were sold were infected and of poor quality.

The papaya production figures of Hawaii bear out the effect that PRSV was having on papaya production (Table I) (Gonsalves *et al.*, 2004b). Production had decreased from 24.0 million kg in 1992 to 12.1 million kg in 1998, a decrease of a half within 6 years. In 1992, Puna produced 95% of Hawaii's papaya, whereas in 1998 its share of total Hawaii production was 75%. Other regions, primarily in the Hamakua district of Hawaii Island, on parts of Oahu Island and on Kauai increased production and so picked up some of the production void left by Puna. Even with the establishment of new production areas, Hawaii's overall papaya production had declined by 36% from 25.3 million kg in 1992 to 16.1 million kg in 1998. Hawaii had to keep up production as much as possible to maintain its markets, especially in mainland United States and Japan.

An important impact of PRSV on the papaya industry was also on the quality of the produce. Much of the papaya that was being harvested in Puna was affected and not of the superior quality, which was the big selling point for the Hawaiian papaya. In fact, the requirement that symptomatic fruit should not be shipped to Japan and the mainland US was lifted as a measure to keep the exports of papaya to Japan and mainland at a reasonably good volume. Another impact was on the

TABLE I
FRESH PAPAYA PRODUCTION IN THE STATE OF HAWAII AND IN THE PUNA DISTRICT FROM 1992 TO 2004

Year	Total (×1000 kg)	Puna (×1000 kg)	Puna (%)
1992[a]	25,340	24,073	95
1993	26,430	25,108	95
1994	25,522	25,215	99
1995	19,028	17,808	94
1996	17,166	15,529	90
1997	16,212	12,629	78
1998[b]	16,167	12,148	75
1999	17,892	11,630	65
2000	22,820	15,417	68
2001	23,614	18,297	77
2002	19,391	16,294	84
2003	18,528	16,228	88
2004	15,533	13,737	88

[a] PRSV was first reported in Puna.
[b] Transgenic seed released.

people associated with the papaya industry. In 1992, Hawaii had eight packing houses but by 1998 only three were in operation and these were not running at full capacity as they were in 1992. It was difficult to find unaffected papaya in Hawaii.

B. Restoring Production in Puna with Transgenic Papaya

Given the tremendous devastation that PRSV had caused in Puna, from a wider perspective, it might be asked why was not Puna abandoned and production moved to other areas of Hawaii Island, Maui and Kauai? The latter two islands were free of PRSV in 1998. In effect, this would have followed the precedent where Hawaii's papaya production was abandoned on Oahu Island and established in Puna in the early 1960s. Aside from the 'moral' obligation to help the farmers in Puna, several factors favoured retaining Puna as the main papaya growing area for the state of Hawaii: (1) Puna has abundant sunshine and rainfall, conditions that favour good papaya production. Overall, it is arguably the ideal place for growing papaya in Hawaii, (2) large tracts of suitable and affordable land for papaya were available for leasing or purchase, (3) many people in Puna were farming on a full- or part-time basis, and growing papaya had become a good way of life for them, (4) the packing houses were located close to Puna and (5) most importantly, the dominant variety Kapoho was selected and adapted for growing in the volcanic soils and rainfall conditions of Puna. In fact, previous efforts to grow Kapoho on Oahu Island and Kauai failed, even in the absence of virus infection.

Based on the farmer and consumer preference for a yellow-flesh papaya, it was expected that Rainbow would be the dominant transgenic cultivar to be grown in Puna. Even though the 1995–1997 field trial had shown that Rainbow and SunUp were highly resistant to PRSV under very severe virus pressure, there was still some trepidation when growers started commercial plantings in mid-1998. Moreover, plantings of the transgenic papaya were not established exclusively in areas where PRSV was absent or some distance away. Many of the plantings were established adjacent to abandoned-infected papaya fields or even within infected fields. The newly planted transgenic papaya showed excellent resistance even under these conditions of severe virus pressure. Within a year of the release of the seeds, it was common to see fields of healthy Rainbow. Numerous observations since 1998 to the present time have shown that resistance of Rainbow and SunUp is durable. Observations suggest that it has been total. Where 'infected' trees were observed in fields of Rainbow or SunUp, subsequent checking of the plants showed that these were in fact non-transgenic papaya.

Despite the observations of durable resistance, it is known that the resistance of Rainbow is narrow. Thus, it is still possible that there are PRSV strains in Hawaii that could overcome the transgenic resistance. Diligence is needed in monitoring for these possibilities. In that sense, personnel in Hawaii are consistently on the lookout for evidence for breakdown of virus resistance. Routinely, isolates are collected and inoculated to Rainbow to assess their resistance to a wide range of PRSV isolates in Hawaii.

C. Reversing the Decrease in Papaya Production Caused by PRSV

In 1992, when PRSV entered Puna, that area produced 95% (24.0 million kg of Hawaii's papaya; but production decreased and by 1998 Puna produced only 75% (12.1 million kg) of Hawaii's production (Table I). Papaya production in Puna had decreased by 50% in 1998, compared to 1992's. Clearly, PRSV was having a devastating affect on papaya production in 1998. Production decreased slightly more in 1999, a year after transgenic seeds were released. Puna papaya production then showed an increase starting in 2000 to 15.4 million kg and 18.3 million kg in 2001. Production in 2002–2004 ranged from 16.3 to 13.7 million kg. Furthermore, Puna production accounted for 84% of the production in 2002, compared to the low of 65% in 1999. In 2004, Rainbow accounted for 88% of the total area of papaya in Puna.

D. Enabling the Production of Non-Transgenic Papaya in Puna

The logical question might be asked: Why does not Hawaii produce only transgenic papaya? It is critical that Hawaii continues to produce non-transgenic papaya to supply the Japanese market, as discussed later. Arguably, one of the major contributions that the transgenic papaya has made to the papaya industry is that of facilitating the economic production of non-transgenic papaya (Gonsalves and Ferreira, 2003). This has occurred in several ways. First, the initial large-scale planting of transgenic papaya in established farms along with the elimination of abandoned virus-infected fields drastically reduced the amount of virus inoculum. This allowed for strategic planting of non-transgenic papaya in areas that no longer had infected sources. In fact, HDOA instituted a plan in 1999 to ensure the continued production of non-transgenic papaya in the Kahuawai area of Puna. Kahuawai was isolated from established papaya fields and prevailing winds in

Kahuawai came from the ocean which borders the area (Gonsalves and Ferreira, 2003). Growers were to monitor for infection and rogue all infected plants quickly. Growers who followed the recommended practices were able to produce Kapoho economically without major losses from PRSV. Second, although definitive experiments have not been carried out, it seems that transgenic papaya can provide a buffer zone to protect non-transgenic papaya that is planted within the confines of the buffer. The reasoning is that viruliferous aphids will probe or feed on transgenic plants and thus lose any virus they are carrying before reaching the non-transgenic plantings within the buffer. This approach also has the advantage of allowing growers to produce transgenic and non-transgenic papaya in relatively close proximity. However, timely elimination of infected trees is necessary to delay large-scale infection of the non-transgenic plants.

E. Allowing the Production of Papaya in a More Limited Area of Puna

In the mid-1990s when PRSV was rampant in Puna, a number of growers moved their production to new land in the Hamakua district of Hawaii Island. This move was not disruptive to the Hamakua area because lots of agricultural land had become available in the area due to the demise of sugarcane production. Furthermore, these areas also became infected with PRSV. Following the introduction of the transgenic papaya, growers have returned to Puna. The transgenic papaya has allowed growers to replant their original land, and also grow non-transgenic papaya, provided it is planted in areas where virus is not present. This in turn has cut down on the amount of new papaya land that has to be cleared in order to grow non-transgenic papaya for the Japanese market. In other words, it has helped to slow down expansion of the industry into new lands simply to escape the virus. This situation helps environmentally by preserving forest and other lands that might otherwise be cleared for papaya plantings to escape PRSV infection.

F. Increase in the Number of Papaya Cultivars Available to Hawaii

A common caution about the use of genetically modified organisms (GMOs) is that it will create a reliance on transgenic crops and not encourage diversification of crop varieties. In fact, the transgenic papaya has had the opposite effect in Hawaii. Before 1992, non-transgenic Kapoho accounted for 95% of the states production. Essentially, the papaya industry had relied on only one variety for many years. The introduction of transgenic papaya has made new distinct varieties available. While SunUp is a new variety, it essentially is a sib of Sunset and Sunrise. However, Rainbow is a new variety since it is an F_1 hybrid of a cross between non-transgenic Kapoho and SunUp. Moreover, the new transgenic cultivar 'Laie Gold', which is a hybrid between 'Rainbow F_2' (selfed Rainbow) and the non-transgenic 'Kamiya', also serves a niche market on Oahu Island. Since 'Rainbow F_2' is not homozygous for the *CP* gene, 'Laie Gold' needs to be micro-propagated to achieve uniformity of production. Micropropagation of the papaya also has the added benefit of ensuring the production of only hermaphrodite plants that are demanded by the market, of earlier and lower bearing trees with initially higher yields, and of providing selected, superior clones that could result in improved quality and yield. In addition to the cultivars mentioned, newer ones developed for niche markets include large-fruited, firm cultivars used as green papaya in South and Southeast Asian cuisine and a red-fleshed 'Laie Gold' progeny called 'Red Kamiya' for cultivation on Oahu Island.

G. Resurgence of Papaya Cultivation on Oahu Island

As noted earlier, the papaya industry originally was centred on Oahu, until production on that island was largely eliminated by PRSV in the 1950s. The availability of PRSV-resistant papaya provided options for papaya growers on Oahu. Prior to the release of transgenic papaya, Oahu growers farmed only small plots of papaya due to the effect of PRSV on production. Growers on Oahu now enjoy a niche market, currently growing 'Rainbow' and 'Laie Gold' papaya for residents in Honolulu and other urban areas of the island. Whereas Oahu grew *c.* 20 ha of papaya in 1960, it now grows *c.* 60 ha of virus-resistant papaya. This would not have happened without the release and utilization of the transgenics.

IX. Challenges Facing the Hawaiian Papaya Industry

Although a major constraint to papaya production in Hawaii was eliminated with the introduction of PRSV-resistant transgenic plants, the industry still faces challenges. Some of these are: gaining market share in Canada and Japan, growing of non-transgenic papaya without severe damage to PRSV, the durability of the resistance in transgenic papaya, concerns by some organic growers that their crops will be contaminated by pollen flow to their orchards, and the general controversy of GMOs. This section has used information from a recent review by the author (Gonsalves *et al.*, 2006).

A. *Canadian and Japanese Markets*

Japan and Canada are large markets for Hawaii papaya. Currently, Japan accounts for 20% of Hawaii's export market and Canada accounts for 11%. Canada approved the import of 'SunUp' and 'Rainbow' transgenic papaya in January 2003 and imports of transgenic papaya continue. In contrast, the application process for sale of transgenic papaya in Japan has not yet been approved. Accordingly, it is critical that papaya shipments to Japan are not contaminated with transgenic papaya fruit. Several steps are being taken to minimize such contamination.

At the request of Japanese importers, HDOA adopted an Identity Preservation Protocol (IPP) that growers and shippers must follow in order to receive an IPP certification letter from HDOA that accompanies the papaya shipment. This is a voluntary programme. Papaya shipments with this certification can be distributed in Japan without delay whilst Japanese officials are conducting spot tests to detect contaminating transgenic papaya. In contrast, papaya shipments without this certificate must remain in custody at the port of entry until Japanese officials complete their spot checks. Completing the tests may take several days or a week, during which time fruits lose quality and marketability.

Some significant features of the IPP are that the non-transgenic papaya must be harvested from papaya orchards that have been approved by HDOA. For approval, every tree in the proposed field must be tested and found negative for the transgenic *GUS* reporter gene that is linked to the virus-resistance gene. Non-transgenic trees must be separated by at least a 4.5 m papaya-free buffer zone and new fields to be certified must be planted with papaya seeds that have been produced in approved non-GMO fields. Tests for detecting transgenic papaya trees in the field are monitored by HDOA and conducted by the applicant who must submit detailed records to HDOA. Before final approval of a field, HDOA will randomly test one fruit from a sample of 1% of the papaya trees present. If approved by HDOA, fruit from these fields can be harvested. Additionally, the applicant must submit the detailed protocols that will be followed to minimize the chance of contamination of non-GMO papaya by GMO papaya. This includes a protocol by the applicant on the random testing of papaya before they are packed for shipment. If the procedures are followed and tests are negative, a letter from HDOA will accompany the shipment stating that the shipment complies with a properly conducted IPP.

The above procedure represents a 'good faith' effort by HDOA and applicants to prevent transgenic papaya contamination in shipments of non-transgenic papaya to Japan. It also illustrates meaningful collaboration between Japan and HDOA. Continued shipment of non-transgenic papaya to Japan with a minimum of delay once they arrive in Japan is allowed while adhering to the policy that transgenic papaya will not commercially enter Japan until it is deregulated by the Japanese government. These efforts, along with the effectiveness of the transgenic papaya in boosting production of non-transgenic papaya, have allowed Hawaii to maintain significant shipments of the latter to Japan.

B. Deregulation of Transgenic Papaya in Japan

Obviously, deregulation of transgenic papaya in Japan will circumvent much of the concern of accidental introduction of transgenic papaya into Japan. Consequently, efforts to allow the transgenic papaya into Japan were initiated by the PAC soon after the transgenic papaya was commercialized in Hawaii. Again, the researchers led in developing the petition. Approval of the transgenic papaya in Japan requires the approval of the Ministry of Agriculture Fisheries and Forestry (MAFF) and the Ministry of Health Labor Welfare (MHLW). The petition to the MAFF was approved in December 2000. The petition process for approval by MHLW is still in progress. An initial petition was submitted to MHLW in April 2003. MHLW then requested more information. A revised and complete dossier was submitted to the Food Safety Sub Committee of MHLW in January 2006. It is hoped that full approval will be obtained and transgenic papaya allowed into Japan in the first quarter of 2007. The transgenic papaya will have to be labelled, as required by the Japanese government.

The implications of receiving approval of the transgenic papaya in Japan would have huge benefits to Hawaii's papaya industry and also advance the cause of transgenic products outside the United States. Although effective, the current IPP process for minimizing the introduction of transgenic papaya to Japan adds significantly to current production costs. Approval will at least mean that shipment of transgenic papaya and mixed batches of non-transgenic and transgenic papaya can be shipped freely to Japan without fear of rejection based on the presence of transgenic papaya. It is very likely that non-transgenic papaya that are shipped to Japan will need to be labelled as such and likely meet a tolerance level of 'potential' contamination with transgenic papaya. The conditions that will be imposed for testing for tolerance levels will be determined during the approval process.

More widely, there is much interest by government agencies, such as Foreign Agricultural Service (FAS), in obtaining approval for transgenic papaya in Japan because of their interests in promoting United States biotechnology products in other countries. Because the transgenic papaya is a fresh product, consumers will have clear side-by-side choice. These conditions will help to answer the questions of consumer acceptance of fresh GMO products outside the United States in a 'real' and 'nonacademic' case scenario. Lastly, since the product was not developed by support from multinational companies, the acceptance arguments, hopefully, will not be about control by large multinational companies. Instead, it is anticipated that product acceptance will be influenced by factors such as quality, price, advertising and the 'philosophy' of the consumer. Also the arguments should not be on whether the transgenic papaya involves major trade issues. This is because papaya is a low-valued transgenic crop when compared to the currently approved transgenic products approved in countries outside the United States. In other words, the transgenic papaya will provide a chance to analyse factors that supposedly concern the consumers in the absence of national media hype on dominance by multinational companies and so on. In this aspect, the transgenic papaya will be a ground-breaking biotechnology product.

C. Co-Existence

The conditions for deregulating transgenic papaya did not restrict the areas where it can be grown. Essentially, it can be grown anywhere in the United States. Since it had been deemed safe to consume and that the product would not harm the environment, the deregulated transgenic product was regarded as any other new conventionally developed variety to be grown in the United States. There are concerns, especially with organic farmers that growing transgenic papaya would contaminate their products and thus not make them eligible for certification or for sale. Although the US organic rules do state that if a grower growing a product that is certified as organic and non-GMO at the time of planting and that the grower has taken reasonable precaution against contamination, the product can be sold as organic. Nevertheless, coexistence in the sense that some want absolute guarantee from cross contamination is a contentious issue, even with the transgenic papaya.

How can gene flow from papaya to papaya be prevented? The commercial Hawaiian papaya is hermaphrodite and largely self-pollinating. Part of the APHIS requirements for the 1995 Kapoho field trial in Puna was that transgene flow from transgenic to non-transgenic papaya should be assessed in the field trial and that an abandoned commercial papaya field that was *c.* 400 m from the field trial should be sampled. Since the solid block of Rainbow was surrounded by several rows of non-transgenic papaya, it was quite easy to assess whether the transgene flowed to non-transgenic papaya growing in adjacent rows. The *GUS* gene helped to make it relatively straight forward to detect transgene flow by testing seeds for the presence of the *GUS* gene using a colourimetric test. The transgene was detected in 43% of the female trees sampled and in 7% of the hermaphrodite trees (Manshardt, 2002). It was not detected in an abandoned orchard 400 m away. In another on-going study, efforts are being made to detect transgene flow in commercial orchards in Puna. Seeds were sampled from border or close to border trees of non-transgenic orchards growing adjacent to Rainbow orchards. So far, the *GUS* transgene has not been detected in any of the 447 non-transgenic trees sampled (Gonsalves, C., unpublished data). Although not yet complete, the results indicate that transgene flow is minimal in non-transgenic papaya orchards growing near to transgenic papaya under commercial conditions in Puna.

The fact, as shown in Section VIII.A, that Hawaii produces transgenic and non-transgenic papaya commercially in Puna clearly shows that their co-existence in the same location is practical and economic. Japan has zero tolerance for transgenic papaya being sold and none has been detected in non-transgenic shipments since the IPP programme began.

D. Durability of Resistance and Potential of New PRSV Strains Reaching Hawaii

As mentioned in Section V.E, studies have shown that the resistance of Rainbow is largely limited to PRSV isolates from Hawaii, or those with very similar sequences to the *CP* gene used to transform Hawaii's transgenic papaya. The resistance of Rainbow has held up to the present time. This durability is likely because the PRSV isolates in Hawaii are homogeneous in their CP sequences. The possibility of new virulent strains developing from recombination of PRSV strains in Puna with the CP transgene of Rainbow is remote. A more realistic danger is the introduction of PRSV strains from outside Hawaii.

SunUp should be resistant to many strains of PRSV that might be introduced into Hawaii (Tennant *et al.*, 2001). So it might be expected that SunUp should be the choice for use in Hawaii. However, growers and customers prefer the yellow-fleshed papaya so it is doubtful that the area of SunUp will increase significantly.

A potential solution is to develop transgenic Kapoho that is resistant to a wide range of strains. This could be used as a 'stand-alone' cultivar, or it could serve as a transgenic parent for creating a new type of 'Rainbow' by crossing the 'transgenic Kapoho' with SunUp. This F_1 hybrid should have the horticultural characteristics of Rainbow but is likely to have much wider resistance than the current Rainbow due to increased *CP* gene dosage. The new 'Rainbow' with a *CP* gene dosage from both the SunUp and the new 'transgenic Kapoho' will probably be sufficient to be resistant to strains of PRSV that might invade Hawaii. Transgenic Kapoho that is resistant to a range of PRSV strains has been developed (Gonsalves and Ferreira, unpublished data) and work began in 2006 to obtain necessary information to petition APHIS, EPA and FDA for deregulation. However, deregulation and commercialization of this transgenic Kapoho is still a few years away. These circumstances highlight the need to guard carefully against the introduction of additional PRSV strains into Hawaii and to maximize the usefulness of existing transgenic cultivars.

As an alternative, Rainbow F_2 plants were backcrossed four times with 'Kapoho' to obtain backcrossed lines that were then self-pollinated. The plants obtained were homozygous for the *CP* gene and had horticultural characteristics similar to those of Kapoho (Fitch and Ferreira, unpublished data). These plants could possibly be used as a transgenic 'Kapoho' parent to cross with 'SunUp', hopefully producing horticultural characteristics equal to those of 'Rainbow', but with increased gene dosage and presumably stronger resistance than 'Rainbow'. Moreover, given that Rainbow has narrow resistance, current practices of collecting PRSV isolates and inoculating them to Rainbow and SunUp should continue.

E. Guarding Against Large-Scale Resurgence of PRSV in Non-Transgenic Papaya in Puna

PRSV still occurs in Puna, even though the transgenic papaya has decreased the incidence of PRSV. If virus inoculum builds up further in Puna, it will become even more difficult to produce non-transgenic papaya economically. Strict attention should be paid to planting non-transgenic papaya in as much isolation as possible, to timely eliminate infected trees, and to remove non-transgenic plantings that are no longer in production. Although important, these simple measures are often not practised when there are no obvious signs of resurgence of PRSV. The tremendous damage caused by PRSV to Hawaii's papaya industry during the period 1992–1998 should not be forgotten.

X. Factors that Influenced the Timely Deployment of Transgenic Papaya in Hawaii

A. *Proactive Research to Develop a Solution to a Potential Problem*

Starting research efforts well before the potential problem was a critical step in the timely deployment of the transgenic solution to the PRSV problem in Puna in 1992. In 1978 research began to develop control measures. It took 6 years of research to develop a resistant R_0 plant and then another 7 years to have it released commercially. If research to develop transgenic papaya had not begun until 1992 Puna would still be over-run by PRSV and unlikely to be producing papaya now. Perhaps, the transgenic papaya would have been ready for release in 2006, much too late to do much good.

B. *Blend of Expertise and Focus of the Research Team*

The practical deployment of biotechnology is a multidisciplinary effort because it involves molecular biology, tissue culture, plant breeding, deregulation and intellectual property rights, among others. Many research groups are not set up to cover the entire span of activities or even willing to do it since many of the other activities would not be within their 'job description'. Our group had a good blend expertise in molecular biology, virology and plant pathology, horticulture and tissue culture. However, the research was not simply done in phases where one group of experts did one thing and then passed their part or product on and subsequently ceased to be involved. Passing the products along did occur but the whole team remained involved and interested in bringing all phases together to complete the goal of the project, which was to develop a practical solution to the PRSV problem of the Hawaii's papaya industry. Lastly, the group was focused on developing a resistant transgenic papaya in a timely manner. This became especially important at the stage of screening for resistance. By moving to propagation and screening of R_0 plants for resistance as the first priority, it opened up the timely identification of the resistant line and setting up the field trial on Oahu Island. The 'conventional wisdom' would have been to obtain R_1 plants and then check these plants for resistance before even contemplating the establishment of a field trial. This would have delayed the establishment of the first field trial by about 18 months.

C. *Deregulation*

Conferences have focused on trying to determine how Universities and small companies might be assisted so that the step of deregulation does not pose an obstacle that would block a biotechnology product from being released. With papaya, part of the research team worked in getting data, consolidating information and submitting the information to EPA, APHIS and FDA. As this process was done by the investigators, the cost was minimal because people and facilities specific for doing the task were not employed. The data were obtained by the team during the research on characterizing the transgenic papaya. It is perhaps fair to say that deregulation of the transgenic papaya would have been delayed or not even contemplated if the investigators had been unwilling to handle these processes. A 'just do it' attitude helps to move the process forward in a timely manner!

D. *Intellectual Property Rights*

The issue of obtaining licences of the intellectual property that was used to create the transgenic papaya was beyond the expertise and authority of the research team, who thus passed the assignment to the PAC. The intellectual property component of the work was handled largely by the lawyer hired by PAC to assist. An important challenge was to persuade growers to move aggressively to obtain the necessary licences. It should be noted that the PAC was composed of papaya farmers who had no previous experience of intellectual property rights issues. The following personal experience might help to illustrate this point. Although PRSV was devastating the industry and the need for the transgenic papaya as a potential solution was urgent, 'rumours' were floating around that the companies who had the intellectual property rights would charge too much for a licence and thus PAC would not be able to commercialize the transgenic papaya. At one of the crucial grower meetings, I told them that they should not simply believe the 'rumours' that it would be too costly to obtain the necessary licences. Instead they should hire a lawyer and approach the companies to negotiate on their behalf. Fortunately, PAC hired a lawyer and moved quite aggressively to get the licences. These were obtained in a timely manner and were affordable.

E. Hawaii Transgenic Papaya was Developed and Commercialized During the 'Days of Innocence' in Relation to the GMO Controversy

Controversy over GMOs was not yet in full swing in 1998 when the transgenic papaya was commercialized. In Hawaii the public was 'rooting' for the investigators to develop a control measure for PRSV. Europe had just started to impose a moratorium on approving transgenic products. In many ways, these were the 'days of innocence'. We were fortunate to do the work during this period. Essentially, we simply had to develop a product that was effective and that would be approved by the three agencies responsible for the deregulation of transgenic products in the United States. In other words, we could focus on developing a product and spend little time over the controversy of GMOs.

XI. Efforts to Transfer Technology to Other Countries

Since PRSV threatens papaya production worldwide, other countries have shown interest in developing this technology for their own use. Thus, a programme was established by my laboratory to develop and transfer the technology to other interested countries. This section summarizes the current status of the programme. Various aspects have been reviewed previously (Fermin and Gonsalves, 2004; Fermin *et al.*, 2004b; Gonsalves, and Fermin, 2004; Gonsalves *et al.*, 2006; Tripathi *et al.*, 2005). Here, the cases of Jamaica, Venezuela and Thailand are described. More details can be obtained on these and other cases in the above reviews.

Starting in 1992, a technology-transfer programme was implemented with various agencies in Brazil, Jamaica, Venezuela and Thailand. It involved students or scientists coming to the host institution (Cornell University) to develop a transgenic papaya that would be used in their own country. Except for Venezuela, initial characterization of the transgenic plants was done at Cornell University before desired lines were transferred back to their own country for further testing and eventual deregulation and commercialization. Since the Hawaiian work had shown that resistance of transgenic Rainbow papaya was largely limited to PRSV strains from Hawaii, the transgenic papaya was targeted for resistance to local virus strains. The *CP* gene from the country of origin was used and transformation was done on papaya cultivars grown in the target country. Emphasis was on developing a transgenic papaya expeditiously and getting it to the target country for timely testing, deregulation and hopefully commercialization.

A. *Jamaica*

This section is covered in more detail elsewhere (Fermin *et al.*, 2004b). Parts of the sections are summarized and some are quoted to better illustrate the challenges experienced in technology transfer. Jamaica is one of the few countries in the Caribbean that has maintained a consistent supply of papaya for its domestic and international markets. 'Sunrise' solo selections originally from Hawaii are exported as fresh produce to the European Community countries, Canada and the United States. Large-fruited varieties, from Floridian and South American selections, are also cultivated mainly for use in the local processing industries. In 1989, a severe infection of PRSV was observed in certain traditional papaya growing regions of Jamaica. Although attempts to eradicate or suppress the virus were undertaken, by 1994 the virus was also observed in other regions with high incidences of PRSV.

Shortly after the major outbreak of PRSV in 1994, a papaya-breeding programme with collaborators at the University of the West Indies (Mona), Cornell University and the private sector organization, the Jamaica Agricultural Development (JADF), was initiated. The major goal of the programme was to develop new papaya varieties with durable resistance to PSRV using *CP* genes of the virus. Transgenic papaya with the translatable and non-translatable versions of the *CP* gene from a PRSV isolate from Jamaica was developed at Cornell University by Paula Tennant, a graduate student sent by JADF. R_0 lines were tested for resistance to the homologous Jamaican isolate at Cornell and kept in greenhouses until Jamaica established biosafety protocols for importing transgenic crops. In 1997, the National Biosafety Committee (NBC) was set up as a subcommittee of the National Commission of Science, and Technology and the Plants (Importation) Control Regulations were passed in Parliament under Section 38 of the Plants (Quarantine) Act to permit the importation and controlled field-testing of the transgenic papaya. Clones of promising R_0 lines were transferred to Jamaica in 1998 and confined field testing was started.

Over the next several years, experiments were conducted on subsequent generations of transgenic papaya for resistance under field conditions (Tennant *et al.*, 2005). Data were obtained on horticultural characteristics, food safety and rat feeding studies were done. Aside from resistance to PRSV, the agronomic characteristics were similar to comparable non-transgenic plants. Furthermore, encouraging data were also obtained in a 12-week dietary study with adult Wistar rats (Fermin *et al.*, 2004b). The rats were divided into three groups: (1) rats fed with normal diet (a marketed laboratory rodent diet recommended for rats, mice, and hamsters), (2) normal diet supplemented with commercial papaya pulp, and (3) normal diet and transgenic papaya pulp. At the end of the study, blood, liver and kidney samples were collected, and analysed for total plasma protein and the activity of transaminases and phosphatases. The levels of the enzymes transaminases (aspartate and alanine transaminases) and phosphatases (acid and alkaline phosphatases) in the plasma are a measure of liver damage; high levels are suggestive of disease or injury of the organ. In the analyses, there were no significant differences ($p < 0.05$) in the levels of acid or alkaline phosphatases in the liver of rats fed with normal rat diet compared to those fed with the different preparations of the transgenic and commercial papaya. A similar trend was observed for the phosphatases in the kidney and plasma. Moreover, the protein content of the plasma was not statistically different among the groups. Overall, the data suggested no negative effects of either of the papaya supplements on tissue or organ integrity.

Below is an excerpt from a review (Fermin *et al.*, 2004b) as written by Tennant. It illustrates the factors that explain delays in implementing the practical transfer of transgenic papaya technology in Jamaica.

The study shown above represents the results of six years of research focused on the development and transfer to the field in Jamaica of PRSV-resistant transgenic papaya lines. The resistance exhibited by the Jamaican transgenic papaya lines (four in particular, 52.2, 52.3, 52.22, 52.24) should be useful for commercial production and efforts are presently being focused on stabilizing interesting lines by continued self-pollination. Moreover, southern data recently obtained in the lab (Tennant et al.*, unpublished data) suggest a single insertion of the CP gene in some of the resistant transgenic lines (e.g. line 52.3). Invariably this will be of benefit in transferring the resistant trait to local papaya varieties, namely the large fruited 'Santa Cruz giant' and 'Cedro' varieties. These varieties are as susceptible to PRSV as the commercial 'Sunrise' solo variety and are grown for the local market, particularly for the* *hotel and baking industries. Given these results, the collaborators put together a time-line for moving the transgenic papaya towards deregulation and commercial release. In 2000, a field trial aimed at further field evaluation and building the seed supply would be set up at Bramptom Farm, in 2001 data would be submitted to the NBC for a decision on deregulation and setting up trials on farmers' orchards, and in 2002 pending a positive ruling from the NBC and the establishment of the necessary regulatory framework, large-scale field release of transgenic seeds to papaya growers.*

However, the project is two years behind the proposed schedule. The third field trial has been set up as planned but the question at hand is whether researchers will be allowed to conduct field-tests on growers' orchards and whether the transgenic papaya will be deregulated and released commercially in Jamaica.

Although Jamaica was ahead of its CARICOM counterparts and Small Island Developing States (SID) in having established a National Biosafety Committee in 1997 and initiated field-testing genetically modified organisms in 1998, the country now appears to be lagging behind. The required regulatory guidelines for the release of genetically modified organisms have not progressed despite formulation in 2000. Discussions on how to facilitate the extension of the field trials to growers' orchards (provided satisfactory review of the data generated by the previous field tests) were initiated last year. Meetings with Parliamentary Counsel representatives advised the NBC on examining the Plant Quarantine Act to accommodate this Post Quarantine stage or confined testing of the transgenic papaya on growers' orchards. No final decisions have been communicated to date.

Moreover, one of the collaborators financing the project (JADF) has requested that the next phase of the project involving field-testing on growers' orchards across the island, be postponed until the NBC has deregulated the transgenic papaya. The organization foresees legal repercussions with some growers should the transgenic materials be released under temporary provisions by the NBC.

However, there is interest among papaya farmers for the completion of the research and release of materials. Growers have visited the field trials and identified transgenic trees exhibiting acceptable commercial traits. Moreover, a recent survey of papaya growers producing on farms for the local (26%) or export market (74%), report that 80% are anxious to receive the transgenic papaya seeds and to participate in setting up experimental plots with the transgenic papaya on their orchards. The other 20% are hesitant of the introduction the material into commerce

(Dawkins et al., *2003, unpublished data). This is because their major markets are in Europe and they fear genetic contamination of their non-transformed materials. Some of these growers have said that their European buyers explicitly stated that fruits from farms with transgenic crops (papaya or otherwise) would not be taken. Another reason given by these growers against adopting the transgenic papaya, is that they may not be safe for consumption and that the transgenic trees actually carry the virus (that is they are 'immunized') and will spread the virus to other trees. Interestingly, these farmers are situated in the western end of the island, where the virus only recently moved in and the disease pressure is very low. They are the ones presently exporting papaya fruit and maintaining the industry.*

In a wider survey of the Jamaican public in the corporate area and outskirts, 40% are willing to try genetically modified products, 49% will not try these products, and 11% were not able to say whether they would purchase these products or not (Pinnock et al., *unpublished data). The respondents felt that genetic engineering could probably have a positive effect on the quality of life (73%) and were in favor of the application of the technology to medicine (73%), the improvement of ornamentals (52%), and plant defenses (69%) rather than the improvement in the taste of foods (39%) and improvement of livestock (32%).*

Thus, there is interest in the project from some papaya farmers and consumers, so the researchers will continue developing a transgenic product with acceptable commercial performance while the regulatory processes 'catch up'. The research will focus into developing homozygous transgenic lines and in transferring the transgenes to other local varieties, as well as developing new varieties using portions of viral genes to elicit resistance by chimeric transgenes (Gonsalves et al., *unpublished results) and safer selection markers. Most importantly, the NBC has mounted on an education program to inform the public on genetic engineering; activities at schools, as well as radio and television broadcasts, have been initiated in order to clear up some of the misconceptions on the technology. But as the old adage goes, only time will tell, whether transgenic papaya will eventually be deregulated in Jamaica.*

B. Venezuela

This section is covered in more detail in a review (Fermin *et al.*, 2004b). It focuses on the emotive and political aspects of technology transfer. Unlike Jamaica and Hawaii, papaya in Venezuela is produced for local consumption. Moreover, the effect of PRSV was not as devastating as in Hawaii or Jamaica. In fact, the project was initiated largely from an academic standpoint. In an effort aimed at developing the infrastructure required to help advance projects in Biotechnology in Venezuela the Inter-American Development Bank provided a grant to Universidad de Los Andes (ULA) for the creation of transgenic papayas resistant to local strains of PRSV. The project in technology transfer was a collaborative effort of Universidad de Los Andes and Cornell University. In 1993 the *CP* gene of two different Venezuelan isolates from Mérida were cloned at Cornell University by Fermin the lead scientist of the project and used to transform a commercially grown papaya in Venezuela known as 'Tailandia Roja' ('Thailand Red'). Transformation of papaya was done in Venezuela, resulting in the recovery of a few transgenic lines in 1995–1996. Female, male and hermaphrodite R_0 plants were inter-crossed or selfed in Mérida to obtain the first seed-derived generation; R_1 and R_2 generations were obtained and analysed by Fermin as part of his PhD graduate studies at Cornell University. The work on analyzing the transgenic plants went well and several resistant lines were characterized (Fermin *et al.*, 2004a).

In parallel with efforts at characterizing the transgenic papaya at Cornell University, the University of Los Andes tried to do confined field tests in Venezuela. At that time Venezuela did not have set rules and regulations governing the import, release and commercialization of GMOs. An excerpt from the review (Fermin *et al.*, 2004b) to illustrate the dramatic events as written by one of the authors (Fermin) of the review is given below.

A small plot of transgenic papayas was set at Lagunillas, Mérida. The area of Lagunillas was chosen for this set of tests for different reasons: one of the virus samples was isolated in the area, papaya is being grown only domestically far from any commercial orchard (and from the experimental field), and the plot is close to the researchers in charge of the field tests. The lack of a national legislation pertaining to the management of GMOs obliged some researchers to look for creative solutions. Lab work and field experiments with GMOs were not expressly prohibited in the country, but a special permit from the Ministry of Health of Venezuela (MHV) was granted to perform the field testing (1999–2000) in a small plot described above (Fig. 5A). The plot was planted with R1 individuals previously selected in the greenhouse as PRSV-resistant. Later on, the plants continued to show very good performance in the field under the local pressure of the virus. When the transgenic, PRSV-resistant papayas were set to flower unexpected problems started to emerge.

Parallel to the development and characterization of the transgenic papayas for Venezuela, open opposition to GMOs from non-governmental organizations (NGO) flourished in Mérida by the end of the 1990s. The appearance of misinformed but well organized opposition groups; the poor preparedness of the public opinion and the lack of supporting legislation on GMOs all conspired against the acceptance of the transgenic papayas, engineered in the cleanest way possible to solve a practical problem. The main arguments against the transgenic papayas were: that the experiments were being performed in secrecy, that the risks associated with the creation and field release of the transgenic papayas were unknown, and that the country lacked the appropriate regulations to deal with GMOs in a responsible and organized manner. Only the third claim had certain truth. The news on the grant obtained to develop transgenic papayas in Mérida was highly publicized nationally. Besides, all the data was publicly available to anyone interested, mainly as monographs from one of the authors of this chapter (G. Fermin), but also through articles in local journals and seminars across the country. The claim of secrecy is probably the most unfair of all. Later on we understood that the accusation of secrecy was just another element of a very well orchestrated campaign to discredit the scientists involved in the experiments and to predispose the public against the transgenic technology as such. The few transgenic papayas tested in the field (Fig. 5A) were held responsible for the increase in mutation rates, abortions, 'contamination' of food and water, and so on; scientists were held as criminals and their names painted on the city walls with degrading epithets. The people of Lagunillas suffered the most, being constantly bombarded with a collection of 'information' aimed at creating panic. Opponents did not accept the legality of the permit issued by MHV and the matter was discussed and solved by the State Legislature. The scientists were allowed to continue the experiments until the plants set fruits, which could be collected and transferred to the lab. The plants, however, should be incinerated after fruits were collected. A serious discussion about GMOs was lacking, the fierce battle ended up exhausting all reasoning, and legal decisions were not accepted, nor respected, by the infuriated people who were convinced by few NGO representatives that transgenic papayas would trigger off serious health problems. The most fanatical opponents of transgenic plants violently attacked the plot's guardian and set the small plot on fire in December 2001 (Fig. 5B).

C. Thailand

The Thailand effort illustrates a case in which technology transfer has progressed, excellent transgenic lines have been obtained, much biosafety data have been obtained on selected lines, but a particular government moratorium is delaying efforts to move the project to the deregulation and commercialization stages. The transgenic effort is an offshoot of a USAID funded project initiated in 1986 with Ms. Vilai Prasartsee of the Thailand Department of Agriculture to help subsistence farmers of northeast Thailand control PRSV by cross-protection and by breeding for papaya lines that are tolerant to PRSV. The efforts with cross-protection did not give practical results, whereas a PRSV-tolerant variety 'Khakdum Tha Pra 2' was developed and released to growers in 1998. The tolerant papaya produces fruit even if the trees become infected. Under severe disease pressure, however, the performance of the tolerant papaya is not very effective, although it is better than that of non-tolerant lines. Much information from this section is from a review (Gonsalves *et al.*, 2006).

In 1994, Thailand Department of Agriculture and my laboratory at Cornell University started a programme for developing PRSV-resistant transgenic papaya for the farmers of northeast Thailand. Two scientists came to work at Cornell. Two cultivars, the popular 'Khakdum' and 'Khaknuan' were targeted for transformation using the translatable or non-translatable *CP* gene of a Thailand isolate of PRSV from northeast Thailand. The project in Cornell worked well and several transgenic resistant lines of *CP* transgenic 'Khakdum' and 'Khaknuan' were identified after greenhouse inoculation. Several potential R_0 lines were delivered to Thailand in July 1997 under proper rules and regulations set by Thailand. By 1999, field trials of the R_1 generation showed excellent results. By 2002, an R_3 line of 'Khaknuan' had been selected and showed excellent PRSV resistance and horticultural characteristics. In comparative tests, the transgenic line showed that 97% of the progeny were resistant under intense disease pressure and yielded 63 kg fruit per tree in the first year, whereas non-transgenic papaya yield only 0.7 kg per tree per year.

Concurrently, molecular characterization, biosafety experiments and analysis of transgenic fruit for food properties and food safety, and intellectual property rights were initiated using material that has been selected for eventual deregulation and commercialization. Nearly all biosafety experiments that are mandated by the national committee on biosafety have been completed. Tests on food safety and other characteristics, such as vitamin, amino acid, soluble solids and other profiles, are being done and should soon be completed. However, an obstacle to deregulation is that the Thailand government has a moratorium for field testing of transgenic crops except in the confined field trials in experiment stations. The field testing in simulated farmer fields is a mandatory part of the deregulation process. Furthermore, some groups including Greenpeace have demonstrated forcefully against the transgenic papaya and controversy has hit national levels.

XII. Final Comments

The pathogen-derived resistance approach for developing virus-resistant transgenic plants is a time-tested and proven technology that works. Numerous reports starting from two decades ago have clearly vouched for its applicability to many viruses and crops. Yet, only three virus-resistant transgenic crops (squash, potato and papaya) have been commercialized in the United States. Viruses clearly cause severe damage to many crops. Although not discussed here, the mechanism of resistance is well established and newer methods for developing even more effective and 'safe' virus-resistant plants have been reported. Yet, these reports are as yet largely academic and products have not been commercialized.

The researchers on the Hawaii papaya project initially never envisioned that they would have had to go beyond their fields of expertise in order get the transgenic papaya commercialized. It is true that the Hawaii papaya project took on increased urgency because it coincided with the devastation that PRSV caused to papaya production. However, my opinion is that this technology should not be used simply as a last resort. Indeed, there will be very few cases where the urgency will reach levels that were encountered by the Hawaii papaya industry in the 1990s. Instead, we should view transgenics as a major practical tool that is available to control viral diseases. Clearly, the issues are not so much the lack of the proper technology; there are other underlying factors for so little commercialization of virus-resistant transgenic crops. It is hoped that presenting these non-technical events here has provided some insights that might help towards the practical use of this powerful technology.

References

Fermin, G., and Gonsalves, D. (2004). Papaya: Engineering resistance against papaya ringspot virus by native, chimeric and synthetic transgenes. *In* "Virus and Virus-like Diseases of Major Crops in Developing Countries" (G. Loebenstein and G. Thottappilly, eds.), pp. 497–518. Kluwer Academic Publishers, The Netherlands.

Fermin, G., Inglessis, V., Garboza, C., Rangel, S., Dagert, M., and Gonsalves, D. (2004a). Engineered resistance against *Papaya ringspot virus* in Venezuelan transgenic papayas. *Plant Dis.* **88:**516–522.

Fermin, G., Tennant, P., Gonsalves, C., Lee, D., and Gonsalves, D. (2004b). Comparative development and impact of transgenic papayas in Hawaii, Jamaica, and Venezuela. *In* "Transgenic Plants: Methods and Protocols" (L. Pena, ed.), pp. 399–430. The Human Press Inc., Totowa, NJ.

Ferreira, S. A., and Mau, R. F. L. (1994). A synopsis of the phased plan for eradicating papaya ringspot disease from the Puna district, Island of Hawaii. *In* "Proceedings of the 30th Annual Hawaii Papaya Industry Association Conference," 23–24 September, Kihei, HI, pp. 7–9.

Ferreira, S. A., Pitz, K. Y., Manshardt, R., Zee, F., Fitch, M., and Gonsalves, D. (2002). Virus coat protein transgenic papaya provides practical control of *Papaya ringspot virus* in Hawaii. *Plant Dis.* **86:**101–105.

Gonsalves, C. V. (2001). Transgenic virus-resistant papaya: Farmer Adoption and Impact in the Puna Area of Hawaii. Master of Art in Liberal Studies, State University of New York, New York.

Gonsalves, C., Lee, D., and Gonsalves, D. (2004a). Transgenic virus resistant papaya: The Hawaiian 'Rainbow' was rapidly adopted by farmers and is of major importance in Hawaii today. *Online. APSnet Feature, American Phytopathological Society, August–September.* http://www.apsnet.org/online/feature/rainbow.

Gonsalves, D. (1998). Control of papaya ringspot virus in papaya: A case study. *Ann. Rev. Phytopathol.* **36:**415–437.

Gonsalves, D., and Ferreira, S. (2003). Transgenic papaya: A case for managing risks of *Papaya ringspot virus* in Hawaii. *Online Plant Health Prog.* doi:10.1094/PHP-2003-1113-03-RV.

Gonsalves, D., and Fermin, G. (2004). The use of transgenic papaya to control papaya ringspot virus in Hawaii and transfer of this technology to other countries. *In* "Handbook of Plant Biotechnology" (P. C. A. H. Klee, ed.), Vol. 2, pp. 1165–1182. John Wiley & Sons, London.

Gonsalves, D., Ferreira, S., Manshardt, R., Fitch, M., and Slightom, J. (1998). Transgenic virus resistant papaya: New hope for control of papaya ringspot virus in Hawaii. *APSnet feature story for September 1998 on world wide web.* Address is: http://www.apsnet.org/education/feature/papaya/Top.htm.

Gonsalves, D., Gonsalves, C., Ferreira, S., Pitz, K., Fitch, M., Manshardt, R., and Slightom, J. (2004b). Transgenic virus resistant papaya: From hope to reality for controlling of papaya ringspot virus in Hawaii. *Online APSnet Feature, American Phytopathological Society, August–September,* http://www.apsnet.org/online/feature/ringspot/July 2004.

Gonsalves, D., Vegas, A., Prasartsee, V., Drew, R., Suzuki, J., and Tripathi, S. (2006). Developing papaya to control papaya ringspot virus by transgenic resistance, intergeneric hybridization, and tolerance breeding. *In* "Plant Breeding Reviews" (J. Janick, ed.), Vol. 26, pp. 35–78. John Wiley & Sons Inc., Hoboken, New Jersey.

Manshardt, R. (2002). Is Organic Papaya Production in Hawaii Threatened by Cross-Pollination with Genetically Engineered Varieties? University of Hawaii College of Tropical Agriculture and Human Resources Bio-1, p. 2.

Powell-Abel, P., Nelson, R. S., De, B., Hoffmann, N., Rogers, S. G., Fraley, R. T., and Beachy, R. N. (1986). Delay of disease development in transgenic plants that express the tobacco mosaic virus coat protein gene. *Science* **232:**738–743.

Sanford, J. C., and Johnston, S. A. (1985). The concept of parasite-derived resistance: Deriving resistance genes from the parasite's own genome. *J. Theor. Biol.* **113:**395–405.

Tennant, P., Fermin, G., Fitch, M. M., Manshardt, R. M., Slightom, J. L., and Gonsalves, D. (2001). *Papaya ringspot virus* resistance of transgenic Rainbow and SunUp is affected by gene dosage, plant development, and coat protein homology. *Eur. J. Plant Pathol.* **107:**645–653.

Tennant, P. F., Gonsalves, C., Ling, K. S., Fitch, M., Manshardt, R., Slightom, J. L., and Gonsalves, D. (1994). Differential protection against papaya ringspot virus isolates in coat protein gene transgenic papaya and classically cross-protected papaya. *Phytopathology* **84:**1359–1366.

Tennant, P. F., Ahmad, M. H., and Gonsalves, D. (2005). Field resistance of coat protein transgenic papaya to *Papaya ringspot virus* in Jamaica. *Plant Dis.* **89:**841–847.

Tripathi, S., Suzuki, J., and Gonsalves, D. (2005). Development of genetically engineered resistant papaya for *Papaya ringspot virus* in a timely manner: A comprehensive and successful approach. *In* "Plant-Pathogen Interactions: Methods and Protocols" (P. Ronald, ed.), Vol. 354, pp. 197–239. The Humana Press, Inc., New Jersey.

Yeh, S.-D., and Gonsalves, D. (1984). Evaluation of induced mutants of papaya ringspot virus for control by cross protection. *Phytopathology* **74:**1086–1091.

Yeh, S.-D., and Gonsalves, D. (1994). Practices and perspective of control of papaya ringspot virus by cross protection. *In* "Advances in Disease Vector Research" (K. F. Harris, ed.), Vol. 10, pp. 237–257. Springer-Verlag, New York.

CASSAVA MOSAIC VIRUS DISEASE IN EAST AND CENTRAL AFRICA: EPIDEMIOLOGY AND MANAGEMENT OF A REGIONAL PANDEMIC

J. P. Legg,[*,†] B. Owor,[‡] P. Sseruwagi,[§] and J. Ndunguru[¶]

[*]International Institute of Tropical Agriculture, Dar es Salaam, Tanzania
[†]Natural Resources Institute, University of Greenwich, Central Avenue Chatham Maritime, Kent ME4 4TB, United Kingdom
[‡]Department of Molecular and Cell Biology, University of Cape Town Private Bag, Rondebosch 7701, South Africa
[§]International Institute of Tropical Agriculture, Kampala, Uganda
[¶]Plant Protection Division, Mwanza, Tanzania

In recent years, the cassava mosaic virus disease (CMD) pandemic in Africa has developed to become one of the most economically important crop diseases. By 2005, it had affected nine countries in East/Central Africa, had covered an area of 2.6 million sq km and was

0065-3527/06 $35.00
DOI: 10.1016/S0065-3527(06)67010-3

causing estimated losses of 47% of production in affected countries equivalent to more than 13 million tonnes (mt) annually, out of an Africa-wide total estimated annual loss of 34 mt. Strategic research investigating the cassava mosaic geminiviruses (CMGs) responsible, their whitefly vector (*Bemisia tabaci*) and interactions with their cassava host have provided the vital insights necessary to monitor the pandemic through regional epidemiological studies. Monitoring and forecasting studies have enhanced the effectiveness of host-plant resistance as a principal component of regional management efforts. Efficient movement of CMD-resistant germplasm into affected countries and regions, using an open quarantine procedure, has been key to the successes achieved to date in mitigating the effects of the pandemic. Novel control tactics, the most important of which is the use of transgenic varieties transformed for virus resistance, offer promise for the future, although transformed plants have yet to be evaluated under field conditions. Set against the promise of current and future control initiatives is the rapidly evolving nature and continued progress of the CMD pandemic. Important new threats include sustained super-abundant populations of *B. tabaci* causing physical damage to cassava and the emergence of cassava brown streak virus disease (CBSD) as a serious problem in Uganda. In conclusion, it is argued that the effective deployment of the whole range of potential control tactics will be required if the CMD pandemic is to be managed effectively and that management efforts should aim to restore the largely benign equilibrium that has characterized the interaction between CMGs and their cassava host for the greater part of their more than century-old shared history.

I. Introduction

Cassava (*Manihot esculenta*) is an important root crop in many parts of the tropics, most particularly in sub-Saharan Africa (SSA). Of the more than 18 million ha of cassava cultivated worldwide, approximately two-thirds are in Africa, producing 110 million tonnes (mt) of tuberous roots annually (FAO, 2006). Cassava is particularly valued both for its diverse potential uses, with both tuberous roots and leaves consumed, and its ability to provide acceptable yields in soils of poor fertility and in areas prone to drought. As such, it plays a vital food security role. Cassava has a relatively recent history in Africa, having been introduced through the Gulf of Guinea region of West Africa in the 16th century by Portuguese seafarers, and to the east coast two centuries later (Jones, 1959). Early cultivation was relatively sparse, however, and the crop did not become more widely grown until early in the 20th century following its active promotion for food security. The earliest report of a disease affecting the crop was made from what is now Tanzania (Warburg, 1894) in which the yellow chlorotic mosaic together with leaf deformation and rugosity were described. However, reports of the occurrence of this disease did not become more widespread until the 1920s, when it was referred to as cassava mosaic disease (CMD). Early epidemics were reported from diverse locations across Africa in the 1920s and 1930s, including Sierra Leone (Deighton, 1926), Uganda (Hall, 1928), Cameroon (Dufrenoy and Hédin, 1929), Ghana (formerly Gold Coast (Dade, 1930)), Ivory Coast (Hédin, 1931), Nigeria (Golding, 1936) and Madagascar (François, 1937). The impact of CMD was so great that large-scale control operations were introduced in several countries. In Uganda, a region-wide programme of phytosanitation was implemented, incorporating bye-laws mandating growers to uproot diseased plantings before disease-free material of partially resistant varieties was introduced after a crop-free period (Jameson, 1964). Breeding programmes were launched independently in both Tanzania (Jennings, 1957) and Madagascar (Cours, 1951; Cours *et al.*, 1997). Considerable success was achieved, and the East African breeding programme formed the basis for the later and more extensive continent-wide efforts to develop and deploy host-plant resistance to CMD.

The breeding work of Jennings and colleagues at Amani in Tanzania followed the earliest comprehensive scientific studies of CMD, its causal pathogen and insect vector led by H. H. Storey (Storey, 1936; Storey and Nichols, 1938). Zimmermann (1906) had earlier proposed that CMD was caused by a viral pathogen. This conclusion was supported by the experimental work of Storey, who also provided the first epidemiological characterization of the disease, confirmed earlier experiments (Kufferath and Ghesquière, 1932) demonstrating the vector to be the whitefly now known as *Bemisia tabaci* (Homoptera: Aleyrodidae) and noted the occurrence of both mild and severe symptom variants of the disease (Storey and Nichols, 1938). Early work on CMD in East Africa was followed by a series of studies in Federal Nigeria on whitefly transmission (Chant, 1958), yield loss and the effects of CMD on metabolism (Chant *et al.*, 1971) and some of the earliest assessments of resistant germplasm introduced from the Amani-based programme in East Africa (Beck, 1982). During the 1970s and 1980s, an ORSTOM-led research programme in Ivory Coast conducted extensive studies on CMD aetiology, epidemiology, yield loss and management, and many of the results were summarized by Fauquet and Fargette (1990). There was little evidence during this period, however, for significant changes in the status of CMD. Although isolated epidemics of severe CMD were reported in the 1990s from parts of Cape Verde and Nigeria (Anon, 1992; Calvert and Thresh, 2002), cassava scientists focused more closely on three pests that had been introduced inadvertently from Latin America in the early 1970s: the cassava mealybug (*Phenacoccus manihoti*), the cassava green mite (*Mononychellus manihoti*) and cassava bacterial blight (*Xanthomonas axonopodis* pv. *manihoti*). Nevertheless, progress was made during this period in confirming the viral aetiology (Bock and Woods, 1983), identifying the principal causal viruses now known as *African cassava mosaic virus* (ACMV) and *East African cassava mosaic virus* (EACMV) (both of Genus: *Begomovirus*; Family: *Geminiviridae*) (Hong *et al.*, 1993) and describing their largely non-overlapping geographic distributions (Swanson and Harrison, 1994). EACMV was mainly restricted to coastal East Africa, while ACMV occurred in all other cassava-growing areas of the continent. Concurrently, progress was made in describing the characteristics of the transmission of the cassava mosaic geminiviruses (CMGs) by their *B. tabaci* whitefly vector (Dubern, 1979, 1994).

This apparently stable disease situation ended in the latter half of the 1980s with the first reports, from north-central Uganda, of an unusually severe and rapidly spreading form of CMD (Otim-Nape *et al.*, 1994a). It was initially thought to be a local phenomenon associated with favourable environmental conditions for the vector. However, it became apparent during the early 1990s that the CMD associated with this epidemic was distinct from that occurring elsewhere in Uganda and the wider East African region (Gibson *et al.*, 1996), and that the epidemic was spreading (Legg, 1995; Legg and Ogwal, 1998; Otim-Nape *et al.*, 1997). By 2005, what had become a pandemic had spread to affect much of the cassava-growing area of East and Central Africa. This chapter provides a detailed description of the character, pattern of spread and impact of the pandemic and reviews management initiatives implemented to mitigate against its effects. This concludes with an assessment of the status of the pandemic and recommendations for its improved management.

II. Development and Spread of the Cassava Mosaic Disease (CMD) Pandemic in East and Central Africa

A. *Biological Characteristics of the Pandemic*

1. *The Virus*

The characteristics and pattern of spread of the severe CMD pandemic have been described in several reviews (Legg, 1999; Legg and Fauquet, 2004; Otim-Nape and Thresh 1998; Otim-Nape *et al.*, 1997,

2000). The most significant early finding was that a novel recombinant virus, referred to as *East African cassava mosaic virus*-Uganda (EACMV-UG), was detected in cassava in the epidemic-affected region of Uganda (Deng *et al.*, 1997; Zhou *et al.*, 1997) and associated with the severely diseased plants that predominated in this region. Recombination of this kind had hitherto not been known for geminiviruses and so this finding represented a significant breakthrough in the understanding of the biology of this virus group. More significant for cassava in East Africa, however, was the fact that the symptoms elicited by the recombinant virus were more severe than those of the previously occurring ACMV, although plants infected with both EACMV-UG and ACMV showed the most severe symptoms (Harrison *et al.*, 1997). The enhanced severity was shown to be the result of a synergistic interaction between ACMV and EACMV-UG (Harrison *et al.*, 1997; Pita *et al.*, 2001a). Subsequent studies have shown that mixed ACMV + 'EACMV-like' virus infections are frequent wherever the severe form of CMD occurs (Berry and Rey, 2001; Ogbe *et al.*, 2003; Were *et al.*, 2004). Synergism between an EACMV-like virus and ACMV has also been reported from Cameroon (Fondong *et al.*, 2000). However, an association between mixed CMG infections and an expanding pandemic of severe CMD has only been demonstrated in East Africa.

2. *The Vector*

A second major distinguishing biological feature of the CMD pandemic is the super-abundance of the *B. tabaci* whitefly vector, particularly at the epidemic 'front' between severely affected and relatively unaffected areas (Colvin *et al.*, 2004; Legg, 1995). Legg and Ogwal (1998) described the spread of the severe CMD epidemic through central and eastern Uganda and noted that *B. tabaci* populations on cassava in the northern epidemic-affected parts of the study area were significantly more abundant than those occurring in the southern as yet unaffected areas. Furthermore, peak populations were recorded at the locations affected by the severe form of CMD. This pattern seems to be a general feature of the CMD pandemic, and it has been hypothesized that numbers of *B. tabaci* decline in areas affected a number of years previously due to a general reduction in the cultivation of cassava that occurred (Legg and Thresh, 2000). There is no definitive evidence to explain why *B. tabaci* populations are boosted in pandemic-affected areas, although both the occurrence of distinct *B. tabaci* genotypes (Legg *et al.*, 2002) and synergistic interactions between *B. tabaci* and severely CMD-diseased cassava hosts (Colvin *et al.*, 2004, this volume, pp. 419–452) have been advanced as possible contributing factors, as discussed in Section V.

3. Cassava Germplasm

A considerable diversity of cassava germplasm exists in East and Central Africa, and before the appearance of the CMD pandemic this was almost entirely composed of local farmer-selected cultivars. Surveys in Uganda recorded 129 cultivars in farmers' fields in 1990–92 and 126 in 1994 (Otim-Nape *et al.*, 2001). Virtually all of these proved to be both susceptible and sensitive to the severe CMD of the epidemic, however, and major yield losses led to the widespread abandonment of cassava cultivation in large areas of eastern and central Uganda (Otim-Nape *et al.*, 1997; Thresh *et al.*, 1994b). Local cultivars were readily infected at a very early stage of growth. Symptoms of infection included the previously described chlorotic mosaic, distortion in shape and reduction in size of the leaves together with a general stunting of plant growth. However, additional symptoms that are particularly characteristic of infected local cultivars in pandemic-affected areas include the 'S' shape and down-turning of petioles immediately above the point of inoculation, the presence of lesions and discoloration on these petioles as well as the drying out and premature abscission of severely infected leaves below affected apices (Fig. 1). The combination of these features in an infected plant has commonly led to the use of the term 'candlestick' or 'paint brush' to describe the overall appearance of the plant. Very high virus titres develop in susceptible local cultivars following dual CMG infection with its concomitant synergistic interaction (Harrison *et al.*, 1997; Pita *et al.*, 2001a). This enhances the further dissemination of these viruses between plants.

B. Cassava Mosaic Geminiviruses in Africa and the CMD Pandemic

1. The CMGs Causing CMD in Africa

It has been recognised for many years that CMD is caused by a number of CMGs. In some of the earliest detailed virological studies conducted in Kenya, two strains with distinct patterns of behaviour in herbaceous test plant hosts were recognised (Bock *et al.*, 1981) and these were shown subsequently through sequence analyses of their DNA-A to be two distinct CMG species and given the names: ACMV and EACMV (Hong *et al.*, 1993). A third species recognised at this time was reported from India and designated as *Indian cassava mosaic virus* (ICMV). Using serology-based

FIG 1. Symptoms of cassava mosaic and cassava brown streak diseases and direct damage caused by the whitefly vector (*B. tabaci*) of the viruses responsible. (A) Moderate leaf mosaic characteristic of ACMV infection, Sangmelima, Cameroon; (B) Severe symptoms caused by mixed EACMV-UG + ACMV infection, Franceville, Gabon; (C) Downturned petioles, stem and petiole lesions associated with EACMV-UG + ACMV infection, Bukoba, Tanzania; (D) Super-abundant *B. tabaci* adults, Namulonge, Uganda; (E) Chlorosis on shoot tip caused by *B. tabaci* adult feeding and sooty mould growth on lower leaves resulting from nymph honeydew secretion; and (F–H) *Cassava brown streak virus* infection in variety TME 204, Namulonge, Uganda ((F) Dry brown necrotic rot in roots; (G) Leaf symptoms; (H) Shoot dieback associated with severe response). (See Color Insert.)

diagnostics through enzyme-linked immunosorbent assay (ELISA), Swanson and Harrison (1994) provided the first distribution map of ACMV and EACMV in Africa. This showed the largely non-overlapping distributions of the two species. ACMV occurred across West Africa, through Central Africa to Uganda and southwards as far as South Africa, whereas EACMV was almost entirely confined to East Africa west of the Rift Valley as well as Malawi and Madagascar.

Following the discovery that the developing pandemic in East Africa was associated with the novel recombinant virus, EACMV-UG, there was increased research interest in CMGs and an extensive series of new surveys was undertaken in different parts of the continent. Significantly, surveys in western Kenya (Ogbe *et al.*, 1996) and north-western Tanzania (Ogbe *et al.*, 1997) provided the first evidence that both ACMV and EACMV can occur in the same region. The nucleic acid-based diagnostic techniques that had facilitated the identification of EACMV-UG in 1997 were subsequently applied more widely and most significantly enabled the detection of mixed CMG infections. This was particularly important for West Africa, where most EACMV-like virus infections were in mixtures with ACMV (Ariyo *et al.*, 2005; Fondong *et al.*, 2000; Ogbe *et al.*, 2003). EACMV-like viruses were identified in this way from several West African countries, although sequence-based characterizations have shown these West African isolates to be a distinct CMG species, *East African cassava mosaic Cameroon virus* (EACMCV) (Fondong *et al.*, 2000). Like EACMV-UG, sequence analyses of both genome components (DNA-A and DNA-B) revealed the presence of recombined portions (in the AC2–AC3 region and in BC1), a feature that seems to be frequent with viruses of the genus *Begomovirus* (Padidam *et al.*, 1999; Pita *et al.*, 2001b). Six species of CMG are recognised from Africa to date (Fauquet and Stanley, 2003). In addition to ACMV, EACMV, EACMCV and EACMV-UG (which is considered to be a strain of EACMV), there are: *East Africa cassava mosaic Malawi virus* (EACMMV) (Zhou *et al.*, 1998), *East African cassava mosaic Zanzibar virus* (EACMZV) (Maruthi *et al.*, 2004a) and *South African cassava mosaic virus* (SACMV) (Berrie *et al.*, 2001). Based on these findings and summarizing existing survey data, Legg and Fauquet (2004) developed an updated distribution map for CMGs in Africa. From the 22 countries included in the dataset, ACMV occurred in 20, EACMV in 10, EACMV-UG in 11, EACMCV in 5, SACMV in 2, EACMZV in 2 and EACMMV in 1. There are also records of SACMV in Madagascar (Ranomenjanahary *et al.*, 2002) and EACMCV in southern Tanzania (Ndunguru *et al.*, 2005a). Senegal and Guinea in West Africa are unique in being the only countries for which extensive surveys have been conducted and where only one virus species has been reported (i.e. ACMV) (Okao-Okuja *et al.*, 2004). This does, however, reflect the generally more limited diversity of CMGs in West Africa compared with eastern and southern Africa.

Sequence characterizations of 13 CMG isolates from Tanzania revealed the occurrence of three species, clear evidence for recombination in at least two of the species (EACMV-UG and EACMCV), and an unprecedented level of diversity based on both sequencing and restriction fragment length polymorphism (RFLP) analyses (Ndunguru, 2005). Based on these findings, East Africa has been proposed as a centre of diversity for CMGs in Africa and a probable source of the begomoviruses that infected cassava and provided the key components for both the historical and evolutionary change in this group of viruses (Ndunguru *et al.*, 2005a). Evolutionary processes have been influencing the function and interactions of CMGs with their plant hosts, whitefly vector and co-infecting CMGs for millenia, including the prolonged period before the first introduction of cassava. The CMD pandemic, however, provides an excellent example both of the relative speed with which CMG evolution can occur and major consequences on plant disease that can arise.

C. Sub-Genomic DNAs and the CMD Pandemic

A number of cassava plants in the pandemic-affected area of northwestern Tanzania were observed with unusual virus-like symptoms during a countrywide CMG survey (Ndunguru, 2005). Detailed laboratory investigation of samples of this material revealed the presence of two novel DNA molecules. These 'small' DNA molecules have been shown to be dependent on geminiviruses for replication and movement within the plant, confirming their status as satellite DNAs (Ndunguru, 2005). Their sizes are 1.0 and 1.2 kbp, respectively, and they are distinct from each other sharing only 23% nucleotide sequence homology. They possess no significant homology with other sequences published in searchable databases, including those of geminiviruses. This clearly raises questions as to their evolutionary origin and the mechanisms underlying their *trans*-replication (Ndunguru, 2005). When these satellite molecules occurred in co-infections with geminiviruses, the satellites caused increased viral accumulation and novel, severe disease symptoms. Moreover, high resistance to geminiviruses in the West African cassava landrace TME 3, which has become an important component of cassava improvement programmes, including the CMD pandemic mitigation effort, can be broken by the satellites. This has raised concern about the impact of these satellites on cassava production and their possible role in the current pandemic of CMD, a question currently under investigation. Most importantly, information is required on their respective distributions and whether or not one or both are consistently associated with the severe symptoms typical of the pandemic. Regardless of whether such an association is found, there is an obvious need to increase understanding of these unusual molecules and how they interact with each of the six African CMG species in order to ensure that control approaches are effective against all potential virus and virus-satellite infections.

D. PCR-Based Diagnostics Aid the Tracking of Pandemic-Associated CMGs

Although ELISA-based diagnostics continue to be useful for detecting many plant viruses, the occurrence of recombinations involving the coat protein and the relatively high frequency of mixed infections for CMGs in Africa mean that this group of viruses is best detected and identified using nucleic acid-based techniques. Specific oligonucleotide primers have been developed for all CMG species known in Africa and they facilitate diagnosis through the polymerase chain reaction (PCR). Zhou *et al.* (1997) developed primers for detecting the principal pandemic zone viruses (ACMV and EACMV-UG) and these have been used widely in subsequent studies. An alternative method combines the use of universal primers to produce near full-length amplicons of DNA-A followed by endonuclease digestion of these amplicons to give RFLPs. This method has been used extensively to enable the diagnosis of single and mixed CMG infections and also to detect unusual virus variants (Ndunguru, 2005; Okao-Okuja *et al.*, 2004; Sseruwagi *et al.*, 2004a,b, 2005a).

Many of the CMD diagnostic surveys conducted between 1997 and 2005 in East and Central Africa have used one or both of these approaches to monitor the spread of the pandemic-associated virus, EACMV-UG (Table I). Surveys conducted over a number of years in similar locations have provided a very clear picture of the pattern of change in virus occurrence as the CMD pandemic spreads through a previously unaffected region or country. Data for Uganda provide the first and one of the clearest examples of this (Table I). In the first surveys (1995 and 1996) conducted during the early period of the epidemic in central-southern Uganda (Harrison *et al.*, 1997; Otim-Nape *et al.*, 1997), ACMV was more frequent than EACMV-UG,

TABLE I

VIRUS SURVEYS IN CMD PANDEMIC-AFFECTED REGIONS OF SUB-SAHARAN AFRICA

Country	Region	Year	Epidemic status	Methods	Number of samples	ACMV	EACMV-UG	ACMV+ EACMV-UG	EACMV	Other	No result	Reference
Uganda	Central	1995	Early epidemic	SP	32	12 (38)	6 (19)	14 (43)	–	–	–	Harrison *et al.*, 1997
	Country	1996	Epidemic	SP	68	22 (32)	21 (31)	25 (37)	–	–	–	Harrison *et al.*, 1997
	Country	1997	Late epidemic	SP	129	36 (31)	62 (53)	19 (16)	–	–	12	Pita *et al.*, 2001a
	Country	2002	Post-epidemic	SP/RFLP	152	28 (18)	97 (64)	27 (18)	–	–	–	Sseruwagi *et al.*, 2004b
Kenya	Western	1999	Late epidemic	SP	>200	(22)	(52)	(17)	3 (n.i.)	1 (n.i.)	–	Were *et al.*, 2004
	Western	2003	Post-epidemic	SP/RFLP	107	6 (6)	75 (74)	8 (8)	12 (12)	–	6	Obiero, H., unpublished data
Tanzania	Northwest	2000	Early epidemic	SP/RFLP	44	26 (59)	11 (25)	7 (16)	–	–	–	Jeremiah, S., unpublished data
	Northwest	2002	Epidemic	SP/RFLP	42	6 (14)	27 (64)	3 (7)	5 (12)	1 (2)	–	Jeremiah, S., unpublished data
Rwanda	Country	2000	Early epidemic	SP	52	26 (79)	6 (18)	1 (3)	–	–	19	Legg *et al.*, 2001
	Country	2001	Early epidemic	SP/RFLP	76	69 (91)	7 (9)	–	–	–	–	Sseruwagi *et al.*, 2005a
	Country	2004	Epidemic	SP	88	32 (38)	35 (41)	18 (21)	–	–	3	Okao-Okuja, G., unpublished data
Burundi	Country	2003	Early epidemic	SP/RFLP	55	34 (62)	17 (31)	3 (5)	1 (2)	–	–	Bigirimana *et al.*, 2004
	Country	2004	Epidemic	SP/RFLP	94	51 (55)	21 (22)	22 (23)	–	–	1	Bigirimana, S., unpublished data
Gabon	Country	2003	Epidemic	SP/RFLP	110	92 (84)	1 (1)	16 (14)	1 (1)	–	–	Legg *et al.*, 2004

Virus data indicate numbers and (percentages) of particular virus species/strains in samples tested.
Abbreviations: SP, specific primers; RFLP, restriction fragment length polymorphism; n.i., not included in percentage calculations.

although there was an increase in the proportion of plants infected with both ACMV and EACMV-UG. During later surveys, however, the proportion of CMD-diseased plants infected with ACMV alone was much reduced while the proportion infected with EACMV-UG alone increased. It was also significant that the proportion of mixed infections decreased over time. Patterns of change over time in virus occurrence are virtually identical for other pandemic-affected countries for which data are available (Table I). These data reveal a general temporal progression in the evolution of CMG-infection patterns associated with the spread of the CMD pandemic, as follows:

1. The virulent EACMV-UG spreads to areas where previously cassava was only infected with ACMV at generally low incidences.
2. Synergistic interactions between EACMV-UG and ACMV in dual-infected plants lead to rapid increases in the incidence of mixed EACMV-UG + ACMV infections.
3. Farmers select vigorous plants to provide cuttings for new planting and reject very severely diseased stems of dual-infected plants. It is then hypothesized that this leads to a reduction in the frequency of dual-infected plants with a concomitant increase in the frequency of single EACMV-UG infected plants. Even if farmers do not select, the trend towards more vigorous single EACMV-UG infected plants is enhanced because mild infections produce many more cuttings than severe ones.

The net result of these processes is the competitive exclusion of ACMV by EACMV-UG, a phenomenon whose validity seems to be confirmed by the CMG survey data (Table I).

The extensive use of monitoring and diagnostic surveys to determine the pattern of distribution of the pandemic-associated EACMV-UG has facilitated the tracking of the spread of this virus over time. Based on results of these and other surveys, efforts have been made at various times to map the extent of coverage of the pandemic (Anon, 1998a,b; Dixon *et al.*, 2003; Legg, 1999; Legg and Fauquet, 2004; Legg and Thresh, 2000). Drawing on these earlier efforts and records of the geographical pattern of spread of EACMV-UG obtained from published data, a diagrammatic representation of the pattern of development of the pandemic from 1997 to 2005 is presented in Fig. 2 and years of first pandemic infection for affected countries are indicated in Table II. It is notable that while the map represents an estimation of the historical behaviour of the pandemic, in many areas it has not been possible to obtain data. This point is most pertinent for the large areas of eastern

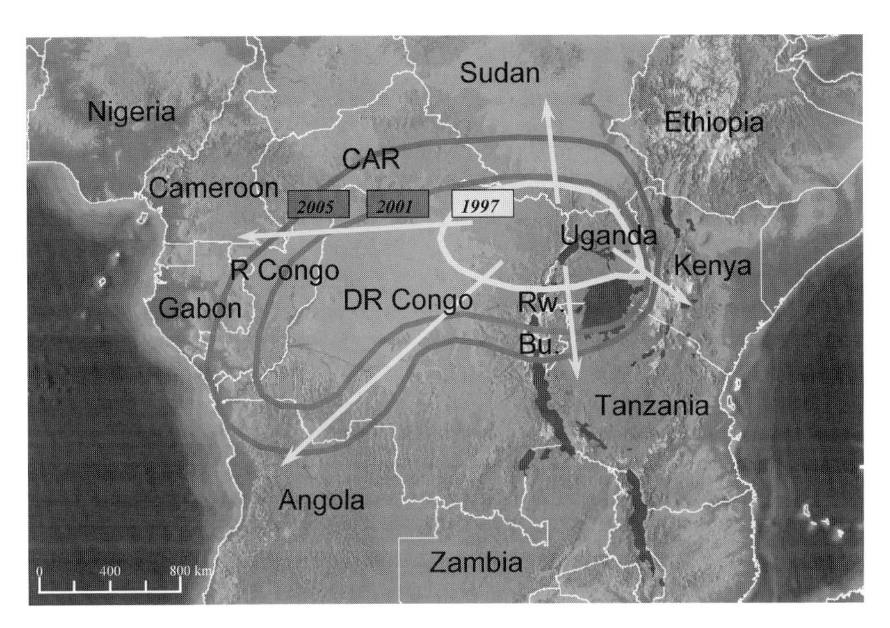

FIG 2. Estimated extent of the epidemic of the CMD pandemic in SSA in years 1997, 2001 and 2005. DR Congo, Democratic Republic of Congo; R Congo, Republic of Congo; Rw., Rwanda; Bu., Burundi. (See Color Insert.)

and central Democratic Republic of Congo (DRC), which are both extremely difficult to access and have been disturbed by internal civil conflict. As such, estimations made for these regions are speculative. This contrasts with the southern and eastern parts of the pandemic-affected zone which have been mapped intensively, particularly since 1998. Using the maps developed in Fig. 2 and by applying simple geographical information system-based techniques, it is possible to estimate that the area affected by the pandemic increased from 520,000 sq km in 1997, when only Uganda and parts of Kenya were affected, to 1,710,000 sq km in 2001 and 2,650,000 sq km in 2005. This information is of considerable importance in estimating the likely impact of the pandemic on cassava production in Africa and targeting control efforts based on knowledge of the proximity of important cassava-growing areas to zones of expected new EACMV-UG spread. These data have been used to develop risk maps which provide an important management tool for pandemic CMD, as discussed in Section III.

TABLE II

First Records of Pandemic Infection for Countries in East and Central Africa

Country	Severe CMD first report	Location	Reference
Uganda	1988	Northern Luwero district	Otim-Nape *et al.*, 1994a
Kenya	1995	Busia district, Western Province	Gibson, R., unpublished
Sudan[a]	1997	Equatoria region, south	Harrison *et al.*, 1997
Rwanda	1997	Umutara prefecture, northeast	Legg, J., unpublished
Tanzania	1998	Kagera region, northwest	Legg, 1999
Republic of Congo	1999	Plateaux region, central	Neuenschwander *et al.*, 2002
Democratic Republic of Congo	2000	Kinshasa Province, southwest	Neuenschwander *et al.*, 2002
Burundi	2003	Kirundo and Muyinga Provinces, northeast	Bigirimana *et al.*, 2004
Gabon	2003	Haute-Ogooué and Ogooué-Ivindo Provinces, east	Legg *et al.*, 2004

[a] Record of EACMV-UG and severe CMD in a single sample.

E. Epidemiology of CMD and the Pandemic

There is extensive literature describing the epidemiology of CMD. Important developments have included:

- The early demonstration of the link between seasonal environmental factors and rates of CMD spread at a locality in Tanzania (Storey, 1938)
- Descriptions of the association between vector abundance and rates of CMD spread (Dengel, 1981; Fargette *et al.*, 1993; Fishpool *et al.*, 1995)
- The occurrence of environmental gradients associated with CMD spread and the primary importance of external inoculum sources compared with internal sources (Fargette *et al.*, 1985, 1990)
- The status of temperature and rainfall as key determinants of cassava growth, vector population increase and subsequent virus spread (Fargette *et al.*, 1993; Fishpool *et al.*, 1995)
- The value of resistant varieties in both delaying and reducing rates of virus spread (Colon, 1984; Hahn *et al.*, 1980; Otim-Nape *et al.*, 1998; Sserubombwe *et al.*, 2001)
- The potential to predict final CMD incidence through a combined assessment of inoculum and the abundance of early vector immigrants (Legg *et al.*, 1997).

The behaviour of CMD at higher-epidemiological levels, as it spreads between fields, regions and countries, has received much less research attention, however, due in part to the inherently greater difficulty of undertaking such extensive studies. Another important reason for the limited treatment of these levels of epidemiology has been that there have been few reports of macro-scale CMD spread over more than a century of the disease's history in Africa, as mentioned in Section I. Two early qualitative descriptions of regional spread were, however, made from Nigeria (Golding, 1936) and Madagascar (Cours, 1951). By contrast, detailed information has been presented describing the regional spread of the CMD pandemic. Much of this relates to Uganda, where the pandemic was first reported. Otim-Nape *et al.* (1997, 2000) and Otim-Nape and Thresh (1998) used qualitative terms to describe the regional dynamics of the CMD epidemic in Uganda. Six zones were defined, representing both a spatial series, running from areas ahead to those behind the 'front' of the epidemic, as well as a temporal series, describing changes in the epidemic as it passed through a given location. 'Zone 1' was defined as the area ahead of the epidemic front in which CMD incidence was low, symptoms were generally mild and where there was little or no disease spread. Whitefly populations were low and local cassava cultivars were being grown sustainably. This zone was referred to as '*Pre-epidemic*'. The '*Epidemic zone*' or 'Zone 3' corresponded to the 'front' of the epidemic where increased whitefly populations were causing rapid spread of CMD, symptoms were severe and local cultivars were becoming entirely diseased. In the '*Recovery zone*' (Zone 5), production of severely diseased local cultivars was being abandoned following the failure of crops where farmers had attempted to replant using cuttings from already severely diseased mother plants. As the intensity of cassava cultivation declined, inoculum levels and whitefly numbers dropped, leading to a general amelioration in the situation that was enhanced by the apparent increase in frequency of plants infected by mild strains.

Legg and Ogwal (1998) used quantitative data collected along two north–south transects in central and eastern Uganda to characterize the changes in CMD and whiteflies between 1992 and 1993. The key variables used to describe these changes were whitefly abundance, CMD incidence and severity and, perhaps most importantly, the relative amounts of cutting and whitefly-borne infection. Current season whitefly-borne infection was recognised by the absence of symptoms on the first-formed leaves at the base of the plant being assessed, whereas cutting-borne infection was apparent from the occurrence of symptoms on first-formed leaves. The relative proportions of these two infection categories allow inferences to be made about the epidemiological situation.

A combined analysis of CMD incidence and infection type using georeferenced data gathered from surveys in the Lake Victoria Basin zone of East and Central Africa was used to map the distributions of different epidemic zones and to highlight threatened areas immediately ahead of the epidemic (Legg *et al.*, 1999a). The '*Zone of epidemic expansion*' was defined as the area where incidence was greater than 70% and the percentage of whitefly-infected plants was commonly more than three times the percentage of cutting-infected plants. At this time, two such zones were identified, the first of which covered Western Province in western Kenya, and the second, the southern Ugandan districts of Masaka and Rakai, just north of the border with Tanzania. Using a similar approach, patterns of change in the epidemiological zones in space and over time within the East and Central African regions were tabulated for the period 1990–1999 (Legg, 1999). An updated and slightly modified analysis of this type is presented in Tables III and IV, which summarizes the pandemic's zones in 2005. It is notable that

TABLE III

STATUS OF CMD-PANDEMIC ZONES IN SUB-SAHARAN AFRICA, 2005 (ADAPTED FROM LEGG, 1999)

Zone	Designation	CMD incidence	Mean CMD severity	Type of infection	*B. tabaci* abundance	Cassava cultivation
1	Pre-epidemic zone	<30%	Low (<2.5)	Mainly cutting	Low (0–1)	Normal
2	Zone of epidemic expansion	30–70%	High (>3)	Mainly whitefly	High (>5)	Normal
3	Mature epidemic zone	>70%	High (>3)	Whitefly and cutting	High (>5)	Sustained but yields reduced
4	Post-epidemic zone	>50%	Moderate (2.5–3)	Mainly cutting	Moderate (1–5)	Reduced or abandoned
5	Zone of recovery	<50%	Low (<2.5)	Mainly cutting	Moderate (1–5)	Being re-established

TABLE IV

EPIDEMIOLOGICAL ZONATION OF PANDEMIC-AFFECTED COUNTRIES IN 2005 (BASED ON ZONES DESCRIBED IN TABLE III)

Country	Zone 1	Zone 2	Zone 3	Zone 4	Zone 5
Uganda	–	–	–	North, south, west	East, centre
Kenya	East, centre	–	South Nyanza Province	North Nyanza Province	Western Province
Rwanda	Southwest	–	Centre, south	Northeast	–
Tanzania	East, centre, south	Kigoma, Mara, Shinyanga regions	Mwanza region	Kagera region	–
Republic of Congo	–	–	–	Entire country	–
Democratic Republic of Congo	Southeast	East	Northeast	Centre	Southwest
Burundi	–	South, west	Centre, north	Northeast	–
Gabon	North, south, centre, west	East	–	–	–

"–": not present.

while five countries have been newly reported as CMD pandemic-affected since 1999, only parts of Uganda, Kenya and the DRC have progressed to Zone 5. The reasons for this apparently limited progress in the management of the pandemic will be discussed in Section IV. However, the comparison between the 1999 and 2005 datasets does highlight the magnitude of the extent of 'new' spread of the pandemic and the scale of the problem that it poses.

An alternative approach that can be taken to collating epidemiological data collected from CMD surveys across Africa is to combine the principal variables affecting CMD epidemiology to develop an '*Epidemic Index*', which gives an overall estimate of the acuteness of the CMD problem in a region. The key variables are indicated in Table V, and in each case three levels or categories have been defined from 'low' (category '1') to 'high' (category '3'). For the three directly scored quantitative measures, higher levels (CMD incidence and CMD-symptom severity) or greater numbers (whitefly abundance) gave higher-category levels. For the CMD-infection factor, derived by dividing the whitefly-borne infection by the incidence of cutting infection, greater values similarly gave higher-category levels. For the virus-diversity factor, two aspects were considered to enhance epidemic severity. The first of these was the presence of EACMV-UG, since it typically has high virulence and the second was the presence of more than one virus species or strain, as virus–virus synergism is known to be a key factor in enhancing CMG spread.

Using country survey data, mean values are calculated for each of the quantitative variables and for each of the regions considered within the survey. The '*Epidemic Index*' value for each of these regions is then calculated by summing the index-category values for each of the five factors. Values can range from a minimum of $5 \times 1 = 5$ (lowest/weakest) to a maximum of $5 \times 3 = 15$ (highest/strongest). Available survey data collected between 1998 and 2003 have been analysed in this way for the 17 countries in SSA shown in Fig. 3, and *Epidemic Index* values obtained have been mapped and are presented in the same figure. Assuming that a region or country with an *Epidemic Index* value of 11 or above is in an '*acute epidemic*' state, the analysis yields 30 regions in 7 countries so affected. On superimposing the region of coverage of the CMD pandemic over this map (Figs. 2 and 3), 28 of the 30 acute zones are within the pandemic-affected area, highlighting this as the only major part of the cassava-growing region of Africa in which CMD is a severe and spreading problem.

TABLE V

FACTORS AND CATEGORIES USED IN CALCULATING THE 'EPIDEMIC INDEX'

Factor	Index category		
	1	2	3
CMD incidence	0–33%	33–67%	67–100%
CMD severity (1–5 scale)	2.0–2.75	2.75–3.5	3.5–5.0
CMD infection[a]	0–0.33	0.33–1.00	>1
Whitefly abundance	0–1	1–5	>5
Virus diversity	Single sp., no EACMV-UG	EACMV-UG present	Mixed spp. + EACMV-UG

[a] CMD infection = ratio of whitefly to cutting infection.

Whitefly infection is the percentage of plants infected by the whitefly vector during the current season and is distinguished by the absence of CMD symptoms on the lowermost first-formed leaves.

Cutting infection is the percentage of plants infected through planting CMD-infected cuttings and is distinguished by the presence of symptoms on the lowermost first-formed leaves.

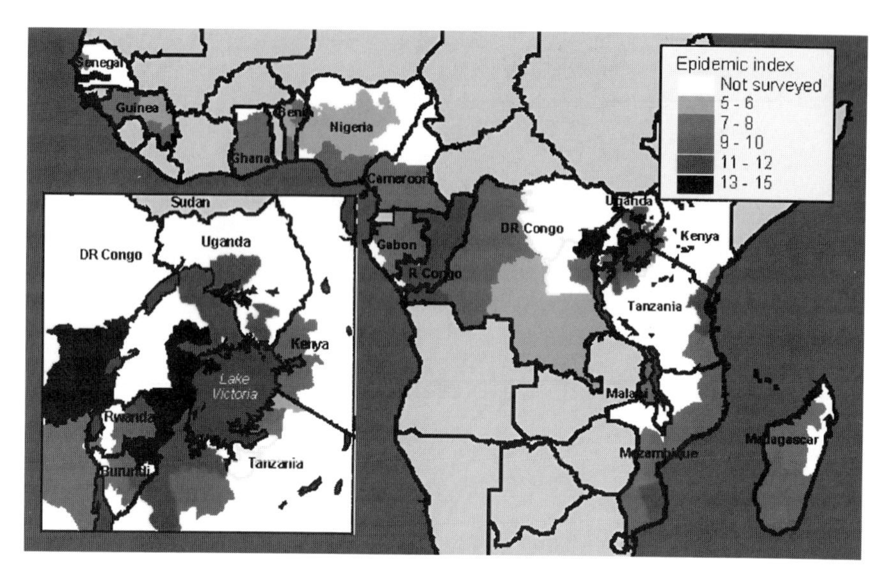

FIG 3. Epidemic index values for cassava mosaic disease in regions of SSA surveyed between 1998 and 2003. (See Color Insert.)

III. Economic and Social Impact of the CMD Pandemic

A. *Introduction and Field-Level Yield Loss Studies*

Because of the severe damage that pandemic CMD has caused to the largely susceptible and sensitive local cultivars grown across the major production zones of East and Central Africa, it is not surprising that there have been diverse impacts both on the production of the crop and, as a consequence of this, on producers and consumers. Considering the basic unit of the plant, losses have been particularly severe where cassava stems from plants infected for the first time by the viruses causing severe pandemic CMD have been used to provide the following season's crop. In Uganda, yield loss estimates for local cultivars measured prior to the pandemic's spread ranged from 20% to 40% (Otim-Nape *et al.*, 1994b). A later study, by collecting planting material from a location within the pandemic-affected zone, quantified losses in tuberous root yield of 66% in CMD-diseased plants of what was then the most commonly grown local cultivar, 'Ebwanateraka' (Byabakama *et al.*, 1999). Similar assessments made during the pandemic in Bukoba district, Kagera region, Tanzania in 2001 and 2002 gave yield losses for three of the most widely grown local cultivars as

72% for Msitu Zanzibar, 85% for Rushura and 90% for Bukalasa Ndogo (=Bukalasa 11 or F279), giving a mean loss of 82% (Ndyetabura, I., unpublished data). Comparable results were obtained for CMD-susceptible varieties in Siaya district, western Kenya, where losses for two local cultivars were estimated at 72% for Karemo and 63% for Bukalasa 11 (mean = 68%) (Mallowa, 2006). However, none of these studies considered the nature of the virus infection in plants whose yields were measured. The only study to date that has attempted to measure yield losses attributable to specific virus infections was in western Uganda from 1999 to 2000 and considered the yield effects of infection by mild and severe forms of EACMV-UG, ACMV and mixed ACMV + EACMV-UG infections on the local cultivar Ebwanateraka (Owor *et al.*, 2004a). As anticipated, losses were greatest for mixed infected plants (82%), intermediate for plants infected with EACMV-UG severe alone (68%) or ACMV alone (42%) and least for EACMV-UG mild infections (12%).

B. *Regional-Level Assessments of Production Losses Due to CMD*

Several regional-level assessments of pandemic-associated cassava production losses have been made. Annual losses in the mid-1990s in Uganda were estimated as being equivalent to the total production of four districts each year, which amounted to an estimated 600,000 t annually, equivalent to a financial cost of US$ 60 million (Otim-Nape *et al.*, 1997). Similar calculations were used to produce an estimated annual loss for western Kenya of US$ 14 million (Legg *et al.*, 1999a). Thresh *et al.* (1997) provided an approximation for continental level losses due to CMD. Their calculation assumed an overall incidence of 50–60% and losses for infected plants of 30–40% and gave a range of 15–24% for losses due to CMD in Africa as a whole. New incidence data allowed this estimate to be updated in 2003, with a revised figure for Africa-wide losses of 19–27 mt (Legg and Thresh, 2004).

Here, we review existing data for pandemic recovery (Uganda), pandemic-affected and as yet unaffected countries and parts of countries (Table VI). The mean loss for CMD-infected plants in pandemic-affected countries has been calculated as 72%, which is the mean value of the loss estimates presented previously of 66% for widely grown cultivars in Uganda (Byabakama *et al.*, 1999), 82% for Tanzania (Ndyetabura, I., unpublished data) and 68% for Kenya (Mallowa, 2006). The comparable mean loss for CMD-infected plants in countries not yet affected by the pandemic is that used previously (Legg and Thresh, 2004; Thresh *et al.*, 1997) of 30–40%, although here we have used the single mean value of 35%.

TABLE VI

CMD-INFECTION CHARACTERISTICS, CASSAVA PRODUCTION AND ESTIMATED LOSSES FOR 17 CASSAVA-PRODUCING COUNTRIES IN AFRICA

Country (reference)	Year	Cassava mosaic disease: Cutting infection (%)	Whitefly infection (%)	Total incidence (%)	Mean symptom severity	*B. tabaci* (no. per shoot)	CMGs detected	Actual production (1000 t)	Estimated loss (1000 t)
Recovery[a]									
Uganda (Okao-Okuja, G., unpublished data; Owor *et al.*, 2004a; Sseruwagi *et al.*, 2005b)	1998, 2003	44	16	60	2.9	25.0	ACMV, EACMV-UG	5500	1048 (16%)
Pandemic-affected[b]									
Western Kenya (Kamau *et al.*, 2005; Obiero, H, unpublished data)	1998, 2005	36	27	63	2.8	6.6	ACMV, EACMV, EACMV-UG	$420^{2/3,d}$	349
Northwestern Tanzania (Jeremiah, S., unpublished data; Ndunguru *et al.*, 2005b)	1998, 2002	18	35	53	3.0	30.6	ACMV, EACMV, EACMV-UG	$2333^{1/3}$	1440
Rwanda (Sseruwagi *et al.*, 2005a)	2001, 2004	27	23	50	3.6	2.4	ACMV, EACMV-UG	782	440
Burundi (Bigirimana *et al.*, 2004; Bigirimana, S., unpublished data)	2003, 2005	36	25	61	3.4	18.3	ACMV, EACMV-UG	710	556
Northern, eastern and western DRC (Tata-Hangy, A., unpublished data)	2002–2003	56	11	67	3.1	41.1	ACMV, EACMV-UG	$9982^{2/3}$	9303
ROC (Ntawuruhunga, P., unpublished data)	2003	82	4	86	3.3	2.0	ACMV, EACMV-UG	900	1464
Eastern Gabon (Legg *et al.*, 2004; Legg, J., unpublished data)	2003	77	7	84	2.8	<1	ACMV, EACMV-UG	$29^{1/8}$	44
Average–pandemic-affected		47	19	66	3.1	14.5			
Total–pandemic-affected								**15,156**	**13,596 (47%)**
Pandemic-unaffected[c]									
Eastern and central Kenya (Kamau *et al.*, 2005; Obiero, H., unpublished data)	1998	24	10	34	2.4	<1	EACMV, EACMZV	$210^{1/3}$	28

Eastern and southern Tanzania (Jeremiah, S., unpublished data; Ndunguru *et al.*, 2005b)	1998	16	10	26	2.9	<1	EACMV, EACMCV	$4667^{2/3}$	467
Southern and central DRC (Tata-Hangy, A., unpublished data)	2002–2003	34	0	34	2.8	<1	No diagnoses	$4991^{1/3}$	674
Western, northern and southern Gabon (Legg *et al.*, 2004; Legg, J., unpublished data)	2003	79	0	79	2.3	1.4	ACMV	$201^{7/8}$	77
Cameroon (Ntonifor *et al.*, 2005)	1998	55	7	62	2.3	3.3	ACMV, EACMCV	1950	538
Nigeria (Echendu *et al.*, 2005)	1998	54	2	56	2.3	2.1	ACMV, EACMCV	38,179	9225
Benin (Gbaguidi *et al.*, 2005)	1998	34	2	36	2.1	2.9	ACMV, EACMCV	3100	447
Ghana (Cudjoe *et al.*, 2005)	1998	56	15	71	2.3	2.3	ACMV, ACMCV	9739	3232
Guinea (Okao-Okuja *et al.*, 2004)	2003	52	10	62	2.4	1.0	ACMV	1350	379
Senegal (Okao-Okuja *et al.*, 2004)	2003	71	12	83	2.3	3.2	ACMV	402	165
Malawi (Theu and Sseruwagi, 2005)	1998	25	17	42	2.8	1.3	EACMV, EACMMV	2600	448
Mozambique (Toko, M., unpublished data)	2003	9	16	25	2.6	17.9	ACMV, EACMV, EACMV-UG	6150	590
Madagascar (Ranomenjanahary *et al.*, 2005)	1998	41	6	47	3.1	5.0	ACMV, EACMV, SACMV	2191	435
Average–pandemic-unaffected		42	8	50	2.5	3.3			
Total–Pandemic-unaffected								**75,730**	**16,705 (18%)**
Unsurveyed area (18% loss estimate)								14,034	3081
Overall Total								**110,420**	**34,430 (24%)**

[a] Yield losses for the recovery country (Uganda) calculated based on an estimated overall production loss due to CMD of 16% (see text for details).
[b] Yield losses for pandemic-affected countries calculated based on a mean loss for CMD-infected plants of 72%.
[c] Yield losses for pandemic-unaffected countries calculated based on a mean loss for CMD-infected plants of 35%.
[d] For countries for which data have been divided between affected and unaffected areas, fractions indicate the estimated proportion of total production in that area.

Uganda has been placed in a distinct pandemic 'recovery' category in view of the extensive dissemination of CMD-resistant varieties (Sserubombwe *et al.*, 2005a) and the post-epidemic amelioration of CMD in local cultivars. The virus survey data show mild and severe strains of EACMV-UG to be the predominant viruses causing CMD (Sseruwagi *et al.*, 2004b). Combining an average for yield losses in a local cultivar infected by EACMV-UG mild (12%) and EACMV-UG severe (68%) (Owor *et al.*, 2004a) with a local cultivar prevalence of 67% (Sserubombwe *et al.*, 2005a) and a CMD incidence of 60% (Okao-Okuja, G., unpublished data), gives an overall estimate for production loss to CMD in Uganda of 16%. This makes the reasonable assumption that losses in CMD-resistant varieties are negligible. For those countries only partially affected by the pandemic (Tanzania, Kenya, DRC and Gabon), loss estimates have been made for affected and unaffected areas. For affected areas, the 72% figure has been used for losses in infected plants, while for unaffected areas, the 35% figure has been used. The regional coverage of the pandemic zone has been used to estimate the fraction of cassava production affected in Tanzania (1/3), DRC (2/3) and Gabon (1/8). In Kenya, the pandemic-affected fraction has been estimated as 2/3, since approximately two-thirds of Kenya's cassava production is in the pandemic-affected western part of the country.

Production loss estimates have been calculated for pandemic-affected and unaffected countries and parts of countries using figures for CMD incidence, percentage loss for infected plants (either 35% or 72%) and the overall production figures for 2005 obtained from the Food and Agriculture Organization (FAO) database (FAO, 2006). The total production of all surveyed countries represents *c.* 87% of total African production. The production loss for the remaining 13% was calculated using the mean percentage loss calculated for non-pandemic countries (=18%). Using this analysis, overall losses for pandemic-affected areas are estimated at 47%, between two and three times the value for those countries not yet affected (18%) and the recovering Uganda (16%). The estimate for total loss due to CMD in Africa of more than 34 mt (equivalent to 24% of total production) is comparable but slightly greater than the previous estimates of Thresh *et al.* (1997) and Legg and Thresh (2004). This is a consequence of the increased figure for yield loss due to CMD in the pandemic zone and the greater area affected. It is recognised that there are deficiencies in the approach used here to estimate yield losses. For countries in the pandemic zone, estimates of the proportion of cassava production affected are imperfect and there is considerable variation in effects on yield depending on the varietal response, stage of infection and the mix of CMG species or strains present. In addition, some countries known to be affected by the pandemic have not been surveyed and, therefore, are not included in the analysis in Table VI (Sudan). Other countries are almost certainly affected but they have not yet been surveyed (Central African Republic (CAR) and Angola). Despite these deficiencies, the results provide a reasonable and relatively conservative reflection of the actual situation, and are probably the best that could be achieved with the data currently available.

C. Social Impact of the CMD Pandemic

The social consequences of the CMD pandemic differed considerably between regions and countries, with the primary determining factor being the importance of cassava to the livelihoods of the affected populations. While the pandemic represents an 'inconvenience' in the high-potential farming areas of parts of southern Uganda, northwestern Tanzania and western Kenya, it poses a significant threat to the survival of more vulnerable communities in drier, less-fertile farming areas such as northeastern Uganda, the Kenya shore of Lake Victoria, the southern and eastern shores of Lake Victoria in Tanzania and eastern areas of both Rwanda and Burundi. In large parts of the humid forest zone of Central Africa, encompassing central and northern DRC, Republic of Congo (ROC) and Gabon, cassava is often the only crop grown, and large quantities of leaves as well as the tuberous roots are consumed. In all such areas, the social impact of the pandemic has been acute.

Otim-Nape *et al.* (2000) listed eight key facets of the pandemic's impact on cassava cultivation in vulnerable farming communities of eastern Uganda. These were: a drastic reduction in cassava production, large decreases in the area of cassava being cultivated, marked increases in market prices for cassava food products and planting material, changes in the relative importance of cassava cultivars being grown, increased trade in cassava planting material as farmers seek out improved varieties, increased cultivation of alternative crops, a rise in thefts of cassava roots and stems (particularly of improved resistant varieties following their introduction) and widespread demands from farmers, farmer groups, local officials and ultimately from politicians for the implementation of urgent mitigation measures. These demands peaked in the early stages of the pandemic's spread through eastern Uganda, which coincided with a sustained drought in the 1993–1994 season and led to widespread food shortages, localized famine and reports of hundreds of hunger-related deaths (Thresh and Otim-Nape, 1994).

Similar situations have continued to occur as the pandemic has spread through East and Central Africa, one of which was a severe food shortage in the provinces of Kirundo and Muyinga in northeastern Burundi in 2004/2005. The combined effects of the CMD pandemic exacerbated by drought led to more than 100 famine-related deaths and threats to thousands more (Anon, 2005a). In an attempt to address the crisis, the Government of Burundi introduced an emergency tax on the salaries of all civil servants and collected one-off payments from all other workers (Anon, 2005b). This was to provide emergency support to 1-million people in the affected provinces. Governments may or may not provide support for affected communities, but individuals impacted have to find ways to cope with the lost cassava production. Some of the most prominent means have included the substitution of cassava by alternative crops, most importantly sweet potato, and search for employment on the farms of more prosperous neighbours. Affected families immediately cut back on all non-essential expenditure, one of the most important of which is the payment of education expenses for children. Other slightly longer-term responses have included migration to urban centres or the removal of some family members to other locations, which are either less severely affected and, therefore, better suited for farming or where there are more opportunities for employment. These negative impacts also have adverse effects on relationships within families which in turn further exacerbate the crisis. At a higher level, the negative social effects of the pandemic have affected politics at both local and national levels. These effects have sometimes been negative, as politicians or political groups have sought to apportion blame for the harmful socio-economic effects of the pandemic rather than making more constructive contributions to addressing the crisis. In some situations, however, awareness amongst national politicians has led to increased governmental commitments to solving the problem. The example of this is Burundi, where the ruling party, whose flag bears an image of a cassava plant, made it a national priority to deal with the production losses arising from the CMD pandemic, following its election victory in mid-2005.

IV. Management of the CMD Pandemic

A. *Introduction: Virus Management Strategies*

A broad range of techniques is known to be of value for the management of plant virus diseases, such as CMD. These have been reviewed extensively (Fauquet and Fargette, 1990; Guthrie, 1988; Thresh and

Cooter, 2005; Thresh and Otim-Nape, 1994; Thresh *et al.*, 1998). As for other plant virus diseases, the strategy for managing CMD focuses on preventing infection, delaying the time of infection, minimizing the effects of infection once it has occurred, or ideally combinations of the three. The key elements of control strategies include the use of phytosanitation (primarily involving the uprooting of diseased plants (roguing) and selection of disease-free stems for new planting) and the development and deployment of host-plant resistance. Largely as a consequence of the success that has been achieved in breeding for resistance to CMGs in cassava, this has been the primary tactic exploited in CMD management programmes. Phytosanitation has mainly been used within the framework of schemes for the multiplication of resistant varieties. Cross protection, which makes use of mild or attenuated virus strains that decrease the effects of infection by related but more virulent strains, has been used for some viruses/virus diseases, but until recently, there was no evidence for the potential value of this approach in controlling CMGs. Various crop management strategies may be used to hinder the spread of CMGs into initially CMD-free crops. These include isolation, field disposition and orientation with respect to inoculum sources, inter-cropping and varietal mixtures in which resistant varieties are used to 'protect' susceptible local cultivars. All of these have their limitations and there is little experimental evidence demonstrating significant benefits, although the varietal mixture approach has been shown to provide some degree of protection for CMD-susceptible material in Uganda (Sserubombwe *et al.*, 2001, 2005b).

Although the CMD pandemic has arisen and spread through the interaction of virus, vector and host plant, surprisingly, little attention has been given to the possible scope and value of vector management. This is in part a result of earlier findings from West Africa in which there was shown to be a weak relationship between field resistance (proportion of infected plants) and vector resistance (Fauquet and Fargette, 1990). However, concern about super-abundant populations of the CMG vector, *B. tabaci*, particularly on CMD-resistant varieties, has led to a renewed interest in possible whitefly-control measures and investigations into both host-plant resistance against whiteflies and biological control through natural enemy augmentation have been initiated. Virus resistance has traditionally been developed using conventional breeding approaches, however, there is an increasing trend towards the development of virus resistance through genetic engineering. Conventional intra- and inter-species crossing is an imprecise means of introgressing resistance genes into a target genotype.

Genetic transformation offers the potential to provide a much more precise method of introducing genes conferring specific beneficial traits. As knowledge of the genomes of crops such as cassava is gained, the prospect increases of 'cut and paste' technologies that will allow the insertion of sets of desirable genes into farmer-preferred local cultivars, thereby assuring acceptability. Experience with the distribution of improved varieties has shown that such a route towards rapid acceptance is highly desirable, as quality characteristics of conventionally bred CMD-resistant materials often differ from those of local farmer-preferred yet CMD-susceptible cultivars, and this mismatch can result in poor uptake of the improved material.

The magnitude of the problems caused by the CMD pandemic has been such that virtually all possible approaches to controlling the disease have been used, including many 'local' methods used by farmers. The following sections detail the characteristics and methods of implementation of the most important of these approaches.

B. *Targeting Control Through Monitoring the CMD Pandemic*

1. *Rationale for Pandemic Monitoring*

The CMD pandemic has been unique, in relation to the history of the disease in Africa, in being dynamic and spreading through a number of countries. This has important implications for the management strategy. Epidemiology has provided information on the rate at which the pandemic is spreading and about the contrasting characteristics of CMD in different 'zones' of the pandemic, and how these change both in space and time. Since the characteristics of CMD at a given location and time determine the most appropriate control approach, detailed epidemiological information is a vital pre-requisite for effective management. Most importantly, epidemiological data, when gathered sufficiently frequently, can identify the areas of real crisis, such as the two examples provided in Section III, namely, eastern Uganda in the early 1990s and northeastern Burundi in 2004/2005. Monitoring and diagnostic surveys have, therefore, become a key component of CMD pandemic mitigation programmes throughout the affected countries and regions (Sseruwagi *et al.*, 2004a).

2. *Qualitative and Quantitative Monitoring Assessments*

The first attempts to monitor the CMD pandemic were undertaken in Uganda in the early 1990s as spread occurred through the central southern part of the country. Observations were made along transects running north–south across the epidemic front and elements of the epidemic assessed included severity, incidence and infection type (by cutting or by whitefly), farmer responses and overall cultivation intensity (Otim-Nape *et al.*, 1997, 2000). From information gathered during these surveys, the zonal characteristics of the spreading epidemic were described and the first attempts were made to estimate rates of spread. Concurrent with the wide-ranging qualitative monitoring surveys, a quantitative study provided further evidence for an annual spread rate of 20–30 km (Legg and Ogwal, 1998), and this information led to the first concerns about spread to neighbouring areas of western Kenya and northwestern Tanzania.

3. *Forecasting Pandemic Spread*

Through the mid-1990s to late 1990s, consistent patterns of CMD spread through Uganda and into the neighbouring countries of Kenya and Tanzania made it clear that patterns of expansion from year to year were predictable. Epidemiological data obtained from surveys in Uganda, Kenya and Tanzania were used to make a forecast of 'threatened' areas in the late 1990s (Legg *et al.*, 1999a) and this approach was modified subsequently to propose a zone, surrounding the known limits of the CMD pandemic, that was expected to be affected within 5 years (Legg, 1999). At this time, monitoring and diagnostic surveys became a routine component within CMD pandemic management programmes (Anon, 1999). Typically, these were conducted annually in the extensive geographical areas in and around the pandemic-affected zone. The most important elements of these surveys were assessments of changes in CMD incidence, symptom severity and pattern of infection and tests for the pandemic-associated virus, EACMV-UG. The principal outputs of these surveys were regional maps that plotted the pattern and speed of spread of the pandemic. Thus, it was possible to delineate newly affected areas and those most severely impacted by the pandemic, with a view to preferentially targeting control activities towards them.

4. *Forecasting Examples from East and West Africa*

Diagnostic survey data have been used to develop risk maps for East Africa. The first step in the construction of such maps is to determine the administrative areas (typically districts or provinces) already affected by the pandemic. Risk to administrative regions beyond the limits of the pandemic-affected zone is then assessed based on the cassava cultivation intensity in the region and the distance of the region from the boundary of the pandemic. Three levels are defined for each of the two factors and combined to determine the level of risk, which can be low, medium or high (Table VII). A risk map is then generated using the results obtained. Figure 4 is a risk map developed in 2001 for use within a regional CMD pandemic mitigation programme. By 2005, all 14 of the regions identified as at high risk in 2001 had been affected by the pandemic, 12 of the 17 regions in the medium-risk category and 6 of the 10 in low-risk regions. The fact that so many of even the lowest risk category regions were affected within 4 years, highlights both the rapidity with which the pandemic has spread in East Africa and the need to increase the distances used in defining the three levels in the distance-based factor used in the risk assessment.

An alternative approach used to locate severe CMD in a previously unsurveyed country for which no data were available was attempted in Gabon in West/Central Africa. Earlier survey work had shown that EACMV-UG occurred in the open, hilly grassland environment of Plateaux Region in central ROC (Neuenschwander *et al.*, 2002). Experience from East Africa had also shown that pandemic spread occurred most readily in open savannah-like environments, such as those of eastern Uganda (Otim-Nape *et al.*, 2000). Vegetation maps of West/Central Africa (Fig. 5) showed clearly that the savannahs of central ROC extended to the east into the eastern part of Gabon. Consequently, it was predicted that EACMV-UG and the pandemic had already

TABLE VII
COMBINATIONS OF FACTORS GIVING THE DIFFERENT RISK LEVELS PLOTTED ON THE RISK ASSESSMENT MAP (FIG. 4)

Distance from pandemic front (km)	Cassava cultivation intensity	Threat/risk level
<50	H	H
	M	H
	L	M
50–150	H	H
	M	M
	L	L
>150	H	M
	M	L
	L	L

Abbreviations: L, low; M, medium; H, high.

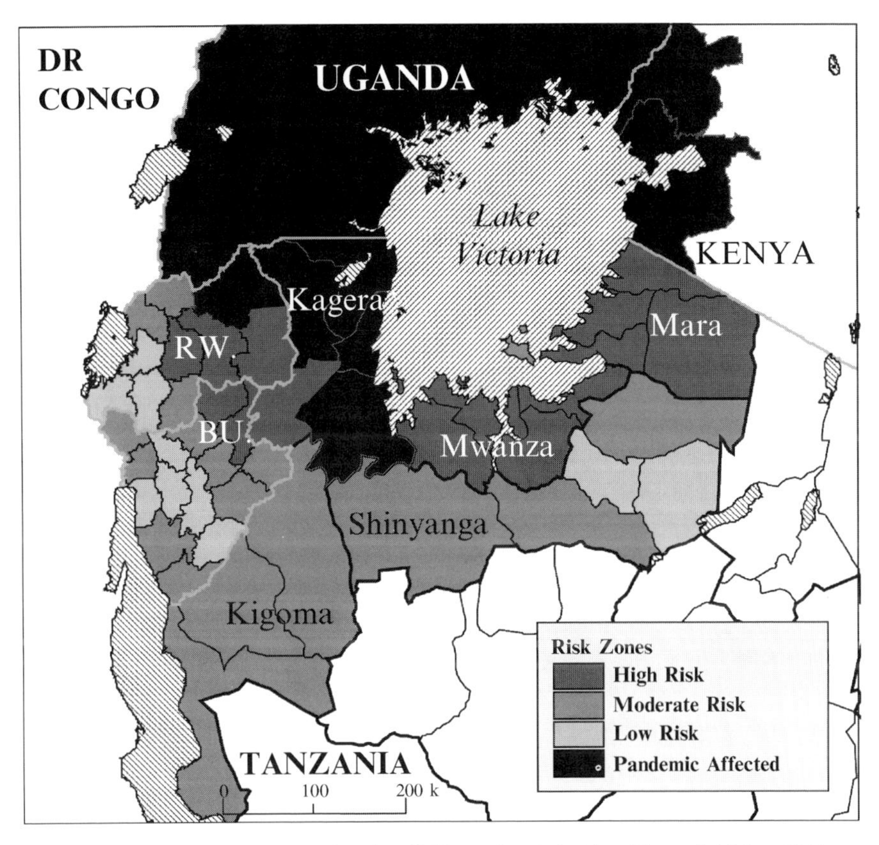

FIG 4. Risk assessment map for the CMD pandemic in East/Central Africa, 2001.

spread into eastern Gabon and that the affected area would be in the grassland zone of central eastern Gabon near to the town of Franceville (Fig. 5). Subsequent surveys confirmed both predictions (Legg *et al.*, 2004) and these findings led directly to the development of recommendations to limit spread of the pandemic through Gabon. This was particularly valuable because at the time of the survey, the incidences of EACMV-UG and severe CMD were very low, offering the possibility of initiating control measures at an early stage of epidemic development. Despite the relative remoteness of this location, the recognition of spread into Gabon represented the earliest stage at which this had been diagnosed in any of the countries affected by the pandemic.

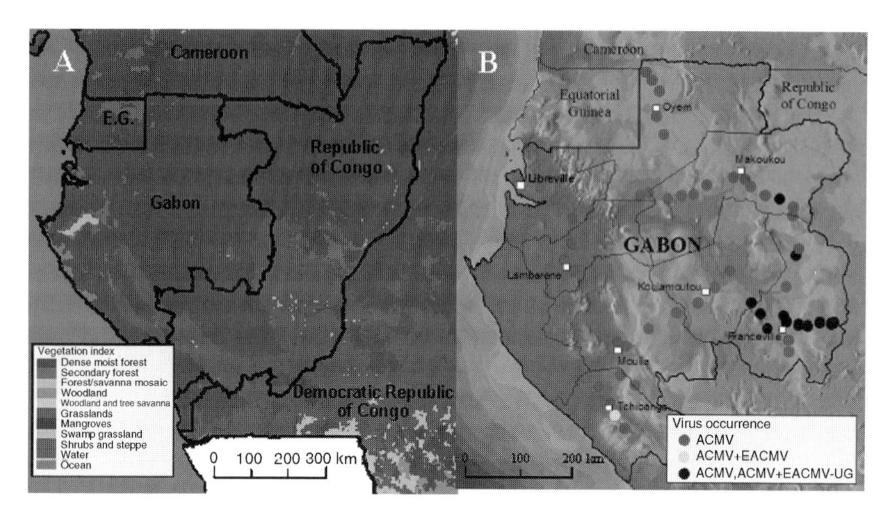

FIG 5. (A) Vegetation map for Central/West Africa used to predict spread of EACMV-UG into Gabon (Anon, 2005d). (B) Occurrence and distribution of cassava mosaic geminiviruses in Gabon, 2003 (Legg *et al.*, 2004). (See Color Insert.)

5. Monitoring Surveys and Impact Assessment

Another key function of monitoring surveys is to collect data that can serve as a pre-intervention baseline, which subsequently can be compared with data collected from the same area in an attempt to assess the impact of control programmes. The most successful example of this is central southern Uganda, where a CMD management programme based on the multiplication and distribution of resistant varieties was accompanied by an annual set of cassava pest and disease assessment surveys between 1998 and 2001 (discussed further in Section H).

C. Host-Plant Resistance Development and Deployment

1. Early Resistance Breeding Initiatives in Tanzania and Madagascar

From the earliest research into CMD, it was recognised that some cassava cultivars were more readily and more severely affected than others. Moreover, there was an understanding that wild relatives of crops are very often more resistant to pest and disease problems than the cultivated forms. With this background, cassava improvement programmes were initiated separately in the 1930s in Madagascar and in what is now Tanzania. Both programmes utilized existing knowledge of breeding to attempt to introgress resistance from both wild cassava relatives and cultivated cassava into target cassava germplasm (Cours, 1951; Jennings, 1957). In the East African regional cassava-breeding programme in what is now Tanzania, greatest success was achieved with crosses with *Manihot glaziovii* that were then triple back-crossed with cultivated cassava to produce progeny that combined the CMD-resistance trait of *M. glaziovii* with the edible storage roots of *M. esculenta*. Although this programme ended in the 1950s, some of the materials produced were carried to Nigeria, where resistance breeding continued under the Federal Research Programme in the 1960s (Beck, 1982). This work was expanded in the major new cassava germplasm development programme launched by the Nigeria-based International Institute of Tropical Agriculture (IITA) from the early 1970s (Hahn *et al.*, 1980).

2. *Africa-Wide Breeding Programme of IITA*

During the early period of the IITA breeding programme in the 1970s, efforts were focused on extending the Amani (Tanzania) work to develop varieties that were resistant to CMD, and also to some of the other key pest and disease constraints, the most important of which at the time was cassava bacterial blight (CBB) (*X. axonopodis* pv. *manihotis*) (Hahn *et al.*, 1980). Significant successes were achieved in deploying resistant germplasm in Nigeria, but progress was slower elsewhere, although many countries received sets of CMD-resistant germplasm in tissue culture form through IITA's continental distribution programme (Manyong *et al.*, 2000). Significantly, resistant varieties had been sent to Uganda in 1984, although this had coincided with a period when pest and disease constraints there were only of moderate importance.

3. *Host-Plant Resistance and the 1990s Epidemic of CMD in Uganda*

As the CMD epidemic first began to affect cassava production in Uganda in the early 1990s, initial attempts were made to address the problem by distributing uninfected material of 'local improved' cultivars, notably 'Ebwanateraka', 'Bao' and 'Aladu Aladu' (Otim-Nape *et al.*, 1997), although it quickly became apparent that these genotypes were susceptible to the severe epidemic CMD. Small collections of IITA-developed varieties were available, however, and these were multiplied rapidly and evaluated with farmers in epidemic-affected areas. Three varieties were selected and released officially in 1994 (Otim-Nape *et al.*, 1994a): TMS 60142 (released as Nase 1), TMS

30337 (Nase 2) and TMS 30572 (Migyera or Nase 3). A rapid programme of multiplication and distribution followed with considerable early success, and Nase 3 in particular was widely adopted by farmers, particularly in the eastern and northwestern districts, where cassava is primarily processed into flour. The Ugandan National Agricultural Research Organization (NARO) released six other varieties in 1999, which were IITA-derived genotypes or locally developed half-sib progeny obtained through seed from CMD-resistant maternal parents. The most important variety from this second release was named SS4 (released as Nase 4), and in view of its very high level of CMD resistance and generally low cyanogenic glucoside content, it was widely promoted in parts of the country where cassava is usually eaten following simple boiling rather than being processed into flour. Based on the successes of TMS 30572 and SS4 in Uganda, these became the main varieties multiplied and distributed in neighbouring Kenya and Tanzania following pandemic spread to these two countries.

4. Nature and Mechanisms of Resistance to CMD

Few quantitative studies of CMD resistance were available prior to the surge in interest in the disease that occurred following the outbreak of the epidemic in Uganda. However, exceptions were the studies of Colon (1984) in Ivory Coast, who assessed a wide range of genotypes, and Hahn *et al.* (1980) in which the performance of IITA-bred germplasm was contrasted to that of local material. Key facets of CMD resistance identified at this time included a reduction in the rate of infection and an overall reduced level of infection. A characteristic feature of the *M. glaziovii*-derived resistance was incomplete systemicity of the virus in infected plants, which led to the observed phenomena of recovery and reversion (Fargette *et al.*, 1994; Pacumbaba, 1985), both of which appeared to be enhanced by high temperatures (Gibson, 1994). Fargette *et al.* (1996) recognised a series of components of resistance to CMG infection. These included: resistance to the vector before infection, reduction in virus replication, inhibition of virus movement within the plant and decreased response of the plant to a given virus content as assessed serologically. Little is known about the molecular processes that underlie these mechanisms of resistance, although there is increasing evidence that other crop plants have developed defence responses to geminivirus infection that involve post-transcriptional gene silencing of virus gene products (Lucioli *et al.*, 2003).

Through germplasm development work at IITA in the early 1990s, it became apparent that many of the local landraces represented in

LEGG *ET AL.*, FIG 1. Symptoms of cassava mosaic and cassava brown streak diseases and direct damage caused by the whitefly vector (*B. tabaci*) of the viruses responsible. (A) Moderate leaf mosaic characteristic of ACMV infection, Sangmelima, Cameroon; (B) Severe symptoms caused by mixed EACMV-UG + ACMV infection, Franceville, Gabon; (C) Downturned petioles, stem and petiole lesions associated with EACMV-UG + ACMV infection, Bukoba, Tanzania; (D) Super-abundant *B. tabaci* adults, Namulonge, Uganda; (E) Chlorosis on shoot tip caused by *B. tabaci* adult feeding and sooty mould growth on lower leaves resulting from nymph honeydew secretion; and (F–H) *Cassava brown streak virus* infection in variety TME 204, Namulonge, Uganda ((F) Dry brown necrotic rot in roots; (G) Leaf symptoms; (H) Shoot dieback associated with severe response).

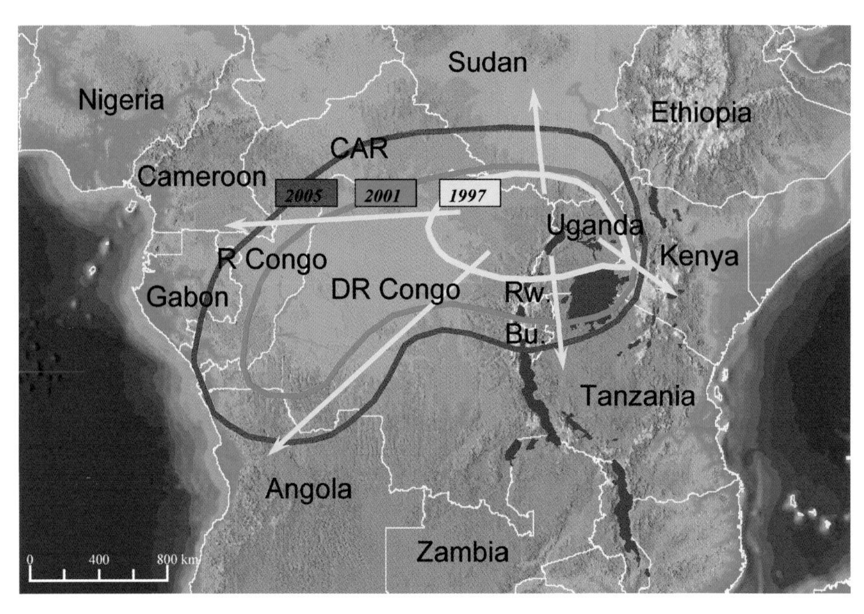

LEGG *ET AL.*, FIG 2. Estimated extent of the epidemic of the CMD pandemic in SSA in years 1997, 2001 and 2005. DR Congo, Democratic Republic of Congo; R Congo, Republic of Congo; Rw., Rwanda; Bu., Burundi.

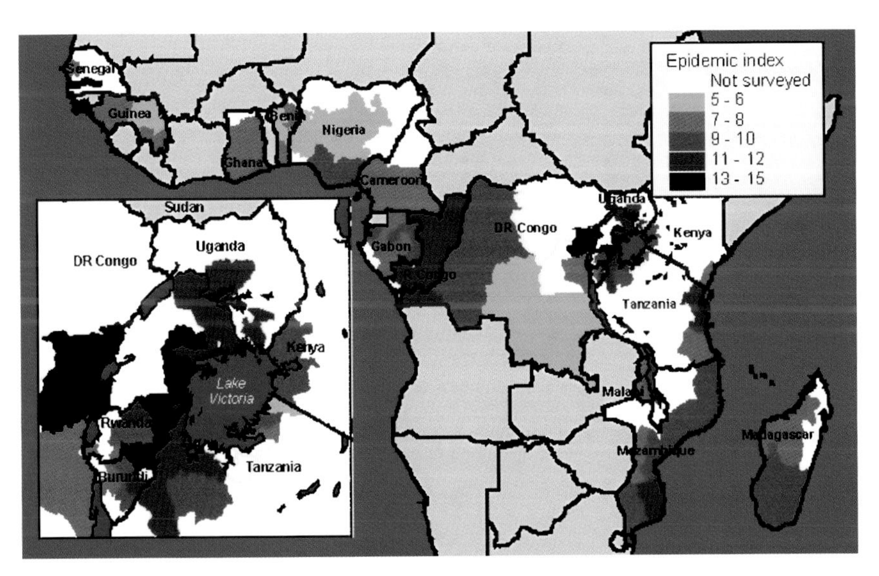

LEGG *ET AL.*, FIG 3. Epidemic index values for cassava mosaic disease in regions of SSA surveyed between 1998 and 2003.

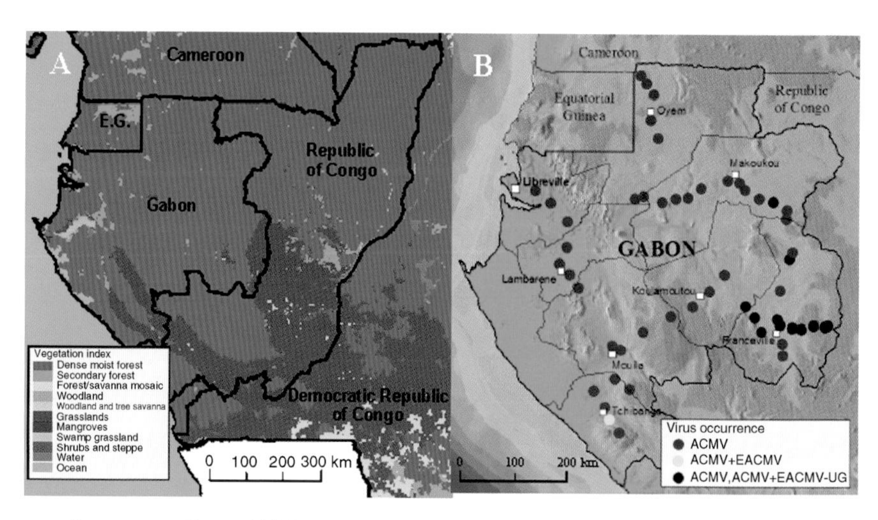

LEGG *ET AL.*, FIG 5. (A) Vegetation map for Central/West Africa used to predict spread of EACMV-UG into Gabon (Anon, 2005d). (B) Occurrence and distribution of cassava mosaic geminiviruses in Gabon, 2003 (Legg *et al.*, 2004).

IITA's germplasm collection, from various parts of West Africa (and designated the Tropical *Manihot esculenta* or TME group) had some resistance to CMD (Mignouna and Dixon, 1997). Significantly, these resistant materials also had many of the quality characteristics preferred by farmers (upright growth habit, 'sweet' taste, mealyness) that were absent in the *M. glaziovii*-derived material (within the bred tropical *Manihot* species or TMS group). Subsequent studies linked resistance in TME genotypes, the type example of which was TME 3, with the presence of a single dominant CMD resistance gene, referred to as *CMD2* (Akano *et al.*, 2002). Much of the breeding work of IITA and its national partners over the last decade has, therefore, involved developing germplasm that has a combination of *CMD2* and *M. glaziovii*-derived resistance (Dixon *et al.*, 2003). Additional sources of resistance continue to be discovered, however, and molecular marker techniques have been used to show that TME 7 has a CMD resistance gene or genes different from, but related to *CMD2* (Lokko *et al.*, 2005). Genomics approaches involving the development of expressed sequence tag (EST) libraries (Anderson *et al.*, 2004) and ultimately the complete sequencing of the cassava genome offer much promise for the identification of further CMD resistance genes and should open up the potential for resistance gene 'pyramiding'. Conventional breeding approaches have and will continue to be used in identifying and combining resistance genes, but increasingly transgenic approaches are likely to be pursued, as discussed further in Section IV.F.

D. Phytosanitation

Phytosanitation describes a number of primarily field-applied techniques that are used to reduce the incidence of disease in a crop. Three of these have been widely applied for the management of CMD in Africa, namely, the use of tissue culture, selection of disease-free stems for new plantings and removing diseased plants from within a crop stand (usually referred to as roguing). The value as well as the potential drawbacks of these methods have been discussed extensively (Thresh and Cooter, 2005; Thresh and Otim-Nape, 1994; Thresh *et al.*, 1998), but there is little published information either confirming or refuting their effectiveness. The main points of concern are that where inoculum pressure is high and susceptible cultivars are being grown, roguing can lead to almost complete removal of the crop, yet where a resistant variety is being grown, the slight yield losses suffered when plants are infected may mean that the losses incurred through roguing would exceed those resulting from infection. Because of these concerns, and the fact that persuading farmers to change their cultural practices is inherently difficult, use of phytosanitation measures has largely been confined to formal germplasm exchange and 'clean stock' multiplication programmes. Tissue culture, involving meristem tip excision, thermotherapy and virus indexing, provides a means to produce virus-tested plantlets that can then be transported across international boundaries without presenting a quarantine hazard (Frison, 1994). This approach has been used by IITA to move virus-tested plantlets of CMD-resistant varieties to CMD-affected countries. The broader strategic use of this method in CMD management programmes is discussed further in Section IV.G. Roguing and selection of CMD-free stems for planting have been widely used in resistant variety multiplication in all of the countries affected by the CMD pandemic (Legg *et al.*, 1999b). Although roguing and selection of planting material by farmers appear to have little role to play in areas affected by the pandemic, evidence from a post-epidemic situation in western Kenya (Mallowa, 2006) suggests that selecting planting material of local cultivars can provide yields that are similar to those of resistant varieties grown under similar conditions. Such a result is of particular significance since resistant variety multiplication programmes typically require 5–10 years before the majority of farmers have access to such planting material. Moreover, where material is available, farmers very often prefer to retain a proportion of their own local cultivars since these have specific desirable quality traits that are absent in the resistant material.

E. *Mild Strain Protection and Virus–Virus Interference*

1. *Mild Strain Protection*

A feature of the post-epidemic condition in parts of East and Central Africa, affected by the CMD pandemic, is the resurgence of local cultivars and a general reduction in the severity of CMD symptoms expressed. Three factors contributing to this effect are: the previously described reduction in frequency of mixed virus infections (Table I), the increased cultivation of more CMD-tolerant local cultivars and decreased virulence of EACMV-UG as mild strains have become frequent (Sseruwagi *et al.*, 2004b). Pita *et al.* (2001a) first demonstrated the occurrence of both mild and severe strains of EACMV-UG in Uganda, and showed that the differences in disease phenotype were due to sequence differences of only a few nucleotides. The practical importance of these differences was confirmed through the virus-specific yield loss studies of Owor *et al.* (2004a) in which EACMV-UG severe infection led to losses of 68% compared with 12% for EACMV-UG mild. Although mildly diseased cassava plants infected with EACMV-UG were frequent in post-epidemic Uganda (Sseruwagi *et al.*, 2004b), it was also noted that initially healthy plants typically expressed severe symptoms when they became infected. This observation suggested that EACMV-UG mild infection provides some kind of cross protection against severe strains that are, by contrast, readily able to infect healthy plants. Early studies on cassava had reported the inability of mild strains to protect against severe strains (Dade, 1930; Storey and Nichols, 1938). To determine if indeed mild strains of EACMV-UG were providing a form of cross protection, an experiment was carried out in central Uganda. Plants grown from initially CMD-free parents and plants initially infected with mild EACMV-UG were grown, and subsequent patterns of infection, symptom expression and tuberous root production were assessed (Owor *et al.*, 2004b). Plants grown from initially CMD-free parents developed more severe disease and yielded less than plants derived from mildly diseased parents. These results were complemented by screenhouse-based studies in which whitefly transmission tests were used to confirm the cross protection effect of mild strains of EACMV-UG (Owor, 2002). These results help to explain the resurgence of local cultivars in post-epidemic areas and also indicate that cross protection could offer an important novel approach for the management of CMGs. An important consideration in implementing such an approach, however, will be to determine where mild strain protection works and where it does not, in view of the diversity of virus occurrence. Further study will also be required to acquire an understanding of the molecular basis for the cross protection effect.

2. Defective Interfering Molecules

Defective interfering (DI) sub-genomic DNAs that arise primarily through initial or serial *in vitro* passage of geminiviruses in test plants such as *Nicotiana benthamiana* can moderate viral damage of the wild-type virus. A naturally occurring truncated form of CMG DNA-A (1525 nts), which upon sequence analysis has been shown to be a defective (df) form derived from EACMV, has been isolated and characterized in Tanzania (Ndunguru *et al.*, 2006). The 'missing' portions include the *AC2* and *AC3* genes on the complementary-sense strand, and the C-terminal portions of *AC1* as well as *AV2* and over 80% of the *AV1* (coat protein gene) on the virion-sense strand. The sub-genomic DNA has, nevertheless, retained all the *cis*-acting elements necessary for its maintenance. Phylogenetic comparisons placed the molecule close to mild and severe isolates of EACMV-UG2 (>95% nucleotide sequence identity). Biolistic inoculation of the infectious df DNA-A clone 15 (df DNA-A 15) with EACMCV showed symptom amelioration as compared to the plants singly infected with EACMCV and there was an accumulation of the df DNA-A 15 in systemically infected leaves 14 days after inoculation. The data indicate that 1.5 kbp df DNA-A 15 is a defective DNA that can modulate symptoms of the wild-type geminivirus. Investigation of a possible trans-encapsidation and transmission of the df DNA-A 15 under field conditions as well as its mechanism of symptom modulation could shed light on its possible role in symptom variability associated with the CMD pandemic and also on any involvement in the mild strain protection phenomenon described previously. Even where these effects are not attributable to sub-genomic molecules, mechanisms of symptom amelioration are likely to be shared, and further study of the molecular mechanisms behind these responses is warranted.

F. Cassava Transformation

In several crops, transgenic plants resistant to an infective virus have been developed by introducing a sequence of the viral genome into the target crop by genetic transformation. Virus-resistant transgenics have been developed in many crops by introducing either viral coat protein or replicase gene-encoding sequences, which interfere with one or more essential steps in the replication cycle of the virus. Replicase (Rep) protein-mediated resistance against a virus in transgenic plants was first shown in tobacco against *Tobacco mosaic virus* in plants containing the 54 kDa putative *Rep* gene (Golemboski *et al.*, 1990). Geminiviruses have been used to study fundamental aspects of RNA interference (RNAi) and virus-induced gene silencing (Chellappan *et al.*, 2005; Turnage *et al.*, 2002). Proof of the concept that RNAi can be engineered to target geminiviruses effectively has been documented in transient assays for ACMV. Vanitharani *et al.* (2003) reported successful generation of a transgenic cassava (line Y85) resistant to ACMV and two other related CMGs. Detection of the transgene-derived siRNAs and extremely low levels of the transgene product (the truncated Rep protein from ACMV) in this line suggested that RNA silencing accounts for the broadly active resistance.

The greatest resistance recorded in experiments done in containment facilities with artificial methods of inoculation has been obtained using the so-called *Rep* gene of ACMV (Chellappan, P., unpublished data). The susceptible genotype TMS 60444 was transformed and near-immune plants were regenerated. In addition, these plants are resistant to other CMG species, including EACMCV and *Sri Lankan cassava mosaic virus* (SLCMV), indicating a wide range of protection, which is a key requirement in view of the molecular variability of known CMGs. Furthermore, these plants have shown a very high level of resistance to the synergistic mixture of ACMV and EACMV-UG, indicating that the strategy employed is very effective against the natural mixture causing the pandemic. It appears that the most resistant plants are using the post-transcriptional gene silencing mechanism and the broad spectrum of protection to other virus species is attributed to common short sequences between their respective Rep protein genes (Chellappan *et al.*, 2004). Plants so transformed are currently being tested under screenhouse conditions in western Kenya.

An alternative approach used anti-sense RNA technology (Zhang *et al.*, 2005) in which targets for the anti-sense interference were the mRNAs of *C1*, *AC2* and *AC3* of ACMV. Assays of virus accumulation in transgenic plants revealed reduced or inhibited replication of ACMV. In a third approach, a hypersensitive response to infection is elicited by transforming TMS 60444 with the bacterial *barnase* and *barstar* genes from *Bacillus amyloliquefaciens*, controlled by the ACMV A bi-directional promoter (Zhang *et al.*, 2003). Reductions of viral replication of 86–99% were demonstrated, when comparing leaves of untransformed and transgenic plants. While this strategy is still at the greenhouse testing stage, both the *Rep* and anti-sense RNA strategies have yielded successfully regenerated cassava lines ready for field testing in areas devastated by the CMD pandemic.

In another development, a toxin gene, *dianthin*, was placed downstream of a transactivatable geminivirus promoter from ACMV. When transgenic *N. benthamiana* plants were inoculated with ACMV, *dianthin* was synthesized only in the virus-infected tissues, where it inhibited virus multiplication (Hong *et al.*, 1996). This approach also warrants further consideration for its potential value in CMD management. Although further refinements of the group of transgene-derived CMG-resistance strategies are required before effective field implementation, these tactics offer great promise. Most importantly, they could provide a valuable complement to conventional control tactics and could effectively allow farmers to continue to grow locally preferred cultivars (albeit transformed for virus resistance). In view of the difficulties that have been experienced in some pandemic-affected areas in addressing farmer quality preferences, this represents a very important development. Biosafety of newly developed transgenic crops remains a concern in much of SSA, however, and legislation governing the introduction, testing and release of transgenic plants is still being developed in many of the pandemic-affected countries. Kenya is currently the only African country to have tested transgenic CMG-resistant varieties of cassava under screenhouse conditions, but even here, field testing of such material has yet to be approved. Despite this initial caution, however, trends in the larger developing countries suggest that early concerns will be overcome and that transgenic technologies will be widely adopted. These novel approaches to CMD management appear, therefore, to have much potential in the near future, a prospect that certainly justifies further investment in the technology.

G. Integrated Management Programmes

1. Early Successes in Uganda

Management of any major disease epidemic covering a large area requires considerable logistical organization and the participation of diverse stakeholders. The first country-level model for the approach to managing the CMD pandemic was developed in Uganda and coordinated by the National Cassava Programme (NCP) of NARO. An informal network of partners was established, including the NCP research team, government extension officers, agricultural workers from non-governmental organizations and farmer groups (Otim-Nape *et al.*, 1994a). Following on-farm evaluations, the best-performing CMD-resistant varieties were identified and multiplied, firstly at institutional sites (mainly comprising research stations, district farm institutes, prisons and other government institutions) and then closer to farming communities at collective multiplication sites. Various approaches were used in different projects that targeted CMD management, and there were varying degrees of success (Otim-Nape *et al.*, 2000). Programmes that focused on the multiplication of CMD-resistant varieties were generally effective, although the degree of adoption by farmers was influenced greatly by the way in which cassava is utilized; results were much better in areas where traditionally cassava is processed to make flour. During the early years of CMD management in Uganda, the TMS varieties were emphasized but some of these were limited in their acceptability by quality issues, and from the late 1990s there has been a greater focus on the promotion of TMS × TME crosses or TME landraces. Widely promoted TME clones include TME 14 and TME 204, while some of the most popular TME × TMS crosses are I92/0067 (=Akena) and I92/0057 (=Omongole). Although these varieties combine the qualities of local farmer-preferred cultivars with good levels of resistance, drawbacks include their apparent attractiveness to whiteflies and the vulnerability of some of them (notably TME 204) to the other main cassava-infecting virus in Africa, *Cassava brown streak virus* (Genus: *Ipomovirus*; Family: *Potyviridae*). These concerns are discussed further in Section V. In addition to the resistant variety multiplication programmes, training of extension workers and farmers and awareness-raising through printed materials and the media (radio and television) were important components of CMD management in Uganda, as described in detail by Otim-Nape *et al.* (2000). The approaches developed there were of crucial importance in developing control programmes elsewhere.

2. *Regional CMD Pandemic Mitigation Programme*

As the CMD pandemic expanded to cover large parts of the cassava-growing areas of the Lake Victoria Basin of East Africa during the second half of the 1990s, it became apparent that a co-ordinating programme was required that would help to transfer some of the lessons learnt in Uganda, provide a regional forum for sharing experiences and information on CMD management and create 'leverage' to seek external support. It was recognised that various forms of such support were needed, including cassava germplasm (through IITA, Nigeria), finance (from donors) and technical 'backstopping' (from centres with specialist expertise). Sustained donor support was essential, and a regional team co-ordinated through IITA, was successful in attracting financial backing for a regional pandemic mitigation programme from the Office of United States Foreign Disaster Assistance (OFDA), which still continues. The regional programme complemented other country-focused CMD management initiatives financed by United States Agency for International Development (USAID), the Crop Protection Programme (CPP) of the UK's Department for International Development (DFID), the Gatsby Charitable Foundation, the Canadian International Development Research Centre (IDRC) and the Rockefeller Foundation. The regional team that was established through the OFDA-supported project linked researchers, extensionists, plant protection and quarantine staff, NGOs and farmers from the participating countries of Uganda, Kenya and Tanzania, although the participation later broadened to include both the ROC in 2001 and Burundi in 2002. The overall strategy of this on-going programme comprises the following principal components: monitoring and forecasting CMD pandemic expansion, germplasm diversification and exchange, multiplication and distribution of CMD-resistant varieties; and training of researchers, extension workers and farmers in CMD management techniques. Country and regional representatives participate in a stakeholder group that co-ordinates the regional CMD mitigation programme through consultative reviews and planning. Within countries, activities are co-ordinated similarly through national steering committees, which have a comparable make-up of participating institutions.

a. Monitoring and Forecasting CMD Pandemic Expansion The programme has been the first to implement a regional CMD-monitoring system. The technical approaches were described earlier (Section IV.B). An important facet of the organization has been the regional-network structure developed, as described in the previous paragraph. This has facilitated the rapid transfer of information, the provision of technical support, where required, to country-level monitoring teams and the channels for the communication of results of the work. This final step has been particularly important, since the early provision of information on newly threatened areas or new problems is vital to the success of controlled implementation work. Efforts were made to use a different approach to monitoring pandemic spread in which farmer groups in advance of the epidemic 'front' were trained in what to expect and asked to provide information on any new spread into their farming areas. Weaknesses in the communication systems from farm, through extension and up to researchers, however, meant that this approach was unworkable. Any future attempt on these lines must pay particular attention to improving communication pathways.

b. Germplasm Diversification and Exchange The rapid provision of appropriate CMD-resistant germplasm for areas affected by the pandemic is key to successful management. The spread of EACMV-UG into countries neighbouring Uganda immediately made this task more complex since known resistant stocks were present in Uganda and movement of germplasm across borders raised phytosanitary concerns. The first affected country to face this issue was Kenya. It was soon realized that the simplest, quickest and most effective way to address pandemic spread there was to introduce the CMD-resistant varieties being used in Uganda in sufficient quantity to facilitate a multiplication and dissemination programme. An 'open quarantine' protocol was developed to introduce substantial quantities of these varieties and yet avoid the risk of inadvertently introducing quarantine pests/pathogens (including EACMV-UG). This enabled the introduction of substantial quantities of cuttings of TMS 30572 and SS4 to a 2 ha fenced open quarantine site at Alupe, western Kenya, in 1997. This plot was maintained under regular surveillance for pests and diseases for a year, after which the stems were harvested and used to initiate the multiplication programme at a series of primary multiplication sites distributed across the pandemic-affected area. This at that time comprised all of Western Province and the northern part of Nyanza Province. The development of the open quarantine site also facilitated the introduction of >600 cassava clones with a background of CMD resistance from the Serere (Uganda)-based regional cassava germplasm development programme of the East African Root Crops Research Network (EARRNET). EARRNET, established as an IITA-executed regional network, subsequently used the Serere-based cassava germplasm collection as the key central 'source' of CMD-resistant germplasm for the entire region, and similar and repeated transfers were made to Tanzania, Rwanda, Burundi and southern Sudan, in addition to Kenya. In addition to the use of open quarantine sites, another key element of the regional germplasm introduction programme was the transport of virus-indexed tissue culture plantlets of CMD-resistant varieties from IITA, Nigeria. As part of this programme:

- Small numbers of tissue-culture plantlets were sent to Kakamega, western Kenya through the Muguga-Nairobi Plant Quarantine Station of the Kenya Plant Health Inspectorate Service.
- 10,000 tissue culture plantlets were sent via Dar es Salaam to Ukiriguru Research Institute, Mwanza, Tanzania, on the southern shore of Lake Victoria, an area ahead of the CMD pandemic.
- 5000 tissue culture plantlets were delivered to Brazzaville, ROC. This represented the first supply of CMD-resistant varieties of known provenance to this country, which by that time was almost entirely affected by severe CMD.

These initiatives had mixed success. The transfer of plantlets to Kenya worked well, although in terms of scale alone, the open quarantine introduction proved to be much more effective and lasting in its impact. Both of the large-scale introductions to Tanzania and ROC suffered substantial losses arising in part from the transport of the fragile plantlets over long distances, and partly also due to difficulties encountered during hardening off and field planting of the plantlets. Nevertheless, where CMD-resistant material was not available previously, as in ROC, the plants that were established provided a vital first set of germplasm for subsequent CMD management.

Following the introduction of CMD-resistant germplasm, evaluation programmes were initiated, starting in many cases from the open quarantine site itself. A few of the best-performing introduced cassava clones were typically then 'fast-tracked' to on-farm evaluations with farmers to accelerate the germplasm selection process, and identify farmer-acceptable CMD-resistant materials for multiplication and dissemination as quickly as possible. Fourteen clones were fast-tracked in Kenya, 10 in Tanzania, 7 in Burundi, 20 in DRC and 17 in ROC.

c. Multiplication and Dissemination of CMD-Resistant Varieties
Efforts to multiply and disseminate cassava germplasm rapidly in pandemic-affected countries of East and Central Africa followed similar approaches to those already implemented successfully in Uganda (Otim-Nape *et al.*, 2000). Typically, a three-tier system was used in which 'nuclear' stocks of resistant material were multiplied initially at primary multiplication sites, before being disseminated to secondary then tertiary sites. Primary sites were mainly at government-owned institutions, such as research stations, district farm institutes and prisons. These sites typically ranged in size from a few to tens of hectares. Secondary sites were commonly established by farmer groups, community-based organizations, or large private farms (usually up to 2 ha in size) and tertiary sites involved individual farmers and tended to be small (<2 ha). Diverse systems were developed for the various components of multiplication, which included the partitioning of costs and labour for management, the handling of planting materials during planting, harvest and distribution, the agronomic methods used and the system of distribution of the harvested cuttings. Although irrigation can be of considerable value in allowing almost continuous production in rapid multiplication systems, this was rarely used because it was seldom available. An exception was the Nyakasanga location near Mwanza, Tanzania, where CMD-resistant germplasm introduced from IITA, Nigeria was multiplied rapidly at an irrigated site. This then enabled the Tanzania team to multiply and disseminate resistant material from a site ahead of the pandemic 'front', while material introduced from Uganda was being multiplied from within the pandemic-affected region.

There are many examples of successful approaches to the multiplication and dissemination of CMD-resistant germplasm, but one of the most effective programmes in Tanzania was that of an NGO, Norwegian Peoples' Aid (NPA) in Ngara District, Kagera Region, bordering Rwanda and Burundi. A 5 ha intensively managed nursery plot has been used to rapidly multiply CMD-resistant material that was then distributed to groups of farmers. Within these groups, up to 20 farmers have planted contiguous multiplication plots of 0.2 ha each and have agreed to contracts with NPA in which they return to NPA two-thirds of the cuttings produced over a two-year period and keep the remaining one-third for themselves. From an initial group of 5 villages in

2003, NPA increased the coverage of the programme to 45 villages by February 2005. The block approach appears to have been more effective than alternatives tried elsewhere, but clearly benefits from the fact that in Ngara suitable land is more readily available than it is in other parts of the region, such as northern Kagera in Tanzania, Burundi and western Kenya.

Multiplication has been greatest in those countries affected earliest. Cassava production in Uganda is currently greater than before the pandemic (FAO, 2006). Similar successes have been achieved in western Kenya, where production in 2004 was only a little less than before the pandemic and almost double that during the worst-affected period in 1997. In Tanzania, more than 1000 ha of resistant varieties had been established by mid-2005. Progress in Burundi, Rwanda and ROC has been slower, however, mainly because activities began later. Problems with civil insecurity initially hindered progress in DRC, but major CMD mitigation programmes being led by IITA and FAO have been operating since 2000, and substantial areas of CMD-resistant varieties have been produced, particularly in the Bas Congo, Kinshasa, Bandundu and South/North Kivu Provinces.

d. Training Increasing the knowledge of all stakeholders affected by the CMD pandemic has been an important component of the regional CMD pandemic management programme. Researchers have been trained in the technical elements of CMD: the ecology of the viruses responsible and their whitefly vector, diagnostic techniques and approaches to CMD management. Practical, field-based training initiatives have been run for agricultural workers of government extension systems and NGOs. Knowledge of affected farmers has also been improved through *in situ* training, typically at multiplication or on-farm evaluation sites, as well as through farmer exchange visits which have involved groups from Uganda, Kenya and Tanzania. Most countries have also attempted to raise awareness by publishing information about the pandemic and ways in which to control the problem via radio and television. Radio has been a particularly important medium since it is almost universally accessible, even in the remotest locations.

3. Pre-Emptive CMD Management Programmes

Much of the CMD pandemic management activity in East and Central Africa has been in response to an existing problem. None of the countries has been successful in disseminating CMD-resistant varieties within communities at risk but as yet unaffected by the pandemic, despite significant efforts to do so in Kenya and Tanzania. A fundamental limiting factor behind this phenomenon is the fact that the generally conservative farming communities are usually not prepared to switch to different cassava varieties, while those that they already have and are accustomed to using continue to provide acceptable yields. The spread of the pandemic into the western part of Central Africa, however, has given rise to concern in Nigeria, Africa's largest cassava producer, of the potential social and economic consequences of the 'arrival' of the pandemic there. To ensure that resistant germplasm is available before such spread occurs, a major 'pre-emptive' pandemic mitigation programme was initiated in 2003 targeting the 12 states considered to be most vulnerable in the south and southeast of the country. In order to overcome potential unwillingness of farmers to adopt new varieties prior to the impact of the pandemic, this programme aims to develop commercial opportunities for processed cassava products that should lead to expanded markets and increased demand. This effort is being enhanced significantly through major governmental interventions. These include a 'Presidential Initiative', which targets the establishment of cassava industries for export promotion and the introduction of a minimum use of 10% cassava flour by bread producers (Anon, 2005c).

H. Impact of Management Initiatives

Considerable success has been achieved through CMD pandemic mitigation activities, although experience from Uganda has shown that the recovery period is protracted and takes at least 10 years. Despite the evident benefits of CMD management, there is currently no impact assessment in the published literature. An examination of production statistics, such as those of FAO (2006), should provide some indications of declines in production associated with spread of the pandemic, and subsequent increases arising from the uptake of CMD-resistant varieties. Unfortunately, the inadequate systems that exist for collecting these data in East and Central Africa mean that these estimates are unreliable and not suitable for use. Monitoring survey data do provide an alternative means of estimating impact, however, as varieties are recorded from surveyed fields, thereby, allowing assessments to be made of changes in frequency of cultivation of particular varieties, including CMD-resistant materials introduced through management programmes. Using this approach, an estimate was made of the impact of CMD-resistant variety multiplication in six central Ugandan districts that were targeted through a USAID-funded project that ran from 1998 to 2001. During this period, annual surveys of more than 270 fields in the 6 districts, revealed a series of changes. Intensity of cultivation increased by 16%, equivalent to an increase in area of more than 11,000 ha, and the proportion of farmers' fields in which CMD-resistant material predominated increased from 17% to 35%. On-farm trials conducted in the project districts gave average yields for local cultivars of 7.4 t/ha compared to 15.5 t/ha for resistant varieties, results which are comparable to data collected from a later survey in 2003 (Sserubombwe *et al.*, 2005a). Considering the benefit to producers alone, these changes give an estimated benefit of more than US$ 22 million, based on a conservative price for fresh cassava of US$ 100 per tonne.

The degree of impact differs greatly between areas, largely due to differences in the acceptability of introduced CMD-resistant varieties to local farming practices and patterns of cassava utilization. Another important determinant is the relative importance of cassava to the affected community. Where cassava is not the main food crop, as in parts of southwest Uganda, Bukoba district in northwestern Tanzania and the high-potential maize-growing areas of western Kenya, adoption of CMD-resistant varieties is slower and farmers often continue to plant diseased local CMD-susceptible cultivars. This is of concern as sustained high incidences of CMGs and the continued presence of high-whitefly populations provide ideal conditions for the emergence of novel and potentially resistance-breaking virus recombinants, which highlights the importance of a sustained focus on CMD control efforts even after the initial crisis of the epidemic condition has passed.

V. New Threats

A. *Super-Abundant* B. tabaci

Increased populations of *B. tabaci* have been a general feature of the CMD pandemic, as described in Section II.A.2. Initial studies seemed to suggest that populations were particularly high at the 'front' of the pandemic and then declined in areas affected earlier, as the area under cassava was reduced (Legg and Ogwal, 1998; Otim-Nape *et al.*, 1997). Experience from Uganda, however, has shown that this decline seems to be temporary, and that subsequently there has been a sustained increase in *B. tabaci* abundance. Data collected from a sequence of 50 × 50 plant arrays of local cultivar 'Bao', each planted in exactly the same manner over the period 1992–1994 and again in 1998 and 1999, illustrate the magnitude of this population change (Fig. 6). Similar

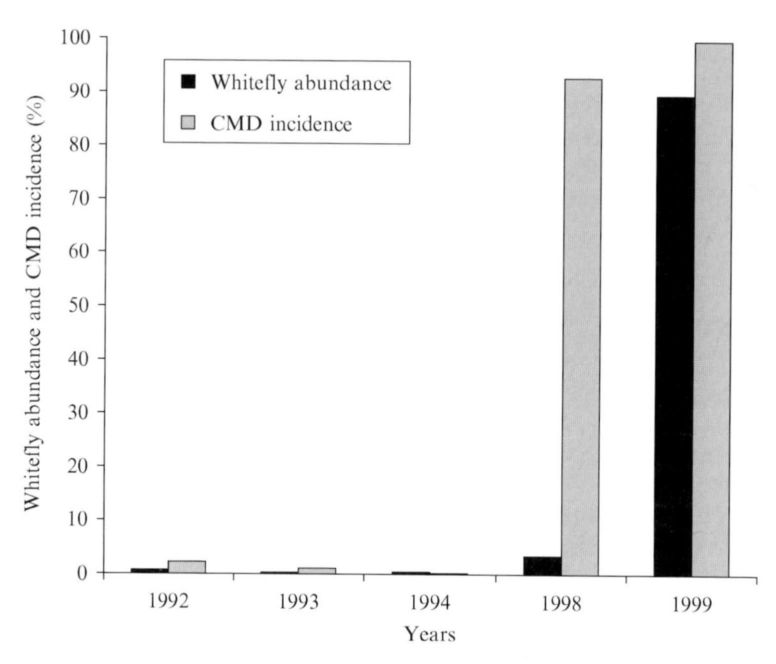

FIG 6. CMD incidence and abundance of *B. tabaci* whiteflies recorded from initially CMD-free plots of the local CMD-susceptible cultivar 'Bao' planted at Namulonge, Uganda from 1992–99 (CMD incidence was the percentage of plants infected with CMD at the end of each trial, 12 months after planting; whitefly abundance was the mean number of *B. tabaci* adults on the top five leaves of a representative shoot of plants sampled weekly over a period of 6 months from 2–8 months after planting).

sustained increases in the abundance of *B. tabaci* have been reported throughout other pandemic-affected parts of East and Central Africa.

In addition to causing rapid spread of CMGs to local susceptible cultivars and threatening resistance breakdown for improved CMD-resistant varieties, super-abundant populations of *B. tabaci* are also causing physical damage to cassava. Symptoms (Fig. 1) include a yellow chlorotic mottling on young newly emerged leaves resulting from feeding by *B. tabaci* adults, 'sooty mould' on lower leaves caused by fungal growth on 'honeydew' excreted by whitefly nymphs and a general reduction in plant growth caused by the compound effect of the damage. Preliminary analysis of yield loss trials indicates that losses resulting from whitefly physical damage alone can be 50% for some of the CMD-resistant varieties being promoted (Legg, J., unpublished data). Fortunately, losses appear to differ considerably between varieties, offering promise for the identification and use within breeding programmes of sources of whitefly resistance.

It is crucial to understand why such a major change has occurred in the numbers of *B. tabaci* on cassava. Experience elsewhere with other pathosystems has shown that increased populations of *B. tabaci* were linked to the appearance of new biotypes or strains of the vector (Bedford *et al.*, 1994; Brown and Bird, 1992). This occurred particularly in areas where *Bemisia* whiteflies were previously unimportant. For example, in the southwestern United States, the B biotype of *B. tabaci*, an introduction from the Middle East (Brown, 2000; Costa and Brown, 1991; Costa *et al.*, 1993) increased in distribution and abundance, and displaced the 'local' A biotype that occurred previously (Costa *et al.*, 1993). The B biotype colonized a large range of plant species (Cock, 1993), leading to outbreaks of previously undescribed begomovirus diseases in the Americas (Brown, 2001; Brown *et al.*, 1995; Morales, this volume, pp. 127–162).

It has been recognised for many years in Africa that genetically distinct populations of *B. tabaci* occur and colonize different crops. Burban *et al.* (1992) in the Ivory Coast, used isozyme-based diagnostic tools to demonstrate the occurrence of a cassava-associated biotype of *B. tabaci* that was more or less restricted to that crop. In Uganda, a similar approach was used by Legg *et al.* (1994) to detect polymorphisms amongst cassava-associated whitefly populations from different locations along a CMD epidemic transect using esterase profiles. However, the high degree of genetic variability in the whiteflies precluded the identification of distinct genotypic groups. Further studies of *B. tabaci* collected from cassava in pandemic-affected parts of Uganda have provided definitive evidence on several key questions. It was shown that *B. tabaci* collected from pandemic-affected regions mated readily and produced fertile offspring with *B. tabaci* collected from cassava elsewhere in Africa (Maruthi *et al.*, 2001). By contrast, mating was not successful with *B. tabaci* collected on sweet potato, even where these populations were sympatric (Maruthi *et al.*, 2004b). Similarly, mating was unsuccessful with *B. tabaci* originating from cassava in India. Furthermore, studies of the transmission of different African CMGs (ACMV and EACMV-UG) by cassava *B. tabaci* populations collected from different parts of Africa revealed relatively minor and insignificant differences (Maruthi *et al.*, 2002). However, none of these findings helps to explain the marked increases in *B. tabaci* abundance associated with the pandemic.

Two general hypotheses have been proposed for the change in *B. tabaci* populations in pandemic-affected areas. The first is that changes in the chemical composition of CMD-infected plants lead to increased populations of whiteflies colonizing them, resulting in more rapid reproduction and a general increase in populations (Colvin *et al.* 2004, this volume, pp. 419–452). Evidence has been adduced for behavioural changes in *B. tabaci* populations on CMD-diseased versus healthy plants (Omongo, 2003), but there is as yet no definitive proof of a causal link between CMD and increased whitefly fitness. This proposal is further confounded by the fact that some of the highest populations of *B. tabaci* are recorded on CMD-free resistant varieties, as observed commonly throughout the pandemic-affected zone. A second hypothesis to explain *B. tabaci* super-abundance is the appearance of a genetically distinct and fitter cassava biotype in pandemic-affected areas. An extension of this view is that the pandemic could be a consequence of the spread of a 'new' fitter *B. tabaci* biotype and that the recombinant virus, EACMV-UG, is simply a fortunate beneficiary. Accordingly, studies were conducted to investigate the genetic variability of *B. tabaci* on cassava in Uganda using the mitochondrial cytochrome oxidase I (mtCOI) marker (Legg *et al.*, 2002). Two genotype clusters were detected along each of three sampling transects that ran north–south across the epidemic 'front'. The clusters were designated Uganda 1 (Ug1) and Uganda 2 (Ug2), and diverged by *c.* 8% (Legg *et al.*, 2002). At the time that Ugandan populations were sampled in 1997, the Ug2 genotypes, which were shown by mismatch analyses to have the characteristics of an 'invasive' population, were associated with the CMD epidemic. The Ug1 genotypes, a non-invasive population that was considered to be indigenous, occurred primarily 'at' and 'ahead' of the epidemic 'front' (Legg *et al.*, 2002). Subsequent observations in 2002 (Sseruwagi, 2005) made after the expansion of the CMD epidemic to all Ugandan cassava-growing areas (Sseruwagi *et al.*, 2004b), confirmed the occurrence of both Ug1 and Ug2 on cassava in the country. Although there had been a distinct epidemic-associated distribution of whitefly genotypes in 1997, this pattern had been lost in 2002, as the two types were more or less randomly distributed with Ug1 predominating, as it made up 83% of the population. The reasons for and mechanisms behind this change remain unclear, although it is recognised that these genotype clusters can interbreed and it is also known that endosymbiotic bacteria can have important effects on *B. tabaci* biology (Zhori-Fein and Brown, 2002), including influencing the pattern of mating success (De Barro, 2005; De Barro and Hart, 2000).

Results indicate that *B. tabaci* genotypes carrying the Ug1 MtCO1 marker now predominate in much of the pandemic-affected zone of East and Central Africa (Legg, J., unpublished data) and retain the super-abundant 'invasive' characteristic wherever they occur throughout this region. This population is less readily distinguished from neighbouring populations in areas yet to be affected by the pandemic, as Ug1 is closely related to southern African genotypes (Legg *et al.*, 2002). However, an important feature retained is the very narrow within-population variability. Collections of *B. tabaci* adults from diverse pandemic-affected locations of Burundi revealed them to be clonal with respect to MtCO1 (Legg, J., unpublished data). Even though genetic uniformity is a consistent feature of super-abundant *B. tabaci* in the pandemic zone, it is difficult to 'track' whitefly genotypes using molecular markers in view of the ready mating and genetic exchange that occurs between populations. This could either be the result of the epidemic-like transfer of a gene or group of genes conferring super-abundance or the consequence of CMD-infected host-plant synergy, or both. Further studies are required to determine which situation pertains.

In view of the major change in the status of *B. tabaci*, control options targeted directly at this whitefly species have become much more important than earlier when populations were relatively low. In Latin America, sources of resistance have been identified against the whitefly, *Aleurotrachelus socialis* (Bellotti and Arias, 2001), which causes direct damage to cassava. Efforts are underway both to test the effectiveness of these resistance sources to African *B. tabaci* and also to introduce wild cassava relatives that may provide new resistance sources. Furthermore, the variability apparent in responses of African cassava germplasm to physical attack from *B. tabaci* suggests that these materials may also have sources of resistance that could be exploited through breeding.

There is also considerable interest in the potential for augmenting existing biological control as a means of reducing *B. tabaci* populations on cassava. A substantial body of knowledge has been generated on the major groups of natural enemies of whiteflies on cassava (Fishpool and Burban, 1994; Legg *et al.*, 2003; Otim *et al.*, 2004, 2006) and some of the work is examining the behaviour of these groups in order to identify potential interventions that could lead to enhanced natural enemy activity and, therefore, improved whitefly control (Otim, M., and Asiimwe, P., personal communication). Given the abundance of *B. tabaci* populations in pandemic-affected areas, no single whitefly control tactic is likely to be effective in providing sustained reductions in population sizes. This highlights the need for an integrated pest management (IPM) approach to address this dual pest-virus vector problem, which should also complement existing virus control strategies.

The deployment of IPM-oriented approaches is currently being promoted through a multi-institutional project targeting whitefly and whitefly-transmitted virus 'hotspots' in Latin America, Africa and Asia and known as the 'Tropical Whitefly IPM Project' (Anderson, 2005; Anderson and Morales, 2005). The primary focus of this project in Africa is the use of IPM to manage whitefly and CMD (Legg, 2005).

B. Cassava Brown Streak Disease

The only virus of economic importance to cassava production in Africa, other than the CMGs, is *Cassava brown streak virus* (CBSV). This RNA *Ipomovirus* causes a disease of cassava that has been known for many years, having first been described from northeastern Tanzania (Storey, 1936). The most economically important symptom of cassava brown streak disease (CBSD) is a corky brown necrotic rot in tuberous roots (Hillocks *et al.*, 1996; Nichols, 1950) (Fig. 1). Surveys conducted in the last decade have confirmed earlier observations that CBSD was largely restricted to coastal East Africa (Tanzania, Kenya, Mozambique) and Malawi (Hillocks *et al.*, 1999, 2002; Legg and Raya, 1998) and that the disease did not spread at altitudes of above *c.* 1000 m a.s.l. (Hillocks *et al.*, 1999). While the disease causes serious yield loss (Hillocks *et al.*, 2001), there was no evidence of the kind of expansion in geographical range that has been such a feature of the current CMD pandemic.

CBSD was first recorded from central Uganda (above 1000 m a.s.l.) in 1935 (Jameson, 1964), but was apparently eradicated through a vigorous phytosanitation campaign and not subsequently observed until 1994 when mild infections at a single location near to Entebbe were reported (Thresh *et al.*, 1994a). In the second half of 2004, however, the first informal reports were made of the widespread occurrence of CBSD-like symptoms in Uganda (Alicai, T., unpublished data). Subsequent laboratory tests confirmed the diagnoses. Significantly, symptoms were most apparent in CMD-resistant varieties of the TME group. Worst affected were TME 204, in which both leaf and root symptoms were seen (Fig. 1), and TME 14, but many other CMD-resistant varieties were also affected to lesser degrees and also some local CMD-susceptible cultivars. Extensive surveys through south-central Uganda showed the problem to be widespread, although largely confined to a few improved varieties. Nevertheless, it was considered that infection was spread over too great an area for a zero-tolerance eradication programme to be feasible. The first major consequence of this development is that it seems that CBSD has become established in

Uganda, and will require routine but sustained management, a situation that is complicated by the fact that even in coastal East Africa, effective and stable sources of host-plant resistance have still to be identified. Of greater concern is the increased potential for the wider regional spread of CBSD throughout the mid-altitude cassava-growing areas of East and Central Africa. Research is required to explain how the apparent altitudinal 'ceiling' of 1000 m a.s.l. for CBSD spread has been broken and to determine the implications for possible future spread throughout the extensive mid-altitude (800–1500 m a.s.l.) cassava-growing zones of the region. There is no evidence to suggest any positive interaction with the pandemic-associated EACMV-UG. Although newly introduced CMD-resistant yet CBSD-susceptible germplasm, may in part, have led to the spread of CBSD in Uganda, a much more plausible cause for increased spread would seem to be the increased abundance of the whitefly vector. Transmission studies have confirmed that *B. tabaci*, vector of the CMGs, also transmits CBSV (Maruthi *et al.*, 2005), although the transmission characteristics have yet to be fully determined. The co-occurrence of super-abundant *B. tabaci* populations, CBSV and EACMV-UG, raises the alarming possibility of a 'dual pandemic', although many research questions will need to be addressed before the likelihood of such a development is known.

VI. Conclusions

Although CMD has been an important constraint to cassava production in Africa for more than a century, changes in the nature of the disease during the last two decades have led to losses on a hitherto unprecedented scale. Strategic epidemiological studies which have traced the development and spread of what is now known as the African CMD pandemic have provided vital insights into the mechanisms and pattern through which this disease is spread and the critical interactions with its whitefly vector, *B. tabaci*. Based on this new knowledge, an effective and wide-ranging management programme has been implemented utilizing each of the principal virus management tools, although the primary focus has been on the deployment of host-plant resistance. Substantial impact has been achieved in areas where management programmes have run for several years. The best example of this is Uganda, where more than a third of the cassava crop is under CMD-resistant varieties (Sserubombwe *et al.*, 2005a). Many challenges remain, however. The first relates to the scale of the problem. Although effective management programmes are running, including innovative 'pre-emptive' control initiatives, the scale of the problem dwarfs the current level of control interventions, and the best current estimate of the affected area exceeds 2.6 million sq km. Of this total, less than a tenth, currently, has adequate control programmes in place (Uganda and western Kenya). In order to improve the management effort significantly, existing programmes must be maintained and expanded, and major programmes are required in more recently affected regions and countries. Particular attention should be given to Central African Republic, Angola and Gabon, where there is currently minimal preparedness for the impact of the pandemic and much needs to be done, if serious losses are to be avoided. A further challenge is the sustained change in abundance of the whitefly, *B. tabaci*, which vectors both CMGs and CBSV. Resurgence of CBSD in Uganda is of considerable concern, raising the possibility of a dual-pandemic throughout the region. The increased interest in controlling *B. tabaci* through both host-plant resistance and biological control is encouraging, but much needs to be done before effective whitefly-management strategies are available for widespread dissemination to pandemic-affected farming communities.

Although the huge scale and continued spread of the African CMD pandemic make it one of the greatest pest/disease challenges now facing agricultural researchers globally, there is considerable hope for successful mitigation. Heightened awareness is leading to increased support from development investors for control programmes. Moreover, experience obtained by multi-stakeholder teams over the past decade has led to the establishment of strong regional networks of personnel active in CMD management. Nevertheless, the CMD pandemic will have a significant negative impact on the continent's cassava production for the foreseeable future. Concerted efforts will be required by researchers, extensionists, farmers and those supporting these groups to ensure that the scale of this impact is minimized, and that in future years CMD reverts to the status of being largely benign. The capacity that the CMGs have demonstrated to 'use' true recombination to exploit new opportunities for spread suggests that elimination is an unrealistic objective, and ambition should, therefore, be restricted to restoring the apparent equilibrium that for long periods seems to have characterized the more than century-old relationship between cassava and CMGs. If this is achieved, it will represent a real success in overcoming what is one of the most pernicious of all crop diseases.

References

Akano, A. O., Dixon, A. G. O., Mba, C., Barrera, E., and Fregene, M. (2002). Genetic mapping of a dominant gene conferring resistance to cassava mosaic disease. *Theor. Appl. Genet.* **105:**521–535.

Anderson, J. V., Delsen, M., Fregene, M. A., Jorge, V., Mba, C., Lopez, C., Restrepo, S., Soto, M., Piegu, B., Verdier, V., Cooke, R., Tohme, J., *et al.* (2004). An EST resource for cassava and other species of Euphorbiaceae. *Plant Mol. Biol.* **56:**527–539.

Anderson, P. (2005). Introduction. *In* "Whiteflies and Whitefly-Borne Viruses in the Tropics: Building a Knowledge Base for Global Action" (P. K. Anderson and F. Morales, eds.), pp. 1–11. Centro Internacional de Agricultura Tropical, Cali, Colombia.

Anderson, P., and Morales, F. (2005). "Whiteflies and Whitefly-Borne Viruses in the Tropics: Building a Knowledge Base for Global Action." Centro Internacional de Agricultura Tropical, Cali, Colombia.

Anon (1992). Quarantine Implications: Cassava Program, 1987–1991 Working document no. 116. CIAT, Cali, Colombia.

Anon (1998a). Cassava mosaic pandemic threatens food security. *Agriforum* **3:**1, 8.

Anon (1998b). Cassava mosaic disease menaces East African food security. Famine Early Warning System Special Report 98–4. http://www.fews.net/, p. 2.

Anon (1999). Fighting cassava mosaic pandemic: Networking critical. *Agriforum* **7:**1, 12.

Anon (2005a). "Burundi: Famine Declared in Two Provinces." Reuters. January 11, 2005. http://www.alertnet.org/.

Anon (2005b). "Une taxe sur les salaires burundais contre la famine." Afrik.com. January 14, 2005.http://www.xn-beaut-fsa.afrik.com/article8032.html.

Anon (2005c). CMD: A blessing in disguise for Nigeria. New Agriculturalist Online. September 1, 2005. http://www.new-agri.co.uk/05–5/focuson/focuson5.html.

Anon (2005d). Vegetation map for Central/West Africa. Central Africa Regional Project for the Environment (CARPE), University of Maryland, USA (http://carpe.umd.edu/) and TREES Project, Joint Research Centre, Ispra, Italy (http://www.jrc.cec.eu.int/).

Ariyo, O. A., Koerbler, M., Dixon, A. G. O., Atiri, G. I., and Winter, S. (2005). Molecular variability and distribution of cassava mosaic begomoviruses in Nigeria. *J. Phytopathol.* **153:**226–231.

Beck, B. D. A. (1982). Historical perspectives of cassava breeding in Africa. *In* "Root Crops in Eastern Africa. Proceedings of a Workshop, Kigali, Rwanda, 1980" (S. K. Hahn and A. D. R. Ker, eds.), pp. 13–18. IDRC, Ottawa, Canada.

Bedford, I. D., Briddon, R. W., Brown, J. K., Rossel, R. C., and Markham, P. G. (1994). Geminivirus transmission and biological characterisation of *Bemisia tabaci* (Gennadius) biotypes from different geographic regions. *Ann. Appl. Biol.* **125:**311–325.

Bellotti, A. C., and Arias, B. (2001). Host plant resistance to whiteflies with emphasis on cassava as a case study. *Crop Prot.* **20:**813–823.

Berrie, L. C., Rybicki, E. P., and Rey, M. E. C. (2001). Complete nucleotide sequence and host range of South African cassava mosaic virus: Further evidence for recombination amongst geminiviruses. *J. Gen. Virol.* **82:**53–58.

Berry, S., and Rey, M. E. C. (2001). Molecular evidence for diverse populations of cassava-infecting begomoviruses in southern Africa. *Arch. Virol.* **146:**1795–1802.

Bigirimana, S., Barumbanze, P., Obonyo, R., and Legg, J. P. (2004). First evidence for the spread of *East African cassava mosaic virus*–Uganda (EACMV-UG) and the pandemic of severe cassava mosaic disease to Burundi. *Plant Pathol.* **53:**231.

Bock, K. R., and Woods, R. D. (1983). The etiology of African cassava mosaic disease. *Plant Dis.* **67:**994–995.

Bock, K. R., Guthrie, E. J., and Figueiredo, G. (1981). A strain of cassava latent virus occurring in coastal districts of Kenya. *Ann. Appl. Biol.* **90:**361–367.

Brown, J. K. (2000). Molecular markers for the identification and global tracking of whitefly vector-begomovirus complexes. *Virus Res.* **71:**233–260.

Brown, J. K. (2001). The molecular epidemiology of begomoviruses. *In* "Trends in Plant Virology" (J. A. Khan and J. Dykstra, eds.), pp. 279–316. The Haworth Press Inc., New York, USA.

Brown, J. K., and Bird, J. (1992). Whitefly-transmitted geminiviruses and associated disorders in the Americas and the Caribbean Basin. *Plant Dis.* **76:**220–225.

Brown, J. K., Frohlich, D. R., and Rosell, R. C. (1995). The sweetpotato/silverleaf whiteflies: Biotypes of *Bemisia tabaci* (Genn.), or a species complex? *Ann. Rev. Entomol.* **40:**511–534.

Burban, C., Fishpool, L. D. C., Fauquet, C., Fargette, D., and Thouvenel, J.-C. (1992). Host-associated biotypes within West African populations of the whitefly *Bemisia tabaci* (Genn.) (Hom., Aleyrodidae). *J. Appl. Entomol.* **113:**416–423.

Byabakama, B. A., Adipala, E., Ogenga-Latigo, M. W., and Otim-Nape, G. W. (1999). The effect of amount and disposition of inoculum on cassava mosaic virus disease development and tuberous root yield of cassava. *Afr. J. Plant Prot.* **5:**21–29.

Calvert, L. A., and Thresh, J. M. (2002). The viruses and virus diseases of cassava. *In* "Cassava: Biology, Production and Utilization" (A. C. Bellotti, R. J. Hillocks, and J. M. Thresh, eds.), pp. 237–260. CABI, Wallingford, UK.

Chant, S. R. (1958). Studies on the transmission of cassava mosaic virus by *Bemisia* spp. (Aleyrodidae). *Ann. Appl. Biol.* **46:**210–215.

Chant, S. R., Bateman, J. G., and Bates, D. C. (1971). The effect of cassava mosaic virus infection on the metabolism of cassava leaves. *Trop. Agr.* **48:**263–270.

Chellappan, P., Masona, M., Vanitharani, R., Taylor, N. J., and Fauquet, C. M. (2004). Broad spectrum resistance to ssDNA viruses associated with transgene-induced gene silencing in cassava. *Plant Mol. Biol.* **56:**601–611.

Chellappan, P., Vanitharani, R., and Fauquet, C. M. (2005). MicroRNA-binding viral protein interferes with *Arabidopsis* development. *Proc. Natl. Acad. Sci. USA* **102:**10381–10386.

Cock, M. J. W. (1993). *Bemisia tabaci*–an update 1986–1992 on the cotton whitefly with an annotated bibliography, p. 78. International Institute of Biological Control, Ascot, UK.

Colon, L. (1984). Contribution à l'étude de la résistance variétale du manioc (*Manihot esculenta* Crantz) vis-à-vis de la mosaïque africaine du manioc. Etude réalisée dans le cadre du programme ORSTOM. Etude de la mosaïque africaine du manioc. ORSTOM, Abidjan, Ivory Coast.

Colvin, J., Omongo, C. A., Maruthi, M. N., Otim-Nape, G. W., and Thresh, J. M. (2004). Dual begomovirus infections and high *Bemisia tabaci* populations drive the spread of a cassava mosaic disease pandemic. *Plant Pathol.* **53:**577–584.

Costa, H. S., and Brown, J. K. (1991). Variation in biological characteristics and in esterase patterns among populations of *Bemisia tabaci* (Genn.) and the association of one population with silverleaf symptom development. *Entomol. Exp. Appli.* **61:**211–219.

Costa, H. S., Brown, J. K., Sivasupramaniam, S., and Bird, J. (1993). Regional distribution, insecticide resistance, and reciprocal crosses between the "A" and "B" biotypes of *Bemisia tabaci. Insect Sci. Appl.* **14:**255–266.

Cours, G. (1951). Le manioc à Madagascar. *Mém. Inst. Sci. Madagascar, Ser. B, Biol. Vég.* **3:**203–400.

Cours, G., Fargette, D., Otim-Nape, G. W., and Thresh, J. M. (1997). The epidemic of cassava mosaic virus disease in Madagascar in the 1930s–1940s: Lessons for the current situation in Uganda. *Trop. Sci.* **37:**238–248.

Cudjoe, A., James, B., and Gyamenah, J. (2005). Whiteflies as vectors of plant viruses in cassava and sweetpotato in Africa: Ghana. *In* "Whiteflies and Whitefly-Borne Viruses in the Tropics: Building a Knowledge Base for Global Action" (P. K. Anderson and F. Morales, eds.), pp. 24–29. Centro Internacional de Agricultura Tropical, Cali, Colombia.

Dade, H. A. (1930). "Cassava Mosaic." Paper No. XXVIII. Yearbook. pp. 245–247. Department of Agriculture, Gold Coast.

De Barro, P. J. (2005). Genetic structure of the whitefly *Bemisia tabaci* in the Asia-Pacific region revealed using microsatellite markers. *Mol. Ecol.* **14:**3695–3718.

De Barro, P. J., and Hart, P. J. (2000). Mating interactions between two biotypes of the whitefly, *Bemisia tabaci* (Hemiptera: Aleyrodidae) in Australia. *Bull. Ent. Res.* **90:**103–112.

Deighton, F. C. (1926). Annual Report of the Lands and Forestry Department, Sierra Leone, pp. 1–2.

Deng, D., Otim-Nape, G. W., Sangare, A., Ogwal, S., Beachy, R. N., and Fauquet, C. M. (1997). Presence of a new virus closely associated with cassava mosaic outbreak in Uganda. *Afr. J. Root Tuber Crops* **2:**23–28.

Dengel, H.-J. (1981). Untersuchengen über das auftreten der imagines von *Bemisia tabaci* (Genn.) auf verschiedenen manioksorten. *Z. Pflanzenkrankh. Pflanzenschutz* **88:**355–366.

Dixon, A. G. O., Bandyopadhyay, R., Coyne, D., Ferguson, M., Ferris, R. S. B., Hanna, R., Hughes, J., Ingelbrecht, I., Legg, J., Mahungu, N., Manyong, V., Mowbray, D., *et al.* (2003). Cassava: From a poor farmer's crop to a pacesetter of African rural development. *Chron. Hort.* **43:**8–14.

Dubern, J. (1979). Quelques propriétés de la Mosaïque Africaine du Manioc. I: La transmission. *Phytopathol. Z.* **96:**25–39.

Dubern, J. (1994). Transmission of African cassava mosaic geminivirus by the whitefly (*Bemisia tabaci*). *Trop. Sci.* **34:**82–91.

Dufrenoy, J., and Hédin, L. (1929). La mosaïque des feuilles du manioc au Cameroun. *Rev. Bot. Appl. Agric. Trop.* **9:**361–365.

Echendu, T. N. C., Ojo, J. B., James, B. D., and Gbaguidi, B. (2005). Whiteflies as vectors of plant viruses in cassava and sweet potato in Africa: Nigeria. *In* "Whiteflies and Whitefly-Borne Viruses in the Tropics: Building a Knowledge Base for Global Action" (P. K. Anderson and F. Morales, eds.), pp. 35–39. Centro Internacional de Agricultura Tropical, Cali, Colombia.

Fargette, D., Fauquet, C., and Thouvenel, J.-C. (1985). Field studies on the spread of African cassava mosaic. *Ann. Appl. Biol.* **106:**285–294.

Fargette, D., Fauquet, C., Grenier, E., and Thresh, J. M. (1990). The spread of African cassava mosaic virus into and within cassava fields. *J. Phytopathol.* **130:**289–302.

Fargette, D., Jeger, M., Fauquet, C., and Fishpool, L. D. C. (1993). Analysis of temporal disease progress of African cassava mosaic virus. *Phytopathology* **84:**91–98.

Fargette, D., Thresh, J. M., and Otim-Nape, G. W. (1994). The epidemiology of African cassava mosaic geminivirus: Reversion and the concept of equilibrium. *Trop. Sci.* **34:**123–133.

Fargette, D., Colon, L. T., Bouveau, R., and Fauquet, C. (1996). Components of resistance of cassava to African cassava mosaic virus. *Eur. J. Plant Pathol.* **102:**645–654.

Fauquet, C., and Fargette, D. (1990). African cassava mosaic virus: Etiology, epidemiology and control. *Plant Dis.* **74:**404–411.

Fauquet, C. M., and Stanley, J. (2003). Geminivirus classification and nomenclature: Progress and problems. *Ann. Appl. Biol.* **142:**165–189.

Fishpool, L. D. C., and Burban, C. (1994). *Bemisia tabaci*: The whitefly vector of African cassava mosaic geminivirus. *Trop. Sci.* **34:**55–72.

Fishpool, L. D. C., Fauquet, C., Thouvenel, J.-C., Burban, C., and Colvin, J. (1995). The phenology of *Bemisia tabaci* populations (Homoptera: Aleyrodidae) on cassava in southern Côte d'Ivoire. *Bull. Ent. Res.* **85:**197–207.

Fondong, V., Pita, J. S., Rey, M. E. C., de Kochko, A., Beachy, R. N., and Fauquet, C. M. (2000). Evidence of synergism between African cassava mosaic virus and the new double recombinant geminivirus infecting cassava in Cameroon. *J. Gen. Virol.* **81:**287–297.

Food and Agriculture Organization (FAO) of the United Nations (2006). Cassava production data 2005. http://www.fao.org.

François, E. (1937). Un grave peril: La mosaïque du manioc. *Agron. Colon.* **26:**33–38.

Frison, E. (1994). Sanitation techniques for cassava. *Trop. Sci.* **34:**146–153.

Gbaguidi, B., James, B., and Saizonou, S. (2005). Whiteflies as vectors of plant viruses in cassava and sweet potato in Africa: Benin. *In* "Whiteflies and Whitefly-Borne Viruses in the Tropics: Building a Knowledge Base for Global Action" (P. K. Anderson and F. Morales, eds.), pp. 30–34. Centro Internacional de Agricultura Tropical, Cali, Colombia.

Gibson, R. W. (1994). Long-term absence of symptoms in heat treated African cassava mosaic geminivirus-infected resistant cassava plants. *Trop. Sci.* **34:**154–158.

Gibson, R. W., Legg, J. P., and Otim-Nape, G. W. (1996). Unusually severe symptoms are a characteristic of the current epidemic of mosaic virus disease of cassava in Uganda. *Ann. Appl. Biol.* **128:**479–490.

Golding, F. D. (1936). Cassava mosaic in southern Nigeria. *Bull. Depart. Agric. Nigeria* **11:**1–10.

Golemboski, D. B., Lomonossoff, G. P., and Zaitlin, M. (1990). Plants transformed with a tobacco mosaic virus non-structural gene are resistant to the virus. *Proc. Natl. Acad. Sci. USA* **87:**6311–6315.

Guthrie, E. J. (1988). African cassava mosaic disease and its control. *In* "Proceedings of the International Seminar on African Cassava Mosaic Disease and its Control, 4–8 May, 1987, Yamoussoukro, Ivory Coast" (C. Fauquet and D. Fargette, eds.), pp. 1–9. CTA, Wageningen, Netherlands.

Hahn, S. K., Terry, E. R., and Leuschner, K. (1980). Breeding cassava for resistance to cassava mosaic disease. *Euphytica* **29:**673–683.

Hall, F. W. (1928). Annual Report. Department of Agriculture, Uganda, p. 35.

Harrison, B. D., Zhou, X., Otim-Nape, G. W., Liu, Y., and Robinson, D. J. (1997). Role of a novel type of double infection in the geminivirus-induced epidemic of severe cassava mosaic in Uganda. *Ann. Appl. Biol.* **131:**437–448.

Hédin, L. (1931). Culture du manioc en Côte d'Ivoire; observations complémentaires sur la mosaïque. *Revue de Botanique Appliquée* **11:**558–563.

Hillocks, R. J., Raya, M., and Thresh, J. M. (1996). The association between root necrosis and above ground symptoms of brown streak virus infection of cassava in southern Tanzania. *Int. J. Pest Man.* **42:**285–289.

Hillocks, R. J., Raya, M. D., and Thresh, J. M. (1999). Factors affecting the distribution, spread and symptom expression of cassava brown streak disease in Tanzania. *Afr. J. Root Tuber Crops* **3:**57–61.

Hillocks, R. J., Raya, M., Mtunda, K., and Kiozia, H. (2001). Effects of brown streak virus disease on yield and quality of cassava in Tanzania. *J. Phytopathol.* **149:**1–6.

Hillocks, R. J., Thresh, J. M., Tomas, J., Botao, M., Macia, R., and Zavier, R. (2002). Cassava brown streak disease in northern Mozambique. *Int. J. Pest Man.* **48:**178–181.

Hong, Y. G., Robinson, D. J., and Harrison, B. D. (1993). Nucleotide sequence evidence for the occurrence of three distinct whitefly-transmitted geminiviruses in cassava. *J. Gen. Virol.* **74:**2437–2443.

Hong, Y., Saunders, K., Hartley, M. R., and Stanley, J. (1996). Resistance to geminivirus infection by virus-induced expression of dianthin in transgenic plants. *Virol.* **220:**119–127.

Jameson, J. D. (1964). Cassava mosaic disease in Uganda. *E. Afr. Agric. J.* **29:**208–213.

Jennings, D. (1957). Further studies in breeding cassava for virus resistance. *E. Afr. Agric. J.* **22:**213–219.

Jones, W. O. (1959). "Manioc in Africa," 315 pp. Stanford University Press, Stanford, USA.

Kamau, J., Sseruwagi, P., and Aritua, V. (2005). Whiteflies as vectors of plant viruses in cassava and sweet potato in Africa: Kenya. *In* "Whiteflies and Whitefly-Borne Viruses in the Tropics: Building a Knowledge Base for Global Action" (P. K. Anderson and F. Morales, eds.), pp. 54–60. Centro Internacional de Agricultura Tropical, Cali, Colombia.

Kufferath, H., and Ghesquière, J. (1932). La mosaïque du manioc. *Compte Rendu Soc. Biol.* **109:**1146–1148.

Legg, J. P. (1995). The ecology of *Bemisia tabaci* (Gennadius) (Homoptera), vector of African cassava mosaic geminivirus in Uganda. Doctoral thesis. University of Reading, UK.

Legg, J. P. (1999). Emergence, spread and strategies for controlling the pandemic of cassava mosaic virus disease in East and Central Africa. *Crop Prot.* **18:**627–637.

Legg, J. (2005). Whiteflies as vectors of plant viruses in cassava and sweet potato in Africa: Introduction. *In* "Whiteflies and Whitefly-Borne Viruses in the Tropics: Building a Knowledge Base for Global Action" (P. K. Anderson and F. Morales, eds.), pp. 15–23. Centro Internacional de Agricultura Tropical, Cali, Colombia.

Legg, J. P., and Fauquet, C. M. (2004). Cassava mosaic geminiviruses in Africa. *Plant Mol. Biol.* **56:**585–599.

Legg, J. P., and Ogwal, S. (1998). Changes in the incidence of African cassava mosaic geminivirus and the abundance of its whitefly vector along south-north transects in Uganda. *J. Appl. Entomol.* **122:**169–178.

Legg, J. P., and Raya, M. (1998). Survey of cassava virus diseases in Tanzania. *Int. J. Pest Man.* **44:**17–23.

Legg, J. P., and Thresh, J. M. (2000). Cassava mosaic virus disease in East Africa: A dynamic disease in a changing environment. *Virus Res.* **71:**135–149.

Legg, J. P., and Thresh, J. M. (2004). Cassava virus diseases in Africa. *In* "Plant Virology in Sub-Saharan Africa Conference Proceedings" (J. d'A. Hughes and B. O. Adu, eds.), pp. 517–552. IITA, Ibadan, Nigeria.

Legg, J. P., Gibson, R. W., and Otim-Nape, G. W. (1994). Genetic polymorphism amongst Ugandan populations of *Bemisia tabaci* (Gennadius) (Homoptera: Aleyrodidae), vector of African cassava mosaic geminivirus. *Trop. Sci.* **34:**73–81.

Legg, J., James, B., Cudjoe, A., Saizonou, S., Gbaguidi, B., Ogbe, F., Ntonifor, N., Ogwal, S., Thresh, J., and Hughes, J. (1997). A regional collaborative approach to the study of ACMD epidemiology in sub-Saharan Africa. *In* "Proceedings of the African Crop Science Conference, 13–17 January, 1997, Pretoria, South Africa" (E. Adipala, J. S. Tenywa, and M. W. Ogenga-Latigo, eds.), pp. 1021–1033. African Crop Science Society, Kampala, Uganda.

Legg, J. P., Sseruwagi, P., Kamau, J., Ajanga, S., Jeremiah, S. C., Aritua, V., Otim-Nape, G. W., Muimba-Kankolongo, A., Gibson, R. W., and Thresh, J. M. (1999a). The pandemic of severe cassava mosaic disease in East Africa: Current status and future threats. *In* "Proceedings of the Scientific Workshop of the Southern African Root Crops Research Network (SARRNET), Lusaka, Zambia, 17–19 August, 1998" (M. O. Akoroda and J. M. Teri, eds.), pp. 236–251. IITA, Ibadan, Nigeria.

Legg, J. P., Kapinga, R., Teri, J., and Whyte, J. B. A. (1999b). The pandemic of cassava mosaic virus disease in East Africa: Control strategies and regional partnerships. *Roots* **6:**10–19.

Legg, J. P., Okao-Okuja, G., Mayala, R., and Muhinyuza, J.-B. (2001). Spread into Rwanda of the severe cassava mosaic virus disease pandemic and associated Uganda variant of *East African cassava mosaic virus* (EACMV-Ug). *Plant Pathol.* **50:**796.

Legg, J. P., French, R., Rogan, D., Okao-Okuja, G., and Brown, J. K. (2002). A distinct, invasive *Bemisia tabaci* (Gennadius) (Hemiptera: Sternorrhyncha: Aleyrodidae) genotype cluster is associated with the epidemic of severe cassava mosaic virus disease in Uganda. *Mol. Ecol.* **11:**1219–1229.

Legg, J. P., Gerling, D., and Neuenschwander, P. N. (2003). Biological control of whiteflies in sub-Saharan Africa. *In* "Biological Control in IPM Systems in Africa" (P. Neuenschwander, C. Borgemeister, and J. Langewald, eds.), pp. 87–100. CABI International, Wallingford, UK.

Legg, J. P., Ndjelassili, F., and Okao-Okuja, G. (2004). First report of cassava mosaic disease and cassava mosaic geminiviruses in Gabon. *Plant Pathol.* **53:**232.

Lokko, Y., Danquah, E. Y., Offei, S. K., Dixon, A. G. O., and Gedil, M. A. (2005). Molecular markers associated with a new source of resistance to the cassava mosaic disease. *Afr. J. Biotech.* **4:**873–881.

Lucioli, A., Noris, E., Brunetti, A., Tavazza, R., Ruzza, V., Castillo, A. G., Bejarano, E. R., Accotto, G. P., and Tavazza, M. (2003). *Tomato yellow leaf curl Sardinia virus* rep-derived resistance to homologous and heterologous geminiviruses occurs by different mechanisms and is overcome if virus-mediated transgene silencing is activated. *J. Virol.* **77:**6785–6798.

Mallowa, S. (2006). Survey and management of cassava mosaic disease in western Kenya with special emphasis on Siaya District. M.Sc. thesis, University of Egerton, Nakuru, Kenya.

Manyong, V. M., Dixon, A. G. O., Makinde, K. O., Bokanga, M., and Whyte, J. (2000). "Impact of IITA-Improved Germplasm on Cassava Production in Sub-Saharan Africa," 16 pp. IITA, Ibadan, Nigeria.

Maruthi, M. N., Colvin, J., and Seal, S. (2001). Mating compatability, life-history traits, and RAPD-PCR variation in *Bemisia tabaci* associated with the cassava mosaic disease pandemic in East Africa. *Entomol. Exp. Appl.* **99:**13–23.

Maruthi, M. N., Colvin, J., Seal, S., Gibson, G., and Cooper, J. (2002). Co-adaptation between cassava mosaic geminiviruses and their local vector populations. *Virus Res.* **86:**71–85.

Maruthi, M. N., Seal, S., Colvin, J., Briddon, R. W., and Bull, S. E. (2004a). *East African cassava mosaic Zanzibar virus*–a recombinant begomovirus species with a mild phenotype. *Arch. Virol.* **149:**2365–2377.

Maruthi, M. N., Colvin, J., Thwaites, R. M., Banks, G. K., Gibson, G., and Seal, S. E. (2004b). Reproductive incompatibility and cytochrome oxidase I gene sequence variability amongst host-adapted and geographically separate *Bemisia tabaci* populations (Hemiptera: Aleyrodidae). *Syst. Entomol.* **29:**560–568.

Maruthi, M. N., Hillocks, R. J., Mtunda, K., Raya, M. D., Muhanna, M., Kiozia, H., Rekha, A. R., Colvin, J., and Thresh, J. M. (2005). Transmission of *Cassava brown streak virus* by *Bemisia tabaci* (Gennadius). *J. Phytopathol.* **153:**307–312.

Mignouna, H. D., and Dixon, A. G. O. (1997). Genetic relationships among cassava clones with varying levels of resistance to African mosaic disease using RAPD markers. *Afr. J. Root Tuber Crops* **2:**28–32.

Ndunguru, J. (2005). Molecular characterization and dynamics of cassava mosaic geminiviruses in Tanzania. Doctoral thesis, 162 pp., University of Pretoria, South Africa.

Ndunguru, J., Legg, J. P., Aveling, T. A. S., Thompson, G., and Fauquet, C. M. (2005a). Molecular biodiversity of cassava begomoviruses in Tanzania: Evolution of cassava geminiviruses in Africa and evidence for East Africa being a center of diversity of cassava geminiviruses. *Virol. J.* **2:**21.

Ndunguru, J., Sseruwagi, P., Jeremiah, S., and Kapinga, R. (2005b). Whiteflies as vectors of plant viruses in cassava and sweetpotato in Africa: Tanzania. *In* "Whiteflies and Whitefly-Borne Viruses in the Tropics: Building a Knowledge Base for Global Action" (P. K. Anderson and F. Morales, eds.), pp. 61–67. Centro Internacional de Agricultura Tropical, Cali, Colombia.

Ndunguru, J., Legg, J., Fofana, B., Aveling, T., Thompson, G., and Fauquet, C. (2006). Identification of a defective molecule derived from DNA-A of the bipartite begomovirus of *East African cassava mosaic virus*. *Plant Pathol.* **55:**2–10.

Neuenschwander, P., Hughes, J. d'A., Ogbe, F., Ngatse, J. M., and Legg, J. P. (2002). The occurrence of the Uganda variant of East African cassava mosaic virus (EACMV-Ug) in western Democratic Republic of Congo and the Congo Republic defines the western-most extent of the CMD pandemic in East/Central Africa. *Plant Pathol.* **51:**384.

Nichols, R. F. J. (1950). The brown streak disease of cassava: Distribution, climatic effects and diagnostic symptoms. *E. Afr. Agric. J.* **15:**154–160.

Ntonifor, N., James, B. D., Gbaguidi, B., and Tumanteh, A. (2005). Whiteflies as vectors of plant viruses in cassava and sweetpotato in Africa: Cameroon. *In* "Whiteflies and Whitefly-Borne Viruses in the Tropics: Building a Knowledge Base for Global Action" (P. K. Anderson and F. Morales, eds.), pp. 40–45. Centro Internacional de Agricultura Tropical, Cali, Colombia.

Ogbe, F. O., Songa, W., and Kamau, J. W. (1996). Survey of the incidence of African cassava mosaic and East African cassava mosaic viruses in Kenya and Uganda using a monoclonal antibody based diagnostic test. *Roots* **3**(1)**:**10–13.

Ogbe, F. O., Legg, J., Raya, M. D., Muimba-Kankolongo, A., Theu, M. P., Kaitisha, G., Phiri, N. A., and Chalwe, A. (1997). Diagnostic survey of cassava mosaic viruses in Tanzania, Malawi and Zambia. *Roots* **4**(2)**:**12–15.

Ogbe, F. O., Thottappilly, G., Dixon, A. G. O., and Mignouna, H. D. (2003). Variants of East African cassava mosaic virus and its distribution in double infections with African cassava mosaic virus in Nigeria. *Plant Dis.* **87:**229–232.

Okao-Okuja, G., Legg, J. P., Traore, L., and Alexandra Jorge, M. (2004). Viruses associated with cassava mosaic disease in Senegal and Guinea Conakry. *J. Phytopathol.* **152:**69–76.

Omongo, C. A. (2003). Cassava whitefly, *Bemisia tabaci*, behaviour and ecology in relation to the spread of the cassava mosaic pandemic in Uganda. Doctoral thesis, University of Greenwich, UK.

Otim, M., Legg, J., Kyamanywa, S., Polaszek, A., and Gerling, D. (2004). Occurrence and activity of *Bemisia tabaci* parasitoids on cassava in different agro-ecologies in Uganda. *Biocontrol* **50:**87–95.

Otim, M., Legg, J., Kyamanywa, S., Polazsek, A., and Gerling, D. (2006). Population dynamics of *Bemisia tabaci* (Homoptera: Aleyrodidae) parasitoids on cassava mosaic disease-resistant and susceptible varieties. *Biocontrol Sci. Techn.* **16:**205–214.

Otim-Nape, G. W., and Thresh, J. M. (1998). The current pandemic of cassava mosaic virus disease in Uganda. *In* "The Epidemiology of Plant Diseases" (G. Jones, ed.), pp. 423–443. Kluwer, Dordrecht, Germany.

Otim-Nape, G. W., Bua, A., and Baguma, Y. (1994a). Accelerating the transfer of improved crop production technologies: Controlling African cassava mosaic virus disease in Uganda. *Afr. Crop Sci. J.* **2:**479–495.

Otim-Nape, G. W., Shaw, M. W., and Thresh, J. M. (1994b). The effects of African cassava mosaic geminivirus on the growth and yield of cassava in Uganda. *Trop. Sci.* **34:**43–54.

Otim-Nape, G. W., Bua, A., Thresh, J. M., Baguma, Y., Ogwal, S., Semakula, G. N., Acola, G., Byabakama, B., and Martin, A. (1997). Cassava mosaic virus disease in Uganda: The current pandemic and approaches to control. Natural Resources Institute, Chatham, UK.

Otim-Nape, G. W., Thresh, J. M., Bua, A., Baguma, Y., and Shaw, M. W. (1998). Temporal spread of cassava mosaic virus disease in a range of cassava cultivars in different agro-ecological regions of Uganda. *Ann. Appl. Biol.* **133:**415–430.

Otim-Nape, G. W., Bua, A., Thresh, J. M., Baguma, Y., Ogwal, S., Ssemakula, G. N., Acola, G., Byabakama, B., Colvin, J., Cooter, R. J., and Martin, A. (2000). The current pandemic of cassava mosaic virus disease in East Africa and its control. Natural Resources Institute, Chatham, UK.

Otim-Nape, G. W., Alicai, T., and Thresh, J. M. (2001). Changes in the incidence and severity of cassava mosaic virus disease, varietal diversity and cassava production in Uganda. *Ann. Appl. Biol.* **138:**313–327.

Owor, B. (2002). Effect of cassava mosaic geminiviruses (CMGs) on growth and yield of a cassava mosaic disease (CMD) susceptible cultivar in Uganda and cross protection studies. M.Sc. thesis, Makerere University, Kampala, Uganda.

Owor, B., Legg, J. P., Okao-Okuja, G., Obonyo, R., and Ogenga-Latigo, M. W. (2004a). The effect of cassava mosaic geminiviruses on symptom severity, growth and root yield of a cassava mosaic virus disease-susceptible cultivar in Uganda. *Ann. Appl. Biol.* **145:**331–337.

Owor, B., Legg, J. P., Obonyo, R., Okao-Okuja, G., Kyamanywa, S., and Ogenga-Latigo, M. W. (2004b). Field studies of cross protection with cassava in Uganda. *J. Phytopathol.* **152:**243–249.

Pacumbaba, P. R. (1985). Virus-free shoots from cassava stem cuttings infected with cassava latent virus. *Plant Dis.* **69:**231–232.

Padidam, M., Sawyer, S., and Fauquet, C. M. (1999). Possible emergence of new geminiviruses by frequent recombination. *Virol.* **265:**218–225.

Pita, J. S., Fondong, V. N., Sangare, A., Otim-Nape, G. W., Ogwal, S., and Fauquet, C. M. (2001a). Recombination, pseudorecombination and synergism of geminiviruses are determinant keys to the epidemic of severe cassava mosaic disease in Uganda. *J. Gen. Virol.* **82:**655–665.

Pita, J. S., Fondong, V. N., Sangare, A., Kokora, R. N. N., and Fauquet, C. M. (2001b). Genomic and biological diversity of the African cassava geminiviruses. *Euphytica* **120:**115–125.

Ranomenjanahary, S., Rabindran, R., and Robinson, D. J. (2002). Occurrence of three distinct begomoviruses in cassava in Madagascar. *Ann. Appl. Biol.* **140:**315–318.

Ranomenjanahary, S., Ramelison, J., and Sseruwagi, P. (2005). Whiteflies as vectors of plant viruses in cassava and sweet potato in Africa: Madagascar. *In* "Whiteflies and Whitefly-Borne Viruses in the Tropics: Building a Knowledge Base for Global Action" (P. K. Anderson and F. Morales, eds.), pp. 72–76. Centro Internacional de Agricultura Tropical, Cali, Colombia.

Sserubombwe, W. S., Thresh, J. M., Otim-Nape, G. W., and Osiru, D. S. O. (2001). Progress of cassava mosaic virus disease and whitefly vector populations in single and mixed stands of four cassava varieties grown under epidemic conditions in Uganda. *Ann. Appl. Biol.* **138:**161–170.

Sserubombwe, W. S., Bua, A., Baguma, Y. K., Alicai, T., Omongo, C. A., Akullo, D., Tumwesigye, S., Apok, A., and Thresh, J. M. (2005a). The relative productivity of local and improved cassava mosaic disease-resistant varieties in Uganda in 1999 and 2003. *Roots* **9:**15–20.

Sserubombwe, W., Thresh, M., Legg, J., and Otim-Nape, W. (2005b). Special topics on pest and disease management: Progress of cassava mosaic disease in Ugandan cassava varieties and in varietal mixtures. *In* "Whiteflies and Whitefly-Borne Viruses in the Tropics: Building a Knowledge Base for Global Action" (P. K. Anderson and F. Morales, eds.), pp. 324–331. Centro Internacional de Agricultura Tropical, Cali, Colombia.

Sseruwagi, P. (2005). Molecular variability of cassava *Bemisia tabaci* and its effect on the epidemiology of cassava mosaic geminiviruses in Uganda. Doctoral thesis, University of Witwatersrand, South Africa.

Sseruwagi, P., Sserubombwe, W. S., Legg, J. P., Ndunguru, J., and Thresh, J. M. (2004a). Methods of surveying the incidence and severity of cassava mosaic disease and whitefly vector populations on cassava in Africa: A review. *Virus Res.* **100:**129–142.

Sseruwagi, P., Rey, M. E. C., Brown, J. K., and Legg, J. P. (2004b). The cassava mosaic geminiviruses occurring in Uganda following the 1990s epidemic of severe cassava mosaic disease. *Ann. Appl. Biol.* **145:**113–121.

Sseruwagi, P., Okao-Okuja, G., Kalyebi, A., Muyango, S., Aggarwal, V., and Legg, J. P. (2005a). Cassava mosaic geminiviruses associated with cassava mosaic disease in Rwanda. *Int. J. Pest Man.* **51:**17–23.

Sseruwagi, P., Legg, J. P., and Otim-Nape, G. W. (2005b). Whiteflies as vectors of plant viruses in cassava and sweetpotato in Africa: Uganda. *In* "Whiteflies and Whitefly-Borne Viruses in the Tropics: Building a Knowledge Base for Global Action" (P. K. Anderson and F. Morales, eds.), pp. 46–53. Centro Internacional de Agricultura Tropical, Cali, Colombia.

Storey, H. H. (1936). Virus diseases of East African plants. VI-A progress report on studies of the disease of cassava. *E. Afr. Agric. J.* **2:**34–39.

Storey, H. H. (1938). Virus diseases of East African plants. VII-A field experiment in the transmission of cassava mosaic. *E. Afr. Agric. J.* **3:**446–449.

Storey, H. H., and Nichols, R. F. W. (1938). Studies on the mosaic of cassava. *Ann. Appl. Biol.* **25:**790–806.

Swanson, M. M., and Harrison, B. D. (1994). Properties, relationships and distribution of cassava mosaic geminiviruses. *Trop. Sci.* **34:**15–25.

Theu, M. P. K. J., and Sseruwagi, P. (2005). Whiteflies as vectors of plant viruses in cassava and sweetpotato in Africa: Malawi. *In* "Whiteflies and Whitefly-Borne Viruses in the Tropics: Building a Knowledge Base for Global Action" (P. K. Anderson and F. Morales, eds.), pp. 68–71. Centro Internacional de Agricultura Tropical, Cali, Colombia.

Thresh, J. M., and Cooter, R. J. (2005). Strategies for controlling cassava mosaic virus disease in Africa. *Plant Pathol.* **54:**587–614.

Thresh, J. M., and Otim-Nape, G. W. (1994). Strategies for controlling African cassava mosaic geminivirus. *Adv. Dis. Vector Res.* **10:**215–236.

Thresh, J. M., Fargette, D., and Otim-Nape, G. W. (1994a). The viruses and virus diseases of cassava in Africa. *Afr. Crop Sci. J.* **24:**459–478.

Thresh, J. M., Otim-Nape, G. W., and Jennings, D. L. (1994b). Exploiting resistance to African cassava mosaic virus. *Asp. Appl. Biol.* **39:**51–60.

Thresh, J. M., Otim-Nape, G. W., Legg, J. P., and Fargette, D. (1997). African cassava mosaic virus disease: The magnitude of the problem. *Afr. J. Root Tuber Crops* **2:**13–19.

Thresh, J. M., Otim-Nape, G. W., and Fargette, D. (1998). The control of African cassava mosaic virus disease: Phytosanitation and/or resistance. *In* "Plant Virus Disease Control" (A. Hadidi, R. K. Khetarpal, and H. Koganezawa, eds.), pp. 670–677. A. P. S. Press, St. Paul., USA.

Turnage, M. A., Muangsan, N., Peele, C. G., and Robertson, D. (2002). Geminivirus-based vectors for gene silencing in *Arabidopsis*. *Plant J.* **30:**107–114.

Vanitharani, R., Chellappan, P., and Fauquet, C. M. (2003). Short interfering RNA-mediated interference of gene expression and viral DNA accumulation in cultured plant cells. *Proc. Natl. Acad. Sci. USA* **100:**9632–9636.

Warburg, O. (1894). Die kulturpflanzen usambaras. *Mitt. Dtsch. Schutzgeb.* **7:**131.

Were, H. K., Winter, S., and Maiss, E. (2004). Occurrence and distribution of cassava begomoviruses in Kenya. *Ann. Appl. Biol.* **145:**175–184.

Zhang, P., Fütterer, J., Frey, P., Potrykus, I., Puonti-Kaerlas, J., and Gruissem, W. (2003). Engineering virus-induced ACMV resistance by mimicking a hypersensitive reaction in transgenic cassava plants. *In* "Plant Biotechnology 2002 and Beyond" (I. K. Vasil, ed.), pp. 143–146. Kluwer Academic Publishers, Dordrecht, Netherlands.

Zhang, P., Vanderschuren, H., Fütterer, J., and Gruissem, W. (2005). Resistance to cassava mosaic disease in transgenic cassava expressing antisense RNAs targeting virus replication genes. *Plant Biotechnol. J.* **3:**385–398.

Zhori-Fein, E., and Brown, J. K. (2002). Diversity of prokaryotes associated with *Bemisia tabaci* (Gennadius) (Hemiptera: Aleyrodidae). *Ann. Ent. Soc. America* **95:**711–718.

Zhou, X., Liu, Y., Calvert, L., Munoz, C., Otim-Nape, G. W., Robinson, D. J., and Harrison, B. D. (1997). Evidence that DNA-A of a geminivirus associated with severe cassava mosaic disease in Uganda has arisen by interspecific recombination. *J. Gen. Virol.* **78:**2101–2111.

Zhou, X., Robinson, D. J., and Harrison, B. D. (1998). Types of variation in DNA-A among isolates of East African cassava mosaic virus from Kenya, Malawi and Tanzania. *J. Gen. Virol.* **79:**2835–2840.

Zimmermann, A. (1906). Die Krauselkrankheit des Maniok. *Pflanzer* **2:**145.

HOST-PLANT VIRAL INFECTION EFFECTS ON ARTHROPOD-VECTOR POPULATION GROWTH, DEVELOPMENT AND BEHAVIOUR: MANAGEMENT AND EPIDEMIOLOGICAL IMPLICATIONS

J. Colvin,* C. A. Omongo,† M. R. Govindappa,‡
P. C. Stevenson,*,§ M. N. Maruthi,* G. Gibson,*
S. E. Seal,* and V. Muniyappa‡

*Natural Resources Institute, University of Greenwich, Central Avenue
Chatham Maritime, Chatham, Kent ME4 4TB, United Kingdom
†National Agricultural Research Organisation, Namulonge Agricultural and Animal
Production Research Institute, Kampala, Uganda
‡Plant Pathology Department, University of Agricultural Sciences, GKVK
Bangalore 560 065, India
§Royal Botanic Gardens Kew, Surrey TW9 3AB, United Kingdom

Single factors, such as the introduction of a non-indigenous vector species and/or plant virus, are often invoked to explain the emergence and spread of plant virus epidemics. This has often been so when an epidemic has been associated with a noticeable increase in the vector population, as frequently occurs during the initial outbreaks and subsequent spread of arthropod vector-borne plant-virus epidemics. Such explanations, however, should be considered more appropriately in the wider context of the complex interactions that occur between the virus(es), host-plant species, the environment and the vector(s). Here, we review evidence for overall positive, negative and neutral effects on the relative population growth of several arthropod vectors when feeding on virus-infected and

0065-3527/06 $35.00
DOI: 10.1016/S0065-3527(06)67011-5

uninfected host plants. The emphasis is on the whitefly, *Bemisia tabaci*, aphids, leafhoppers, mites and thrips. The *B. tabaci*-borne cassava mosaic disease pandemic in sub-Saharan Africa and the tomato leaf curl virus epidemics in the Indian subcontinent are then considered in more detail and experimental data are presented to show, for both of these cases, that a mutually beneficial relationship between the virus and the vector is an important component of the mechanism driving these epidemics. This mechanism is discussed in the context of associated changes in the behaviour of the whitefly vector as well as biochemical changes within the plant.

I. Introduction

The increased movement of plant material globally has resulted in the frequent introduction of plant viruses and vectors into new regions. This increase in global traffic has occurred together with a trend towards intensification of agriculture arising from the need to produce more food from a decreasing land area and with fewer resources. These factors have combined to alter ecosystems and favour the rapid spread of plant virus diseases, particularly those that are transmitted by arthropod vectors such as whiteflies, aphids, leafhoppers, thrips and eriophyid mites (Bos, 1992; Thresh, 1980).

Single causative factors, such as the arrival of a non-indigenous vector species and/or a plant virus, can sometimes provide an appealingly simple explanation for the appearance and rapid spread of plant virus epidemics. This has generally been so when epidemics are associated with an easily identifiable characteristic such as an increased disease severity and/or a marked increase in the vector population. More appropriately, however, the mechanisms driving arthropod-borne plant virus epidemics may be considered to encompass a suite of complex interactions of differing importance, which occur at several levels between the vector(s), the virus(es), the host-plant species and the environment.

Here, we review the evidence for positive, negative and neutral effects on the population growth of several arthropod vector species when feeding on virus-infected compared with uninfected host plants. The emphasis is on whiteflies and aphids as these have received the greatest attention. In part, this has been due to the dramatic increase, in the last 10–15 years, in agricultural and horticultural problems caused particularly by the whitefly, *Bemisia tabaci,* and the many viruses it transmits (Boulton, 2003; Jones, 2003; Morales and Anderson, 2001; Polston and Anderson, 1997; Morales, this volume, pp. 127–162). Two important examples are the *B. tabaci*-borne cassava mosaic disease (CMD) pandemic, which continues to devastate cassava production in large areas of sub-Saharan Africa (SSA), and the tomato leaf curl disease epidemics in the Indian subcontinent, which have had equally serious implications for tomato production in this region. These problems are considered in detail and experimental data are included to show, for both pathosystems, that the *B. tabaci* colonising virus-infected host plants have significantly higher-population growth rates compared to those colonising virus-free hosts. This effect is discussed in the context of the significantly higher densities of *B. tabaci* present on symptomatic cassava and the behavioural changes associated with this effect. Data are also presented to show that the concentrations of four amino acids were significantly higher in the phloem sap of CMD-infected cassava plants. These interacting effects are considered in relation to probable mechanisms contributing to the rapid spread of these epidemics.

II. Vector–Virus–Host Plant Interactions

A. *Aphids*

Experiments on *Beet mosaic virus* (BtMV; Genus: *Potyvirus*; Family: *Potyviridae*) provided some of the first evidence that the population growth rate of an arthropod vector could be affected by the health status of the host plants (Kennedy, 1951). *Aphis fabae* reproduced approximately 1.5 times faster on virus-infected than on uninfected plants, which resulted in the aphid colonies on infected plants becoming overcrowded more rapidly. This, in turn, resulted in emigration of viruliferous adults beginning more quickly from the infected plants which, it was proposed, would lead to further virus spread.

Baker (1960) reported a similar phenomenon and showed that three aphid species bred more rapidly and lived longer on sugar beet leaves infected with virus yellows than on the green leaves of healthy plants. Other researchers also subsequently demonstrated that the English grain aphid, *Sitobion avenae* (formerly *Macrosiphum granarium*), had an increased rate of development, lived longer, had a longer reproductive period and, therefore, produced more progeny on *Barley yellow dwarf virus* (BYDV; Genus: *Luteovirus*; Family: *Luteoviridae*)-infected oats than on healthy plants (Miller and Coon, 1964). A significantly greater fecundity and reduced developmental period was also demonstrated for

S. avenae when feeding on BYDV-infected wheat compared to those feeding on healthy plants (Fereres *et al.*, 1989).

The performance of the aphid, *Myzus persicae*, was also reported to improve on sugar beet plants infected with *Beet yellows virus* (BYV; Genus: *Closterovirus*; Family: *Closteroviridae*) compared to those colonising healthy plants. This effect was so rapid that nymphs born at the time of virus inoculation benefited as much as those produced later (Williams, 1995).

The groundnut rosette virus disease complex (Naidu *et al.*, 1998) provides another example in which symptomatic groundnut plants were more attractive to the aphid vector and populations colonising them developed more quickly. Also, higher numbers of winged adults were produced than on healthy plants (Réal, 1955). In other work, *Aphis craccivora* populations increased significantly faster on the infected compared to healthy plants of the groundnut varieties ICG12991 and JL24, which was due to an increase in aphid fecundity (Willekens, 2003).

As well as affecting the reproductive rate of vectors, virus-infected host plants have been shown to affect the physiological development of vectors. The cereal grain aphids, *S. avenae* and *Rhopalosiphum padi*, for instance, when reared from birth on BYDV-infected barley or oats, produced more than twice as many alates (winged adults) compared to those reared on otherwise comparable healthy plants. This effect was also apparent when field-collected first and second instar nymphs were transferred to BYDV-infected plants (Gildow, 1983; Montllor and Gildow, 1986). It was proposed that changes in host-plant physiology had affected aphid nutrition and development, leading to an increased ratio of winged to non-winged progeny. These interactions were considered adaptive in that they favoured viruliferous aphid dispersal by flight and, thus, the spread of viruses within and between crops (Gildow, 1983; Sohi and Swenson, 1964). Power (1992) later summarised the then available data for the BYDV pathosystem, and suggested that it involved an indirect form of mutualism between the vectors and virus mediated through the host plant.

The most recent work on this system involves the complex interactions that occurred when the bird cherry–oat aphid, *R. padi,* was allowed to colonise transgenic and untransformed wheat challenged with BYDV (Jiménez-Martínez *et al.*, 2004a,b). The results indicated that transgenic virus resistance in wheat directly influenced *R. padi* settling preference, total fecundity, length of reproductive period and intrinsic rate of increase. Healthy transgenic plants were superior hosts for *R. padi* compared to the untransformed parental variety

Lambert. The situation was reversed, however, when Lambert plants were infected with BYDV and, in this situation, the transgenic plants became the inferior hosts.

For a different system, Markkula and Laurema (1964) found that the reproduction of the pea aphid, *Acyrthosiphon pisum*, increased on *Bean yellow mosaic virus* (BYMV; Genus: *Potyvirus*; Family: *Potyviridae*)-infected red clover (*Trifolium pratense*) expressing moderate viral symptoms compared with healthy plants, but that it decreased on plants with severe symptoms. They also found that the reproduction of *R. padi* increased with increasing concentration of the free amino acids on oats infected with BYDV but that the reproduction of two other aphid species remained unchanged, demonstrating that the effects of a virus on a particular host-plant species may not always result in an increased fecundity and survival of all virus vectors.

This point was also illustrated when *M. persicae* was reared on healthy and *Cucumber mosaic virus* (CMV; Genus: *Cucumovirus*; Family: *Bromoviridae*)-infected seedlings of *Nicotiana tabacum, Gomphrena globosa* and *Zinnia elegans*. After 8–12 days, populations were 2.0–4.8 times larger on the healthy than on diseased seedlings of each species (Lowe and Strong, 1962). In CMV-infected *N. tabacum*, the amount of glutamic acid had decreased after 182 h, which was proposed as the explanation for the poorer performance of aphids on CMV-infected plants.

The amino acid content of the phloem of diseased plants typically increases in response to virus infection (Matthews, 1981) and Selman *et al.* (1961) suggested that some viruses affect the biochemistry of host plants in a unique way. BYDV-infected spring wheat, for example, had an increased total amino acid content, and both alanine and glutamine occurred consistently in greater amounts than in healthy leaves. At the oldest growth stage, for instance, there was approximately 2.5 times more alanine in the infected plants' phloem sap than in that of uninfected plants (Ajayi, 1986).

In BYV-infected sugar beet plants, aspartic and glutamic acids decreased, whereas other amino acids either increased, such as citrulline and threonine by three- and fourfold, respectively, or remained unchanged (Fife, 1961). These changes may account for the improved performance of *M. persicae* on BYV-infected plants compared to those colonising healthy sugar beet (Williams, 1995).

Much of the variability in the influence of virus-infected plants on vectors has been attributed to differences in the nitrogenous compounds present in these plants (Power, 1992), as these form an important constituent of the food of phloem-feeding insects (Auclair,

1963; Kunkel, 1977). In most infections, the amount of virus protein synthesised within an infected plant is probably less than to be expected from the quantities of precursors available. Increased concentrations of free amino acids could potentially benefit both the virus and the vector, and further instances in which higher-nitrogen levels in the host plant led to increased vector populations are provided by Barker and Tauber (1951) and Coon (1959). However, there is a dearth of data on the specific amino acid requirements of vectors and so it is difficult to relate their performance to the effect of the virus infection on the concentration of individual amino acids within the host plant.

BYDV infection has also been shown to increase carbohydrate content of wheat tissues and it was suggested that this may account for the increased fecundity of aphids observed on such plants (Fereres *et al.*, 1990).

Symptom expression, or otherwise, by virus-infected plants is subject to various selection pressures and, therefore, may also be a component of the mechanism that influences vector dispersal and thus virus spread. Many viruses produce a marked yellowing of infected tissue, yellow mosaics or yellow vein mosaics (Ajayi and Dewar, 1983; Muniyappa, 1980; Watson and Healy, 1953). Yellow is an attractive colour for many aphid, whitefly and other vector species (Dixon, 1985; Kennedy *et al.*, 1961; Mound, 1962) and, in the field, this may increase the 'contact rate' between the virus and the vector, and may thus increase the proportion of viruliferous vectors in the population as a whole (Ajayi and Dewar, 1983).

The aphid *M. persicae* is the principal vector of *Potato leaf roll virus* (PLRV; Genus: *Polerovirus*; Family: *Luteoviridae*) and it transmits the virus in a persistent manner (Harrison, 1984). *M. persicae* grows faster, has higher fecundity and settles preferentially on cultivated potato, *Solanum tuberosum,* infected by PLRV than on uninfected potato plants (Castle and Berger, 1993; Castle *et al.*, 1998). Eigenbrode *et al.* (2002) have since demonstrated that this preferential colonisation is influenced by volatile emissions from PLRV-infected plants.

Aphid species have plant physiological preferences which, combined with the tendency to move more frequently when feeding on a food source of poor quality, result in aphids colonising parts of plants where they can achieve the highest growth and developmental rates (Ibbotson and Kennedy, 1950; Kennedy *et al.*, 1950). Other aphid species induce plant galls that provide them diverse benefits including a richer food supply than non-galled leaves, extension of the usual period for which high-quality food is available and protection from weather and predators (Forrest, 1971; Kennedy, 1951). Although they do not induce plant galls, aphid species including *Schizaphis graminum* and *Diuraphis noxia* induce chlorotic lesions and increase the amino acid concentrations of the phloem sap when feeding on grasses possibly in a nutritionally advantageous manner (Sandström *et al.*, 2000).

Aphids also demonstrate a strong preference for the best feeding sites on plants (Whitham, 1979), but the effect of viral disease symptoms on the suitability of these feeding sites is generally unknown.

B. Leafhoppers, Thrips and Mites

Information on the effects of plant viruses on their leafhopper vectors is relatively sparse and various authors have reported contradictory results. Several reviews, however, have concluded that plant viruses which multiply within their insect vector, such as those transmitted by delphacid leafhoppers, have not been shown unequivocally to harm the vectors (Nault, 1994; Nault and Ammar, 1989; Sinha, 1981). A general difficulty with experiments investigating this possibility, however, is the practical problem of separating the direct effects of the virus on the vector from those caused by the vector feeding on virus-infected host-plant tissue.

Fife (1956) reported that sugar beet plants infected with *Beet curly top virus* (BCTV; Genus: *Curtovirus*; Family: *Geminiviridae*) transmitted by the beet leafhopper, *Neoaliturus tenellus* (formerly *Circulifer tenellus*), contained twice the concentration of amino acids that occurred in equivalent healthy plants. Other work showed that sugar beet leafhopper populations were highest in fields where plant growth had been retarded in the early stages of development. These fields also had the highest incidence of BCTV and it was suggested that the disruption in leaf canopy produced sunnier and lower humidity conditions within the infected crop, which suited the beet leafhopper (Bennett, 1971). In the last decade, epidemics of BCTV have been occurring with increasing frequency in New Mexico, resulting in substantial losses (Creamer *et al.*, 2003). Although the plant–virus–vector interactions driving these epidemics have yet to be elucidated, it seems likely that there is an indirect positive effect of host-plant virus infection on *N. tenellus* population growth.

An example of the feeding behaviour of a leafhopper being affected by the virus it transmits is provided by *Cicadulina storeyi*, which spent more time ingesting from mesophyll and tissues other than the phloem when host plants were infected with *Maize streak virus* (MSV; Genus: *Mastrevirus*; Family: *Geminiviridae*) (Mesfin and Bosque-Pérez, 1998).

MSV virions are usually located in the nuclei of cells from the epidermis to the phloem parenchyma (Pinner *et al.*, 1993) and so this altered behaviour increases the likelihood of virus acquisition.

One of the earliest reports of a plant virus affecting the population dynamics of its vector involved *Thrips tabaci*—the vector of yellow spot virus disease of pineapple in Hawaii (Carter, 1939). On its alternative weed host, *Emilia sonchifolia*, which is also susceptible to the virus, thrips populations were consistently higher on diseased than on healthy plants. It was suggested that the diseased plants also lived longer and that their mass of curled leaves provided better shelter to the thrips.

In a series of experiments, the direct and indirect (through the host plant) effects of *Tomato spotted wilt virus* (TSWV; Genus: *Tospovirus*; Family: *Bunyaviridae*) on its main vector, the western flower thrips, *Frankliniella occidentalis*, were quantified (Belliure *et al.*, 2005; Maris *et al.*, 2004). There was both a positive direct effect of TSWV on *F. occidentalis* and an indirect effect in which *F. occidentalis* juvenile survival and developmental rates were lower on healthy pepper plants than on those that had been either infected by thrips or by mechanical inoculation. It was suggested, therefore, that TSWV had evolved a mechanism to overcome the plant defences against the thrip vectors and thus assist its spread.

Another example of viral infection inducing changes in host-plant morphology is provided by *Blackcurrant reversion virus* that infects blackcurrant, *Ribes nigrum*, bushes. Infected plants have a reduced density of hairs on their leaves, flowers and stems and produce more shoots of a shorter length, which greatly increases the availability, accessibility and vulnerability of buds to the eriophyid gall mite vector, *Cecidophyopsis* (formerly *Phytoptus*) *ribis* (Thresh, 1967).

In the pigeonpea sterility mosaic virus (PPSMV; ungrouped) pathosystem (Jones *et al.*, 2004), a mutually beneficial interaction also occurs between the eriophyid mite, *Aceria cajani*, and the virus. This mite is highly host specific and is largely confined to pigeonpea and its wild relatives. The mites predominantly inhabit the lower surface of symptomatic leaves of PPSMV-infected plants. Multiplication of *A. cajani* on cultivated pigeonpea was much greater on PPSMV-infected plants than on healthy plants of the same genotype (Kulkarni *et al.*, 2002; Muniyappa and Nangia, 1982; Reddy and Nene, 1980).

Another example of this effect is provided by the eriophyid mite, *Phyllocoptes fructiphylus*, the vector of rose rosette disease (RRD: causal agent yet to be determined) in multiflora rose, *Rosa multiflora*. From July to September, field populations of *P. fructiphylus* were up to

17 times higher on RRD-symptomatic *R. multiflora* than on non-symptomatic plants (Epstein and Hill, 1999).

C. The Whitefly, B. tabaci

The whitefly, *B. tabaci*, has become an extremely serious agricultural and horticultural pest both by causing direct damage and as a vector of more than 110 plant viruses (Jones, 2003; Polston and Anderson, 1997). Several studies have reported effects of virus infection of the host plants on *B. tabaci* reproduction rate and behaviour. Berlinger (1986), for instance, reported a significant preference by *B. tabaci* for healthy tomato plants compared to those infected with *Tomato yellow leaf curl virus* (TYLCV; Genus: *Begomovirus*; Family: *Geminiviridae*), and the effect was apparent even before the virus symptoms had developed. Negative effects of TYLCV on *B. tabaci* were also reported subsequently by Rubenstein and Czosnek (1997), where viruliferous *B. tabaci* exhibited a 17–23% reduction in adult life expectancy and produced 40–50% fewer eggs compared to TYLCV-free individuals. It was suggested that TYLCV, which had previously been considered to infect plants only, had features reminiscent of an insect pathogen.

In another study (Costa *et al.*, 1991), adult *B. tabaci* from a colony that had been reared on a succession of pumpkin plants for more than 5 years were exposed in clip cages to six plant species infected with one of four whitefly-transmitted plant viruses. Groups of 10 adult females were confined on them for 48 h and then the number of eggs was counted and their survival through to adulthood was assessed 3 to 4 weeks later. The mean number of eggs oviposited on healthy pumpkin compared to pumpkin infected with the watermelon curly mottle strain of *Squash leaf curl virus* (WCMoV/SLCV; Genus: *Begomovirus*; Family: *Geminiviridae*) was not significantly different. Survival to adulthood, however, was significantly greater on the diseased than on the healthy pumpkin and total free amino acid concentrations were significantly higher in virus-infected pumpkin, lettuce, tomato, zucchini squash, cotton and cantaloupe melon than in healthy plants of the same species. No simple relationship could be detected between total free amino acid levels and the oviposition or survival rates of *B. tabaci*, however, which suggested that *B. tabaci* did not assess host suitability for offspring survival and that, for any given host plant/virus/*B. tabaci* combination, there was not necessarily a positive effect on vector oviposition or survival (Costa *et al.*, 1991).

More work has demonstrated clearly that the performance of *B. tabaci* populations can be affected positively by the health status of the plants they feed on. Colvin *et al.* (1999) reported that the fecundity of cassava *B. tabaci* was greater on cassava infected with *East African cassava mosaic virus*-Uganda [Namulonge] (EACMV-UG [NAM]; Genus: *Begomovirus*; Family: *Geminiviridae*) than on otherwise comparable healthy plants. Mayer *et al.* (2002) and McKenzie *et al.* (2002) also reported a similar effect with *Tomato mottle virus* (ToMoV; Genus: *Begomovirus*; Family: *Geminiviridae*), where a *B. tabaci* biotype B population produced 2.5-fold more eggs on infected plants than those on healthy plants. This study is particularly interesting as the *B. tabaci* biotype B is continuously expanding its geographic range and has arrived in South India (Banks *et al.*, 2001).

III. Studies on the Tomato Leaf Curl Pathosystem in India

Tomato leaf curl viruses (ToLCVs) are the most important viral pathogens of tomato in India and infected plants show a variety of symptoms including leaf curling, stunting and partial or complete sterility (Chowda Reddy *et al.*, 2005; Ramappa *et al.*, 1998; Saikia and Muniyappa, 1989). *B. tabaci* is the only known vector of ToLCVs (Vasudeva and Sam Raj, 1948) and both the vector and the viruses have a wide host-plant range, and many species act as a reservoir for both the viruses and the vectors throughout the year (Chowda Reddy *et al.*, 2005; Ramappa *et al.*, 1998).

Populations of *B. tabaci* fluctuate significantly each year in India and in the south they are highest during the hot season, which lasts from February to May (Ramappa *et al.*, 1998; Saikia and Muniyappa, 1989). ToLCV disease is highly positively correlated with the size of the *B. tabaci* population and, in ToLCV-susceptible tomato not sprayed with insecticide, the disease incidence frequently reaches 100%, resulting in complete yield loss if plants are infected as seedlings (Holt *et al.*, 1999; Ramappa *et al.*, 1998; Saikia and Muniyappa, 1989).

A. B. tabaci *Population Growth on Healthy and ToLCV-Infected Plants*

In order to investigate possible reasons for the positive correlation between the size of the *B. tabaci* population and ToLCV-disease incidence, the common weed hosts *Euphorbia geniculata, Parthenium hysterophorus, Acanthospermum hispidum, Ageratum conyzoides* and tomato (*Lycopersicon esculentum* cv. Arka Vikas) were selected as experimental species as these are important sources of both *B. tabaci* and *Tomato leaf curl Bangalore virus*-[Bangalore 4] (ToLCBV-[Ban4]; Genus: *Begomovirus*; Family: *Geminiviridae*) (Chowda Reddy *et al.*, 2005; Muniyappa *et al.*, 2000; Ramappa *et al.*, 1998).

Seeds of the five plant species were collected, planted and seedlings were transplanted individually into earthenware pots at 7 days after germination. Plants were kept for a further 7 days after transplanting to allow establishment and further growth before use. The transplanted seedlings were maintained under *B. tabaci*-proof nylon netting to ensure they remained free of whiteflies and were not infected accidentally with ToLCVs.

The *B. tabaci* colony was an indigenous strain (analysed by sequencing of the mitochondrial cytochrome oxidase I gene (mtCOI), GenBank no. AMϕ4ϕ595) that had been maintained on cotton at the Hebbal Campus of the University of Agricultural Sciences, Bangalore. The virus isolate used in the experiment was ToLCBV-[Ban4]. Cohorts of newly emerged adults were collected as follows. A single cotton plant was selected that had a high proportion of *B. tabaci* 'pupae' present on the leaves. All the adult insects were removed from the plant at 16:00 h on a particular day, and the plant was then left undisturbed overnight in a separate cage. The following day at 11:00 h, newly emerged adults were collected and approximately equal numbers were allowed to feed for 24 h on healthy or ToLCV-[Ban4]-infected tomato seedlings. The virus-free and the potentially viruliferous groups of *B. tabaci* were then collected separately and sexed. Initial populations of 10 male and 10 female insects were used to initiate either ToLCV-[Ban4]-free or ToLCV-[Ban4]-infected colonies on the host-plant seedlings. The seedlings with *B. tabaci* adults were enclosed in netting bags that prevented the adult *B. tabaci* from escaping, without allowing the build-up of condensation that might otherwise trap insects in water droplets or affect behaviour due to the possible impact of high humidity. Eighteen days after the release of the colonising *B. tabaci*, any adults present were collected and counted. The leaves of each plant were then examined under a microscope, and the numbers of eggs and nymphs were recorded. The adults were then released back onto the same plant and the process was repeated after a further 18 days. Data were analysed using a repeated measurements analysis of variance (GenStat 8.1, 2005).

The *B. tabaci* population growth on the ToLCV-[Ban4]-infected plants was significantly higher for all host-plant species, although the extent of the effect depended on the plant species (Fig. 1 and Table I). *B. tabaci* successfully colonised all of the plant species and populations increased significantly over time regardless of plant health status (Fig. 1 and Table I). The *B. tabaci* population increase was lowest on healthy

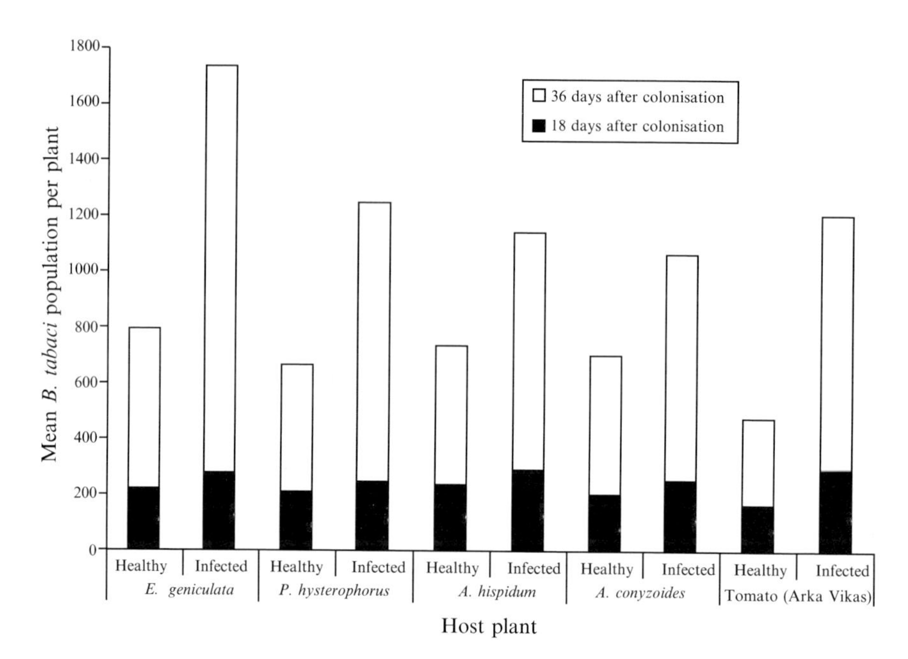

FIG 1. The population (adults, nymphs and eggs) growth of Indian *B. tabaci* on ToLCBV-[Ban4]-infected and uninfected host plants of tomato (*Lycopersicon esculentum* cv. Arka Vikas) and the common weed hosts *E. geniculata, Parthenium hysterophorus, Acanthospermum hispidum, Ageratum conyzoides.*

TABLE I

REPEATED MEASUREMENTS ANALYSIS OF VARIANCE FOR LN (*B. TABACI* POPULATION COUNT) AT 18 AND 36 DAYS AFTER GROUPS OF 10 MALE AND 10 FEMALE NEWLY EMERGED ADULTS PER PLANT WERE ALLOWED TO COLONISE FIVE DIFFERENT PLANT SPECIES

Source of variation	Wald statistic	d.f.	Wald/d.f.	Chi (*P*)
Time	21 929.24	2	10 964.62	<0.001
Plant species	63.27	4	15.82	<0.001
Plant health[a]	353.01	1	353.01	<0.001
Time × plant species	67.05	8	8.38	<0.001
Time × plant health	287.56	2	143.78	<0.001
Plant species × plant health	38.68	4	9.67	<0.001
Time × plant species × plant health	32.98	8	4.12	<0.001

[a] Plants either remained healthy or became infected with ToLCBV-[Ban4].

tomato plants and highest on infected *E. geniculata*, which was consistent with the observation that tomato is not considered a preferred host plant of South Indian *B. tabaci* (Ramappa *et al.*, 1998).

IV. Studies on the Cassava Mosaic Pathosystem in Africa

A. *Current Cassava Mosaic Pandemic in Africa*

Cassava is grown widely in SSA as a staple food crop although, in the last decade, production has been affected severely by a pandemic of an unusually severe form of CMD (Otim-Nape *et al.*, 2000; Thresh *et al.*, 1997). This disease is caused by whitefly-borne cassava mosaic viruses (Genus: *Begomovirus*; Family: *Geminiviridae*). Data collected as the CMD pandemic spread into new regions of southern Uganda were consistent with the rapid spread of the novel EACMV-UG (Zhou *et al.*, 1997) into areas where formerly only *African cassava mosaic virus* (ACMV) occurred. This resulted in a predominance of cassava plants with dual infections that expressed symptoms that were more severe than those caused by either virus alone (Colvin *et al.*, 2004; Harrison *et al.*, 1997). The spread of the pandemic was associated with significantly increased *B. tabaci* populations, which migrated into new areas ahead of a CMD 'front', thus causing its rapid spread (Colvin *et al.*, 2004; Legg and Ogwal, 1998; Otim-Nape *et al.*, 2000).

One hypothesis to explain the increased *B. tabaci* numbers proposed that a putative 'invader' *B. tabaci* genotype cluster, identified by a single mtCOI gene sequence present in *B. tabaci* adults collected from cassava in 1997 and 1999, was spreading southwards with EACMV-UG. It was proposed that this invader *B. tabaci* population might be better adapted to cassava, more highly fecund and able to transmit EACMV-UG with greater efficiency than the 'local' Ugandan *B. tabaci* genotypes that were present previously (Legg *et al.*, 2002). Various experiments designed to test for associations between fitness traits of the 'invader biotype' and the presence of the pandemic, however, proved unsuccessful (Colvin *et al.*, 2004; Maruthi *et al.*, 2001, 2002, 2004a). Moreover, in a separate analysis of *B. tabaci* samples collected from within and outside the CMD-affected zone in 1997, there was no apparent association between the presence of EACMV-UG and the two mtCOI haplotypes (Fig. 2 and Table II). Furthermore, in the separate study reported here, *B. tabaci* populations (UgCas-Ss1 and UgCas-Ss2) collected in 1997 with the so called invader mtCOI sequence were found in areas of Ssanji, southern Uganda, which were, at that time,

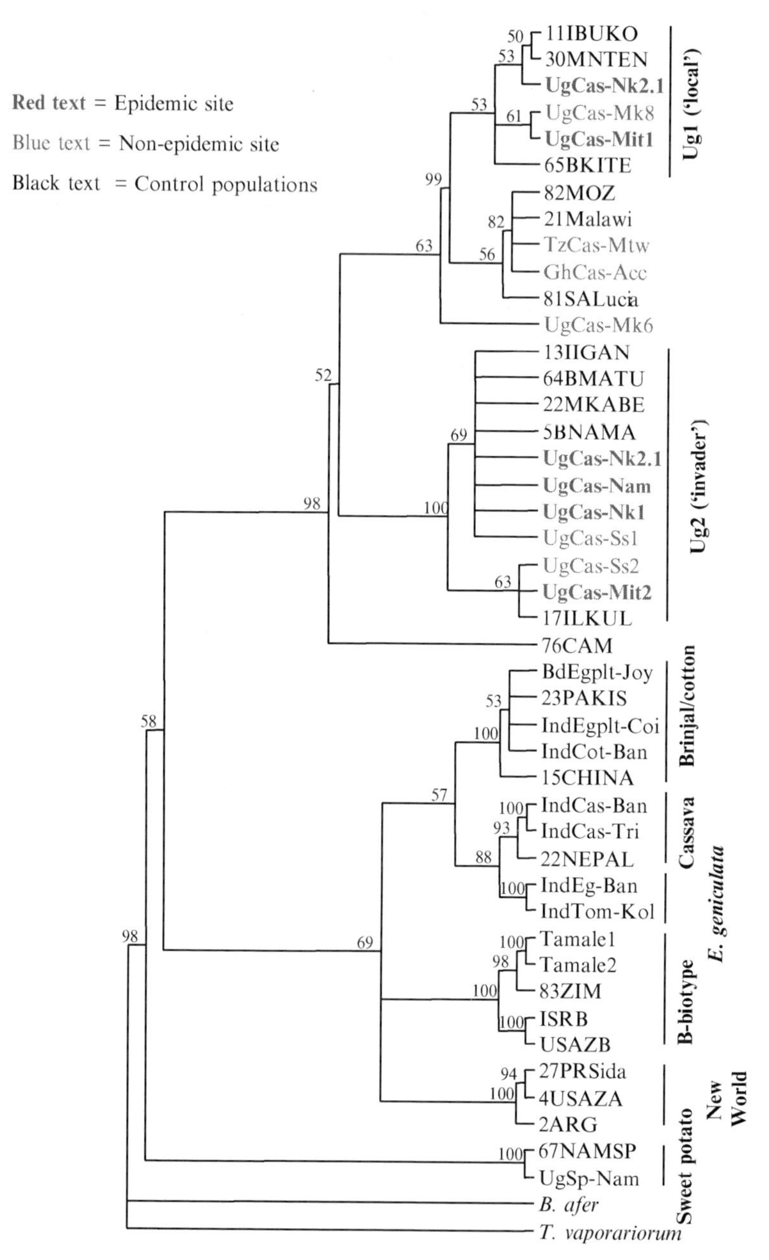

FIG 2. Clustering of partial mtCOI sequences of *B. tabaci* associated with CMD in Africa and ToLCVD in India. Samples in red and blue originate from the CMD epidemic and pre-epidemic zones in Uganda, respectively. (See Color Insert.)

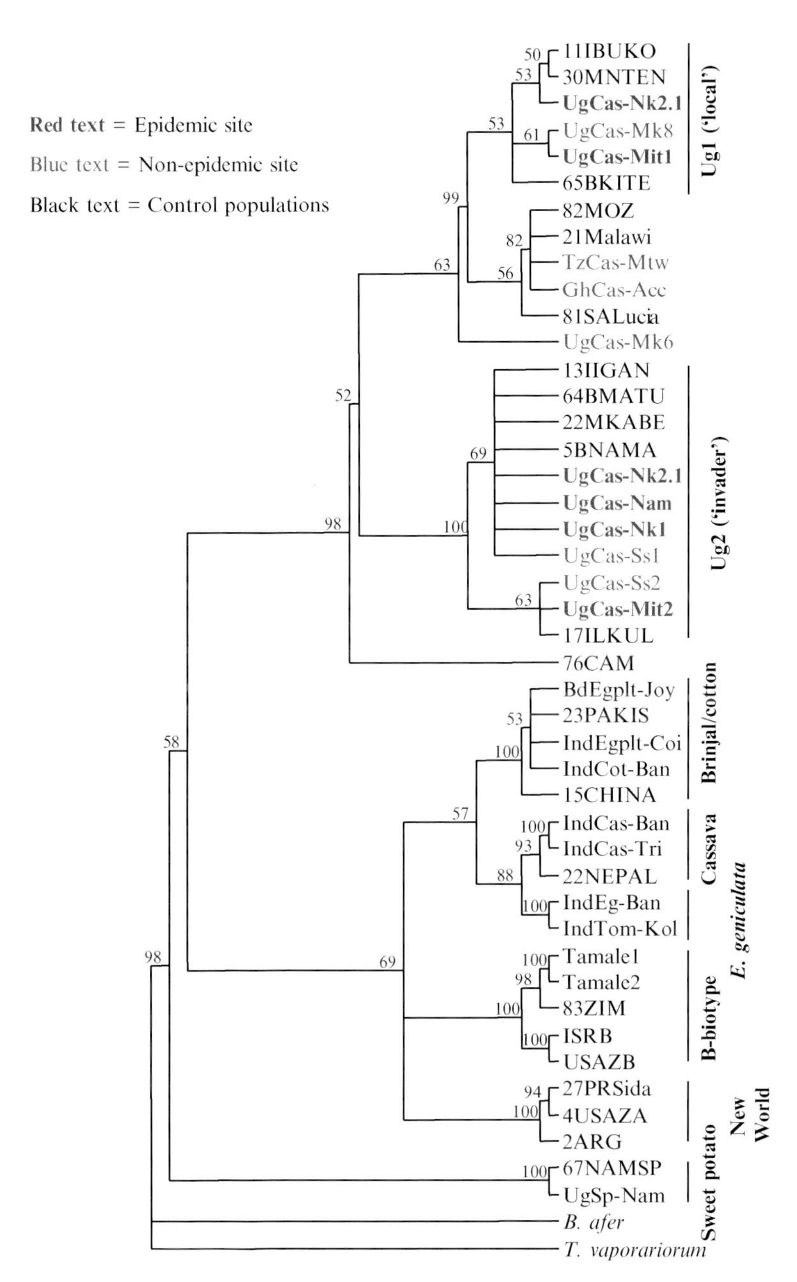

COLVIN *ET AL.*, FIG 2. Clustering of partial mtCOI sequences of *B. tabaci* associated with CMD in Africa and ToLCVD in India. Samples in red and blue originate from the CMD epidemic and pre-epidemic zones in Uganda, respectively.

TABLE II

THE *B. TABACI* PARTIAL MTCOI GENE SEQUENCES OBTAINED IN THIS STUDY FROM THE DIFFERENT LOCATIONS, MARKED IN BOLD SCRIPT, AND THOSE TAKEN FROM THE DATABASE, THEIR ACCESSION NUMBERS AND ABBREVIATIONS

Source plant, country	Location	Abbreviation	Accession numbers
Cassava, Uganda		17IIkul	AY057158
		13IIgan	AY057154
		64BMatu	AY057204
		22MKabe	AY057163
		5BNama	AY057145
		11IBuko	AY057151
		30MNten	AY057171
		65BKite	AY057205
	Mukono site 6	UgCas-Mk6	**AM040598**
	Mukono site 8	UgCas-Mk8	**AM040599**
	Mityana site 1	UgCas-Mit1	**AM040600**
	Mityana site 2	UgCas-Mit2	**AM040601**
	Nkosi site 1	UgCas-Nk1	**AM040602**
	Nkosi site 2	UgCas-Nk2.1	**AM040603**
	Nkosi site 2	UgCas-Nk2.2	**AM040604**
	Namulonge	UgCas-Nam2	**AM040605**
	Ssanji	UgCas-Ss1	**AM040606**
	Ssanji	UgCas-Ss2	**AM040607**
Cassava, Ghana	Tamale site 1	GhCas-Tam1	**AM040608**
	Tamale site 2	GhCas-Tam2	**AM040609**
	Accra	GhCas-Acc	AF418668
Cassava, Mozambique		82Moz	AF344278
Cassava, Malawi		21Malawi	AY057162
Cassava, Tanzania	Mtwara	TzCas-Mtw	AF418667
Cassava, South Africa	St Lucia	81SA Lucia	AF344259
Cassava, Cameroon		76Cam	AF344247
Zimbabwe		83Zim	AF344285
Sweet potato, Uganda	Namulonge	UgSp-Nam	AF418665
Cassava, India	Bangalore	IndCas-Ban	AF418666
	Trivandrum	IndCas-Tri	AF418670
Tomato	Kolar	IndTom-Kol	AF321928
E. geniculata	*Bangalore*	*IndEg-Ban*	AF418664
Cotton	Bangalore	IndCot-Ban	**AM040595**
Egg plant	Coimbatore	IndEgg-Coi	**AM040596**
Bangladesh	Joydebpur	BdEg-Joy	**AJ748400**

(continues)

TABLE II (*continued*)

Source plant, country	Location	Abbreviation	Accession numbers
China		15China	AF342777
Nepal		22Nepal	AF342779
Pakistan		23Pakis	AF342778
Tel Aviv, Israel	Cabbage	IsCab-Tel	AF418671
Arizona, USA		USAZ-B	AY057140
Puerto Rico	*Sida* spp.	27PRSida	AY057134
Arizona, USA		4USAZ-A	AY057122
Argentina		2Arg	AF340213
Namulonge, Uganda	Sweet potato	67NamSp	AY057207
Entebbe, Uganda	Cassava	*B. afer*	AF418673
Bangalore, India	Phyllanthus emblica L.	T. vaporariorum	AF418672

unaffected by the pandemic (Fig. 2). A further problem with the invader hypothesis arises from a survey of the *B. tabaci* cassava populations in post-epidemic areas of Uganda, which failed to detect invader nymphs developing on cassava (Sseruwagi, 2005), although several other known biotypes were identified to be present for the first time (Sseruwagi *et al.*, 2005). Thus, there is presently no evidence of a link between the spread of the pandemic and the presence of a *B. tabaci* invader population with innately superior fitness attributes.

B. Increased B. tabaci *Populations on Cassava Mosaic Disease-Affected Plants*

A possible alternative explanation for the high-*B. tabaci* populations associated with the epidemic was investigated by an experiment to determine whether plant health status influenced the population growth rate of cassava *B. tabaci*. Thirty-five, 3 weeks old, healthy cassava plants (var. Ebwanateraka) were inoculated using either virus-free or EACMV-UG-infective, *B. tabaci* collected from the pre-epidemic area. Ten days after inoculation, any *B. tabaci* nymphs were removed mechanically from the plants. Five male and five female, 1-day-old, non-viruliferous, pre-epidemic, *B. tabaci* adults were introduced onto each plant, which was then covered with a perforated plastic bag. Nymph and adult numbers were recorded both 3 and 6 weeks later as was the presence of any CMD symptoms. On the latter date, the top three leaves from each plant were removed for polymerase chain reaction (PCR) diagnostic analysis for geminivirus. Amino acids were also extracted from a randomly selected sub-sample of freeze-dried leaves that were placed in 50% methanol and fractionated by Dowex 50 ion-exchange chromatography (BDH Ltd, UK) before derivatisation with FMOC-Cl and subsequent analysis by high-performance liquid chromatography (Carratu *et al.*, 1995; Worthen and Liu, 1992). Amino acids were identified and quantified by comparison of retention times and peak area with authentic standards (Sigma Ltd). Control plants for the biochemical analysis were EACMV-UG–infected and healthy plants that were free of *B. tabaci*. PCR and biochemical work was conducted using a blind-analysis protocol.

The *B. tabaci* populations increased significantly more rapidly on EACMV-UG–infected than on equivalent healthy plants (Fig. 3 and Table III) and the concentrations of asparagine, glutamine, tryptophan and tyrosine increased significantly in the phloem sap of the diseased compared with healthy plants (Table IV).

C. Increased Field-Population Densities of B. tabaci *on CMD-Symptomatic Plants*

Many arthropod vector species have been reported to aggregate on plants infected with the viruses that they transmit (Bautista *et al.*, 1995; Eigenbrode *et al.*, 2002; Maris *et al.*, 2004) and it has been proposed that this can be explained if plant pathogens in general have evolved mechanisms to overcome plant defences against their vectors, thus promoting pathogen spread (Belliure *et al.*, 2005).

The severe chlorosis that is produced by cassava plants dually infected with both EACMV-UG and ACMV causes a big reduction in plant size, total leaf area and distribution of green leaf tissue (Maruthi *et al.*, 2002; Omongo, 2003). When total *B. tabaci* numbers per plant were counted, no significant differences were found between populations on CMD-symptomatic and healthy plants (Gibson *et al.*, 1996). This assessment, however, failed to capture the complexity of the situation, as it did not consider the preference of *B. tabaci* to feed and oviposit on the non-chlorotic areas of symptomatic leaves (Gibson et al*., 1996; Omongo, 2003).*

Measurements of leaf area of the top five leaves of representative plants showed that the total leaf area of mildly and severely affected plants was reduced to 71% and 30% of the leaf area of symptomless plants, respectively. For plants exhibiting mild symptoms, 80% of the total leaf area was green tissue compared with only 30% for plants

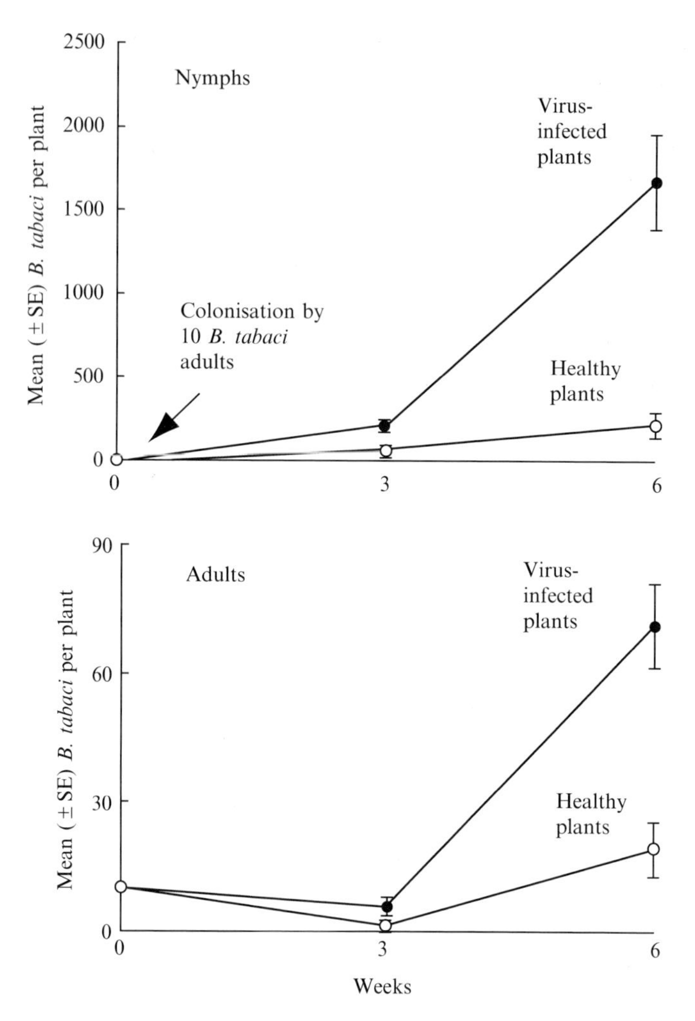

FIG 3. Population growth of African cassava *B. tabaci* on virus-infected and uninfected host plant species.

exhibiting severe symptoms. These factors were considered when investigating the densities of adult and immature stages of *B. tabaci* colonising blocks of cassava (cv. Bao) grown at the Namulonge Research Station, near Kampala, in Uganda. These data show clearly that the densities of *B. tabaci* adults and nymphs breeding on symptomatic diseased cassava leaves are significantly greater than on the

TABLE III

REPEATED MEASUREMENTS ANALYSIS OF VARIANCE FOR LN (*B. TABACI* NYMPH COUNT + 1) AND LN (*B. TABACI* ADULT COUNT + 1) AT 21 AND 42 DAYS AFTER GROUPS OF FIVE MALE AND FIVE FEMALE NEWLY EMERGED ADULTS PER PLANT WERE ALLOWED TO COLONISE CASSAVA PLANTS

Source of variation	Wald statistic	d.f.	Wald/d.f.	Chi (*P*)
Nymphs				
Time	1301.89	2	650.95	<0.001
Plant health[a]	44.53	1	44.53	<0.001
Time × plant health	65.84	2	32.92	<0.001
Adults				
Time	83.70	2	41.85	<0.001
Plant health[a]	32.53	1	32.53	<0.001
Time × plant health	21.79	2	10.89	<0.001

[a] Plants either remained healthy or became infected with EACMV-UG[Nam].

TABLE IV

THE QUANTITIES OF FOUR AMINO ACIDS IN HEALTHY (*N* = 26) AND INFECTED CASSAVA (*N* = 13)

Amino acid[a]	Healthy cassava mg/g dry leaf weight ± SE	CMD-affected cassava mg/g dry leaf weight ± SE	d.f.	*P*
Asparagine	0.028 ± 0.003	0.136 ± 0.045	37	<0.001
Glutamine	0.135 ± 0.021	0.293 ± 0.101	37	=0.045
Tryptophan	0.037 ± 0.005	0.061 ± 0.009	37	=0.018
Tyrosine	0.039 ± 0.007	0.070 ± 0.016	37	=0.043

[a] Quantities of the 16 other amino acids tested were not significantly different. Data for each amino acid were analysed separately as a two factor ANOVA (healthy or infected, with or without *B. tabaci*). In the above four cases, both the effect of *B. tabaci* and the interaction effect (*B. tabaci* × plant-health status) were not significant and therefore data for plant-health status treatments were pooled.

leaves of healthy plants (Fig. 4), reflecting the greater attractiveness and suitability of these plants to whitefly. In addition, virus content of leaves has also been shown to increase with increasing symptom severity (Fargette *et al.*, 1987), which suggests that these viruses have probably evolved mechanisms to overcome the cassava plant's defences against *B. tabaci*, thus, aiding their acquisition and subsequent inoculation.

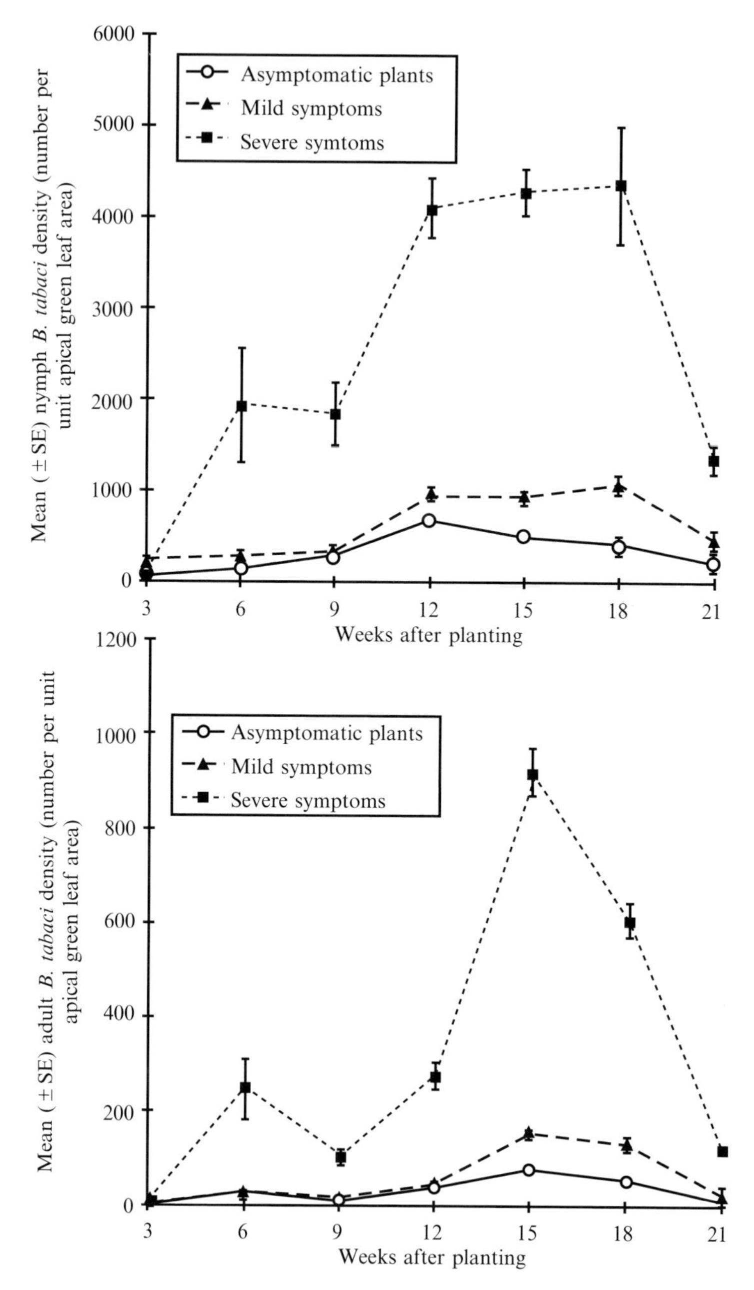

FIG 4. Densities of cassava *B. tabaci* on plants in the field expressing different levels of severity in CMD symptoms.

V. Concluding Remarks

Previous publications and the new experimental data presented here provide an increasing body of evidence to suggest that vector–virus–host plant interactions are both widespread and complex, and that they play an important role in the mechanisms driving many important arthropod-borne plant virus disease epidemics. Of the large range of potential interactions, the one that best explains the increased populations of viruliferous vectors that are associated with some plant virus epidemics is the greater population growth rate of vectors colonising and feeding on virus-infected compared with healthy host plants.

It has also often been assumed that an increase in vector numbers, at least initially, should result in an increased incidence of disease, although, for various reasons, this has not always been easy to demonstrate in the field. Nevertheless, there are several examples of significant correlations particularly between aphid numbers and virus spread within different pathosystems and from various parts of the world (Alper and Loebenstein, 1966; Dickenson *et al.*, 1956; Schwarz, 1965; Watson and Healy, 1953).

For *B. tabaci*, field experiments by several researchers have shown that CMD incidence was correlated positively with the size of the population of adult vectors about one month earlier. This delay corresponds with the expected latent period between infection and CMD symptom expression (Fargette *et al.*, 1990, 1994; Fishpool *et al.*, 1995; Leuschner, 1978; Otim-Nape, 1993). After crops reached 5 months old, however, the relationship was less clear probably because the crop became less attractive to whiteflies and less susceptible to infection (Colvin *et al.*, 1998; Fishpool *et al.*, 1995; Robertson, 1987).

Fauquet *et al.* (1988) found less consistent relationships between the size of the *B. tabaci* population and subsequent disease spread and attributed this to differences in the potency, prevalence and distribution of nearby sources of infection. High-disease incidences occurred, however, even in fields several kilometres from infected cassava, which suggested movement of viruliferous vectors over considerable distances (Fauquet *et al.*, 1988). In more recent studies in Uganda, the extent and proximity of sources of infection was shown to be important (Legg and Ogwal, 1998), and at the epidemic front there was a highly significant association between the size of the cassava *B. tabaci* population and CMD spread (Colvin *et al.*, 2004).

Swenson (1968) noted that an effective relationship between a virus and its vector(s) necessitates a dependable means of transfer to new susceptible hosts. For several of the pathosystems described here, this is achieved partially through the increased vector population growth rates on virus-infected plants which, in turn, leads to crowding and the production of migrant viruliferous vectors (Byrne and Blackmer, 1996; Lees, 1966; Zhang *et al.*, 2000). This mechanism is clearly an important component driving the CMD pandemic in East and Central Africa, where cassava plants that expressed the most severe symptoms generally contained both ACMV and EACMV-UG and generated large populations of viruliferous *B. tabaci* adults (Colvin *et al.*, 2004). We suggest that it is highly probable that the improved performance of the *B. tabaci* populations on infected cassava is the result of the higher-amino acid concentrations in the phloem sap, which is generally poor in these constituents that are essential for the growth of other phloem-feeding species such as aphids (Dixon, 1985).

The severe chlorosis associated with the top leaves of these diseased cassava plants also reduces the acceptable areas of green tissue available for *B. tabaci* feeding and development. These two effects result in significantly increased population densities on symptomatic plants that, in turn, promote the movement of viruliferous adults and thus enhance disease spread.

The combination of ACMV and EACMV-UG in cassava plants clearly produces a range of interrelated effects that are highly efficient at ensuring the continuing spread of both the viruses and the vector. This mechanism can explain how the pandemic is effectively 'self-perpetuating' and why physical barriers, such as rivers and mountains, or seasonal reductions in *B. tabaci* populations present no apparent obstacle to its continuing progress (Otim-Nape *et al.*, 2000). All that is required is the arrival of adult cassava *B. tabaci*, that are carrying EACMV-UG, into a new area where susceptible cassava varieties are being grown and where ACMV is already endemic.

The concept that the cassava viruses act to overcome the plant's defences against the *B. tabaci* may also explain the presence of the so called invader *B. tabaci* population on cassava during the CMD pandemic in Uganda from 1997 to 1999 (Legg *et al.*, 2002). Rather than being a cassava *B. tabaci* population with inherently superior fitness attributes, the available data suggest that it is more likely to be an indigenous population whose normal niche probably includes alternative plant species to cassava as its preferred hosts. The breakdown in vector resistance in the severely diseased cassava plants may, therefore, have temporarily increased the acceptability of cassava to this population, thereby enabling colonisation to take place.

A mutually beneficial interaction between viruses and *B. tabaci* populations is clearly also present in the ToLCV-pathosystem in South India. Data presented here show that significantly greater *B. tabaci* populations were generated on several ToLCV-infected weed species, which occur extremely commonly throughout most of South India and often cover much of the open ground around tomato fields (Ramappa *et al.*, 1998). This mechanism explains the rapid build-up of a large reservoir of ToLCV and *B. tabaci* during the summer season and why, when tomato was planted in a new area where it had not been grown previously, ToLCV-infected plants still appeared very rapidly and the final incidence was extremely high (Ramappa *et al.*, 1998).

In 1999 in South India, an unusually severe epidemic of ToLCV disease occurred in Karnataka state, which resulted in the complete failure of the tomato crop. This was associated with strikingly high *B. tabaci* populations exceeding 1000 adults per tomato plant (Banks *et al.*, 2001). The population associated with the ToLCV epidemic was identified subsequently by mtCOI gene sequence analysis and by the squash silver leaf assay as biotype B of *B. tabaci* (Banks *et al.*, 2001; Rekha *et al.*, 2005). This population had probably been introduced into southern India through the importation of horticultural material. Since then, the B biotype has spread rapidly throughout Karnataka, and into the neighbouring states of Andhra Pradesh and Tamil Nadu, undoubtedly assisted by the activities of nurserymen raising tomato seedlings who transport these to considerable distances. The arrival of the B biotype into new areas has also been associated with the appearance of new epidemics of *B. tabaci*-borne viruses causing cotton leaf curl, okra yellow vein mosaic and pumpkin yellow vein mosaic diseases. Completely new diseases caused by geminiviruses, such as potato leaf curl, have also appeared (Maruthi *et al.*, 2003, 2004b; Rekha *et al.*, 2005). However, the virus–vector–host plant interactions driving these new epidemics remain unknown, and research in this area represents a considerable opportunity and challenge for the future.

Certain aphid species alight on both host and non-host plants and host selection occurs only after arrival (Kennedy *et al.*, 1959a,b). Migratory behaviour takes precedence over host-plant finding behaviour and alighting on a host plant does not usually terminate flight (see Reynolds *et al.*, this volume, pp. 453–517). The aphids stay longer on host plants, feed and reproduce, and so their accumulation on them is mostly a result of differential departure rates, rather than preferential alightment (Kennedy *et al.*, 1959a,b). The non-discriminatory alighting and probing of aphids, the initial dominance of migratory over host-finding behaviour, and the intensity and duration of aphid dispersal are all activities that facilitate the spread of non-persistent viruses (Kennedy, 1960). These factors ensure a dependable means of virus spread even into crops that have no colonising vector species as occurs, for example, with BYMV (Swenson, 1957; Swenson and Nelson, 1959) and melon mosaic viruses (Dickenson *et al.*, 1949). This phenomenon may also occur with other vectors which make transitory visits to the crop, for example, leafhoppers that transmit MSV in the forest regions of Nigeria were considered to come from infection sources located outside the field and further spread occurred as the result of the transient leafhoppers spreading the virus from plant-to-plant within the field (Asanzi, 1991).

More studies on the aphid *R. padi*, which along with 25 other aphid species in North America transmits BYDV in a persistent-circulative manner, showed that they respond to the volatile cues emitted by infected plants and settle preferentially on them (Eigenbrode *et al.*, 2002; Jiménez-Martínez *et al.*, 2004b). Behavioural responses that influence vector distribution and movement are particularly important because of their potential effect on the spread of virus in the field. Such responses also include the enhanced visual attraction of alate vectors to the yellowed leaves of BYDV-infected cereals (Ajayi and Dewar, 1983).

For cassava *B. tabaci*, field data show clearly that adults colonise severely symptomatic cassava plants most rapidly (Omongo, 2003), although it remains to be discovered which cues are important in this effect. In addition, it is unknown whether the yellow mosaic symptoms caused by many geminiviruses increases the attractiveness of affected plants to the *B. tabaci* biotypes that transmit the associated geminiviruses.

The data presented here have implications for the management of cassava mosaic geminiviruses, which is the subject of a review by Thresh and Cooter (2005). For this pathosystem, the main reservoirs of both the vector and the viruses are the cassava crop (Burban *et al.*, 1992) and, in addition, our results suggest that infected plants are attractive to *B. tabaci* for oviposition, growth and development. Area-wide phytosanitation methods that remove these plants as soon as they begin to express symptoms (roguing) should, therefore, reduce the inoculum pressure to a disproportionately large degree. In the previous CMD epidemics in Uganda in the 1940s and 1960s, removal of infected plants was enforced by law, which together with resistant varieties successfully reduced losses due to CMD (Jameson, 1964). Currently, the inoculum pressure at, and immediately after, the pandemic front is so high that a control policy of only roguing is impractical, as it would result in the removal of all but the most resistant varieties. Any roguing policy, therefore, clearly needs to be combined with the supply to farmers of healthy stocks of CMD-resistant planting material.

The situation is further complicated because many of the new cassava varieties are tolerant to disease in the sense that, even though they become infected, they still produce what farmers consider to be a satisfactory yield. Farmers, therefore, are understandably unwilling to remove and destroy these plants, and so a reservoir of virus and *B. tabaci* is maintained. This situation is probably one of the main reasons that cassava *B. tabaci* populations have remained high in the post-epidemic regions of Uganda.

Other recommendations made in the 1960s (Jameson, 1964) still hold, in that varieties that keep for more than one year in the ground should probably be avoided as these become reservoirs of infection. However, varieties with this characteristic are also valued as an important food reserve in the event that the rains fail, and so farmers may be reluctant to stop growing them. The emphasis in future control methods, however, should remain on phytosanitary measures that reduce both virus incidence and vector numbers (Colvin *et al.*, 2003). Strict quarantine regulations on the movement of plant material between countries should also be enforced.

As well as the dissemination and distribution of tolerant cassava varieties that may be highly susceptible to *B. tabaci* even when virus-free, a potentially valuable strategy for tackling the breakdown in *B. tabaci* resistance associated with the CMD pandemic in Africa is to breed cassava varieties that incorporate both virus and vector resistance originating from different sources. Parental material with high levels of *B. tabaci* resistance have been identified in Africa (Omongo and Colvin, unpublished data) and in South America (Bellotti, 2002; Bellotti and Arias, 2001), and an international project has been initiated to produce new varieties with agronomic traits that are acceptable to farmers.

The evidence from the literature and the data presented here for various *B. tabaci* populations and viruses provides an increasing body of evidence demonstrating that co-adaptation and co-evolution between viruses and their arthropod vectors is even more prevalent than thought previously and that it has created several mechanisms that drive arthropod-borne plant virus disease epidemics.

Research was carried out to investigate whether or not various viral infections affected aphid performance differently and, if so, whether any pattern was apparent according to the type of virus-vector relationship. Plants infected with PLRV, a circulative virus highly dependent on *M. persicae* for dispersal and transmission, caused the greatest intrinsic rate of increase (r_m) in the aphid population. Plants infected with *Potato virus Y* (Genus: *Potyvirus*; Family: *Potyviridae*), a non-circulative virus less dependent on *M. persicae* for dispersal, caused intermediate r_m values and plants infected with *Potato virus X* (Genus: *Potexvirus*), a non-vectored virus independent of *M. persicae*, were least suitable hosts and these aphid colonies produced the lowest r_m values (Castle and Berger, 1993). This work and the additional examples described here suggest that improved vector fitness may be found most commonly where there is a close relationship between the virus and the vector, and the virus is transmitted in a persistent or semi-persistent manner. This concept is consistent with the cassava mosaic pathosystem in both Africa and India, where the cassava whitefly populations are specialised on cassava and do not interbreed with sympatric *B. tabaci* populations on other host species nearby (Maruthi *et al.*, 2001, 2004a).

The continued survival of a vector-dependent plant virus requires that the number of infected plants should never fall so low that transfer of virus to another susceptible host becomes unlikely (Swenson, 1968). The characteristics that determine transfer are therefore undoubtedly subject to selective pressures and so vectors must have a consistent relationship with their host plant(s) even if only to the extent of alighting and probing by a non-colonising species during migration. In such cases, the virus may be transmitted by several host plant-specific aphid species, but the virus itself may not be host-plant specific. Also, the fitness and abundance of diseased plants can be severely reduced by infection and their elimination is obviously not advantageous to the vector. It has been suggested that this negative interaction may explain the lack of specificity in some systems (Barbosa, 1991).

The pathosystems described in this article all involve interactions between plant, viruses and their vector(s) that result in complex and varied disease cycles of huge economic significance. BYDV, for instance, is recognised as one of the most serious viral diseases of wheat, barley, oats, grasses and other cereal crops throughout the world (Jedlinski, 1981) and the CMD pandemic continues to contribute to the food insecurity in a large region of SSA (Legg *et al.*, this volume, pp. 355–418). There is, therefore, a continuing need to study these interactions both to gain a better understanding of the level of co-adaptation and co-evolution in the pathosystems and, subsequently, to use this information to develop rational, practical, improved and sustainable pest and disease management strategies.

Acknowledgments

We are grateful to Prof. J. M. Thresh and Dr F. Kimmins for constructive criticism of this article. We are also grateful to Dr G. W. Otim-Nape and Dr A. Bua for support and encouragement provided for the experimental work and to Mr Poro Severin for technical assistance. This publication is an output from a research project funded by the United Kingdom Department for International Development (DFID) for the benefit of developing countries [DFID project codes R8247, R8425 and R8303 Crop Protection Programme]. The views expressed are not necessarily those of DFID.

References

Ajayi, O. (1986). The effect of barley yellow dwarf virus on the amino acid composition of spring wheat. *Ann. Appl. Biol.* **108:**145–149.

Ajayi, O., and Dewar, A. M. (1983). The effect of barley yellow dwarf virus on field populations of the cereal aphids, *Sitobion avenae* and *Metopolophium dirhodum*. *Ann. Appl. Biol.* **103:**1–11.

Alper, M., and Loebenstein, G. (1966). Trapping of flying aphids in several areas of Israel, as a guide to growing virus free potato and iris stocks. *Isr. J. Agr. Res.* **16:**143–146.

Asanzi, C. M. (1991). Studies of epidemiology of maize streak virus and its Cicadulina leafhopper vectors in Nigeria. Ph.D. thesis. The Ohio State University, USA.

Auclair, J. L. (1963). Aphid feeding and nutrition. *Annu. Rev. Entomol.* **8:**439–490.

Baker, P. F. (1960). Aphid behaviour on healthy and on yellows-virus-infected sugar beet. *Ann. Appl. Biol.* **48:**384–391.

Banks, G. K., Colvin, J., Chowda Reddy, R. V., Maruthi, M. N., Muniyappa, V., Venkatesh, H. M., Kiran Kumar, M., Padmaja, A. S., Beitia, F. J., and Seal, S. E. (2001). First report of the *Bemisia tabaci* B biotype in India and an associated tomato leaf curl virus disease epidemic. *Plant Dis.* **85:**231; published on-line as D-2000–1218–01N, 2000.

Barbosa, P. (1991). Plant pathogens and non-vector herbivores. *In* "Microbial Mediation of Plant-Herbivore Interactions" (P. Barbosa, V. A. Krischik, and C. G. Jones, eds.), pp. 341–382. Wiley, New York.

Barker, J. S., and Tauber, O. E. (1951). Fecundity of and plant injury by the pea aphid as influenced by nutritional changes in the garden pea. *J. Econ. Entomol.* **44:**1010–1012.

Bautista, R. C., Mau, R. F. L., Cho, J. J., and Custer, D. M. (1995). Potential of tomato spotted wilt tospovirus plant hosts in Hawaii as reservoirs for transmission by *Frankliniella occidentalis* (Thysanoptera: Thripidae). *Phytopathology* **85:**953–958.

Belliure, B., Janssen, A., Maris, P. C., Peters, D., and Sabelis, M. W. (2005). Herbivore arthropods benefit from vectoring plant viruses. *Ecol. Lett.* **8:**70–79.

Bellotti, A. C. (2002). Arthropod pests. *In* "Cassava: Biology, Production and Utilization" (R. J. Hillocks, A. Bellotti, and J. M. Thresh, eds.), pp. 209–235. CAB International, Wallingford, UK.

Bellotti, A. C., and Arias, B. (2001). Host plant resistance to whiteflies with emphasis on cassava as a case study. *Crop Prot.* **20:**813–823.

Bennett, C. W. (1971). The curly top disease of sugar beet and other plants. *Monograph 7, American Phytopathology Society.* St. Paul, Minnesota.

Berlinger, M. J. (1986). Host plant resistance to *Bemisia tabaci*. *Agr. Ecosys. Environ.* **17**:69–82.

Bos, L. (1992). New plant virus problems in developing countries: A corollary of agricultural modernization. *Adv. Virus Res.* **38**:349–407.

Boulton, M. I. (2003). Geminiviruses: Major threats to world agriculture. *Ann. Appl. Biol.* **142**:143.

Burban, C., Fishpool, L. D. C., Fauquet, C., Fargette, D., and Thouvenel, J.-C. (1992). Host associated biotypes within West African populations of the whitefly, *Bemisia tabaci* (Genn.) (Homoptera: Aleyrodidae). *J. Appl. Entomol.* **113**:416–423.

Byrne, D. N., and Blackmer, J. L. (1996). Examination of short range migration by *Bemisia*. *In* "*Bemisia:* 1995 Taxonomy, Biology, Damage, Control and Management" (G. Gerling and R. T. Mayer, eds.), pp. 17–28. Intercept Ltd.

Carratu, B., Boniglia, C., and Bellomonte, G. (1995). Optimization of the determination of amino-acids in parenteral solutions by high-performance liquid-chromatography with precolumn derivatization using 9-fluorenylmethyl chloroformate. *J. Chromatogr.* **708**:203–208.

Carter, W. (1939). Populations of *Thrips tabaci*, with special reference to virus transmission. *J. Anim. Ecol.* **8**:261–271.

Castle, S. J., and Berger, P. H. (1993). Rates of growth and increase of *Myzus persicae* on virus infected potatoes according to type of virus-vector relationship. *Entomol. Exp. et Applic.* **69**:51–60.

Castle, S. J., Mowry, T. M., and Berger, P. H. (1998). Differential settling by *Myzus persicae* (Homoptera: Aphididae) on various virus-infected host plants. *Ann. Entomol. Soc. Am.* **91**:661–667.

Chowda Reddy, R. V., Colvin, J., Muniyappa, V., and Seal, S. E. (2005). Diversity and distribution of begomoviruses infecting tomato in India. *Arch. Virol.* **150**:845–867.

Colvin, J., Fishpool, L. D. C., Fargette, D., Sherington, J., and Fauquet, C. (1998). *Bemisia tabaci* (Hemiptera: Aleyrodidae) trap catches in a field in Côte d'Ivoire in relation to environmental factors and the distribution of African cassava mosaic disease. *Bull. Entmol. Res.* **88**:369–378.

Colvin, J., Otim-Nape, G. W., Holt, J., Omongo, C., Seal, S., Stevenson, P., Gibson, G., Cooter, R. J., and Thresh, J. M. (1999). Factors driving the current epidemic of severe cassava mosaic disease in East Africa. *In* "VIIth International Plant Virus Epidemiology Symposium—Plant Virus Epidemiology: Current Status and Future Prospects," 11–16 April, 1999, Aguadulce (Almeria), Spain, pp. 76–77. International Society of Plant Pathology.

Colvin, J., Adolph, B., and Muniyappa, V. (2003). Participatory development and uptake of whitefly-transmitted virus management technologies in subsistence farming systems. *Australasian Plant Pathol.* **32**:435–439.

Colvin, J., Omongo, C. A., Maruthi, M. N., Otim-Nape, G. W., and Thresh, J. M. (2004). Dual begomovirus infections and high *Bemisia tabaci* populations: Two factors driving the spread of a cassava mosaic disease pandemic. *Plant Pathol.* **53**:577–584.

Coon, B. F. (1959). Aphid population on oats grown in various nutrient solutions. *J. Econ. Entomol.* **51**:624–625.

Costa, H. S., Brown, J. K., and Byrne, D. N. (1991). Life history traits of the whitefly, *Bemisia tabaci* (Homoptera: Aleyrodidae), on six virus-infected or healthy plant species. *Environ. Entomol.* **20**:1102–1107.

Creamer, R., Carpenter, J., and Rascon, J. (2003). Incidence of the beet leafhopper, *Circullifer tenellus* (Homoptera: Cicadellidae) in New Mexico Chile. *Southwest. Entomol.* **28**:177–182.

Dickenson, R. C., Swift, J. E., Anderson, L. D., and Middleton, J. T. (1949). Insect vectors of cantaloupe mosaic in California's desert valleys. *J. Econ. Entomol.* **42:**770–774.
Dickenson, R. G., Johnson, M. M., Flock, R. A., and Laird, E. F., Jr. (1956). Flying aphid populations in southern California citrus groves and their relation to the transmission of the tristeza virus. *Phytopathology* **46:**204–210.
Dixon, A. F. G. (1985). Aphid Ecology, p. 157. Blackie & Son Ltd, London.
Eigenbrode, S. D., Hongjian Ding Shiel, P., and Berger, P. H. (2002). Volatiles from potato plants infected with potato leafroll virus attract and arrest the virus vector, *Myzus persicae* (Homoptera: Aphididae). *Proc. R. Soc. Lond. B.* **269:**455–460.
Epstein, A. H., and Hill, J. H. (1999). Status of rose rosette disease as a biological control for multiflora rose. *Plant Dis.* **83:**92–101.
Fargette, D., Thouvenel, J.-C., and Fauquet, C. (1987). Virus content of cassava infected by African cassava mosaic virus. *Ann. Appl. Biol.* **110:**65–73.
Fargette, D., Fauquet, C., Grenier, E., and Thresh, J. M. (1990). The spread of African cassava mosaic virus into and within cassava fields. *J. Phytopathol.* **130:**289–302.
Fargette, D., Jeger, M., Fauquet, C., and Fishpool, L. D. C. (1994). Analysis of temporal disease progress of African cassava mosaic virus. *Phytopathology* **84:**91–98.
Fauquet, C., Fargette, D., and Thouvenel, J.-C. (1988). Some aspects of the epidemiology of African cassava mosaic geminivirus in Ivory Coast. *Trop. Pest Manag.* **34:**92–96.
Fereres, A., Lister, R. M., Araya, J. E., and Foster, J. E. (1989). Development and reproduction of the English grain aphid (Homoptera: Aleyrodidae) on wheat cultivars infected with barley yellow dwarf virus. *Environ. Entomol.* **18:**388–393.
Fereres, A., Araya, J. E., Housley, T. L., and Foster, J. E. (1990). Carbohydrate composition of wheat infected with barley yellow dwarf virus. *J. Plant Dis. Prot.* **97:**600–608.
Fife, J. M. (1956). Changes in the concentration of amino acids in the leaves of sugar beet plants infected with curly top. *J. Am. Soc. Sugar Beet Technol.* **9:**207–211.
Fife, J. M. (1961). Changes in the concentration of amino acids in sugar beet plants induced by virus yellows. *J. Am. Soc. Sugar Beet Technol.* **11:**327–333.
Fishpool, L. D. C., Fauquet, C., Thouvenel, J.-C., Burban, C., and Colvin, J. (1995). The phenology of *Bemisia tabaci* (Homoptera: Aleyrodidae) populations on cassava in southern Côte d'Ivoire. *Bull. Entmol. Res.* **85:**197–207.
Forrest, J. M. S. (1971). The growth of *Aphis fabae* as an indicator of the nutritional advantages of galling to the apple aphid, *Dysaphis devecta*. *Entomol. Exp. et Applic.* **14:**477–483.
GenStat Release 8.1. (2005). *Lawes Agricultural Trust (Rothamsted Experimental Station).*
Gibson, R. W., Legg, J. P., and Otim-Nape, G. W. (1996). Unusually severe symptoms are a characteristic of the current epidemic of mosaic disease of cassava in Uganda. *Ann. Appl. Biol.* **128:**479–490.
Gildow, F. E. (1983). Influence of barley yellow dwarf-infected oats and barley on morphology of aphid vectors. *Phytopathology* **73:**1196–1199.
Harrison, B. D. (1984). Potato leaf roll virus. *In* "CMI/AAB Description of Plant Viruses," p. 291. Association of Applied Biologists, Warwick, UK.
Harrison, B. D., Zhou, X., Otim-Nape, G. W., Liu, Y., and Robinson, D. J. (1997). Role of a novel type of double infection in the geminivirus-induced epidemic of severe cassava mosaic in Uganda. *Ann. Appl. Biol.* **131:**437–448.
Holt, J., Colvin, J., and Muniyappa, V. (1999). Identifying control strategies for tomato leaf curl virus disease using an epidemiological model. *J. Appl. Ecol.* **36:**1–10.

Ibbotson, A., and Kennedy, J. S. (1950). The distribution of aphid infestation in relation to leaf age II. The progress of *Aphis fabae* Scop. infestations on sugar beet in pots. *Ann. Appl. Biol.* **37:**680–696.

Jameson, J. D. (1964). Cassava mosaic disease in Uganda. *East Afr. Agric. For. J.* **29:**208–213.

Jedlinski, H. (1981). Rice root aphid, *Rhopalosiphum rufiabdominalis*, a vector of barley yellow dwarf virus in Illinois and the disease complex. *Plant Dis.* **65:**975–978.

Jiménez-Martínez, E. S., Bosque-Pérez, N. A., Berger, P. H., and Zemetra, R. S. (2004a). Life history of the bird cherry-oat aphid, *Rhopalosiphum padi* (Homoptera: Aphididae), on transgenic and untransformed wheat challenged with *Barley yellow dwarf virus*. *J. Econ. Entomol.* **97:**203–212.

Jiménez-Martínez, E. S., Bosque-Pérez, N. A., Berger, P. H., Zemetra, R. S., Hongjian, Ding, and Eigenbrode, S. D. (2004b). Volatile cues influence the response of *Rhopalosiphum padi* (Homoptera: Aphididae) to barley yellow dwarf virus-infected transgenic and untransformed wheat. *Environ. Entomol.* **33:**1207–1216.

Jones, D. R. (2003). Plant viruses transmitted by whiteflies. *Eur. J. Plant Pathol.* **109:**195–219.

Jones, A. T., Kumar, P. L., Saxena, K. B., Kulkarni, N. K., Muniyappa, V., and Waliyar, F. (2004). Sterility mosaic disease—the "Green Plague" of pigeonpea. *Plant Dis.* **88:**436–445.

Kennedy, J. S. (1951). Benefits to aphids from feeding on galled and virus-infected leaves. *Nature* **168:**825.

Kennedy, J. S. (1960). The behavioural fitness of aphids as field vectors of viruses. *Rept. VIIth Commonw. Entomol. Congr.,* London **7:**165–168.

Kennedy, J. S., Ibbotson, A., and Booth, C. O. (1950). The distribution of aphid infestation in relation to leaf age. I. *Myzus persicae* (Sulz.) and *Aphis fabae* Scop. on spindle trees and sugar beet plants. *Ann. Appl. Biol.* **37:**651–679.

Kennedy, J. S., Booth, C. O., and Kershaw, W. J. S. (1959a). Host finding by aphids in the field I. Gynoparae of *Myzus persicae* (Sulzer). *Ann. Appl. Biol.* **47:**410–423.

Kennedy, J. S., Booth, C. O., and Kershaw, W. J. S. (1959b). Host finding by aphids in the field II. *Aphis fabae* Scop. (gynoparae) and *Brevicoryne brassicae* L.; with a reappraisal of the role of host finding behaviour in virus spread. *Ann. Appl. Biol.* **47:**424–444.

Kennedy, J. S., Booth, C. O., and Kershaw, W. J. S. (1961). Host finding by aphids in the field. III. Visual attraction. *Ann. Appl. Biol.* **49:**1–21.

Kulkarni, N. K., Kumar, P. L., Muniyappa, V., Jones, A. T., and Reddy, D. V. R. (2002). Transmission of pigeonpea sterility mosaic virus by the eriophyid mite, *Aceria cajani* (Acari: Arthropoda). *Plant Dis.* **86:**1297–1302.

Kunkel, H. (1977). Membrane feeding systems in aphid research. *In* "Aphids as Virus Vectors" (K. F. Harris and K. Maramorosch, eds.), pp. 311–338. Academic Press.

Lees, A. D. (1966). The control of polymorphism in aphids. *Adv. Insect Physiol.* **3:**207–277.

Legg, J. P., and Ogwal, S. (1998). Changes in the incidence of African cassava mosaic virus disease and the abundance of its whitefly vector along south-north transects in Uganda. *J. Appl. Entomol.* **122:**169–178.

Legg, J. P., French, R., Rogan, D., Okao-Okuja, G., and Brown, J. K. (2002). A distinct *Bemisia tabaci* (Gennadius) (Hemiptera: Sternorrhyncha: Aleyrodidae) genotype cluster is associated with the epidemic of severe cassava mosaic disease in Uganda. *Mol. Ecol.* **11:**1219–1229.

Leuschner, K. (1978). "Whiteflies: Biology and Transmission of African Cassava Mosaic Disease. Proceedings of the Cassava Protection Workshop" (T. Brekelbaum, A. Bellotti, and J. C. Lozana, eds.), pp. 51–58. 7–12 November 1977, Centro Internacional de Agricultura Tropical, Cali, Colombia, CIAT Series, CE-14.

Lowe, S., and Strong, F. E. (1962). The unsuitability of some viruliferous plants as host for the green peach aphid, *Myzus persicae*. *J. Econ. Entomol.* **56**:307–309.

Maris, P. C., Joosten, N. N., Goldbach, R. W., and Peters, P. (2004). Tomato spotted wilt tospovirus improves host suitability for its vector, *Frankliniella occidentalis*. *Phytopathology* **94**:706–711.

Markkula, M., and Laurema, S. (1964). Changes in the concentration of free amino acids in plants induced by virus diseases and the reproduction of aphids. *Ann. Agric. Fenn.* **3**:265–271.

Maruthi, M. N., Colvin, J., and Seal, S. E. (2001). Mating compatibility, life-history traits and RAPD-PCR variation in *Bemisia tabaci* associated with the cassava mosaic disease pandemic in East Africa. *Entomol. Exp. et Applic.* **99**:13–23.

Maruthi, M. N., Colvin, J., Seal, S. E., Gibson, G., and Cooper, J. (2002). Co-adaptation between cassava mosaic geminiviruses and their local vector populations. *Virus Res.* **86**:71–85.

Maruthi, M. N., Colvin, J., Briddon, R. W., Bull, S. E., and Muniyappa, V. (2003). Pumpkin yellow vein mosaic virus: A novel begomovirus infecting cucurbits. *J. Plant Pathol.* **85**:64–65.

Maruthi, M. N., Colvin, J., Thwaites, R. M., Banks, G. K., Gibson, G., and Seal, S. (2004a). Reproductive incompatibility and cytochrome oxidase I gene sequence variability amongst host-adapted and geographically separate *Bemisia tabaci* populations. *Syst. Ent.* **29**:560–568.

Maruthi, M. N., Rekha, A. R., Kiran Kumar, M., Chowda Reddy, R. V., Muniyappa, V., and Colvin, J. (2004b). Diversity of *Bemisia tabaci* populations in Karnataka, South India, and emerging problems associated with the spread of the B-biotype. *In* "Proceedings of the 2nd European Whitefly Symposium," pp. 15–16, 5–9 October 2004.

Matthews, R. E. F. (1981). "Plant Virology," 2nd Ed., Academic Press, New York.

Mayer, R. T., Moshe Inbar, McKenzie, C. L., Shatters, R., Borowicz, V., Albrecht, U., Powell, C. A., and Doostdar, H. (2002). Multitrophic interactions of the silverleaf whitefly, host plants, competing herbivores and phytopathogens. *Arch. Insect Biochem. Physiol.* **51**:151–169.

McKenzie, C. L., Shatters, R. G., Jr., Doostdar, H., Lee, S. D., Moshe Inbar, and Mayer, R. T. (2002). Effect of geminivirus infection and *Bemisia* infestation on accumulation of pathogenesis-related proteins in tomato. *Arch. Insect Biochem. Physiol.* **49**:203–214.

Mesfin, T., and Bosque-Pérez, N. A. (1998). Feeding behaviour of *Cicadulina storeyi* China (Homoptera: Cicadellidae) on maize varieties susceptible or resistant to maize streak virus. *Afr. Entomol.* **6**:185–191.

Miller, J. W., and Coon, B. F. (1964). The effect of barley yellow dwarf on the biology of its vector the English grain aphid, *Macrosiphum granarium*. *J. Econ. Entomol.* **57**:970–974.

Montllor, C. B., and Gildow, F. E. (1986). Feeding responses of two grain aphids to barley yellow dwarf virus-infected oats. *Entomol. Exp. et Applic.* **42**:63–69.

Morales, F. J., and Anderson, P. K. (2001). The emergence and dissemination of whitefly-transmitted geminiviruses in Latin America. *Arch. Virol.* **146**:415–441.

Mound, L. A. (1962). Studies on the olfaction and colour sensitivity of *Bemisia tabaci* (Genn.) (Homoptera: Aleyrodidae). *Entomol. Exp. et Applic.* **5**:99–104.

Muniyappa, V. (1980). Whiteflies. *In* "Vectors of Plant Pathogens" (K. F. Harris and K. Maramorosch, eds.), pp. 39–85. Academic Press, Inc.

Muniyappa, V, and Nangia, N. (1982). Pigeonpea cultivars and selections for resistance to sterility mosaic in relation to prevalence of eriophyid mite *Aceria cajani* Channabasavanna. *Trop. Grain Legume. Bull.* **25:**28–30.

Muniyappa, V., Venkatesh, H. M., Ramappa, H. K., Kulkarni, R. S., Zeidan, M., Tarba, C.-Y., Ghanim, M., and Czosnek, H. (2000). Tomato leaf curl virus from Bangalore (ToLCV-Ban4): Sequence comparison with Indian ToLCV isolates, detection in plants and insects, and vector relationships. *Arch. Virol.* **145:**1583–1598.

Naidu, R. A., Bottenberg, H., Subrahmanyam, P., Kimmins, F. M., Robinson, D. J., and Thresh, J. M. (1998). Epidemiology of groundnut rosette virus disease: Current status and future research needs. *Ann. Appl. Biol.* **132:**525–548.

Nault, L. R. (1994). Transmission biology, vector specificity and evolution of planthopper transmitted viruses. *In* "Planthoppers: Their Ecology and Management" (R. F. Denno and T. J. Perfect, eds.), pp. 429–448. Chapman & Hall Inc.

Nault, L. R., and Ammar, E. D. (1989). Leafhopper and planthopper transmission of plant viruses. *Ann. Rev. Entomol.* **34:**503–529.

Omongo, C. A. (2003). Cassava whitefly, *Bemisia tabaci*, behaviour and ecology in relation to the spread of the cassava mosaic epidemic in Uganda. Ph.D. thesis, p. 227. Natural Resources Institute, University of Greenwich, UK.

Otim-Nape, G. W. (1993). Epidemiology of the African cassava mosaic geminivirus disease (ACMD) in Uganda. Ph.D. thesis, p. 256. University of Reading, UK.

Otim-Nape, G. W., Bua, A., Thresh, J. M., Baguma, Y., Ogwal, S., Semakula, G. N., Acola, G., Byabakama, B., Colvin, J., Cooter, R. J., and Martin, A. (2000). *Cassava mosaic virus disease in Uganda: The current pandemic and approaches to control*. University of Greenwich, PSTC28.

Pinner, M. S., Medina, V., Plaskitt, K. A., and Markham, P. G. (1993). Viral inclusions in monocotyledons infected by maize streak and related geminiviruses. *Plant Pathol.* **42:**75–87.

Polston, J. E., and Anderson, P. K. (1997). The emergence of whitefly-transmitted geminiviruses in tomato in the western hemisphere. *Plant Dis.* **81:**1358–1369.

Power, A. G. (1992). Patterns of virulence and benevolence in insect-borne pathogens of plants. *Crit. Rev. Plant Sci.* **11:**351–372.

Ramappa, H., Muniyappa, V., and Colvin, J. (1998). The contribution of tomato and alternative host plants to tomato leaf curl virus inoculum pressure in different areas of South India. *Ann. Appl. Biol.* **133:**187–198.

Réal, P. (1955). Le cycle annuel du puceron de l'arachide (*Arachis leguminoseae* Theobald), en Afrique Noir francaise et son determinisme. *Revue de Pathologie Vegetal* **34:**3–122.

Reddy, M. V., and Nene, Y. L. (1980). Influence of sterility mosaic resistant pigeonpeas on multiplication of the mite vector. *Indian Phytopathol.* **33:**61–63.

Rekha, A. R., Maruthi, M. N., Muniyappa, V., and Colvin, J. (2005). Occurrence of three genotypic clusters of *Bemisia tabaci* and the rapid spread of the B-biotype in South India. *Entomol. Exp. et Applic.* **117:**221–233.

Robertson, L. A. D. (1987). The role of *Bemisia tabaci* Gennadius in the epidemiology of ACMV in East Africa: Biology, population dynamics and interaction with cassava varieties. *In* "Proceedings International Seminar on the African Cassava Mosaic Disease and its Control," pp. 57–63, 4–8 May 1987, Yamoussoukro, Ivory Coast.

Rubenstein, G., and Czosnek, H. (1997). Long-term association of tomato yellow leaf curl virus with its whitefly vector, *Bemisia tabaci:* Effect on the insect transmission capacity, longevity and fecundity. *J. Gen. Virol.* **78:**2683–2689.

Saikia, A. K., and Muniyappa, V. (1989). Epidemiology and control of tomato leaf curl virus in southern India. *Trop. Agric.* **66:**350–354.

Sandström, J., Telang, A., and Moran, N. A. (2000). Nutritional enhancement of host plants by aphids—a comparison of three aphid species on grasses. *J. Insect. Physiol.* **46:**33–40.

Schwarz, R. E. (1965). Aphid-borne virus diseases of citrus and their vectors in South Africa. B. Flight activities of citrus aphids. *S. Afr. J. Agric. Sci.* **8:**931–940.

Selman, I. W., Brierley, M. R., Pegg, G. F., and Hill, T. A. (1961). Changes in the free amino acids and amides in tomato plants inoculated with tomato spotted wilt virus. *Ann. Appl. Biol.* **49:**601–615.

Sinha, R. C. (1981). Vertical transmission of plant pathogens. *In* "Vectors of Disease Agents" (J. J. McKelvey, B. F. Eldridge, and K. Maramorosch, eds.), pp. 109–121. Praeger, New York.

Sohi, S. S., and Swenson, K. G. (1964). Pea aphid biotypes differing in bean yellow mosaic virus transmission. *Entomol. Exp. et Appl.* **7:**9–14.

Sseruwagi, P. (2005). Molecular variability of cassava *Bemisia tabaci* and its effect on the epidemiology of cassava mosaic geminiviruses in Uganda. Ph.D. thesis. The University of the Witwatersrand, Johannesburg, Republic of South Africa.

Sseruwagi, P., Legg, J. P., Maruthi, M. N., Colvin, J., Rey, M. E. C., and Brown, J. K. (2005). Genetic diversity of *Bemisia tabaci* (Gennadius) (Hemiptera: Aleyrodidae) populations and the presence of the B biotype and a non-B biotype that can induce silverleaf symptoms in squash, in Uganda. *Ann. Appl. Biol.* **147:**253–265.

Swenson, K. G. (1957). Transmission of bean yellow mosaic virus by aphids. *J. Econ. Entomol.* **50:**727–731.

Swenson, K. G. (1968). Role of aphids in the ecology of plant viruses. *Ann. Rev. Phytopathol.* **6:**351–374.

Swenson, K. G., and Nelson, R. L. (1959). Relation of aphids to the spread of cucumber mosaic virus in gladiolus. *J. Econ. Entomol.* **52:**421–425.

Thresh, J. M. (1967). Increased susceptibility of blackcurrant bushes to the gall-mite vector (*Phytoptus ribis* Nal.) following infection with reversion virus. *Ann. Appl. Biol.* **60:**455–467.

Thresh, J. M. (1980). The origins and epidemiology of some important plant virus diseases. *Ann. Appl. Biol.* **5:**1–65.

Thresh, J. M., and Cooter, R. J. (2005). Strategies for controlling cassava mosaic virus disease in Africa. *Plant Pathol.* **54:**587–614.

Thresh, J. M., Otim-Nape, G. W., Legg, J. P., and Fargette, D. (1997). African cassava mosaic disease: The magnitude of the problem. *Afr. J. Root Tuber Crops* **2:**13–19.

Vasudeva, R. S., and Sam, Raj (1948). Leaf curl disease of tomato. *Phytopathology* **18:**364–369.

Watson, M. A., and Healy, M. J. R. (1953). The effect of beet yellows and beet mosaic viruses in the sugar-beet crop II. The effects of aphid numbers on disease incidence. *Ann. Appl. Biol.* **40:**38–59.

Whitham, T. G. (1979). Territorial behaviour of *Pemphigus* gall aphids. *Nature* **279:**324–325.

Willekens, J. (2003). Mechanism of vector resistance in groundnut to control *Groundnut rosette virus* disease in sub-Saharan Africa. Ph.D. thesis, p. 199. The University of Greenwich, Natural Resources Institute, UK.

Williams, C. T. (1995). Effects of plant age, leaf age and virus yellows infection on the population dynamics of *Myzus persicae* (Homoptera: Aphididae) on sugarbeet in field plots. *Bull. Entmol. Res.* **85:**557–567.

Worthen, H. G., and Liu, H. (1992). Automatic pre-column derivatization and reversed-phase high-performance liquid-chromatography of primary and secondary amino-acids in plasma with photodiode array and fluorescence detection. *J. Liq. Chromatogr.* **15:**3323–3341.

Zhang, X.-S., Holt, J., and Colvin, J. (2000). A general model of plant-virus disease infection to incorporate vector aggregation. *Plant Pathol.* **49:**435–444.

Zhou, X., Liu, Y., Clavert, L., Munoz, C., Otim-Nape, G. W., Robinson, D. J., and Harrison, B. D. (1997). Evidence that DNA-A of a geminivirus associated with severe mosaic disease in Uganda has arisen by interspecific recombination. *J. Gen. Virol.* **78:**2101–2111.

ADVANCES IN VIRUS RESEARCH, VOL 67

THE MIGRATION OF INSECT VECTORS OF PLANT AND ANIMAL VIRUSES

D. R. Reynolds,* J. W. Chapman,† and R. Harrington†

*Natural Resources Institute, University of Greenwich
Chatham Maritime, Kent ME4 4TB, United Kingdom
†Plant and Invertebrate Ecology Division, Rothamsted Research
Harpenden, Hertfordshire AL5 2JQ, United Kingdom

Many plant viruses are transmitted by insect vectors, and the pattern of spread of a particular virus will depend critically on the characteristic flight activity of its vector. In most cases flight is over

0065-3527/06 $35.00
DOI: 10.1016/S0065-3527(06)67012-7

short distances, but in this chapter the focus is on plant virus vectors (particularly some species of Homoptera), which can migrate tens or hundreds of kilometres and cause disease outbreaks far from the source of infection. Parallels are drawn with the atmospheric transport of vectors of animal and human diseases, particularly small blood-feeding Diptera such as blackflies, biting midges and mosquitoes. The chapter first summarizes the conceptual framework that distinguishes migration from other types of movement, and then briefly outlines some of the methodologies used to study long-range migration of insect vectors. Several case studies of vector migration, particularly ones illustrating recent progress, are then presented. Emphasis is given to the interactions between migratory flight behaviour and atmospheric processes, which together control the ascent, horizontal movement, aerial concentration or dispersion, and the eventual landing of the vectors. Attention is also given to aspects of local post-migratory host-seeking flights that result in the inoculation of the virus into new hosts. Finally, we suggest priorities for further research on insect vectors, which is needed for the formulation or improvement of disease control strategies.

I. Introduction

An important factor in the potential spread of many disease-causing viruses of plants and vertebrate animals is that they are vectored by insects, some of which can move distances of tens or even hundreds of kilometres, often within periods ranging from a few hours up to day or so, but occasionally longer. This raises the possibility that uninfected plant or animal populations that are far from any previous sources of inoculum may suddenly (and sometimes unexpectedly) become infected. Some vectors are very common in the air, and there may be a veritable 'rain' of viruliferous insects landing on susceptible hosts which, even in very isolated or ephemeral habitats, are at risk of infection.

As most vectors are relatively small insects, movements over any substantial distance will necessarily entail the power of the wind to transport individuals much further and faster than would be possible by self-propelled flight alone. The modern view of these long-range movements is that they are seldom 'accidental' but are *actively* initiated and maintained through specialized behaviour patterns occurring during a distinct migratory phase in the life-cycle (Dingle, 1996). Often this occurs soon after the adult becomes flight-capable, but before sexual maturity (i.e. migration is post-teneral and pre-reproductive) (Johnson,

1969). Migration will usually be initiated by sustained climbing flight taking the individual out of its 'flight boundary layer' (a layer of air, extending a variable distance up from the ground, where the wind speed is lower than the insect's flight speed (Taylor, 1974)), thus facilitating long-distance, windborne transport (Drake and Gatehouse, 1995; Gatehouse, 1997; Johnson, 1969).

Earlier work on long-range windborne migration by plant virus vectors has been reviewed in Thresh (1983, 1985), Taylor (1986), Irwin and Thresh (1988), Berger and Ferriss (1989) and Irwin and Kampmeier (1989), and by vectors of animal viruses in Sellers (1980) and Pedgley (1983). We concentrate mainly on studies published since these reviews, but they should still be consulted as many of the ecological principles expounded, and epidemiological issues raised, are still highly relevant today. Aspects of the terminology and mechanisms of the arthropod transmission of viruses have been reviewed by Nault (1997) and Gray and Banerjee (1999); here transmission mechanisms are discussed only where they have a direct bearing on the infectivity of the immigrant vector.

A. *Definition and Characteristics of Migration*

One of the most significant advances of recent years has been the improvement in the conceptual definition of various movement behaviours, particularly following Kennedy's influential (1985) paper, in which he removed the confusion caused by the confounding of behavioural and ecological definitions of migration. It is best to consider migration as a *behavioural* process that has *ecological* consequences, and the most complete and satisfactory definition is that of Kennedy (1985), which has been widely accepted (see Byrne and Blackmer, 1996; Clements, 1999; Dingle 1996; Drake *et al.*, 1995; Gatehouse, 1987, 1997; Hardie, 1993; Woiwod *et al.*, 2001). Kennedy wrote:

> Migratory behaviour is persistent and straightened-out movement effected by the animal's own locomotory exertions or by its active embarkation on a vehicle. It depends upon some temporary inhibition of station-keeping responses but promotes their eventual disinhibition and recurrence.

Migratory locomotion thus tends to be more continuous than other types of movement and the migrant's track over the ground becomes straightened out, causing an individual to be transported away from a localised area (home-range). Migratory flight in most insects uses the wind as a vehicle, and 'active embarkation' is generally achieved by the migrant ascending high into the air. An important characteristic of

Kennedy's definition is that during the migratory phase, responses to 'vegetative' stimuli (e.g. food, shelter, mates, egg-laying sites) are temporarily inhibited to some extent. Inhibition of the normal 'station-keeping' flight and settling responses will, in itself, allow the migrant to escape from a localized emergence or diapause site, and to cover new ground. As migration continues, there is a gradual increase in responsiveness to vegetative stimuli that will eventually halt it, and in some aphids (e.g. *Aphis fabae, Rhopalosiphum padi*) at least, there is a complex system of antagonistic inhibitory and excitatory interactions between migratory and vegetative behaviours (see summary of Kennedy's work on *A. fabae* in Dingle, 1996; also Nottingham and Hardie, 1989; Nottingham *et al.*, 1991). There is accumulating evidence that migratory behaviour is only one of a suite of co-evolved traits (the '*migration syndrome*'), which also include morphological, physiological and biochemical characteristics (Dingle, 1996, 2001).

In contrast to migration, 'vegetative' (= foraging or 'appetitive') movements are directed towards the exploitation of resources, particularly those required for growth and reproduction, and these movements are readily interrupted by an encounter with the resource items in question (food, a mate, and so on) (Dingle, 1996; Hardie *et al.*, 2001; Kennedy, 1985). Vegetative movements often tend to be station-keeping and retain the animal within a locality, which can be equated with the 'home-range' of vertebrates (Dingle, 1996). However, if the required resource is not available, or is not found within a given time, an individual may switch to a different mode of behaviour that may allow it to explore a wider area than its local habitat patch or home-range. These movements, sometimes called 'ranging', will still tend to cease if the resource is found—there is no prolonged inhibition of the vegetative responses such as occurs with migration.

Some migrations cover rather short distances (Solbreck, 1985), and sometimes these may be difficult to distinguish from the dispersal that occurs as a side-effect of various 'vegetative' movements. Local movements of vectors are certainly important in spread of virus diseases, as has been emphasised in several reviews (Byrne and Blackmer, 1996; Garrett and MacLean, 1983; Loxdale *et al.*, 1993), but as these movements take place relatively close to the ground, they can usually be studied by standard entomological techniques (short-range mark–recapture experiments and various forms of ground-based trapping). In this chapter, the focus is mainly on longer-range migrations, that is, at least several kilometres or tens of kilometres in extent. The fact that migrants usually ascend high into the air means that these movements are intrinsically difficult to study. There may be so few data on the timing, scale and intensity of vector migration that the process is virtually ignored when formulating disease management strategies. This is unfortunate because with a virulent disease, the immigration of just a few infective individuals may lead to an outbreak in a susceptible host population.

B. Extended Foraging

Although the vast majority of long-range displacements of insect vectors probably occur due to migration, some powerfully flying insects may move fairly long distances (hundreds of metres, or even kilometres) during foraging flights. As mentioned previously, there is no inhibition of vegetative responses during these flights as there is in true migration, but if the resource (e.g. a suitable host in the case of a blood-feeding dipteran) is not encountered, the flights may become quite extended. Examples might include the movement of tsetse flies (vectors of the protozoan *Trypanosoma*; Hargrove, 2000), tabanids (vectors of the nematode causing loiasis; Chippaux *et al.*, 2000), and calliphorid or muscid flies that can be non-specific mechanical vectors of viruses and other pathogens, which have contaminated their mouthparts or bodies (Carn, 1996; Foil and Gorham, 2000). These movements are not considered in detail here.

II. Techniques for the Study of Vector Migration

Techniques used to monitor insect migration have been described by Reynolds *et al.* (1997) and Osborne *et al.* (2002), and a detailed review is inappropriate here where the focus is on some recent methodological developments. All the techniques have advantages and disadvantages, and large campaigns to study important vectors may use a combination of methods including: aerial sampling, monitoring by field surveys or networks of ground traps, physiological studies, various forms of artificial and natural marking, remote sensing (particularly entomological radar) and trajectory analysis (see, e.g. the multi-disciplinary investigation of the migration of aphids into Illinois (Section IV.C)).

A. Aerial Sampling

It is clearly important to be able to catch vectors during high-altitude migration, in order to confirm the identity of the migrants and also to test for the presence of viruses in the migrating population.

By contrast, vectors caught in ground traps may have acquired the virus post-migration. Samples of migrants may also be required for 'genetic finger-printing' or the assessment of flight-fuel depletion (Irwin and Thresh, 1988).

Well-funded projects may permit the aerial sampling of virus vectors using nets attached to aircraft, either conventionally piloted (Berry and Taylor, 1968; Dung, 1981; Glick, 1939; Greenstone *et al.*, 1991; Hollinger *et al.*, 1991) or remote-controlled (Gottwald and Tedders, 1986; Shields and Testa, 1999). Powered aircraft have the advantage of being able to sample relatively large volumes of air in a short time, at a range of altitudes, and to sample in lower wind speeds than is possible with netting from balloons and kites (see later). The Illinois aphid migration project (Hendrie *et al.*, 1985), used a specially designed aerial sampler mounted on a helicopter (Hollinger *et al.*, 1991). This sampler incorporated features to: (1) ensure that air entered the sampler in straight streamlines (avoiding spillage at the mouth of the device), (2) separate samples with respect to time, altitude, and so on, and (3) reduce damage to the catch by decreasing the velocity of impact in the collection sieve.

If resources do not permit aircraft sampling, a relatively inexpensive alternative is netting from tethered helium-filled blimps (kytoons) (Chapman *et al.*, 2004) or, if the wind is sufficiently constant, from parafoil kites (Farrow and Dowse, 1984). There may, however, be restrictions on the flight heights of tethered balloons due to air traffic safety regulations. For example, in the United Kingdom, special permission is required from the Civil Aviation Authority to fly a tethered balloon above 60 m (or less than this in controlled airspace near airfields). One would often need to sample above this height, especially in stably stratified conditions at night, when insect activity is often concentrated in layers near the top of the temperature inversion at heights of *c.* 200–400 m above ground, and this may necessitate operating the balloon in an aircraft exclusion zone (Chapman *et al.*, 2004).

Aerial sampling can be particularly effective if radar can be used to position the net within layered concentrations of migrants (Hendrie *et al.*, 1985; Riley *et al.*, 1991; and see later), because densities there may be two or three orders of magnitude greater than those above or below the layer. If the insect layers are thought to be associated with local maxima in the temperature or wind speed (Hendrie *et al.*, 1985; Isard *et al.*, 1990), the altitude of these meteorological features can be determined by radiosonde ascents or perhaps by meteorological sodars (Riley *et al.*, 1995a), and the net positioned accordingly.

B. Networks of Ground-Based Traps

The sampling of vectors high in the air, valuable as it is, usually only takes place during short 'campaigns' of a few weeks, and in a relatively restricted area. For season-long monitoring of vector activity over large areas (country- or even region-wide), it is usually necessary to rely on a network of ground traps or surveys (Robert, 1987). The Rothamsted Insect Survey network of 12.2-m high suction-traps has been monitoring aphid populations over the United Kingdom since the 1960s (Woiwod and Harrington, 1994). It has been shown that aphids caught at this height are engaged in flights lasting, on average, ~2 h, suggesting, but not proving, that the flights are predominantly migratory rather than vegetative (Taylor and Palmer, 1972). Certainly, aphids at this height generally show 'cumuliform drift' (Taylor, 1986), that is, they are being circulated to varying heights in the atmosphere by convection so that they become randomly distributed. The suction-trap data are used for fundamental studies on factors affecting the dynamics of aphid populations and to provide information that aids aphid control decisions. The original United Kingdom trap network has been expanded to cover about 20 European countries and, under the European Union EXploitation of Aphid Monitoring systems IN Europe (EXAMINE) project, the data have been collated on a single database to facilitate national and pan-European analyses (Harrington *et al.*, 2004). The whole European dataset has been used to study the potential impacts of climatic and land use changes on aphid phenology and abundance (e.g. Cocu *et al.*, 2005; Harrington *et al.*, 2007), but most work so far done on plant-virus epidemiology and forecasting has been at a national or more local scale (Northing *et al.*, 2004; Qi *et al.*, 2004; see Section VI).

It is rarely possible for data from such networks to provide direct evidence of how far individual aphids have flown. This is because most aphids that are common enough to be sampled regularly by this method also have a wide distribution, and it is difficult to eliminate the possibility of a local source, which may even be imported plant material. Furthermore it is never possible to determine whether the aphids caught were undergoing migratory or vegetative flight. However, there is occasionally strong circumstantial evidence for long-distance movement. Loxdale *et al.* (1993) provide a useful review of this evidence and conclude that such movement is exceptional. Nonetheless, such exceptions may have important consequences with respect to the introduction of viruses to an area. Heathcote (1984) provided evidence from suction trap samples that large numbers of *Myzus persicae* (peach–potato aphid), an important vector of potato and sugar beet viruses in the United Kingdom, arrived in eastern England from the Netherlands at the end of July 1982, albeit too late in the cropping season to cause significant problems as a result of any viruses they may have been carrying, but demonstrating the possibility of such problems arising. Conversely, the record numbers of *Metopolophium dirhodum* (rose–grain aphid) found in eastern England in 1979, were shown to have been 'home grown' in the United Kingdom (Heathcote, 1980). Apparent mass migrations of aphids from southern Europe northwards are often the result of increasingly late local development as a result of the relationship between crop phenology and latitude (Dewar *et al.*, 1980), demonstrating the care needed in the interpretation of data.

Networks of aphid-sampling suction traps have also been set up in some states in the United States (e.g. Quinn *et al.*, 1991). Among other long-standing networks for monitoring highly migratory vectors are the extensive systems of standardized crop searches, light-trapping and net trapping used to forecast the rice planthoppers *Nilaparvata lugens, Sogatella furcifera* and *Laodelphax striatellus* in East Asia (Kisimoto and Yamada, 1998; Tang *et al.*, 1994; Watanabe, 1995; Zhou *et al.*, 1995).

Finally, it should be noted that catches far from possible sources in inhospitable places, such as deserts (Dickson, 1959) or ships at sea (Hardy and Cheng, 1986; and references in Kisimoto and Sogawa, 1995), demonstrate that at least a proportion of a vector population has the potential for long-range flight. Observations at ocean weather station 'Tango', south of Japan, were highly influential in demonstrating that *S. furcifera* and *N. lugens* can migrate to Japan, at a time when a published report was advocating that they overwintered there as eggs (Kisimoto, 1981).

C. Radar

Radar offers a powerful means of observing the migration of many high-flying insects, and the timing, altitude, density and direction of flight of the migrants can be established. The precise height of flight is often critical in determining the origin of migrant populations (Irwin and Thresh, 1988, and see Section II.G). However, entomological radars usually operate at X-band (3.2-cm wavelength) and many virus vectors are too small to be detected individually within the working range of these systems. The new vertical-looking radars (VLRs) (Chapman *et al.*, 2003b) are a slight improvement in this respect over the older azimuthally scanning radars where the beam was directed at a slant angle, but insect targets still have to weigh ~2 mg or more to be detected at the shortest available VLR range (~150 m).

Useful information can sometimes be obtained from observations of collective echo from concentrations of small insects, for example, the observations on layers of migrating aphids made with the 10-cm wavelength 'CHILL' meteorological radar by Hendrie *et al.* (1985). In this case, the species composition of insects forming the layers was determined by aerial sampling from a helicopter. Nieminen *et al.* (2000) utilised a Doppler C-band (5.33-cm wavelength) weather radar to document mass migrations of aphids (including *R. padi*) into the Helsinki area of Finland (see Section IV.C).

To detect *individual* small insects like aphids over most of their heights of migratory flight a shorter wavelength is required instead of the 3.2-cm normally used by entomological radars. A high-frequency (8-mm wavelength) entomological *scanning* radar has, in fact, been built (Riley, 1992) and was used successfully to observe individual brown planthoppers, *N. lugens* (which weigh ~2 mg) at heights up to 1000 m above ground (Riley *et al.*, 1987, 1991, 1994). Simultaneous sampling by a balloon-supported net confirmed the identity of the planthoppers, and also enabled testing for the presence of viruses in the migrating population (Riley *et al.*, 1994).

However, in contrast to X-band technology, there is a comparative lack of inexpensive millimetric radar components available, which could be used to construct a robust and reliable high-frequency VLR system, and some components would probably have to be custom made at considerable cost. A more feasible alternative, particularly for the study of nocturnal migrants flying at high aerial densities, might be to use an unsophisticated vertical-pointing radar sounder to reveal the altitudinal location of the mass of small targets, allowing these to be sampled efficiently by aerial trapping. A simple X-band sounder (without the rotating dipole or nutating beam configuration included in our VLR system) could probably be constructed for ~£7000 (=$12,000).

D. Artificial Markers

There is a plethora of methods of artificially tagging insects including labels, mutilation marks, dyes, paints, fluorescent substances, rare elements and radio-isotopes (Akey, 1991; Reynolds *et al.*, 1997; Southwood and Henderson, 2000). However, a major problem is the very low probability of re-capturing marked individuals after long-distance migration, because of the immense dilution of the released population.

Nevertheless, a large multi-institute Chinese study in the 1970s did succeed in recapturing some marked planthoppers after long-range migration (Department of Plant Protection, Nanjing Agricultural College *et al.*, 1981). Generally, however, these methods are more appropriate for studying movement over distances of tens of metres up to a few kilometres. For example, fluorescent dusts have been used in studies of migration in the whitefly *Bemisia tabaci* (Byrne and Blackmer, 1996; Cohen *et al.*, 1988), and rubidium marking of the planthoppers *N. lugens* and *S. furcifera* was used as an aid in interpreting trap catches (Padgham *et al.*, 1984).

E. Natural Markers

Some ways in which insects become naturally marked (e.g. by pollen, algae, phoretic mites, or gut contents (Reynolds *et al.*, 1997)) have had little application in the study of vector migration. Among methods that have had more utility are elemental composition and, especially, molecular markers.

1. Elemental Composition

The origin of a putative migrant can sometimes be identified from the rarer elements that have accumulated in the body of the insect during its growth stage (its 'chemoprint') (Turner and Bowden, 1983), assuming that the distribution of elements is characteristic of the geographical area in which the insect developed. However, some of the studies on vector aphids have not produced clear source-related chemoprints (e.g. a study on *R. padi* in Britain showed differences between different morphs and host plants rather than source areas; Sherlock *et al.*, 1986).

2. Molecular Markers

The use of molecular markers in entomology has been reviewed by Loxdale and Lushai (1998) and Hoy (2003), and specifically in relation to aphids by Loxdale and Lushai (2007). These techniques are sometimes used to detect migration *directly* by 'genetic finger-printing', where a genotype known from one location is detected at another from which it was previously absent (Foster *et al.*, 1998). More commonly, however, the degree of migration is *inferred* from studies of gene flow between sub-populations assessed by comparing allele and genotype frequencies. Earlier studies often used protein markers, especially allozymes, but more recent workers have tended to employ various types of DNA marker. These provide more variation per locus, and (particularly if non-coding sequences, such as microsatellites, are used) are likely to be selectively neutral (Osborne *et al.*, 2002).

a. Allozymes These are variant forms of an enzyme that are coded for by different alleles at the same locus of a gene. The allelic variation can be detected relatively easily by electrophoresis and staining. Allozymes have been used to make deductions about the gene-flow between sub-populations of aphids (Loxdale and Brookes, 1990; Loxdale *et al.*, 1985) (see later).

b. DNA-Based Markers A variety of nuclear and mitochondrial DNA-based markers is now available for investigations of the genetic structure of natural insect populations (Loxdale and Lushai, 2001, 2007; Osborne *et al.*, 2002). Among the markers used in migration-related studies, are: microsatellite repeats (Llewellyn *et al.*, 2003; Simon *et al.*, 1999); random amplified polymorphic DNA (RAPDs) (De Barro *et al.*, 1995) and amplified fragment length polymorphisms (AFLPs) (Simon *et al.*, 2003). As far as the molecular ecology of virus vectors is concerned, aphids (Loxdale and Lushai, 2007) are the most studied group, but there have also been studies of mosquitoes (Lourenço de Oliveira *et al.*, 2003) and planthoppers (Mun *et al.*, 1999).

Studies of microsatellite polymorphisms in alates of the grain aphid, *Sitobion avenae*, caught in suction traps along a north–south transect across Britain, revealed similar allele frequencies (Llewellyn *et al.*, 2003), and this supports other evidence (Chapman *et al.*, 2004; Hardy and Cheng, 1986) that the species can be highly migratory. There was a degree of genetic differentiation between British and Spanish populations of *S. avenae* (Loxdale *et al.*, 1985), but surprisingly little considering the distance (800 km) and geographical barriers between the two countries. Overall, this suggests high gene-flow/aerial movements of this aphid species over large areas of Western Europe. Other species common in 12.2-m high suction traps, such as *R. padi* and the sycamore aphid, *Drepanosiphum platanoidis*, show homogenous allele frequency patterns over large spatial scales (>100 km), indicating high levels of migration (Delmotte *et al.*, 2002; Wynne *et al.*, 1994).

In contrast, a study of another species of *Sitobion, S. fragariae*, using allozymes suggested restricted gene flow and low dispersal between local populations (Loxdale and Brookes, 1990). Other examples where there was good evidence for only short-range migration (~15–30 km) include the damson-hop aphid, *Phorodon humuli*, and the tansy aphid, *Macrosiphoniella tanacetaria* (see Loxdale and Lushai, 2007).

It should be mentioned that some studies using molecular markers to measure gene-flow produce results that are difficult to reconcile with other information on migration biology (Peterson *et al.*, 2001), underlining the continuing need for direct assessment of migration.

F. Flight Mills and Flight Chambers

Laboratory methods of assessing the relative flight potential of insects, and the factors influencing it, have been reviewed by Reynolds *et al.* (1997) and Reynolds and Riley (2002). 'Flight mill' systems, where the insect is tethered to a rotating arm, are often intended for large insects such as moths, but mills sensitive enough to be used with small-sized vector species have been developed for *Simulium* blackflies (Cooter, 1982), and *Cicadulina* spp. and *Nephotettix virescens* leafhoppers (Cooter *et al.,* 2000; Riley *et al.*, 1997). Among other effects, Cooter *et al.* (2000) found that there was a higher proportion of 'fliers' among *N. virescens* adults feeding for a short period on tungro-diseased rice compared with healthy rice, and this behaviour is likely to facilitate virus spread. Tethered-flight studies of the aphid *R. padi* showed individuals carrying *Barley yellow dwarf virus* (BYDV) had shorter flight durations and a reduced age at which maiden flights started, compared with non-viruliferous aphids (Levin and Irwin, 1995).

Because most insect vectors are small sized it is possible to investigate the partitioning between migratory and vegetative behaviours during free flight in a vertical wind-tunnel resembling that originally developed by J. S. Kennedy (Kennedy and Booth, 1963) for his classic studies on *A. fabae*. A computerised version of the Kennedy flight chamber has been developed (David and Hardie, 1988), and a rather similar system has been employed to investigate the flight of *Bemisia* whiteflies (Blackmer and Byrne, 1993a,b; Blackmer *et al.*, 1995). Upward flight of the test insect towards a white light located in the ceiling of the chamber (accompanied by disregard of plant-mimicking visual cues, initially at least) is considered to be evidence of migratory flight.

It is now possible to fly tethered insects in ingenious 'virtual-reality' flight simulators where they experience visual stimuli as if under free-flight conditions (Dickinson and Lighton, 1995). Flight forces and wing-beat frequencies are sensed and used to provide appropriate visual feedback and other visual stimuli via panoramas on the arena wall produced by, for example, computer-controlled banks of light-emitting diodes (LEDs). If the test species has good vision and flight-control (like fruitflies) it can even be tracked by a stereo video system in a free-flight arena (the 'Fly-O-Rama') with visual panorama or odour cues provided (see http://www.dickinson.caltech.edu/research_flyorama.html).

All the above systems require investment in equipment development, but useful results can also be obtained by suspending the test insect by a stick attached to its thorax, and recording the duration of wing-beating (Dingle, 1985; Rose, 1972). It is disappointing that these very simple and inexpensive flight-assessment methods have not been used more widely for virus vectors where currently there is a lack of any direct information on flight potential.

G. *Trajectories*

When there is reason to believe that a significant migration event has occurred, it is often of interest to determine likely source areas (in the case of an invasion), or sink areas (in the case of an observed emigration), or even the complete migration 'pathway' (*sensu* Drake *et al.*, 1995). This can be achieved by the trajectory models, which utilize meteorological information at various levels of sophistication to estimate the path taken by a hypothetical parcel of air carried in the windfield (Reynolds *et al.*, 1997). Where information on the timing of an invasion is imprecise (to within a period of a few days, say) only a general suggestion of the weather system responsible for transporting the vectors may be warranted. For example, the spread of African horse sickness by *Culicoides* midges to Spain and Cyprus, respectively, probably occurred on spells of winds unusual for the time of year (Sellers *et al.*, 1977). Where more information is available, such as the dates and approximate times of arrival, forward- or back-trajectories may be constructed manually, for example, by the streamline–isotach method used by Rosenberg and Magor (1983) to determine the migration routes of rice planthoppers in East Asia. However, earlier studies sometimes had to rely on meteorological data from a very restricted number of altitudes [e.g. only the surface (10 m) and 850 mbar (*c.* 1500 m)] and these may not necessarily be the most appropriate heights for the migration in question. Nocturnal migrants, for instance, may be concentrated at the height of a boundary layer wind-speed maximum at *c.* 200–400 m above ground, and be travelling much faster than would be suggested by winds at standard altitudes (such as 850 mbar). Observations of migration in progress, especially those made with radar, are very useful in this respect, because information on the actual height of flight can be obtained. In areas of the world with dense networks of meteorological recording stations, it is worth using computerized trajectory routines based on objective analyses of the windfield at multiple altitudes in the atmosphere (McCorcle and Fast, 1989; Scott and Achtemeier, 1987). The Scott and Achtemeier method took standard 12-h data, and estimated wind speed and direction for every 3 h at 7 altitudes using a grid composed of 100 km squares of interpolated points. This method has been used to determine the sources of *Rhopalosiphum maidis* aphids migrating into the state of Illinois, USA (Hendrie *et al.*, 1985; Irwin and Hendrie, 1985).

The most recent advance is the development of trajectory models associated with increasingly sophisticated and powerful numerical models of the atmosphere (Otuka *et al.*, 2005; Turner *et al.* 1999). For example, recent back-trajectory models for rice planthoppers invading Japan (Otuka *et al.*, 2005) employed windfield analyses from the 'MM5' mesoscale model developed by Pennsylvania State University and the National Center for Atmospheric Research. Another example is the Numerical Atmospheric-dispersion Modelling Environment model (NAME), which uses 3D analyses of windfields (including mesoscale effects) from the United Kingdom Meteorological Office's 'Unified Model'. This is being used to investigate the trajectories of *Culicoides* midges, vectors of blue tongue (Gloster, John, personal communication). It should be noted that general-purpose trajectory models (such as NAME) are intended primarily for tracking passive particles, such as atmospheric pollution, and they may well need modification to portray accurately the movements of insects. Although the *horizontal* self-powered speeds of vectors are slow enough (<1 ms^{-1}) for them to be ignored in trajectory models, even small insects have a considerable degree of control over the timing and height of their migratory flights. In the absence of wing-beating, the fall speeds of *A. fabae* are *c.* 0.8–1.8 ms^{-1} (Thomas *et al.*, 1977), and this would allow migrants to terminate migratory flight and return to earth in less than 1 h under most conditions except, perhaps, in the strong updrafts associated with thunderstorms. Incidentally, there is increasing observational data from meteorological radars that small insects actively resist being carried to great heights by convection (Geerts and Miao, 2005). Contrary to the impression given by some writers, small insects (including vectors) cannot be regarded as particles ('aerial plankton') which are carried passively in the atmosphere.

III. Ascent, Transmigration and Landing Phases of Vector Migration

The increase in knowledge of the interactions between high-flying migrant insects and the atmospheric environment has very largely resulted from observations with entomological radars over the last

30 years or so. However, with the exception of the previously mentioned millimetric radar used to observe small rice insects in the Philippines and China, most of the studies have been on taxa of relatively large insects (e.g. moths and grasshoppers), and the equipment used would not resolve *individuals* of small vector species such as planthoppers, leafhoppers and aphids. While not ideal, useful information can be acquired from observations of the *collective* echo from clouds of small insects, and insights can sometimes be obtained through parallels with the high-altitude migration of larger species.

Migrating insects are largely confined to the first 2 km of the atmosphere, with higher flights (*c.* 3 km) under certain circumstances, for example, in hot seasons in the tropics. This range of flight altitudes largely coincides with the *atmospheric (or planetary) boundary layer* (ABL) (Drake and Farrow, 1988) within which the air is directly affected by the sun's heating of the Earth's surface, and by the friction due to surface roughness. The interactions between flying insects and meteorological factors that lead to the various phenomena observed by radar (layers, line-echoes, cellular patterns, and so on) cannot be discussed here, but they have been reviewed in Drake and Farrow (1988) and Burt and Pedgley (1997).

A. Ascent

The take-off and initial ascent phase of vector migration is comparatively easy to study using suction traps at various heights from the ground. In fact, many of the factors controlling ascent into the atmosphere by migrant aphids were elucidated by C. G. Johnson, L. R. Taylor and their colleagues in the 1950s and 1960s (Johnson, 1969).

Diel flight periodicity distributions have been investigated by means of time-segregating suction traps (Lewis and Taylor, 1964) and at least some of this flight activity will represent emigration. The condition of the ABL in which a migrant will find itself is of course critically dependent on the timing of take-off. For example, dusk emigration in fine weather will tend to take the insect into a stable and highly stratified nocturnal ABL, while day-time emigrants will enter an unstable ABL where the air is being mixed vertically by thermal convection (Drake and Farrow, 1988). One of the most obvious and consistently observed features in radar studies is mass emigration at particular times of day (Gatehouse, 1997). These emigrations can be categorised as follows.

1. Mass take-off around dusk, stimulated by declining illumination levels, which may be followed by prolonged windborne movement over periods of several hours, and occasionally all night. This mode of emigration is adopted by, for example, rice planthoppers and leafhoppers (Riley *et al.,* 1987, 1991), as well as many other taxa. Average ascent rates of *c.* 0.2 ms^{-1} up to heights of 700 m or more were observed by radar for *N. lugens* in China (Riley *et al.*, 1991) and this species, and other small insects emigrating at dusk, can evidently reach high altitudes unassisted by convective updrafts. As noted in a later section, *N. lugens* take-off may occur earlier in the afternoon (at higher-illumination levels) as autumn progresses, presumably so that migrants are not 'grounded' by low air temperatures at dusk.

2. Mass take-off around dawn (Riley *et al.,* 1987, 1991), apparently stimulated by similar illumination values to those at dusk. Where a species can emigrate at dusk and at dawn, the dusk movement is generally the more important due to the higher densities of insects involved, and because the subsequent flight durations are longer than those following dawn take-off (e.g. Riley *et al.,* 1991).

3. Take-off and ascent over a more extended period during the day, commencing after the sun has started to warm the surface giving rise to convective activity. This is the mode of emigration found in aphids, but in this case the diel emigration curve is typically bimodal due to the temporary lack of flight-mature individuals during the middle of the day (Johnson, 1969). Daytime migrants can utilize convective up-currents to assist in their ascent—an option that is not generally available to dusk and dawn emigrants. Aphid ascent rates measured in laboratory flight chambers include values of 0.22 ms^{-1} for *A. fabae* autumn migrants (gynoparae) after the first 10 min of flight, but the summer migrants (virginoparae) had lower rates of climb (0.14 ms^{-1}) (David and Hardie, 1988).

B. *Transmigration*

This phase of migration, sometimes called 'horizontal translocation', occurs after the initial ascent phase and before the final descent prior to landing. Horizontal translocation seems more descriptive of migration in the stable ABL that occurs at night in fine weather rather than migration under convective conditions during the day. Nocturnally migrating insects, after their initial climbing flight, may continue for many hours at essentially the same altitude. For example, they may form a layer in the warm air at the top of a temperature inversion

(Drake and Farrow, 1988; Gatehouse, 1997), or they may form a 'ceiling layer' at higher altitudes (Riley *et al.*, 1991). In contrast, insects flying in the convective ABL on a sunny day will be circulated around in the atmosphere by up- and down-drafts, perhaps taking $\sim$1–3 h to complete a circuit (Johnson, 1969). Taylor (1986) terms this type of migratory flight 'cumuliform downwind migration'.

On days when atmospheric convection is absent, aphids do not show this cumuliform vertical distribution. Instead, aerial density profiles are stratified, with aphids often being restricted to within a few tens of metres of the surface (Johnson, 1969), or confined by stable or neutral layers aloft (Isard *et al.*, 1990).

Diurnally migrating aphids may, under certain conditions (which include air temperatures remaining above the flight threshold), continue in flight for long periods during the night (Berry and Taylor, 1968; Chapman *et al.*, 2004; Reynolds *et al.*, 1999; Riley *et al.*, 1995a; Taylor, 1986). The ability of aphids to maintain migration under illumination levels which would inhibit take-off is not well understood—possibly in the dark the migrants are cut off from the normal sensory cues from vegetation, which would promote descent (Berry and Taylor, 1968). Additionally, if a strong nocturnal temperature inversion develops, aphids may well be inhibited from descending into the layer of comparatively cold air near the ground. In any event, the continuation of flight after dark is obviously important in extending the range of migration, and if the migrants encounter the low-level wind jets which frequently develop near the top of a temperature inversion, displacement distances may be further extended.

C. Descent and Landing

Of the three phases of migration discussed here, the termination of migration is the least understood, particularly in nocturnal migrants. Unlike emigration that is often synchronised and therefore appears on the radar as a spectacular mass ascent, there is usually no obvious mass descent at the end of migration. Rather, the concentration of migrants (in a layer, say) generally seems to gradually thin out and eventually dissipates, presumably due to the fall-out of individuals over an extended period. [It is well known that the frequency distribution of flight durations in a population is highly variable, with a long right-hand tail (Davis, 1980; Johnson, 1976).] It should be noted however, that in day-flying *A. fabae* at least, the end of migration is not due to simple exhaustion but to the return of responsiveness to 'vegetative' cues such as long wavelength ($<$500 nm) light reflected from plants

(Dingle, 1996; Kennedy, 1985). Nocturnal migrations in temperate regions are no doubt terminated in some cases by air temperatures falling below the threshold needed to maintain wing-beating.

There are many observations of small insects landing in inappropriate areas (e.g. *N. lugens* landing on lawns and vegetable plots; Kisimoto and Sogawa, 1995), and this suggests that migrants have little detailed information on the suitability of the habitat below them until they have descended more or less to the height of the vegetation canopy and undertaken 'alighting flights' (see later). On the other hand, Favret and Voegtlin (2001) found that migrating aphids showed different arrival rates in immediately adjacent (non-host) habitats; however the range at which the selection was made (i.e. during the descending phase of migratory flight or post-migratory 'alighting flight') is still unclear.

An exception to the gradual fall-out of a migrating population occurs when insects are concentrated by strong mesoscale convergence (such as rainstorm outflows) followed by 'deposition' by down-drafts or precipitation (Dickison *et al.,* 1983; Drake and Farrow, 1988). For example, thunderstorm activity along fronts in the Great Lakes area of North America can terminate aphid and leafhopper migration up the 'Mississippi flyway' (Johnson, 1969; Thresh, 1983; Zeyen *et al.*, 1987; and see later). Mesoscale wind convergence can easily increase aerial densities by an order of magnitude or more in a short time (Drake and Farrow, 1988; Gatehouse, 1997), and the subsequent mass deposition may therefore result in dense ground populations. Entomological radars cannot distinguish small insects from rain, so the exact sequence of events during deposition is not easily observable. Generally, however, the air seems relatively clear of insects after an area of rain has passed over the radar, and there is evidence that moderate or heavy rain causes larger insects, at least, to descend to near ground level (Riley *et al.*, 1983). Landing during heavy rain can be detrimental to small insects, however, with large mortalities due to drowning.

Topographical effects may also influence the landing places of small insects. In a nationwide survey in Japan of places that seemed to be vulnerable to large influxes of *N. lugens*, it was found that one such location was on the eastern side of hills exposed to westerly winds near the coast (Noda and Kiritani, 1989). Various eddies or vortices may form in the lee of hills, which might concentrate migrant populations and/or promote landing due to the lighter winds (Burt and Pedgley, 1997). Another favoured type of landing place for *N. lugens* was at the head of windward-facing valleys (Noda and Kiritani, 1989).

D. Post-Migratory Alighting Flights

There is little information on factors influencing flight immediately after a migrant has descended into its 'flight boundary layer', except for aphids and to a lesser extent, whiteflies. Aphids make controlled 'alighting flights', which appear to be mainly guided visually, towards potential hosts (Hardie, 1989; Nottingham *et al.*, 1991). After landing on a plant an aphid will investigate it by probing the epidermal cells, but even on a preferred host, the aphid is often too 'excitable' to settle, and it will often take off again and later land on another plant. There may be a succession of such landings and take-offs, and this behaviour facilitates aphids rapidly acquiring and spreading *non-persistent* viruses (see Section IV.C). When they eventually settle down and feed from the phloem, any *persistent* viruses that they have carried with them during migration may be transmitted to the new host.

Because vectors, such as aphids and whiteflies have low self-powered flight speeds, it is often easier for them to make host-orientated alighting flights in the shelter of windbreaks or even breaks in the crop canopy (Johnson, 1969), and the subsequent pattern of virus incidence reflects this (Thresh, this volume, pp. 89–125).

IV. Vectors of Plant Viruses

Table I lists some insect vectors of plant viruses for which there is evidence of windborne migration. Much is known about the migration biology of a few of these vectors but, in other cases, information is fragmentary or only suggestive. An example of the latter is *Toya propinqua*, a vector of *Cynodon chlorotic streak virus*. This planthopper has been caught 100 m above ground in central India (Reynolds and Wilson, 1989), which suggests that it can be a windborne migrant, and it has also been caught consistently albeit in small numbers on the East China Sea (Kisimoto, 1987), which shows that some individuals have considerable potential for flight. Pedgley (1999) lists 16 types of circumstantial evidence that may indicate long-distance insect flight.

In the following sections, some case studies of well-documented long-range migration by plant virus vectors are presented, with emphasis on recent research findings.

A. Nilaparvata lugens

The macropterous (long-winged) form of the brown planthopper, *Nilaparvata lugens* (Homoptera: Delphacidae) is a noted long-distance

TABLE I

PLANT-VIRUS VECTORS: SOME INSECT VECTORS FOR WHICH THERE IS EVIDENCE OF WINDBORNE MIGRATION (NB: ONLY A SMALL SELECTION OF THE APHID VECTORS CAN BE INCLUDED HERE, SEE BLACKMAN AND EASTOP (2000) FOR A MORE COMPREHENSIVE LISTING)

Order	Family	Species	Virus	Reference to migration of the insect vector
Heteroptera	Piesmidae	*Piesma quadratum*	Beet leafcurl rhabdovirus	*P. maculatum* in Freeman, 1945
Homoptera	Cicadellidae	*Nephotettix virescens*	Rice tungro viruses, rice transitory yellowing rhabdovirus, rice gall dwarf phytoreovirus, rice bunchy stunt phytoreovirus	Riley *et al.*, 1987
		N. nigropictus	Rice tungro viruses, rice transitory yellowing rhabdovirus, rice dwarf phytoreovirus, rice gall dwarf phytoreovirus	Riley *et al.*, 1987
		N. cincticeps	Rice tungro viruses, rice transitory yellowing rhabdovirus, rice dwarf phytoreovirus, rice gall dwarf phytoreovirus, rice bunchy stunt phytoreovirus	Kisimoto, 1981
		Recilia dorsalis	Rice tungro viruses, rice dwarf phytoreovirus, rice gall dwarf phytoreovirus	Reynolds *et al.*, 1999; Riley *et al.*, 1987
		Neoaliturus (=Circulifer) tenellus	Beet curly top curtovirus	Johnson, 1969; Thresh, 1983
		Graminella nigrifons	Maize chlorotic dwarf waikavirus	Lopes *et al.*, 1995; Sparks *et al.*, 1986
		Cicadulina mbila and others	Maize streak mastrevirus	Rose, 1978
		Macrosteles fascifrons	Oat blue dwarf marafivirus	Johnson, 1969

Delphacidae	*Javasella pellucida*	Wheat European striate mosaic tenuivirus, oat sterile dwarf fijivirus, maize rough dwarf fijivirus	Denno and Perfect, 1994
	Delphacodes kuscheli	Maize rough dwarf fijivirus	Grilli and Gorla, 1999
	Nilaparvata lugens	rice grassy stunt tenuivirus, rice ragged stunt oryzavirus	Kisimoto, 1981, 1987; Reynolds *et al.*, 1999; Riley *et al.*, 1991
	Sogatella vibix (=*S. longifurcifera*)	Maize rough dwarf fijivirus	Kisimoto, 1981
	Sogatella kolophon	Digitaria striate rhabdovirus	Sparks *et al.*, 1986
	Tagosodes orizicolus (=*Sogatodes orizicola*)	Rice hoja blanca tenuivirus	Thresh, 1983, 1985
	Tagosodes cubanus	Rice hoja blanca tenuivirus	Thresh, 1983
	Laodelphax striatellus	Rice stripe tenuivirus, rice black-streaked dwarf fijivirus, maize rough dwarf fijivirus	Kisimoto, 1981, 1987
	Perkinsiella saccharicida	Sugarcane Fiji disease reovirus	Thresh, 1983
	Toya propinqua	Cynodon chlorotic streak rhabdovirus	Kisimoto, 1987; Reynolds and Wilson, 1989
	Agallia constricta	Potato yellow dwarf rhabdovirus, clover wound tumour phytoreovirus	Glick, 1939
Aleyrodidae	*Bemisia tabaci*	Many, including cassava mosaic viruses, tomato yellow leaf curl begomovirus, sweet potato mild mottle ipomovirus	Byrne and Blackmer, 1996; Isaacs and Byrne, 1998; Thresh, 1983
	Trialeurodes abutilonea	Abutilon yellows crinivirus	Kampmeier (quoted in Byrne and Blackmer, 1996
	Trialeurodes vaporariorum	Tomato infectious chlorosis crinivirus	Coombe, 1982
		Beet pseudo-yellows crinivirus	

(*continues*)

TABLE I (*continued*)

Order	Family	Species	Virus	Reference to migration of the insect vector
Homoptera	Aphididae	*Myzus persicae*	Many, including barley/cereal yellow dwarf luteoviruses, beet mild yellowing luteovirus, beet yellows closterovirus, potato Y potyvirus	Hardy and Cheng, 1986; Heathcote, 1984; Thresh, 1983; Wiktelius, 1977
		Sitobion avenae	Barley/cereal yellow dwarf luteoviruses	Chapman *et al.*, 2004; Hardy and Cheng, 1986
		Sitobion miscanthi (=Macrosiphum miscanthi)	Barley yellow dwarf luteovirus	Close and Tomlinson, 1975
		Aphis craccivora	Groundnut rosette viruses, subterranean clover stunt nanavirus, Johnsongrass mosaic potyvirus, yam mosaic potyvirus	Gutierrez *et al.*, 1974; Johnson, 1957; Rainey (in discussion after Thresh, 1983)
		Rhopalosiphum maidis	Barley/cereal yellow dwarf luteoviruses, maize dwarf mosaic potyvirus, sugarcane mosaic potyvirus	Berry and Taylor, 1968; Irwin and Thresh, 1988; Medler, 1962
		Rhopalosiphum padi	Barley/cereal yellow dwarf luteoviruses	Chapman *et al.*, 2004; Hardy and Cheng, 1986; Medler, 1962; Nieminen *et al.*, 2000; Wiktelius, 1984
		Schizaphis graminum	Barley/cereal yellow dwarf luteoviruses	Irwin and Thresh, 1988
		Lipaphis erysimi	Cauliflower mosaic caulimovirus, turnip mosaic potyvirus, sugarcane mosaic potyvirus	Reynolds *et al.*, 1999; Riley *et al.*, 1995
		Toxoptera citricida	Citrus tristeza closterovirus	Loxdale *et al.*, 1993

	Pseudococcidae	*Pseudococcus njalensis* (First-instar 'crawlers')	Cacao swollen shoot badnavirus	Cornwell, 1961
Thysanoptera	Thripidae	*Frankliniella* spp.	Tomato spotted wilt tospovirus	Glick, 1939; Lewis, 1973
		Thrips tabaci	Tomato spotted wilt tospovirus	North and Shelton, 1986
Coleoptera	Chrysomelidae	*Diabrotica undecimpunctata howardi* (=*D. duodecimpunctata*)	Maize chlorotic mottle machlomovirus, squash mosaic comovirus, cowpea severe mosaic comovirus, bean pod mottle comovirus, and others	Glick, 1939; Johnson, 1969; Sparks *et al.*, 1986
		Diabrotica virgifera	Maize chlorotic mottle machlomovirus, cowpea severe mosaic comovirus	Coats *et al.*, 1986; Grant and Seevers, 1989; Isard *et al.*, 2001
		Ceratoma trifurcata	Bean pod mottle comovirus, southern bean mosaic sobemovirus	Boiteau *et al.*, 1979; Glick, 1939; Thresh, 1985
		Chaetocnema pulicaria	Maize chlorotic mottle machlomovirus (also Stewart's bacterial wilt of corn)	Esker *et al.*, 2002; Glick, 1939
		Colaspis brunnea	Bean pod mottle comovirus	Glick, 1939
		Phyllotreta spp.	Erysimum latent tymovirus, turnip yellow mosaic tymovirus, turnip crinkle carmovirus, radish mosaic comovirus	Freeman, 1945; Glick, 1939; Kocourek *et al.*, 2002; Moreton, 1945; Thresh, 1985
	Curculionidae	*Sitona* spp.	Broad bean mottle bromovirus, broad bean stain comovirus, broad bean true mosaic comovirus	Fisher and O'Keeffe, 1979; Freeman, 1945

migrant. The species cannot overwinter in the temperate zone, and infestations of this species over temperate eastern Asia are initiated every year from the tropics and subtropics. *N. lugens* is a vector of *Rice ragged stunt virus* and *Rice grassy stunt virus*, both of which are transmitted in a persistent manner, with multiplication in the vector, but no passage to the progeny (Hibino, 1996). These viruses mainly cause disease in rice in the tropics (Hibino, 1996), but the sporadic outbreaks in northern areas, such as Japan, are very probably related to annual variations in the numbers of infective immigrants from afar (Kisimoto, 1981).

A few *N. lugens* can survive the winter in areas of China south of the 12°C mean January isotherm, but the major sources of migrants are probably located further south (in Indochina or the south of Hainan Island) below the 19°C January isotherm where the planthoppers can breed all-year-round (see review by Kisimoto and Sogawa, 1995). Progressive northeastward windborne migrations by successive generations of *N. lugens* (Cheng *et al.*, 1979; Kisimoto and Sogawa, 1995) eventually allow the species to colonize and exploit virtually all of the vast summer rice-growing area in East Asia. These flights by *N. lugens* (and another rice delphacid, *S. furcifera*) take place at high altitude—planthoppers were caught by aircraft trapping in July in East Central China at heights of 300–2500 m, with the largest numbers being taken between 1500 and 2000 m (Dung, 1981). The annual planthopper invasion of Korea and Japan from mid-latitude regions of China in the 'Bai-u' rainy season (June–July) has been studied by trajectory analysis using increasingly sophisticated atmospheric models to simulate the windfields (Otuka *et al.*, 2005; Turner *et al.*, 1999; Watanabe, 1995). Migration to Japan is often associated with a low-level jet that develops to the south of the Bai-u front.

Several studies have attempted to determine which parts of the permanent-breeding zone of *N. lugens* constitute the ultimate sources of the waves of planthoppers that annually invade central and northern China, Korea and Japan. The consensus, based on historical changes and current distribution of 'biotypes' (Wada *et al.*, 1994) is that the most likely source is the Red River Delta of northern Vietnam. A study using a mitochondrial DNA marker (Mun *et al.*, 1999) provided evidence that, as expected, the Korean population originates from China. However, some of the other results of the study were surprising, particularly the lack of genetic differentiation across Southeast Asian countries from Malaysia to Bangladesh, suggesting regular gene flow across this extensive permanent-breeding area. This seems contrary to evidence of rather limited movements occurring within some areas of the tropics (Riley *et al.*, 1987), as well as the high degree of brachyptery (Nagata and Masuda, 1980) and the geographical variation in traits such as insecticide resistance and 'biotype' in tropical *N. lugens* populations (Nagata and Masuda, 1980; Wada *et al.*, 1994).

Detailed studies with radar and aerial netting in eastern China have revealed the nature of the southward 'return' migrations of *N. lugens* in the autumn (Riley *et al.*, 1991, 1994). The main emigration occurred during the late afternoon–dusk period, when there was a large increase in both the number of *N. lugens* caught in the aerial net and in the aerial density of planthopper-sized insects recorded by radar. Emigration tended to begin earlier in the afternoon as autumn progressed, and this seemed to be an adaptation to allow flight to start before the air temperature at ground level fell below the threshold for take-off. After taking off the planthoppers climbed actively at a rate of *c.* 0.2 ms^{-1}, and flew for several hours during the evening, often forming dense concentrations 400–1000 m above ground, which produced a dramatic display on the radar screen (Fig. 1A and B). These layers often had a well-defined ceiling, which corresponded to an air temperature of *c.* 16°C, close to the minimum for sustained flight in *N. lugens*. The mean migration height was well *above* the top of the surface temperature inversion, so most of the planthoppers did not fly at the height where the air was warmest. Layers usually lasted until about midnight, but sometimes persisted until dawn, showing that some individuals remained in continuous flight for about 12 h. Most of the planthoppers forming the dense layer concentrations overflying the radar site near Nanjing in September 1988 (Riley *et al.*, 1991) had probably emigrated from outbreaks in central Jiangsu Province, up to 200 km away to the northeast. The presence of these outbreaks upwind of the study site was almost certainly responsible for the high aerial densities of *N. lugens* detected by the radar. A second period of mass take-off was usually observed at dawn, and although insect layers sometimes formed at altitude, they did not last longer than about 1.0–1.5 h. Little flight occurred on some mornings, probably because low temperatures inhibited take-off.

The mass movements of *N. lugens* observed near Nanjing would take the migrants to areas in south Anhui Province or north Jiangxi Province if they flew all night (Fig. 2), and here they could infest late rice crops (Riley *et al.*, 1991, 1994). A further radar study in northern Jiangxi Province in October (Riley *et al.*, 1995b) indicated that the next generation of migrants may then have been able to reach overwintering areas near the coast of Guangdong Province. This would mean that the

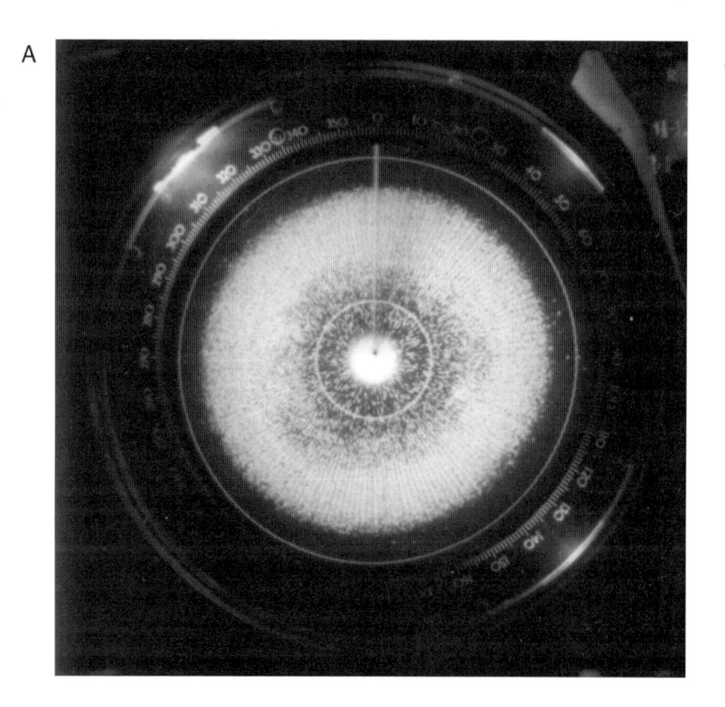

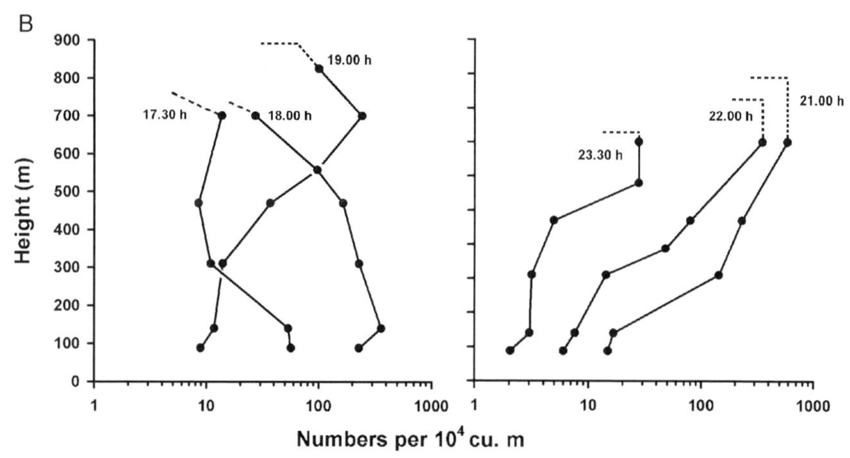

FIG 1. (A) Photograph of the display of the 8-mm wavelength scanning radar at 22.04 hrs on the evening of 28 September 1988 at Jiangpu (near Nanjing) China, showing a dense layer comprised mainly of brown planthoppers (*Nilaparvata lugens*) overflying the radar site between about 380 and 825 m above ground level. *N. lugens* comprised 90% of the radar-detectable insects in an aerial netting sample taken from the bottom of this layer. (Radar beam elevation = 44°; distance to outer range-ring = 0.75 nautical miles (1390 m); the heading marker points north). (From Riley *et al.*, 1991. Reprinted with

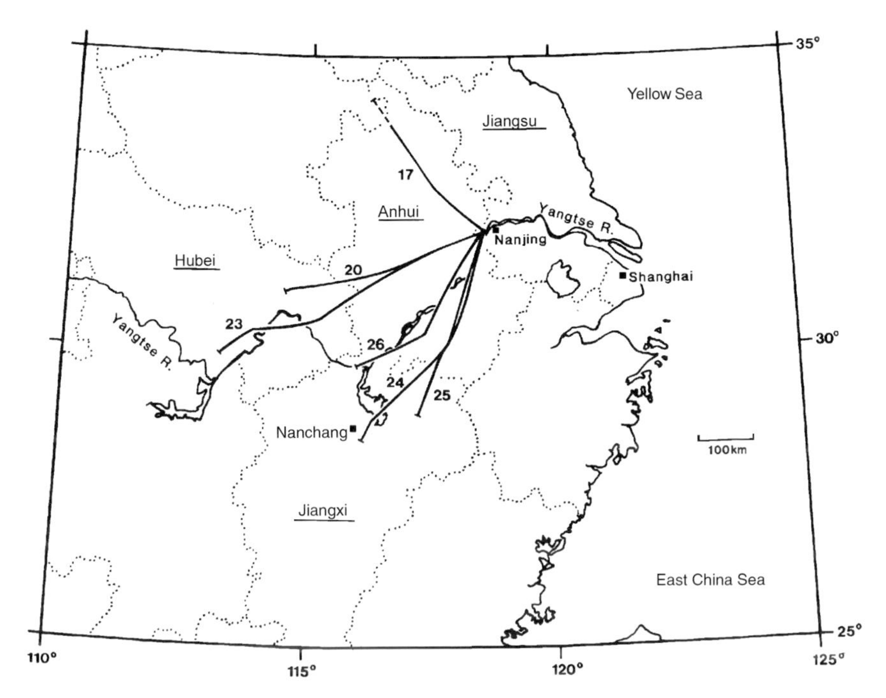

FIG 2. Examples of 12-h forward trajectories for *Nilaparvata lugens* observed emigrating from around a radar site at Jiangpu, near Nanjing, China on nights in the second half of September 1988. The trajectories are for 18.00 hrs departures on 17, 20, 23, 24 and 25 September 1988 (at 900 m altitude) and 26 September (at 600 m). The names of provinces are underlined. It can be seen that all-night flights in late September can take migrant planthoppers from southern Jiangsu as far as southern Hubei or northern Jiangxi Provinces. (From Riley *et al.*, 1994. Reprinted with permission from Blackwell Publishing Ltd., Oxford.)

permission from Blackwell Publishing Ltd., Oxford.) (*Note*: As the radar antenna rotates around the vertical with a selected elevation angle, the beam sweeps through a cone-shaped section of the sky. A horizontal layer is thus detected at the same slant ranges all around the radar and it therefore appears as an annulus on the radar display.)

(B) Radar-derived vertical profiles of insect density during the evening of 28 September 1988 at Jiangpu. The profiles at 17.30 and 18.00 hrs reflect the mass take-off and ascent of small insects, particularly *N. lugens*, from around, and a few kilometres upwind of, the radar site. From 19.00 hrs onwards, the insects were mainly concentrated in a high-altitude layer with a well-defined upper boundary. Densities in the layer were particularly great at *c*. 21.00–22.00 hrs, due to the mass overflight of *N. lugens* from outbreaks extending *c*. 200 km to the northeast of the radar. (From Riley *et al.*, 1991. Reprinted with permission from Blackwell Publishing Ltd., Oxford.)

migration pathways of *N. lugens* populations in East Asia are 'closed circuit', at least for individuals colonizing rice in mid-latitude China. Any successful 'return' migrants would, of course, help to maintain highly migratory genotypes in the permanently breeding tropical populations. Judging by the catches in autumn at weather station 'Tango' in the Pacific south of Japan, it seems likely that many of the progeny of individuals invading that country are lost at sea, and so migrations into the *northern* latitudes of East Asia might well represent 'pied piper' population movements (Walker, 1980) from which there is no return.

Overall, the radar results demonstrated that mass, long-range windborne movements of macropterous *N. lugens* occur regularly in East Central China, and that these intense and prolonged migrations contrasted strikingly with the short movements observed in earlier studies in the Philippines, where flights at altitude were largely confined to periods of *c.* 30 min at dawn and dusk (Holt *et al.*, 1996; Riley *et al.*, 1987).

Aerial netting studies in Central India and Bengal have shown that long-distance movements by *N. lugens* also occur there (Reynolds and Wilson, 1989; Reynolds *et al.*, 1999; Riley *et al.*, 1995a). The aerial densities of *N. lugens* were generally highest at dusk and then declined slowly through the night. However, the fact that some individuals of this species were still flying after midnight (Riley *et al.*, 1995a) suggests that long-distance movements are more common in India than in the Philippines.

Given these facts, it seems reasonable to conclude that long-distance migration of *N. lugens* is most adaptive in temperate continental areas, moderately adaptive in tropical continental areas (especially those with a more pronounced dry season) and least adaptive in archipelagos (e.g. Philippines) or peninsulas (e.g. Malaysia) in the humid tropics. In the latter region, short-range migrations are sufficient for colonization of new rice fields and longer movements entail a high risk of loss at sea.

B. Laodelphax striatellus *and* Rice stripe virus

The small brown planthopper (*L. striatellus*) is the major vector of *Rice stripe virus* (RSV), which is transmitted in a persistent manner with multiplication in the vector, and with transovarial passage to the progeny (Hibino, 1996). *L. striatellus* is caught regularly over the seas around Japan (Kisimoto, 1981, 1987) although it shows shorter flight durations in laboratory tests than do *N. lugens* and *S. furcifera*

(Ohkubo quoted in Kisimoto, 1987). In contrast to the latter two species, *L. striatellus* can overwinter in the temperate zone as a diapausing fourth instar nymph on weeds, and therefore long-distance movements from the tropics are not essential for establishing populations of this species in East Asia. In the south of the temperate zone and in the sub-tropics, the nymphal diapause becomes progressively less important and *L. striatellus* passes through more generations per year (Kisimoto and Yamada, 1998).

L. striatellus has a wide host range, occurring on rice and many other graminaceous plants. RSV problems generally arise in Japan when many infective first-generation adults move from ripening wheat or barley to young or transplanted rice plants in late May–early June (Hibino, 1996; Kisimoto and Yamada, 1998). So, although *L. striatellus* can migrate far (and some migrants caught at sea show antibody reactions indicating the presence of RSV), spread of the virus would normally require only short-range movements (a few kilometres). There may be occasional exceptions to this scenario, however, as when a severe outbreak of RSV occurred on Ishigaki Island (*c.* 200 km east of Taiwan) in 1987 following the invasion of infective populations of *L. striatellus* (Kisimoto and Yamada, 1998).

Management of RSV disease in Japan has been reviewed by Kisimoto and Yamada (1998) (and see Section VI). Monitoring vector density and the proportion of infective *L. striatellus* has shown that if RSV occurrence was boosted to a high level by unusual increases in numbers of vectors, population infectivity did not decrease even when vector density was subsequently reduced (by pesticide application, say). The main defence against RSV epidemics will therefore depend on RSV-resistant rice cultivars rather than insecticidal control (Kisimoto and Yamada, 1998).

C. Aphids

Insect-borne viruses are often classified as non-persistent, semi-persistent or persistent, according to their relationship with their vectors (Berger and Ferriss, 1989; Nault, 1997), and this relationship has a critical impact on the likelihood of transmission occurring after long-distance transport. Non-persistent viruses are transmitted almost solely by aphids. They occur mainly in the epidermal cells of plants and are acquired rapidly (in seconds) by the probing aphid. Once acquired the viruses can be inoculated immediately, but are retained only for relatively short periods on the aphid stylets and are soon lost in subsequent probing. They can be transmitted by aphids that are not feeding but are merely probing a plant to check its suitability as a food source, and hence they have a wide potential vector range. Persistent viruses, in contrast, occur in the phloem and can only be acquired and inoculated by aphids that have identified the plant concerned as a suitable food source and whose stylets have hence penetrated to the phloem in order to begin feeding. This greatly limits the number of vector species. Once acquired, the virus circulates in the host, ending up in the accessory salivary glands ready for inoculation after a period of a few days following acquisition.

Intuitively, persistent viruses would seem to have a greater potential to be transported over long distances by their vectors than do non-persistent viruses, although there is insufficient evidence to substantiate this. Some outbreaks of *non-persistent* viruses have been reported far from known sources. For example in 1977, *Maize dwarf mosaic virus* (MDMV) was transported to Minnesota, where infection is unusual, from the southern US plains. It was thought to have been retained in its vector for 19 h, an unusually long period for a non-persistent virus (Berger *et al.*, 1987; Zeyen *et al.*, 1987). MDMV is common in the southern plains states and in spring 1977 temperatures were unusually high, resulting in large populations of vector aphids. A severe drought followed, causing deterioration in the palatability of the crop to aphids and the consequent production of large numbers of alates. In late June, southerly low-level jet winds formed between the southern plains and Minnesota, and these probably transported the aphids over 1000 km northwards until they reached a cold front and associated thunderstorms where the aphids were deposited (Zeyen *et al.*, 1987).

A severe constraint on the spread of non-persistent viruses is that, once acquired, transmission can only occur to the first few plants visited by the aphid, typically after rather short flights (Kennedy, 1960) and these plants may not be susceptible to the virus. Consequently, the long-range spread of MDMV, mentioned above, may have been due to the long uninterrupted nature of the migratory flight (and the consequent inability of the aphids to land and probe) rather than any exceptional retention per se of this non-persistent virus in the aphid.

Semi-persistent viruses, as the name suggests, are intermediate in terms of persistence in the aphid and hence in the range of distances over which they are likely to be transported. The more extensive distribution of the persistent *Beet mild yellowing virus* (BMYV) compared with the semi-persistent *Beet yellows virus* (BYV) may be due partly to its greater persistence (Russell, 1963). Wiktelius (1977) was able to predict the incidence of BMYV and BYV in the south of Sweden in September on the basis of the incidence of southerly airflow in July. He suggested that neither the virus nor the main vector, *M. persicae*, could overwinter outdoors in Sweden and that southerly airflows brought viruliferous aphids over the southern part of the Baltic Sea. Aphid catches were greatest on days when a cold front passed over the trapping site (Wiktelius, 1984).

Several aphid species are recorded as long-distance migrants in North America, including *S. avenae* (English grain aphid), *R. maidis* (corn leaf aphid), *R. padi* (bird cherry–oat aphid) and *Schizaphis graminum* (greenbug) (see reviews in Berger and Ferriss, 1989; Irwin and Thresh, 1988; Johnson, 1995; Loxdale *et al.*, 1993). Each spring the re-establishment of these species in the northern Mid-West states of the United States and the Prairie Provinces of Canada depends largely upon an influx from cereal fields in the south. In seasons with appropriate conditions, the influx of cereal aphids can result in a high incidence of virus-infected plants. The northward movement of aphids and other migrant insects up the Mississippi River Drainage Basin (MRDB) in spring is facilitated by particular synoptic weather situations. Typically, a mid-latitude depression with an associated cold front approaches from the west, while a high-pressure area lies near, or just off, the Atlantic Coast (Johnson, 1969, 1995) (Fig. 3). This produces a northward flow of warm tropical air over the MRDB, and low-level jets (with wind speeds up to 80 km/h) may form in the warm sector of the depression. Invasions of aphids do not always coincide with presence of low-level jet winds, but the migrants are still transported northwards each spring by weaker, less sustained southerly winds.

Southward movements in late summer and autumn by later generations of aphids are more difficult to detect as any migrants will be moving into areas that already contain populations. However, synoptic weather systems that would transport aphids southwards certainly exist in August–September (Johnson, 1995), and there is good evidence for such 'return' movements in the leafhopper *Empoasca fabae* (Johnson, 1995; Taylor and Reling, 1986).

1. *Migration of Aphid Vectors into Illinois, USA*

Movement of aphids, particularly *R. maidis* into the state of Illinois was studied in the 1980s in the multi-disciplinary 'Pest and Weather Project' (Hendrie *et al.*, 1985; Irwin and Thresh, 1988), which employed a variety of techniques including high-altitude aerial sampling, X-band

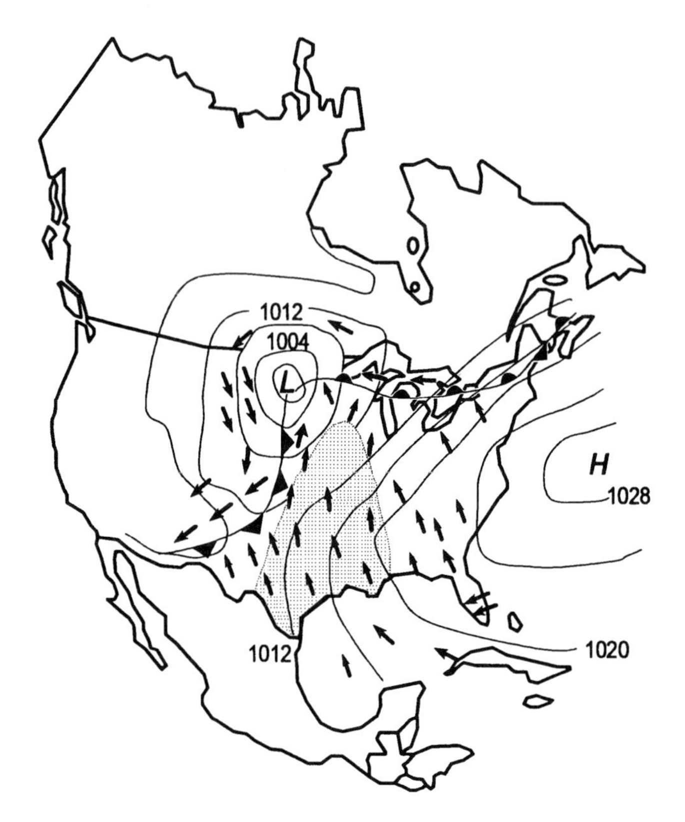

FIG 3. Typical synoptic weather situation transporting aphids, leafhoppers and other migrant insects northwards in North America in spring or early summer, showing the anti-clockwise wind flow around a mid-latitude depression. The shaded area shows the region where low-level jets occur. (From Johnson, 1995. Reprinted with permission from Cambridge University Press, Cambridge.)

and S-band (10-cm wavelength) radar observations, back-trajectory construction, genetic markers and flight-fuel analysis.

Analysis of an immigration of aphids on the morning of 9 August 1984 demonstrates graphically the importance of information on the height of flight. Aphids were observed on the radar to be flying in two layers, 500–650 m and 900–1200 m above ground, respectively, but these two groups had radically different source areas. Flight duration estimates based on lipid utilisation analyses of captured specimens indicated that aphids in the lower layer had probably been flying for 2.0–5.5 h, and that therefore they had probably taken off earlier that morning from relatively nearby sources (50–80 km to the west–southwest). Aphids in the upper layer had been in flight for more than 5 h, which implied that they must have taken off sometime on the day before. Back-trajectory analyses suggested putative sources between southern Missouri and northern Texas, 400–1100 km to the southwest.

2. *Mass Migration of Aphids into Finland*

During fine weather, masses of insect targets are sometimes detected on weather radars, and a project to investigate the nature of this clear-air echo on the Doppler C-band radar at the University of Helsinki in 1988 was well-placed to record some mass migrations of aphids into Finland in spring of that year (Nieminen *et al.*, 2000; http://www.helsinki.fi/~mleskine/engl/migra.html). The aphid species involved were identified from specimens caught in nets and suction traps placed on high buildings, and in light traps, and some of the influxes included large numbers of *R. padi*, many of which were carrying *Barley yellow dwarf virus*. During one of the *R. padi* migrations (on 30 May), radar observations made about 4 h after sunrise revealed an echo layer, which was approximately uniform over a depth of nearly 2 km. The aphids were being carried by the wind at *c.*10 ms^{-1}, and probably had been in flight all night. Trajectory modelling of the various migration events indicated sources in Estonia, Latvia and western Russia south of St Petersburg. The huge populations of immigrant *R. padi* (numbers per hectare in cereal fields were over 100 times higher than typical immigrant densities), along with locally bred populations, later necessitated 'the most extensive insecticide sprayings ever performed in Finland' (Nieminen *et al.*, 2000).

D. Bemisia tabaci

Following the greatly increased prevalence of whitefly borne viruses in many parts of the tropics and subtropics in recent years (Brown, 1994; Polston and Anderson, 1997), there has been a surge of interest in whitefly (Homoptera: Aleyrodidae) dispersal, particularly of the sweet potato whitefly, *B. tabaci* (also referred to as the tobacco whitefly). Work by Byrne and co-authors over several years (Byrne and Blackmer, 1996) has led to a re-evaluation of movement in what was previously considered a non-migratory species. Most individuals tested in a flight chamber did undertake only 'vegetative' flights, responding more or less immediately to a cue simulating their host plants, but a small percentage (*c.* 6%) were clearly migratory in the sense of Kennedy (1985), that is, their response to the host cue was inhibited until they had flown for some time (at least 15 min, and over 2 h in a few individuals) (Blackmer and Byrne, 1993a; Blackmer *et al.*, 1995). Interestingly, the whiteflies did not exhibit a clear-cut oogenesis-flight syndrome (where the young, pre-reproductive adults migrate, and resources are only diverted into reproductive activities like egg-production after completion of the migratory phase (Johnson, 1969)). The longer flying individuals had, if anything, higher levels of egg proteins (Blackmer *et al.*, 1995). On the other hand, mean egg-load decreased with the height at which whiteflies were trapped, which suggests some sort of 'trade-off' between flight and oogenesis (Isaacs and Byrne, 1998).

As would be expected, mark–recapture studies in the field showed that most *B. tabaci* individuals will only move very short distances to neighbouring hosts before settling. However, the distribution of white-flies trapped at different distances from the source field was bimodal, indicating that a proportion of individuals migrated over distances of a few kilometres (Byrne and Blackmer, 1996) reflecting the behaviour observed in the laboratory. Whiteflies were also caught on traps several metres into the air—evidence that some individuals were actively ascending out of their flight boundary layer (Isaacs and Byrne, 1998), and there was some fragmentary evidence of flight at high altitudes which may indicate that dispersal occurs over much longer distances. Confirmation that high-altitude flight occurs in whiteflies, and that wind convergence can increase population densities, has been obtained by aerial trapping in Sudan (see Thresh, 1983). Longer-range wind-borne movement of *B. tabaci* may also be suggested by the sequential spread of *Tomato yellow leaf curl virus* (TYLCV) and other gemini-viruses from island to island in the Caribbean since the late 1980s (Polston and Anderson, 1997).

Marking *B. tabaci* with fluorescent dusts helped to elucidate the epidemiology of TYLCV in the Jordan Valley, where the whiteflies survive on weeds (particularly *Cyananchum acutum*) outside the tomato-growing season (Cohen *et al.*, 1988). The numbers of viruliferous *B. tabaci*, produced on *C. acutum* growing on the banks of the Jordan, peaked at the same time as the tomato transplanting season, and it was shown that marked whiteflies were migrating from the weeds into the main tomato cropping area, 7 km away.

B. tabaci is the vector of cassava mosaic viruses, and new and virulent strains of these have spread across Uganda and into Kenya and Tanzania, often as a clearly defined front, advancing at an overall mean rate of *c.* 20–30 km per year (Otim-Nape *et al.*, 1997). However, the rate of spread was sporadic, with periods of rapid progress and others of relative quiescence. Spread into western Kenya was particularly rapid, and affected a large area almost simultaneously, suggesting the eastward migration of *B. tabaci* over a considerable distance from sites near the Uganda border (Legg, J. P., and Thresh, J. M., personal communication).

E. Beetles

A number of leaf-feeding Coleoptera, particularly in the Chrysomelidae but with examples in the Curculionidae, Coccinellidae and Meloidae, are vectors of plant viruses (Fulton *et al.*, 1987; Thresh, 1985). However, with the possible exceptions of the chrysomelids *Diabrotica undecimpunctata howardi* (Johnson, 1969) and *D. virgifera virgifera* (Coats *et al.*, 1986; Grant and Seevers, 1989; Isard *et al.*, 2001; Naranjo, 1990), there is little published information on long-distance movements of coleopteran vectors (Table I) and still less on the implications for virus epidemiology. Interest in the *Diabrotica* spp. is primarily due to direct damage caused to maize crops in North America, but the beetles also transit *Maize chlorotic mottle virus* that acts with other viruses to cause corn lethal necrosis. The spread of *D. virgifera* in Europe after its introduction into Serbia in 1992 is currently causing concern (MacLeod *et al.*, 2004).

There seems little doubt that *D. virgifera* undertakes windborne movements because large numbers of beetles have been found along the shores of Lake Michigan after the passage of synoptic-scale cold fronts (Grant and Seevers, 1989). Migrant beetles flying at altitude were presumably swept up by the front (with aerial densities probably being increased by wind convergence) and they were later deposited in the lake by downdrafts (or landed en-masse for some other reason). Laboratory studies of *D. virgifera* show that mated females can fly for sustained periods of up to 4 h (Coats *et al.*, 1986). A careful assessment of the evidence of *D. virgifera* spread in the United States and in Europe, for a pest risk analysis (MacLeod *et al.*, 2004), suggested that a slow rate of spread in this univoltine species might be typically 20 km/year (maximum 40 km/year) while a fast rate of spread might be typically 80 km/year (maximum 100 km/year). The partitioning of migratory and vegetative flight over the diel period still requires clarification, however, because some authors (Coats *et al.*, 1986) found that sustained flights occurred mainly at night, while others (Isard *et al.*, 2000) reported that flight activity peaked in the morning or late afternoon and none was detected at night.

Ceratoma trifurcata, another chrysomelid, is the primary vector of *Bean pod mottle virus* that can cause serious damage to soybean crops in the United States (Giesler *et al.*, 2002). In a trapping study, most *C. trifurcata* undertook short flights (<30 m), but there was some evidence that longer flights occurred during the autumn movements to overwintering sites in wooded or shrubby areas away from the crop (Boiteau *et al.*, 1979). A few beetles were caught at heights of 50–60 m at this time (Boiteau *et al.*, 1979; Glick, 1939).

Flea beetles of the genus *Phyllotreta* (Chrysomelidae) transmit several viruses of brassicaceus plants (Table I), and sometimes these insects can be seen flying over crops in huge numbers on warm days in spring or earlier summer (Moreton, 1945). Trapping in the Czech Republic indicated that the seasonal flight peaks of *Phyllotreta* spp. were due to overwintering adults migrating onto crops after temperatures had risen sufficiently in spring (May–early June), and to movements of the new generation of beetles in early July (Kocourek *et al.*, 2002). There was presumed to be an autumn flight to overwintering (diapausing) sites, but this was not detected by the yellow water traps used in the study. Day-time flight of another species of *Phyllotreta* (*P. consobrina*) has, however, been detected by suction trapping in autumn (Lewis and Taylor, 1964).

There are casual references on horticultural websites to the 'migrations' of other coleopteran vectors such as *Acalymma trivittatum* (Chrysomelidae), *Epilachna varivestis* (Coccinellidae) and *Apion* spp. (Curculionidae), but there appears to be little detailed information on their movements.

V. Vectors of Pathogenic Agents of Animals

The most important group of insects involved in the spread of pathogens of vertebrates are the Diptera (true flies), and haematophagous dipterans have been implicated in transporting viruses over long distances (Pedgley, 1983; Sellers, 1980). However, the occurrence of long-range migration in the vectors of animal pathogens is less well established than for plant virus vectors, and the traditional viewpoint of many researchers is that the most important vectors (mosquitoes and biting midges) are too weak-flying to be involved in the long-distance spread (Sellers, 1980). Thus it was generally assumed that spread was caused by migratory movements of bird or mammal hosts, or by the accidental transfer of the insect vector onboard human vehicles. To some extent this attitude has persisted, and some medical/veterinary entomologists regard the presence of vectors, such as blackflies and mosquitoes, high in the air as accidental and non-adaptive (Crosskey, 1990; Service, 1997). However, certain vector species are commonly caught flying hundreds of metres above the ground (see Table II) while other species are seldom or never found there, and this indicates that some taxa have a propensity for windborne migration, while other (possibly closely related) taxa do not. Furthermore, there are many cases of animal disease outbreaks where there is circumstantial evidence to implicate the windborne movement of infective vectors into a previously uninfected area as the initial cause of the epidemic (see reviews in Pedgley, 1983; Sellers, 1980). Diseases may reappear in an area that has been disease-free for some time due to persistence in host populations, migration of infected vertebrate hosts, or windborne movements of infective vectors. In some cases a combination of these processes may be responsible, and it may be very difficult to disentangle which is the most important. Nonetheless, the windborne transport of infective vectors should not be overlooked, and a greater knowledge of vector migration may aid the forecasting of pathogen spread.

The most straightforward scenario for long-distance disease spread by insects is of vectors migrating after they have had a blood meal from an infected host, and directly spreading the virus or other pathogen to an uninfected host population through further blood meals. Insect migration is generally associated with newly emerged, nulliparous adults (the 'oogenesis flight syndrome'; Johnson, 1969) and thus it has been argued by Service (1997) that the windborne transport of *older* (i.e. blood-fed) female mosquitoes is not true migration, but 'accidental' dispersal. However, infective vectors are known to migrate after a blood meal in at least one group of Diptera (*Simulium* blackflies; see later), and this behaviour may be widespread amongst other vectors, although good data on migration are lacking in most groups. Even if post-feeding migration is rare or absent in some vectors, there are two additional routes by which the long-range migration of vectors may result in disease outbreaks. First, the mass arrival of vectors in an area may cause increased transmission of pathogens that had persisted at a low frequency in the host population, including transmission to additional hosts. Second, transovarial transmission of virus may be possible in some vector species, such as Japanese encephalitis in *Culex tritaeniorhynchus* (Rosen *et al.*, 1980), and thus it is conceivable that virus may be carried long distances by newly emerged vectors before their first blood meal. There is a lack of direct evidence that either of these routes has led to disease outbreaks, but clearly research into these hypotheses is necessary before they can be discounted.

TABLE II

SOME INSECT VECTORS OF VIRUSES OR PATHOGENS CAUSING HUMAN AND ANIMAL DISEASES FOR WHICH THERE IS EVIDENCE OF WINDBORNE MIGRATION

Order	Family	Species	Pathogen	Reference to migration of the insect vector
Hemiptera	Reduviidae	*Triatoma infestans*	Chagas disease (caused by a flagellate: *Trypanosoma cruzi*)	Pedgley, 1983; Schofield *et al.*, 1992
Diptera	Culicidae	*Culex tritaeniorhynchus*	*Japanese encephalitis virus*	Ming *et al.*, 1993; Reynolds *et al.*, 1996
		Culex gelidus	*Japanese encephalitis virus*	Reynolds *et al.*, 1996
		Culex visnui complex	*Japanese encephalitis virus, West Nile virus, Nairobi sheep disease virus* (Ganjam)	Reynolds *et al.*, 1996
		Culex annulirostris	*Japanese encephalitis virus, Murray Valley encephalitis virus, Ross River virus, Barmah Forest virus*	Kay and Farrow, 2000; Ritchie and Rochester, 2001
		Culex tarsalis	*Western equine encephalitis virus, St. Louis encephalitis virus*	Johnson, 1969; Sellers and Maarouf, 1988, 1993
		Culex pipiens	*West Nile virus, Rift Valley fever virus*	Pedgley, 1983
		Culex quinquefasciatus	*West Nile virus, Dengue virus*	Reynolds *et al.*, 1996
			Bancroftian filarisis (caused by the nematode *Wuchereria bancrofti*)	
		Aedes spp.	*Rift Valley fever virus*	Pedgley 1983; Sellers *et al.*, 1982
		Aedes sollicitans	*Venezuelan equine encephalitis virus, Eastern equine encephalitis virus*	Johnson, 1969; Sellers, 1980

	Aedes taeniorhynchus	*Venezuelan equine encephalitis virus*	Johnson, 1969; Sellers, 1980
	Aedes vigilax	*Murray Valley encephalitis virus*	Johnson, 1969; Sellers, 1980
	Anopheles pharoensis	Malaria (caused by *Plasmodium* spp.—a protozoan)	Garret-Jones, 1950 (quoted in Johnson, 1969)
	Anopheles pulcherrimus	Malaria	Daggy, 1959
	Anopheles vagus	Bancroftian filarisis (caused by the nematode *Wuchereria bancrofti*)	Reynolds *et al.*, 1996
	Mansonia spp.	Brugian filarisis (caused by *Brugia* spp. nematodes)	Reynolds *et al.*, 1996
Psychodidae (Phlebotominae)	*Phlebotomus* and *Lutzomyia* spp.	Pappataci fever virus Also mucocutaneous and visceral leishmaniasis caused by a protozoan *Leishmania*	Killick-Kendrick *et al.*, 1984
Simuliidae	*Simulium damnosum*, complex	Onchocerciasis (caused by a nematode *Onchocerca volvulus*)	Baker *et al.*, 1990; Garms and Walsh, 1987
Chironomidae	*Chironomus* spp.	Cholera (caused by a bacterium *Vibrio cholerae)*	Broza *et al.*, 2005
Ceratopogonidae	*Culicoides imicola* and other spp.	*Bluetongue virus, African horse sickness virus*	Pedgley, 1983; Purse *et al.*, 2005
	Culicoides brevitarsis	*Bluetongue virus, Akabane virus, Sathuperi virus* (Douglas virus) and others	Murray and Kirkland, 1995
	Culicoides spp.	*Bovine ephemeral fever virus*	Pedgley, 1983
Muscidae	*Stomoxys calcitrans*	Capripox viruses, *African swine fever virus, Equine infectious anemia virus*	Isard *et al.*, 2001

A. Blackflies (Simulidae)

Blackflies of the *Simulium damnosum* complex [vectors of *Onchocerca volvulus*, a nematode that causes human onchocerciasis (river blindness)] are discussed first because they provide one of the best documented examples of long-distance migration in an insect that transmits an animal pathogen. Due to the progressive eradication of the aquatic larvae of *Simulium* over huge areas of West Africa, and the efforts made to detect reinvasion of treated areas by the adult flies, the Onchocerciasis Control Programme (OCP) was well-placed to study the range of movement of various species in the *damnosum* complex (Fig. 4). Some forest species were found to move relatively short distances (e.g. ~5 km in *S. yahense*, whose larvae inhabit permanently flowing small rivers), while some of the savanna species migrated long distances on the southwesterly monsoon winds (e.g. at least 500 km in *S. damnosum* sensu strictu and *S. sirbanum*, where the larvae live in waterways which do not flow for part of the year) (Garms and Walsh, 1987). In contrast to many migrant insects, these long distances are not covered in one sustained flight soon after emergence, but in a succession of shorter movements taking up to 4–5 weeks for the whole journey (Baker *et al.*, 1990). Migratory flights occur after blood feeding (but before full egg maturation) in each gonotrophic cycle throughout the life of the fly (Garms and Walsh, 1987). This means that flies infected soon after emergence can carry the infective *O. volvulus* larvae over vast distances, and because of this the OCP had to extend the treated area considerably to eliminate the sources of invasion (Garms and Walsh, 1987). Migrants tend to bite close to their oviposition sites, and so intense transmission rates will only occur in human settlements located on riverbanks—this fact has led to the disease being given the English name of 'river blindness' (Garms and Walsh, 1987). Finally, it may be noted that the OCP is an excellent example of a preventive pest management strategy against a highly mobile disease vector, and it has allowed millions of children in West Africa to grow up free from the risk of river blindness.

B. Mosquitoes

Long-range migration in mosquitoes is a contentious issue, as some researchers maintain that windborne movements are accidental and non-adaptive, and thus should not even be termed migrations (Service, 1997). However, certain species of mosquito are found much more frequently than others at high altitudes, and the notion that ascent

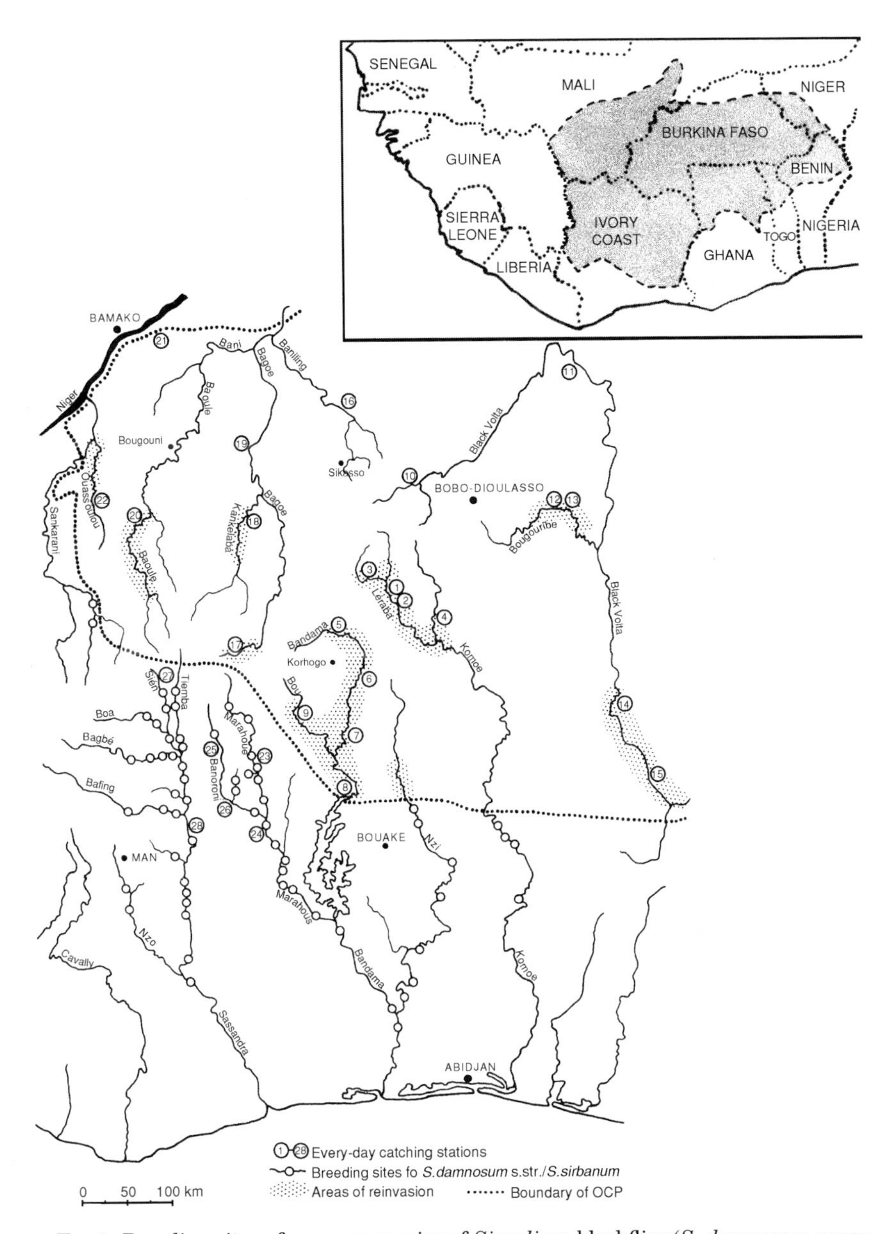

FIG 4. Breeding sites of savanna species of *Simulium* blackflies (*S. damnosum* sensu str. and *S. sirbanum*) south and southwest of Onchocerciasis Control Programme (OCP), the areas of re-invasion, and location of fly catching stations. (After Garms, Walsh and Davies, 1979. Reprinted with permission from Georg Thieme Verlag, Stuttgart.). The inset shows the location in West Africa of the original OCP area.

out of their 'flight boundary layer' might be one manifestation of an evolved migration syndrome cannot simply be dismissed without further investigation. The sustained ascent necessary to reach the observed height of flight, usually at night in the absence of vertical air movements, requires behaviour of a quite different type to the low-altitude host-seeking flights. Maintenance of flight at high altitude for extended periods, under conditions where cessation of wing-beating would produce a fairly rapid descent to the surface, certainly satisfies some of the characteristics of migration (persistent undistracted locomotion, straightened-out track; see Section I.A; Clements, 1999), even if it is not clear whether there has been any inhibition of station-keeping responses (another migratory criterion), although this might be implied by the initial sustained climbing flight. Obviously more work is needed on the nature of long-range movements in mosquitoes, but it is apparent that some species have a predisposition for high-altitude windborne movements at a certain stage in their life cycle, while other species are much less prone to be carried long distances by the wind. In this section, an example of a virus (*Japanese encephalitis virus*) that is spread by a group of highly migratory *Culex* species is discussed first, followed by *Dengue virus* that is spread by *Aedes* species with much lower propensities for windborne movements.

1. Culex *and* Japanese encephalitis virus

Japanese encephalitis (JE) is a well-known viral disease of humans that causes a serious public health problem in East, Southeast and South Asia (Rosen, 1986). The primary hosts are birds and pigs, but the disease appears regularly in humans during epidemics (Sellers, 1980). In Asia the principal vector is *Culex tritaeniorhynchus*, but other species including *Cx. gelidus* and the *Cx. vishnui* group have also been implicated in its spread (Reynolds *et al.*, 1996). In temperate regions JE dies out during the winter, but in some areas it reappears each spring, year after year. This persistence may be explained by the virus surviving in a host population, or by the annual migration of infected birds into the area. However, there is convincing evidence that *Cx. tritaeniorhynchus* cannot survive over winter in some regions that have disease problems, for example, Japan, northern and central China, and South Korea (Ito *et al.*, 1986; Ree *et al.*, 1976; Zhang *et al.*, 1985). Therefore, long-range migrations of the vectors from permanent-breeding sites in tropical and subtropical regions of southern Asia into temperate areas further north are essential for the re-invasion of the virus, even if infected mosquitoes are not actually involved in the flights. There is persuasive evidence that *Cx. tritaeniorhynchus* is a migratory species that regularly ascends high into the air and can travel long distance on the wind—for example, this species was trapped at heights of 80–380 m above ground in China (Ming *et al.*, 1993) and at heights of 150 m in India (Reynolds *et al.*, 1996). The mosquitoes are not carried accidentally to great heights on convective up-drafts, as the high-altitude movements of *Cx. tritaeniorhynchus* occur at night, and so the flights are probably true migrations. This species does *not* fly high during non-migratory, searching flights—maximum numbers assessed by suction trap and biting catches in West Africa occurred within a metre of the ground (Clements, 1999, p. 310), in accordance with a species that feeds on large mammals. Windborne vector migration is therefore critical for the persistence of JE in temperate regions of eastern Asia, and examples of how the warm southwesterly winds associated with the Bai-u rainy season are responsible for its spread northwards are discussed by Sellers (1980).

JE has spread via Papua New Guinea (PNG) and the Torres Strait into the Cape York peninsular in mainland northern Australia, where there have been a few human cases (Hanna *et al.*, 1999). The principal vector in Australia is *Cx. annulirostris*, which is considered to be the morphological and biological counterpart of *Cx. tritaeniorhynchus* (Kay and Farrow, 2000). Because of their similarity, *Cx. annulirostris* might be expected to be highly migratory, and in fact there is evidence that the spread of JE to Australia is due to windborne migration by mosquitoes from PNG, rather than from the migration of infected birds. Kay and Farrow (2000) demonstrated that *Cx. annulirostris* is caught commonly hundreds of metres in the air in New South Wales, and they estimated that potential flight distances of ~200 km were long enough for aerial carriage across the Torres Strait. Ritchie and Rochester (2001) carried out trajectory analyses, and confirmed that *Cx. annulirostris* could have crossed the Torres Strait in single flights during periods of sustained strong northerly winds from PNG to Cape York, during 1995 and 1998, immediately preceding large incursions of the virus. Furthermore, a study of genetic differentiation in *Cx. annulirostris* populations in PNG, the Torres Strait and Cape York, established that a panmictic population occurs throughout the region, strongly indicating the presence of frequent, widespread dispersal (Chapman *et al.*, 2003a). Thus it seems clear that windborne invasions of *Cx. annulirostris* are implicated in the spread of JE to Australia, but to prove that migrant *Cx. annulirostris* are directly responsible for this invasion it is necessary to catch engorged, gravid or parous mosquitoes engaged in active migration across the Torres Strait. Johansen *et al.* (2003) attempted to do this, by aerial sampling on an island in the

Torres Strait, but unfortunately wind speeds and directions during their study were not conducive for mosquito migration, and they failed to catch any female *Culex*. Further work is necessary therefore, but all the evidence gathered so far strongly indicates that windborne movements of mosquitoes are an integral component of the dynamics of JE infections.

2. Aedes *and Dengue Fever*

Two species of *Aedes* are the principal vectors of dengue worldwide, namely *Ae. aegypti* (originally an African species) and *Ae. albopictus* (a native of Southeast Asia). Dengue is a serious disease of humans in many tropical regions of the world, and it often spreads in explosive epidemics (Reiter *et al.*, 1995). In contrast to the dispersal ability of *Culex* species, many *Aedes* species are considered to be poor dispersers, and are thought to travel only a few hundred metres at most during their lifetime (PAHO, 1994; Sellers, 1980). Therefore human-aided transport of the vectors is generally deemed to be the most important factor in the spread of dengue and other viruses transmitted by *Aedes* spp. Direct evidence for the poor dispersal ability of *Ae. aegypti* has been collected by Harrington *et al.* (1995), who showed that movement between neighbouring villages in Thailand ($\sim$500 m apart) happened only rarely. Indirect evidence for low-dispersal rates has also been found in *Ae. albopictus* by Lourenço de Oliveira *et al.* (2003), who found that populations in Florida tended to become more differentiated as geographic distances separating them increased, confirming the short-range migration ability of this species. However, other studies have indicated that it is wrong to assume that control operations can be based on very short dispersal distances (e.g. 50 m), as gravid females have been recorded flying distances up to 850 m (Liew and Curtis, 2004; Reiter *et al.*, 1995). Thus, even for vectors with suspected low-dispersal ability, it is important to characterize maximum dispersal distances fully if control programmes are to be successful.

C. Culicoides *Biting Midges (Ceratopogonidae)*

Ceratopogonidae of the genus *Culicoides* are important vectors of arboviruses worldwide, where they serve as biological vectors for the viruses causing bluetongue, African horse sickness, Akabane, Ibaraki, epizootic hemorrhagic disease and bovine ephemeral fever (Mellor *et al.*, 2000). Numerous studies have related the spread of *Culicoides*-borne viruses to the windborne transport of vectors, and early studies are reviewed extensively in Sellers (1980), Pedgley (1983) and Mellor *et al.* (2000). Here the focus is on the role of windborne migration in the spread of *Bluetongue virus* (BTV), and in particular on studies carried out since the earlier reviews mentioned above.

BTV infects all ruminants, but severe disease symptoms usually occur only in sheep and deer, and it is the occurrence of bluetongue disease in the former that is of economic importance (Mellor *et al.*, 2000). BTV is distributed throughout tropical and subtropical regions of the Americas, Africa, Asia and Australasia. The principal vector species vary throughout the world, and include: the *C. variipennis* group in North America; *C. insignis* and *C. pusillus* in Central and South America; *C. imicola* throughout Africa, Asia and southern Europe; *C. bolitinos* in South Africa; and *C. brevitarsis* and *C. fulvus* in eastern Asia and Australasia (Mellor *et al.*, 2000). Historically, BTV did not exist permanently in Europe, but at times it made incursions, sometimes resulting in many sheep deaths—for example, in the 1956–1960 epizootic, 180,000 sheep died in Spain and Portugal (Mellor *et al.*, 2000). These incursions have tended to be brief, sporadic and only into the fringes of Europe. However, since 1998, BTV has spread across 12 countries, and 800 km further north into Europe than it had previously, resulting in a continuous and unprecedented epizootic in the Mediterranean region and the death of 1 million sheep (Purse *et al.*, 2005). This epizootic has arisen from the almost simultaneous invasion of six different viral strains, arriving from at least two different directions (namely, via Turkey and Greece, and from North Africa via Corsica, Sicily and the Balearics) (Purse *et al.*, 2005). It appears from the work of Purse *et al.* (2005) that the spread of BTV throughout southern Europe is driven by climate change, acting through a number of mechanisms including the northward range expansion of the principal vector, *C. imicola*. The vector's rapid range expansion, in response to temperature increases north of its historical range, has been made possible by this species' propensity for long-distance windborne migration. It also seems probable that the sudden and widespread influx of BTV into southern Europe was due to mass migration of infective vectors, perhaps facilitated by more frequent warm, southerly winds suitable for transport, although there is a lack of direct evidence for this hypothesis. However, there is convincing evidence that once BTV had reached the fringes of southern Europe during the current epidemic, it did indeed spread mainly via windborne dispersal of infected *C. imicola*. For example, the virus spread from Rhodes and Lesvos to mainland Greece and Bulgaria in 1999 (Purse *et al.*, 2005), and it also reached Sicily in 2000, probably from Sardinia (Torina *et al.*, 2004). The northward expansion of the largely subtropical *C. imicola* into areas such as mainland Italy, the Balkans and European Turkey has serious implications for the spread of the disease further north, beyond the direct impact that this species has on infection rates. This is because climate amelioration in these regions has increased the spatio-temporal overlap of *C. imicola* with a suite of Palaearctic *Culicoides* species, such as the *C. obsoletus* and *C. pulicaris* groups, which have the potential to become novel, temperate-region vectors (Purse *et al.*, 2005). *Culicoides* spp. are certainly transported by the wind over Britain, because small numbers have been caught regularly by sampling at 200 m above ground (Chapman *et al.*, 2004; Reynolds, D., and Carpenter, S., unpublished data) and *C. pulicaris* has also been caught far from the coast over the North Sea (Hardy and Cheng, 1986). There is undoubtedly a very real risk that BTV will reach northern Europe, and an understanding of windborne migration in all the potential vectors is essential for any attempt to forecast the spread of this disease.

VI. Modelling the Migration of Plant Virus Vectors

Simulation models and 'expert systems' that take account of long-range migration have been developed for forecasting and management of rice planthoppers in East Asia (Cheng and Holt, 1990; Holt *et al.*, 1990; Tang *et al.*, 1994; Zhou *et al.*, 1995). However, these models were conceived to aid management of direct damage by the planthoppers rather than damage caused as virus vectors. There has also been much work on forecasting aphid invasions and subsequent seasonal population growth from trap catches and weather variables using various methods ranging from regression and multi-variate analysis (Masterman *et al.*, 1996) to artificial neural networks (Worner *et al.*, 2001). Again, however, the virus–vector system was not considered explicitly in these models.

Models that consider virus incidence as well as vector dynamics have been reviewed by Jeger *et al.* (2004). These include empirical statistical models (Raccah *et al.,* 1988) or complex process-based simulation models (Kisimoto and Yamada, 1986; Ruesink and Irwin, 1986; Sigvald, 1998) of specific insect-borne virus systems. There has been some success in modelling propagative planthopper- or leafhopper-borne viruses of rice, such as RSV and *Rice dwarf virus*, where there is transovarial transmission to the vector's progeny (Kisimoto and Yamada, 1986; Nakasuji *et al.*, 1985). For example, RSV epidemics over areas the size of a prefecture in Japan could be described by combining *L. striatellus* density estimates (from trap catches or counts) and assessments of the population infectivity of each generation of the vector resulting from two factors—the rate of transovarial transmission, and the rate at which uninfected individuals acquired the virus from infected plants (Kisimoto and Yamada, 1986).

A weather-driven stochastic simulation model has been developed for forecasting of secondary spread of *Barley yellow dwarf virus* in crops in the United Kingdom, and this model forms the basis of a prototype Web-based 'decision-support system' (DSS), which provides virus incidence risk assessments for regions within Britain (Northing *et al.*, 2004). The numbers of the main BYDV vector, *R. padi*, caught in suction traps in autumn are known to be closely correlated with numbers found in cereal crops, and trap catches can hence be used to assess numbers per unit area of crop. Together with estimates of the proportion of flying aphids that are carrying BYDV and the proportion of these that will transmit the virus to the crop, the trap-derived data are used to initialize the simulation model. This distributes the virus according to algorithms describing the influence of weather variables on aphid population development and movement. It should be noted that there is no implication of long-distance migration in the model. Similar Web-based DSSs are under development in Western Australia (http://www.agric.wa.gov.au/bydv/) and in New Zealand (http://www.aphidwatch.com//bydv/).

There has also been interest in more analytical and general 'SEIR' models (which divide the host population into Susceptible, Exposed (latent), Infectious and Recovered individuals) to investigate the invasion and persistence of insect-borne viruses in crops (Jeger *et al.*, 1998, 2004; Madden *et al.*, 2000). These models have vector immigration and emigration terms to allow for different levels of infectivity in arriving vectors, but they are mainly concerned (explicitly or implicitly) with short-range movements. However, results can be dependent on interactions between virus and vector, particularly the degree of persistence of the virus. For example, in the case of non-persistently transmitted plant viruses, the local activity rather than density of virus vectors was shown to be an important factor in models of the dynamics of virus transmission (Madden *et al.*, 2000), because the virus remains in the insect for a short time, and the rate of probing activity is important. The model accounts for the well-known fact that epidemics of a non-persistent virus are difficult to control by in-field monitoring and insecticidal application, particularly if the vector mobility is high. In contrast, reducing the vector-population density by insecticidal control has greater potential in the case of persistent viruses.

Many of the earlier models ignored the spatial aspects of insect-borne plant virus epidemic spread to concentrate on temporal aspects of the disease in a single population (although some temporal models do allow for a random or clumped spatial distribution of new inoculations within a field (Ruesink and Irwin, 1986)). Berger and Ferriss, however, developed a stochastic simulation model to examine the general effect of transmission characteristics on the spread of arboviruses between plants in a spatial lattice (Berger and Ferriss, 1989; Ferriss and Berger, 1993). Zhang *et al.* (2000) used a model incorporating vector aggregation to improve understanding of the dynamics of cassava mosaic disease in Uganda. Increases in spatial patchiness (which were a consequence of the preference of whiteflies to feed on diseased plants) reduced the disease incidence within a field but caused increased vector emigration to other fields, thus enhancing disease spread on the larger scale.

A major constraint in the successful modelling of long-range vector movement, where both the virus and the vector are explicitly considered, is the lack of accurate input data needed to run the model, especially information on the infectivity of particular populations of migrants. In summary, the tremendous complexity and subtlety of interactions between virus, vector, hosts and the environment, mean that models taking account of post-arrival, in-field movements are difficult enough. Consideration of infective vectors in putative source areas and their long-range windborne transport, dependent as it is on virus retention times and the vagaries of vector migration behaviour and the weather (Berger and Ferriss, 1989), will be even more challenging.

VII. Discussion

The widespread acceptance of Kennedy's behavioural definition of migration, and the development of a generalised conceptual model of a migration system by Drake *et al.* (1995), mean that there is now a solid theoretical framework for migration research. We believe it is time to discard the view, still persisting in some quarters, that windborne migration by vectors and other small insects is largely accidental and 'non-adaptive'. There is of course a need to use appropriate techniques to investigate whether a particular vector species is a long-range migrant—in many cases the answer may well be 'no', but the possibility cannot be dismissed just because the migratory process is not immediately obvious to a ground-based observer.

We also believe that new techniques (such as radar, netting from kytoons, numerical trajectory models, molecular marking, and so on), combined with more traditional entomological trapping and survey methods, now make it possible to study virtually all aspects of the long-range migration of insect vectors. (A possible exception may be field observations of the termination of migration, particularly in nocturnal migrants.) There are also new field or laboratory tests for vector infectivity utilising nucleic acid-based methods (e.g. nucleic acid hybridisation and PCR) (Candresse *et al.*, 1998), which are more sensitive than the traditional methods using serology or transmission tests. Further investigations on vector movement are, therefore, largely dependent on funding and research priorities rather than technical constraints per se. Some of the techniques are undoubtedly expensive (e.g. the development of high-frequency entomological radars) or have high running costs (e.g. aircraft trapping), but ultimately the highest costs are probably those of establishing and retaining a highly skilled multi-disciplinary team (ecologists, molecular biologists, meteorologists and engineers) for a suitably long period. Currently, there seem to be very few ambitious projects comparable to those carried out in the 1970s and 1980s on migratory insect pests (such as the previously described Illinois ‘Pest and Weather Project’ or the radar observations of *N. lugens* in China; also Dickison *et al.*, 1983; Wolf *et al.*, 1990). This suggests that funding for ecological studies of insect migration has become less fashionable and is unattractive to funding agencies. Hopefully, the activities of the Europe-wide ‘EXAMINE’ consortium (see Section II.B) and initiatives to develop a vertical-looking radar system to monitor small insects in China (Dr. Bao-ping Zhai, personal communication) provide counter-examples to this view.

A. *Future Research Priorities*

Thresh (1983) listed six conditions that should be satisfied in order to provide evidence for long-distance spread of insect-borne viruses, and he identified some important research questions that required attention. Since that time there have been distinct increases in understanding of the migration of some vectors, particularly in North America, Europe, and East and Southeast Asia, as detailed in this review. However, knowledge of the migration biology of other important vector species has shown disappointingly little advance (as for example with *Tagosodes* spp., the planthopper vectors of *Rice hoja blanca virus* in South/Central America), and in many cases basic data are still not available on the flight abilities of the vector population. As a first step, laboratory assays using simple tethered flight techniques would indicate whether there is any *prima facie* evidence of prolonged flight durations. Experiments in vertical flight chambers (Section II.F) could then be performed, to see whether the flights include a migratory phase or whether they are purely resource-finding (i.e. vegetative) and are halted when the resource is encountered. If there is evidence of migration, it may be worth undertaking aerial sampling from tall masts, kites or balloons near prolific source areas to see if some dispersal takes place at altitude. Sampling near ground level alone is not sufficient, because even if a large percentage of dispersers are caught in low-altitude traps, a small proportion of migrants ascending to higher altitudes may be important for long-range spread. It is perhaps worth stressing that plant viruses can generally reach nearby hosts without difficulty and this enables efficient local exploitation. It is, however, much more difficult for them to reach and colonize new habitats, but this may be crucial where host plants are annuals growing in highly seasonal climates. It is therefore not surprising that viruses infecting crops are often vectored by rather mobile, 'r-selected', colonizing species of insects such as aphids (Thresh, 1980, 1983). Such organisms, whether the crop-infesting viruses themselves or the insect vectors, trade-off the high risks of dispersal-related mortality against the 'rewards' of being among the first to reach new (and often relatively unexploited) habitats.

Turning to interactions between the vector and the viruses it carries, progress appears to have been slow, and more work is required on, for example, effects of virus latent period and persistence during long-distance flights (Nault, 1997). Long-range spread is mainly thought to be a feature of persistent viruses that are retained for considerable periods in the vector and not lost by probing non-hosts. However, non-persistent viruses may occasionally be spread long distances if aphid vectors are transported quickly and uninterruptedly by favourable weather, as in the case of *Maize dwarf mosaic virus* in North America mentioned in Section IV.C. More information is also required on viral uptake by aphids before migration. In particular, do individuals that have shed non-persistent virus from the stylets during the moult from the final nymphal instar, feed or probe before flight and thereby become infective? If not, long-distance spread of non-persistent viruses would not occur.

Finally, there is much investigative work to be done on the many direct and indirect affects of the virus on its vector (Belliure *et al.*, 2005). Plant viruses apparently do not generally cause direct deleterious effects on their vectors (although there are exceptions, e.g. *Rice dwarf virus*; Nakasuji and Kiritani, 1970), but there are many effects caused by induced changes to host plant physiology (Colvin *et al.*, this volume, pp. 419–452), and there is increasing evidence that pathogens have evolved mechanisms to assist their vector in overcoming host plant defences (Belliure *et al.*, 2005). Some of these plant-mediated indirect effects affect vector flight. Examples include the influence of virus infection of the host on the production of winged forms of the vector (Blua and Perring, 1992; Fiebig *et al.*, 2004; Gildow, 1980; Montllor and Gildow, 1986), and on flight performance itself (such as the above-mentioned effect of BYDV producing a shorter time 'window' for long-distance flight in *R. padi* (Levin and Irwin, 1995)). Vector preference for diseased plants, and the higher fecundity of vectors breeding on them, can lead to overcrowding and, consequently, to increased emigration (Colvin *et al.*, this volume, pp. 419–452; Zhang *et al.*, 2000).

B. The Difficulties of Long-Term Vector and Virus Monitoring

Where vector migration is well established, samples are needed from actual migrant populations to determine percentage infectivity in any given season (Kisimoto and Yamada, 1998). Ideally, an on-going monitoring system should be established in which the size of the vector population *and* the infectivity levels in migrants are assessed, and these inputs would be used to drive models providing forecasts of the timing and severity of incipient plant disease outbreaks. This would often require considerable practical co-operation between organisations in the source and sink regions, however, and these may well be in different provinces/states or even in different countries making co-operation difficult. Furthermore, it would be naïve to assume that funding would be available for the regular collection of such data over periods of many years except, perhaps, in the case of dangerous insect-borne viruses of humans (http://medent.usyd.edu.au/projects/arbovirus%20surveillance.htm) or livestock, or where long-range movements of vectors are likely to increase with climate change, potentially bringing new diseases to previously unaffected areas [e.g. the predicted arrival of bluetongue in Northern Europe (Purse *et al.*, 2005)]. Financial administrators are notoriously wary of long-term funding commitments, and there are several examples of even long-running monitoring schemes lapsing, for example, the routine collection of BYDV infectivity data that was maintained at Rothamsted in the United Kingdom for many years (Plumb *et al.*, 1986). The movement towards privatized services rather than ones provided by the state sector may mean that existing large-scale vector monitoring systems (such as those against rice pests in East Asia) might receive less support in the future. Ultimately, it is up to farmers and other stakeholders, in consultation with their virological advisors, to decide which operational monitoring schemes might be worthwhile, and to lobby politicians and funding bodies for their long-term support.

Acknowledgments

We thank Professor J. M. Thresh for his comments on a draft. Rothamsted Research receives grant-aided support from the UK Biotechnology and Biological Sciences Research Council.

References

Akey, D. H. (1991). A review of marking techniques in arthropods and an introduction to elemental marking. *Southwest Entomol. Suppl.* **14:**1–8.

Baker, R. H. A., Guillet, P., Seketeli, A., Boakye, D., Wilson, M. D., and Bissan, Y. (1990). Progress in controlling the reinvasion of windborne vectors into the western area of the Onchocerciasis Control Programme in West Africa. *Phil. Trans. Roy. Soc. Lond. B* **328:**731–750.

Belliure, B., Janssen, A., Maris, P. C., Peters, D., and Sabelis, M. W. (2005). Herbivore arthropods benefit from vectoring plant viruses. *Ecol. Lett.* **8:**70–79.

Berger, P. H., and Ferriss, R. S. (1989). Mechanisms of arthropod transmission of plant viruses: Implications for the spread of disease. *In* "Spatial Components of Plant Disease Epidemics" (M. J. Jeger, ed.), pp. 40–84. Prentice Hall, New Jersey.

Berger, P. H., Zeyen, R. J., and Groth, J. V. (1987). Aphid retention of maize dwarf mosaic virus (potyvirus): Epidemiological implications. *Ann. Appl. Biol.* **111:**337–344.

Berry, R. E., and Taylor, L. R. (1968). High-altitude migration of aphids in maritime and continental climates. *J. Anim. Ecol.* **37:**713–722.

Blackman, R. L., and Eastop, V. F. (2000). "Aphids on the World's Crops," 2nd Ed. John Wiley & Sons, New York.

Blackmer, J. L., and Byrne, D. N. (1993a). Flight behavior of *Bemisia tabaci* in a vertical flight chamber. *Physiol. Entomol.* **18:**223–232.

Blackmer, J. L., and Byrne, D. N. (1993b). Environmental and physiological factors influencing phototactic orientation in *Bemisia tabaci* (Gennadius). *Physiol. Entomol.* **18:**336–342.

Blackmer, J. L., Byrne, D. N., and Tu, Z. (1995). Behavioral, morphological, and physiological traits associated with migratory *Bemisia tabaci* (Homoptera: Aleyrodidae). *J. Insect Behav.* **8:**251–267.

Blua, M. J., and Perring, T. M. (1992). Alatae production and population increase of aphid vectors on virus-infected host plants. *Oecologia* **92:**65–70.

Boiteau, G., Bradley, J. R., Jr., and Van Duyn, J. W. (1979). Bean leaf beetle: Flight and dispersal behavior. *Ann. Entomol. Soc. Am.* **72:**298–302.

Brown, J. K. (1994). The status of *Bemisia tabaci* Genn. as a pest and vector in world agroecosystems. *FAO Plant Prot. Bull.* **42:**3–32.

Broza, M., Gancz, H., Halpern, M., and Kashi, Y. (2005). Adult non-biting midges: Possible windborne carriers of *Vibrio cholerae*. Non-O1 Non-O139. *Environ. Microbiol.* doi:10.1111/j.1462–2920.2005.00745.x Ref.

Burt, P. J. A., and Pedgley, D. E. (1997). Nocturnal insect migration: Effects of local winds. *Adv. Ecol. Res.* **27:**61–92.

Byrne, D. N., and Blackmer, J. L. (1996). Examination of short-range whitefly migration. *In* "*Bemisia*: 1995. Taxonomy, Biology, Damage, Control and Management" (D. Gerling and R. T. Mayer, eds.), pp. 17–28. Intercept Press, Andover, UK.

Candresse, T., Hammond, R. W., and Hadidi, A. (1998). Detection and identification of plant viruses and viroids using polymerase chain reaction (PCR). *In* "Plant Virus Disease Control" (A. Hadidi, R. K. Khetarpal, and H. Koganezawa, eds.), pp. 399–416. APS Press, St. Paul, Minnesota.

Carn, V. M. (1996). The role of dipterous insects in the mechanical transmission of animal viruses. *Br. Vet. J.* **152:**377–393.

Chapman, H. F., Hughes, J. M., Ritchie, S. A., and Kay, B. H. (2003a). Population structure and dispersal of the freshwater mosquitoes *Culex annulirostris* and *Culex palpalis* (Diptera: Culicidae) in Papua New Guinea and Northern Australia. *J. Med. Entomol.* **40:**165–169.

Chapman, J. W., Reynolds, D. R., and Smith, A. D. (2003b). Vertical-looking radar: A new tool for monitoring high-altitude insect migration. *Bioscience* **53:**503–511.

Chapman, J. W., Reynolds, D. R., Smith, A. D., Smith, E. T., and Woiwod, I. P. (2004). An aerial netting study of insects migrating at high-altitude over England. *Bull. Entomol. Res.* **94:**123–136.

Cheng, J. A., and Holt, J. (1990). A systems analysis approach to brown planthopper control on rice in Zhejiang Province, China. I. Simulation of outbreaks. *J. Appl. Ecol.* **27:**85–99.

Cheng, S.-N., Chen, J.-C., Si, H., Yan, L.-M., Chu, T.-L., Wu, C.-T., Chien, J.-K., and Yan, C.-S. (1979). Studies on the migrations of brown planthopper *Nilaparvata lugens* Stål. *Acta Entomol. Sinica* **22:**1–21.

Chippaux, J.-P., Bouchité, B., Demanou, M., Morlais, I., and Le Goff, G. (2000). Density and dispersal of the loaiasis vector *Chrysops dimidiata* in southern Cameroon. *Med. Vet. Entomol.* **14:**339–344.

Clements, A. N. (1999). "The Biology of Mosquitoes. Sensory Reception and Behaviour," Vol. 2. Chapman & Hall, London.

Close, R. C., and Tomlinson, A. I. (1975). Dispersal of the grain aphid *Macrosiphum miscanthi* from Australia to New Zealand. *N. Z. Entomol.* **6:**62–65.

Coats, S. A., Tollefson, J. J., and Mutchmor, J. A. (1986). Study of migratory flight in the western corn rootworm (Coleoptera: Chrysomelidae). *Environ. Entomol.* **15:**620–625.

Cocu, N., Harrington, R., Rounsevell, M. D. A., Worner, S. P., and Hullé, M., and the EXAMINE project participants (2005). Geographical location, climate and land use influences on the phenology and numbers of the aphid, *Myzus persicae*, in Europe. *J. Biogeog.* **32:**615–632.

Cohen, S., Kern, J., Harpaz, I., and Ben-Joseph, R. (1988). Epidemiological studies on the tomato yellow leaf curl virus (TYLCV) in the Jordan Valley, Israel. *Phytoparasitica* **16:**259–270.

Coombe, P. E. (1982). Visual behaviour of the greenhouse whitefly, *Trialeurodes vaporariorum*. *Physiol. Entomol.* **7:**243–251.

Cooter, R. J. (1982). Studies on the flight of black-flies (Diptera: Simuliidae). I. Flight performance of *Simulium ornatum* Meigen. *Bull. Entomol. Res.* **72:**303–317.

Cooter, R. J., Winder, D., and Chancellor, T. C. B. (2000). Tethered flight activity of *Nephotettix virescens* (Hemiptera: Cicadellidae) in the Philippines. *Bull. Entomol. Res.* **90:**49–55.

Cornwell, P. B. (1961). Movements of the vectors of virus dieases cacao in Ghana. II. Wind movements and aerial dispersal. *Bull. Entomol. Res.* **51:**175–201.

Crosskey, R. W. (1990). "The Natural History of Blackflies." John Wiley & Sons, New York.

Daggy, R. H. (1959). Malaria in oases of eastern Saudi Arabia. *Am. J. Trop. Med. Hyg.* **8:**223–291.

David, C. T., and Hardie, J. (1988). The visual responses of free-flying summer and autumn forms of the black bean aphid, *Aphis fabae*, in an automated flight chamber. *Physiol. Entomol.* **13:**277–284.

Davis, M. A. (1980). Why are most insects short flyers? *Evol. Theory* **5:**103–111.

De Barro, P. J., Sherratt, T. N., Brookes, C. P., David, O., and Maclean, N. (1995). Spatial and temporal variation in British field populations of the grain aphid *Sitobion avenae* (F.) (Hemiptera: Aphididae) studied using RAPD-PCR. *Proc. Roy. Soc. London. B* **262:**321–327.

Delmotte, F., Leterme, N., Gauthier, J. P., Rispe, C., and Simon, J.-C. (2002). Genetic architecture of sexual and asexual populations of the aphid *Rhopalosiphum padi* based on allozyme and microsatellite markers. *Molecular Ecol.* **11:**711–723.

Denno, R. F., and Perfect, T. J. (eds.) (1994). "Planthoppers: Their Ecology and Management." Chapman and Hall, New York.

Department of Plant Protection, Nanjing Agricultural College, and others (1981). Test on the releasing and recapturing of marked planthoppers *Nilaparvata lugens* and *Sogatella furcifera* [in Chinese, with English summary]. *Acta Ecologia Sinica* **1:**49–53.

Dewar, A. M., Woiwod, I., and Choppin de Janvry, E. (1980). Aerial migrations of the rose-grain aphid, *Metopolophium dirhodum* (Wlk.), over Europe in 1979. *Plant Pathol.* **29:**101–109.

Dickison, R. B. B., Haggis, M. J., and Rainey, R. C. (1983). Spruce budworm moth flight and storms: Case study of a cold front system. *J. Climate Appl. Meteor.* **22:**278–286.

Dickinson, M. H., and Lighton, J. R. B. (1995). Muscle efficiency and elastic storage in the flight motor of *Drosophila*. *Science* **268:**87–90.

Dickson, R. C. (1959). Aphid dispersal over southern California deserts. *Ann. Entomol. Soc. Am.* **52:**368–372.

Dingle, H. (1985). Migration. *In* "Comprehensive Insect Physiology, Biochemistry and Pharmacology" (G. A. Kerkut and L. I. Gilbert, eds.), Behaviour, Vol. 9, pp. 375–415. Pergamon Press, Oxford.

Dingle, H. (1996). "Migration: The Biology of Life on the Move." Oxford University Press, Oxford.

Dingle, H. (2001). The evolution of migratory syndromes in insects. *In* "Insect Movement: Mechanisms and Consequences" (I. P. Woiwod, D. R. Reynolds, and C. D. Thomas, eds.), pp. 159–181. CABI Publishing, Wallingford, UK.

Drake, V. A., and Farrow, R. A. (1988). The influence of atmospheric structure and motions on insect migration. *Ann. Rev. Entomol.* **33:**183–210.

Drake, V. A., and Gatehouse, A. G. (eds.) (1995). "Insect Migration: Tracking Resources through Space and Time." Cambridge University Press, Cambridge, UK.

Drake, V. A., Gatehouse, A. G., and Farrow, R. A. (1995). Insect migration: A holistic conceptual model. *In* "Insect Migration: Tracking Resources Through Space and Time" (V. A. Drake and A. G. Gatehouse, eds.), pp. 427–457. Cambridge University Press, Cambridge, UK.

Dung, W. S. (1981). A general survey on seasonal migrations *Nilaparvata lugens* (Stål) and *Sogatella furcifera* (Horvath) (Homoptera: Delphacidae) by means of airplane collections [in Chinese with English summary]. *Acta Phytophylacica Sinica* **8:**73–81.

Esker, P. D., Obrycki, J., and Nutter, F. W. (2002). Temporal distribution of *Chaetocnema pulicaria* (Coleoptera: Chrysomelidae) populations in Iowa. *J. Econ. Entomol.* **95:**739–747.

Farrow, R. A., and Dowse, J. E. (1984). Method of using kites to carry tow nets in the upper air for sampling migratory insects and its application to radar entomology. *Bull. Entomol. Res.* **74:**87–95.

Favret, C., and Voegtlin, D. J. (2001). Migratory aphid (Hemiptera: Aphididae) habitat selection in agricultural and adjacent natural habitats. *Environ. Entomol.* **30:**371–379.

Ferriss, R. S., and Berger, P. H. (1993). A stochastic simulation model of epidemics of arthropod-vectored plant diseases. *Phytopathology* **83:**1269–1278.

Fiebig, M., Poehling, H.-M., and Borgemeister, C. (2004). BYDV, wheat and *Sitobion avenae*: A case of trilateral interactions. *Entomol. Exp. Appl.* **110:**11–21.

Fisher, J. R., and O'Keeffe, L. E. (1979). Seasonal migration and flight of the pea leaf weevil *Sitona lineatus* (Coleoptera: Curculionidae) in northern Idaho and eastern Washington. *Entomol. Exp. Appl.* **26:**189–196.

Foil, L. D., and Gorham, J. R. (2000). Mechanical transmission of disease agents by arthropods. *In* "Medical Entomology: A Textbook on Public Health and Veterinary Problems Caused by Arthropods" (B. F. Eldridge and J. D. Edman, eds.), pp. 461–514. Kluwer Academic, Dordrecht, The Netherlands.

Foster, S. P., Denholm, I., Harling, Z. K., Moores, G. D., and Devonshire, A. L. (1998). Intensification of insecticide resistance in UK field populations of the peach-potato aphid, *Myzus persicae* (Hemiptera: Aphididae) in 1996. *Bull. Entomol. Res* **88:**127–130.

Freeman, J. A. (1945). Studies in the distribution of insects by aerial currents. The insect population of the air from ground level to 300 feet. *J. Anim. Ecol.* **14:**128–154.

Fulton, J. P., Gergerich, R. C., and Scott, H. A. (1987). Beetle transmission of plant viruses. *Ann. Rev. Phytopathol.* **25:**111–123.

Garms, R., and Walsh, J. F. (1987). The migration and dispersal of blackfies: *Simulium damnosum* s.l., the main vector of human onchocerciasis. *In* "Black Flies: Ecology, Population Management and Annotated World List" (K. C. Kim and R. W. Merritt, eds.), pp. 201–214. Pennsylvania State University, University Park, Pennsylvania.

Garms, R., Walsh, J. F., and Davies, J. B. (1979). Studies on the reinvasion of the Onchocerciasis Control Programme in the Volta River basin by *Simulium damnosum* s.l. with emphasis on the south-western areas. *Tropenmed. Parasitol.* **30:**345–362.

Garrett, R. G., and MacLean, G. D. (1983). The epidemiology of some aphid-borne viruses in Australia. *In* "Plant Virus Epidemiology" (R. T. Plumb and J. M. Thresh, eds.), pp. 199–209. Blackwell Scientific Publications, Oxford.

Gatehouse, A. G. (1987). Migration: A behavioural process with ecological consequences. *Antenna* **11:**10–12.

Gatehouse, A. G. (1997). Behavior and ecological genetics of wind-borne migration by insects. *Ann. Rev. Entomol.* **42:**475–502.

Geerts, B., and Miao, Q. (2005). Airborne radar observations of the flight behaviour of small insects in the atmospheric convective boundary layer. *Environ. Entomol.* **34:**361–377.

Giesler, L. J., Ghabrial, S. A., Hunt, T. E., and Hill, J. H. (2002). Bean pod mottle virus: A threat to US soybean production. *Plant Dis.* **86:**1280–1289.

Gildow, F. E. (1980). Increased production of alate by aphids reared on oats infected with barley yellow dwarf virus. *Ann. Entomol. Soc. Am.* **73:**343–347.

Glick, P. A. (1939). "The Distribution of Insects, Spiders and Mites in the Air," Technical Bulletin No. 673, USDA, Washington, D.C.

Gottwald, T. R., and Tedders, W. L. (1986). MADDSAP-1, a versatile remotely piloted vehicle for agricultural research. *J. Econ. Entomol.* **79:**857–863.

Grant, R. H., and Seevers, K. (1989). Local and long range movement of adult western corn rootworm (*Diabrotica virgifera*) as evidenced by washup along southern Lake Michigan shores. *Environ. Entomol.* **18:**266–272.

Gray, S. M., and Banerjee, N. (1999). Mechanisms of arthropod transmission of plant and animal viruses. *Microb. Molecul. Biol. Rev.* **63:**128–148.

Greenstone, M. H., Eaton, R. R., and Morgan, C. E. (1991). Sampling aerially dispersing arthropods: A high-volume, inexpensive, automobile and aircraft-borne system. *J. Econ. Entomol.* **84:**1717–1724.

Grilli, M. P., and Gorla, D. E. (1999). The distribution and abundance of Delphacidae (Homoptera) in central Argentina. *J. Appl. Entomol.* **123:**13–21.

Gutierrez, A. P., Havenstein, D. E., Nix, H. A., and Moore, P. A. (1974). The ecology of *Aphis craccivora* Koch and subterranean clover stunt virus. III. A regional perspective of the phenology and migration of the cowpea aphid. *J. Appl. Ecol.* **11:**21–35.

Hanna, J. N., Ritchie, S. A., Phillips, D. A., Lee, J. M., Hills, S. L., van den Hurk, A. F., Pyke, A. T., Johansen, C. A., and Mackenzie, J. S. (1999). Japanese encephalitis in north Queensland, Australia, 1998. *Med. J. Aust.* **170:**533–536.

Hardie, J. (1989). Spectral specificity for targeted flight in the black bean aphid, *Aphis fabae*. *J. Insect. Physiol.* **35:**619–626.

Hardie, J. (1993). Flight behaviour in migrating insects. *J. Agric. Entomol.* **19:**239–245.

Hardie, J., Gibson, G., and Wyatt, T. D. (2001). Insect behaviours associated with resource finding. *In* "Insect Movement: Mechanisms and Consequences" (I. P. Woiwod, D. R. Reynolds, and C. D. Thomas, eds.), pp. 87–109. CABI Publishing, Wallingford, UK.

Hardy, A. C., and Cheng, L. (1986). Studies in the distribution of insects by aerial currents. III. Insect drift over the sea. *Ecol. Entomol.* **11:**283–290.

Hargrove, J. W. (2000). A theoretical study of the invasion of cleared areas by tsetse flies (Diptera: Glossinidae). *Bull. Ent. Res.* **90:**201–209.

Harrington, L. C., Scott, T. W., Lerdthusnee, K., Coleman, R. C., Costero, A., Clark, G. G., Jones, J. J., Kitthawee, S., Kittayapong, P., Sithiprasasna, R., and Edman, J. D. (1995). Dispersal of the dengue vector *Aedes aegypti* within and between rural communities. *Am. J. Trop. Med. Hyg.* **72:**209–220.

Harrington, R., Verrier, P., Denholm, C., Hullé, M., Maurice, D., Bell, N., Knight, J., Rounsevell, M., Cocu, N., Barbagallo, S., Basky, Z., Coceano, P.-G., *et al.* (2004). 'EXAMINE' (EXploitation of Aphid Monitoring in Europe): An EU Thematic Network for the study of global change impacts on aphids. *In* "Aphids in a New Millennium (Proceedings 6th International Aphid Symposium, Rennes, France, 3–7 September 2001)" (J. C. Simon, C. A. Dedryver, C. Rispe, and M. Hullé, eds.), pp. 45–49. INRA Editions Versailles.

Harrington, R., Clark, S. J., Welham, S. J., Verrier, S. J., Denholm, C. H., Hullé, M., Maurice, D., Rounsevell, M. D. A., Cocu, N., and EU EXAMINE Consortium (2007). Environmental change and the phenology of European aphids. *Global Change Biol.* (in press).

Heathcote, G. D. (1980). Were we invaded by greenfly? *Trans. Suffolk Nat. Soc.* **18:**144–147.

Heathcote, G. D. (1984). Did aphids come to Suffolk from the Continent in 1982? *Trans. Suffolk Nat. Soc.* **20:**45–51.

Hendrie, L. K., Irwin, M. E., Liquido, N. J., Ruesin, W. G., Mueller, W. M., Voegtlin, D. J., Achtemeier, G. L., Steiner, W. M., and Scott, R. W. (1985). Conceptual approach to modeling aphid migration. *In* "The Movement and Dispersal of Agriculturally Important Biotic Agents" (D. R. MacKenzie, C. S. Barfield, G. C. Kennedy, R. D. Berger, and D. J. Taranto, eds.), pp. 541–582. Claitor's Publishing Division, Baton Rouge, Louisiana.

Hibino, H. (1996). Biology and epidemiology of rice viruses. *Ann. Rev. Phytopathol.* **34:**249–274.

Hollinger, S. E., Sivier, K. R., Irwin, M. E., and Isard, S. A. (1991). A helicopter-mounted isokinetic aerial insect sampler. *J. Econ. Entomol.* **84:**476–483.

Holt, J., Cheng, J. A., and Norton, G. A. (1990). A systems analysis approach to brown planthopper control in Zhejiang Province, China. III. An expert system for making recommendations. *J. Appl. Ecol.* **27:**113–122.

Holt, J., Chancellor, T. C. B., Reynolds, D. R., and Tiongco, E. R. (1996). Risk assessment for rice planthopper and tungro disease outbreaks. *Crop Prot.* **15:**359–368.

Hoy, M. A. (2003). "Insect Molecular Genetics: An Introduction to Principles and Applications," 2nd Ed. Academic Press/Elsevier, San Diego.

Irwin, M. E., and Hendrie, L. K. (1985). The aphids are coming. *Illinois Res.* **27:**11.

Irwin, M. E., and Kampmeier, G. E. (1989). Vector behavior, environmental stimuli and the dynamics of plant virus epidemics. *In* "Spatial Components of Plant Disease Epidemics" (M. J. Jeger, ed.), pp. 14–39. Prentice Hall, New Jersey.

Irwin, M. E., and Thresh, J. M. (1988). Long range aerial dispersal of cereal aphids as virus vectors in North America. *Phil. Trans. Roy. Soc. Lond. B* **321:**421–446.

Isaacs, R., and Byrne, D. N. (1998). Aerial distribution, flight behavior and eggload: Their inter-relationship during dispersal by the sweetpotato whitefly. *J. Anim. Ecol.* **67:**741–750.

Isard, S. A., Irwin, M. E., and Hollinger, S. E. (1990). Vertical distribution of aphids (Homoptera: Aphididae) in the planetary boundary layer. *Environ. Entomol.* **19:**1473–1484.

Isard, S. A., Spencer, J. L., Nasser, M. A., and Levine, E. (2000). Aerial movement of western corn rootworm (Coleoptera: Chrysomelidae): Diel periodicity of flight activity in soybean fields. *Environ. Entomol.* **29:**226–234.

Isard, S. A., Kristovich, D. A. R., Gage, S. H., Jones, C. J., and Laird, N. F. (2001). Atmospheric motions systems that influence the redistribution and accumulation of insects on the beaches of the Great Lakes in North America. *Aerobiologia* **17:**275–291.

Ito, S., Buei, K., Yoshida, M., and Nakamura, J. (1986). Ecological studies on the overwintering of mosquitoes, especially of *Culex tritaeniorhynchus* Giles in Osaka Prefecture. 2. Physiological age composition of overwintered females and population density of adults and larvae in autumn. *Jap. J. Sanit. Zool.* **37:**341–347.

Jeger, M. J., van den Bosch, F., Madden, L. V., and Holt, J. (1998). A model for analysing plant virus transmission characteristics and epidemic development. *IMA J. Math. Appl. Biol. Med.* **14:**1–18.

Jeger, M. J., Holt, J., van den Bosch, F., and Madden, L. V. (2004). Epidemiology of insect-transmitted plant viruses: Modelling disease dynamics and control interventions. *Physiol. Entomol.* **29:**291–304.

Johansen, C. A., Farrow, R. A., Morrisen, A., Foley, P., Bellis, G., van den Hurk, A. F., Montgomery, B., Mackenzie, J. S., and Ritchie, S. A. (2003). Collection of wind-borne haematophagous insects in the Torres Strait, Australia. *Med. Vet. Entomol.* **17:**102–109.

Johnson, B. (1957). Studies on the dispersal by upper winds of *Aphis craccivora* Koch in New South Wales. *Proc. Linn. Soc. N.S.W* **82:**191–198.

Johnson, C. G. (1969). "Migration and Dispersal of Insects by Flight." Metheun & Co, London.

Johnson, C. G. (1976). Lability of the flight system: A context for functional adaptation. *In* "Insect Flight" (R. C. Rainey, ed.), pp. 217–234. Blackwell Scientific, Oxford.

Johnson, S. J. (1995). Insect migration in North America: Synoptic-scale transport in a highly seasonal environment. *In* "Insect Migration: Tracking Resources Through Space and Time" (V. A. Drake and A. G. Gatehouse, eds.), pp. 31–66. Cambridge University Press, Cambridge, UK.

Kay, B. H., and Farrow, R. A. (2000). Mosquito (Diptera: Culicidae) dispersal: Implications for the epidemiology of Japanese and Murray valley encephalitis in Australia. *J. Med. Entomol.* **37:**797–801.

Kennedy, J. S. (1960). The behavioural fitness of aphids as field vectors of viruses. *In* "Report of the Seventh Commonwealth Entomological Conference, 6th–15th July 1960," pp. 165–168. Eastern Press Ltd, London.

Kennedy, J. S. (1985). Migration, behavioural and ecological. *In* "Migration: Mechanisms and Adaptive Significance" (M. A. Rankin, ed.), *Contrib. Marine Sci.* **27**(Suppl.), pp. 5–26.

Kennedy, J. S., and Booth, C. O. (1963). Free flight of aphids in the laboratory. *J. Exp. Biol.* **40:**67–85.

Killick-Kendrick, R., Rioux, J. A., Bailly, M., Guy, M. W., Wilkes, T. J., Guy, F. M., Davidson, I., Knetchli, R., Ward, R. D., Guilvard, E., Perieres, J., and Dubois, H. (1984). Ecology of leishmaniasis in the south of France. 20. Dispersal of *Phlebotomus ariasi* Tonnoir, 1921 as a factor in the spread of visceral leishmaniasis in the Cevennes. *Ann. Parasit. Hum. Comp.* **59:**555–572.

Kisimoto, R. (1981). Development, behaviour, population dynamics and control of the brown planthopper, *Nilaparvata lugens* Stål. *Rev. Plant Prot. Res.* **14:**26–58.

Kisimoto, R. (1987). Ecology of planthopper migration. *In* "Proceedings of 2nd International Workshop on Leafhoppers and Planthoppers of Economic Importance, Provo, Utah, USA, 28 July–1 August 1986" (M. R. Wilson and L. R. Nault, eds.), pp. 41–54. CAB International Institute of Entomology, London.

Kisimoto, R., and Sogawa, K. (1995). Migration of the brown planthopper *Nilaparvata lugens* and the white-backed planthopper *Sogatella furcifera* in East Asia: The role of weather and climate. *In* "Insect Migration: Tracking Resources Through Space and Time" (V. A. Drake and A. G. Gatehouse, eds.), pp. 67–91. Cambridge University Press, Cambridge, UK.

Kisimoto, R., and Yamada, Y. (1986). A planthopper-rice virus epidemiology model: Rice stripe and small brown planthopper, *Laodelphax striatellus* Fallén. *In* "Plant Virus Epidemics: Monitoring, Modelling and Predicting Outbreaks" (G. D. McLean, R. G. Garrett, and W. G. Ruesink, eds.), pp. 327–344. Academic Press, Sydney, Australia.

Kisimoto, R., and Yamada, Y (1998). Present status of controlling rice stripe virus. *In* "Plant Virus Disease Control" (A. Hadidi, R. K. Khetarpal, and H. Koganezawa, eds.), pp. 470–483. APS Press, St. Paul, Minnesota.

Kocourek, F., Láska, P., and Jarošík, V. (2002). Thermal requirements for flight of six species of flea beetle of the genus Phyllotreta (Coleoptera: Chrysomelidae). *Plant Prot. Sci. (Prague)* **38:**76–80.

Lewis, T. (1973). "Thrips, Their Biology, Ecology and Economic Importance." Academic Press, London, New York.

Lewis, T., and Taylor, L. R. (1964). Diurnal periodicity of flight by insects. *Trans. R. Ent. Soc. Lond.* **116:**393–479.
Levin, D. M., and Irwin, M. E. (1995). Barley yellow dwarf luteovirus effects on tethered flight duration, wingbeat frequency, and age of maiden flight in *Rhopalosiphum padi* (Homoptera: Aphididae). *Environ. Entomol.* **24:**306–312.
Liew, C., and Curtis, C. F. (2004). Horizontal and vertical dispersal of dengue vector mosquitoes, *Aedes aegypti* and *Aedes albopictus*, in Singapore. *Med. Vet. Entomol.* **18:**351–360.
Llewellyn, K. S., Loxdale, H. D., Harrington, R., Brookes, C. P., Clark, S. J., and Sunnucks, P. (2003). Migration and genetic structure of the grain aphid (*Sitobion avenae*) in Britain related to climate and clonal fluctuation as revealed using microsatellites. *Molecular Ecol.* **12:**21–34.
Lopes, J. R. S., Nault, L. R., and Phelan, P. L. (1995). Periodicity of diel activity of *Graminella nigrifrons* (Homoptera: Cicadellidae) and implications for leafhopper dispersal. *Ann. Entomol. Soc. Am.* **88:**227–233.
Lourenço de Oliveira, R., Vazeille, M., De Filippis, A. M. B., and Failloux, A.-B. (2003). Large genetic differentiation and low variation in vector competence for dengue and yellow fever viruses of *Aedes Albopictus* from Brazil, the United States, and the Cayman Islands. *Am. J. Trop. Med. Hyg.* **69:**105–114.
Loxdale, H. D., and Brookes, C. P. (1990). Genetic stability within and restricted migration (gene flow) between local populations of the blackberry-grain aphid *Sitobion fragariae* in south-east England. *J. Anim. Ecol.* **59:**495–512.
Loxdale, H. D., and Lushai, G. (1998). Molecular markers in entomology (Review). *Bull. Ent. Res.* **88:**577–600.
Loxdale, H. D., and Lushai, G. (2001). Use of genetic diversity in movement studies of flying insects. *In* "Insect Movement: Mechanisms and Consequences" (I. P. Woiwod, D. R. Reynolds, and C. D. Thomas, eds.), pp. 361–386. CABI Publishing, Wallingford, UK.
Loxdale, H. D., and Lushai, G. (2007). Population genetic issues: The unfolding story using molecular markers. *In* "Aphids as Crop Pests" (H. F. van Emden and R. Harrington, eds.). CABI Publishing, Wallingford, UK (in press).
Loxdale, H. D., Tarr, I. J., Weber, C. P., Brookes, C. P., Digby, P. G. N., and Castañera, P. (1985). Electrophoretic study of enzymes from cereal aphid populations. III. Spatial and temporal genetic variation of populations of *Sitobion avenae* (F.) (Hemiptera: Aphididae). *Bull. Entomol. Res.* **75:**121–141.
Loxdale, H. D., Hardie, J., Halbert, S., Foottit, R., Kidd, N. A. C., and Carter, C. I. (1993). The relative importance of short- and long range movement of flying aphids. *Biol. Rev.* **68:**291–311.
MacLeod, A., Baker, R., Holmes, M., Cheek, S., Cannon, R., Mathews, L., and Agallou, E. (2004). *Diabrotica virgifera virgifera*. Pest Risk Analysis. May 2004. Central Science Laboratory, York, UK. http://www.defra.gov.uk/planth/pra/diab.pdf.
Madden, L. V., Jeger, M. J., and van den Bosch, F. (2000). A theoretical assessment of the effects of vector-virus transmission mechanism on plant virus disease epidemics. *Phytopathology* **90:**576–594.
Masterman, A. J., Foster, G. N., Holmes, S. J., and Harrington, R. (1996). The use of the Lamb daily weather types and the indices of progressiveness, southerliness and cyclonicity to investigate the autumn migration of *Rhopalosiphum padi*. *J. Appl. Ecol.* **33:**23–30.
McCorcle, M. D., and Fast, J. D. (1989). Prediction of pest distribution in the Corn Belt: A meteorological analysis. "Proceedings of the Ninth Conference on Biometeorology and

Aerobiology, Charleston, South Carolina, 7–10 March 1989," pp. 298–301. American Meteorological Society, Boston.

Medler, J. T. (1962). Long-range displacement of Homoptera in the central United States. *Proc. XI Int. Congr. Entomol. Vienna (1960)* **3:**30–35.

Mellor, P. S., Boorman, J., and Baylis, M. (2000). *Culicoides* biting midges: Their role as arbovirus vectors. *Ann. Rev. Entomol.* **45:**307–340.

Ming, J.-G., Jin, H., Riley, J. R., Reynolds, D. R., Smith, A. D., Wang, R.-L., Cheng, J.-Y., and Cheng, X.-N. (1993). Autumn southward 'return' migration of the mosquito *Culex tritaeniorhynchus* in China. *Med. Vet. Entomol.* **7:**323–327.

Montllor, C. B., and Gildow, F. E. (1986). Feeding responses of two grain aphids to barley yellow dwarf virus-infected oats. *Entomol. Exp. Appl.* **42:**63–69.

Moreton, B. D. (1945). On the migration of flea beetles (*Phyllotreta* spp.) (Col., Chrysomelidae) attacking *Brassica* crops. *Entomol. Mon. Mag.* **81:**59–60.

Mun, J. H., Song, Y. H., Heong, K. L., and Roderick, G. K. (1999). Genetic variation among Asian populations of rice planthoppers, *Nilaparvata lugens* and *Sogatella furcifera* (Hemiptera: Delphacidae): Mitochondrial DNA sequences. *Bull. Entomol. Res.* **89:**245–253.

Murray, M. D., and Kirkland, P. D. (1995). Bluetongue and Douglas virus activity in New South Wales in 1989: Further evidence for long-distance dispersal of the biting midge *Culicoides brevitarsis. Aust. Vet. J.* **72:**56–57.

Nagata, T., and Masuda, T. (1980). Insecticide susceptibility and wing-form ratio of the brown planthpper, *Nilaparvata lugens* (Stål) and the white backed planthopper, *Sogatella furcifera* (Horvath) (Hemiptera: Delphacidae) of Southeast Asia. *Appl. Ent. Zool.* **13:**10–19.

Nakasuji, F., and Kiritani, K. (1970). Ill-effects of rice dwarf virus upon its vector, *Nephotettix cincticeps* Uhler (Hemiptera: Deltocephalidae), and its significance for changes in relative abundance of infected individuals among vector populations. *Appl. Entomol. Zool.* **5:**1–12.

Nakasuji, F., Miyai, S., Kawamoto, H., and Kiritani, K. (1985). Mathematical epidemiology of rice dwarf virus transmitted by green rice leafhoppers: A differential equation model. *J. Appl. Ecol.* **22:**839–847.

Naranjo, S. E. (1990). Comparative flight behavior of *Diabrotica virgifera virgifera* and *Diabrotica barberi* in the laboratory. *Entomol. Exp. Appl.* **55:**79–90.

Nault, L. R. (1997). Arthropod transmission of plant viruses: A new synthesis. *Ann. Entomol. Soc. Am.* **90:**521–541.

Nieminen, M., Leskinen, M., and Helenius, J. (2000). Doppler radar detection of exceptional mass-migration of aphids into Finland. *Int. J. Biomet.* **44:**172–181.

Noda, T., and Kiritani, K. (1989). Landing places of migratory planthoppers, *Nilaparvata lugens* (Stal) and *Sogatella furcifera* (Horvath) (Homoptera: Delphacidae) in Japan. *Appl. Ent. Zool.* **24:**59–65.

Nottingham, S. F., and Hardie, J. (1989). Migratory and targeted flight in seasonal forms of the black bean aphid, *Aphis fabae. Physiol. Entomol.* **14:**451–458.

Nottingham, S. F., Hardie, J., and Tatchell, G. M. (1991). Flight behaviour of the bird cherry aphid, *Rhopalosiphum padi. Physiol. Entomol.* **16:**223–229.

North, R. C., and Shelton, A. M. (1986). Ecology of Thysanoptera within cabbage fields. *Environ. Entomol.* **15:**520–526.

Northing, P., Walters, K., Barker, I., Foster, G., Harrington, R., Taylor, M., Tones, S., and Morgan, D. (2004). Use of the internet for provision of user specific support for decisions on the control of aphid-borne viruses. *In* "Aphids in a New Millennium (Proceedings 6th International Aphid Symposium, Rennes, France, 3–7 September

2001)" (J. C. Simon, C. A. Dedryver, C. Rispe, and M. Hullé, eds.), pp. 331–336. INRA Editions, Versailles.

Osborne, J. L., Loxdale, H. D., and Woiwod, I. P. (2002). Monitoring insect dispersal: Methods and approaches. *In* "Dispersal Ecology: The 42nd Symposium of the British Ecological Society Held at the University of Reading, 2–5 April 2001" (J. M. Bullock, R. E. Kenward, and R. S. Hails, eds.), pp. 24–49. Blackwell Science Malden, USA.

Otim-Nape, G. W., Bua, A., Thresh, J. M., Baguma, Y., Ogwal, S., Semakula, G. N., Acola, G., Byabakama, B., and Martin, A. (1997). "Cassava Mosaic Virus Disease in Uganda: The Current Pandemic and Approaches to Control." Natural Resources Institute, Chatham, UK.

Otuka, A., Dudhia, J., Watanabe, T., and Furuno, A. (2005). A new trajectory analysis method for migratory planthoppers, *Sogatella furcifera* (Horváth) (Homoptera: Delphacidae) and *Nilaparvata lugens* (Stål), using an advanced weather forecast model. *Agric. Forest Entomol.* **7:**1–9.

Padgham, D. E., Cook, A. G., and Hutchison, D. (1984). Rubidium marking of the rice pests *Nilaparvata lugens* (Stål) and *Sogatella furcifera* (Horváth) (Hemiptera: Delphacidae) for field dispersal studies. *Bull. Entomol. Res.* **74:**379–385.

PAHO (1994). "Guidelines for the Prevention and Control of Dengue and Dengue Haemorrhagic Fever in the Americas." Pan American Health Organization, Washington, DC.

Pedgley, D. E. (1983). The windborne spread of insect-transmitted diseases of animals and man. *Phil. Trans. R. Soc. Lond. B* **302:**463–470.

Pedgley, D. E. (1999). Weather influences on pest movement. *In* "Handbook of Pest Management" (J. R. Ruberson, ed.), pp. 57–78. Marcel Dekker, New York.

Peterson, M. A., Denno, R. F., and Robinson, L. (2001). Apparent widespread gene flow in the predominantly flightless planthopper *Tumidagena minuta*. *Ecol. Entomol.* **26:**629–637.

Plumb, R. T., Lennon, E. A., and Gutteridge, R. A. (1986). Forecasting barley yellow dwarf virus by monitoring vector populations and infectivity. *In* "Plant Virus Epidemics: Monitoring, Modelling and Predicting Outbreaks" (G. D. McLean, R. G. Garrett, and W. G. Ruesink, eds.), pp. 387–398. Academic Press, Sydney, Australia.

Polston, J. E., and Anderson, P. K. (1997). The emergence of whitefly-transmitted geminiviruses in tomato in the western hemisphere. *Plant Disease* **81:**1358–1369.

Purse, B. V., Mellor, P. S., Rogers, D. J., Samuel, A. R., Mertens, P. P. C., and Baylis, M. (2005). Climate change and the recent emergence of bluetongue in Europe. *Nat. Rev. Microbiol.* **3:**171–181.

Qi Aiming, Dewar, A. M., and Harrington, R. (2004). Decision making in controlling virus yellows of sugar beet in the UK. *Pesticide Manag. Sci.* **60:**727–732.

Quinn, M. A., Halbert, S. E., and Williams, L. (1991). Spatial and temporal changes in aphid (Homoptera: Aphididae) species assemblages collected with suction traps in Idaho. *J. Econ. Entomol.* **84:**1710–1716.

Raccah, B., Pirone, T. P., and Madden, L. V. (1988). Correlation between the incidence of aphid species and the incidence of two nonpersistent viruses in tobacco. *Agric. Ecosyst. Environ.* **21:**281–292.

Ree, H. I., Wada, Y., Jolivet, P. H. A., Hong, H. K., Self, L. S., and Lee, K. W. (1976). Studies on overwintering of *Culex tritaenioryhnchus* Giles in the Republic of Korea. *Cahiers ORSTROM: Serie Entomologie medicale et Parasitologie* **14:**105.

Reiter, P., Amador, A. M., Anderson, R. A., and Clark, G. G. (1995). Short report: Dispersal of *Aedes aegypti* in an urban area after blood-feeding as demonstrated by rubidium-marked eggs. *Am. J. Trop. Med. Hyg.* **52:**177–179.

Reynolds, D. R., and Riley, J. R. (2002). Remote-sensing, telemetric and computer-based technologies for investigating insect movement: A survey of existing and potential techniques. *Comp. Electron. Agric.* **35:**271–307.
Reynolds, D. R., and Wilson, M. R. (1989). Aerial samples of micro-insects migrating at night over central India. *J. Plant Prot. Tropics* **6:**89–101.
Reynolds, D. R., Smith, A. D., Mukhopadhyay, S., Chowdhury, A. K., De, B. K., Mondal, S. K., Nath, P. S., Das, B. K., and Mukhopadhyay, S. (1996). Atmospheric transport of mosquitoes in Northeast India. *Med. Vet. Entomol.* **10:**185–186.
Reynolds, D. R., Riley, J. R., Armes, N. J., Cooter, R. J., Tucker, M. R., and Colvin, J. (1997). Techniques for quantifying insect migration. *In* "Methods in Ecological and Agricultural Entomology" (D. R. Dent and M. P. Walton, eds.), pp. 111–145. CAB International, Wallingford, UK.
Reynolds, D. R., Mukhopadhyay, S., Riley, J. R., Das, B. K., Nath, P. S., and Mandal, S. K. (1999). Seasonal variation in the windborne movement of insect pests over Northeast India. *Int. J. Pest Manag.* **45:**195–205.
Riley, J. R. (1992). A millimetric radar to study the flight of small insects. *Electron. Comm. Eng. J.* **4:**43–48.
Riley, J. R., Reynolds, D. R., and Farmery, M. J. (1983). Observations of the flight behaviour of the armyworm moth, *Spodoptera exempta*, at an emergence site using radar and infra-red optical techniques. *Ecol. Entomol.* **8:**395–418.
Riley, J. R., Reynolds, D. R., and Farrow, R. A. (1987). The migration of *Nilaparvata lugens* (Stal) (Delphacidae) and other Hemiptera associated with rice during the dry season in the Philippines: A study using radar, visual observations, aerial netting and ground trapping. *Bull. Entomol. Res.* **77:**145–169.
Riley, J. R., Cheng, X.-N., Zhang, X.-X., Reynolds, D. R., Xu, G.-M., Smith, A. D., Cheng, J.-Y., Bao, A. D., and Zhai, B.-P. (1991). The long distance migration of *Nilaparvata lugens* (Stål) (Delphacidae) in China: Radar observations of mass return flight in the autumn. *Ecol. Entomol.* **16:**471–489.
Riley, J. R., Reynolds, D. R., Smith, A. D., Rosenberg, L. J., Cheng, X.-N., Zhang, X.-X., Xu, G.-M., Cheng, J.-Y., Bao, A.-D., Zhai, B.-P., and Wang, H.-K. (1994). Observations on the autumn migration of *Nilaparvata lugens* (Homoptera: Delphacidae) and other pests in east central China. *Bull. Ent. Res.* **84:**389–402.
Riley, J. R., Reynolds, D. R., Mukhopadhyay, S., Ghosh, M. R., and Sarkar, T. K. (1995a). Long-distance migration of aphids and other small insects in Northeast India. *Eur. J. Entomol.* **92:**639–653.
Riley, J. R., Reynolds, D. R., Smith, A. D., Edwards, A. S., Zhang, X.-X., Cheng, X.-N., Wang, H.-K., Cheng, J.-Y., and Zhai, B.-P. (1995b). Observations of the autumn migration of the rice leaf roller *Cnaphalocrocis medinalis* (Lepidoptera: Pyralidae) and other moths in eastern China. *Bull. Entomol. Res.* **85:**397–414.
Riley, J. R., Downham, M. C. A., and Cooter, R. J. (1997). Comparison of the performance of *Cicadulina* leafhoppers on flight mills with that to be expected in free flight. *Entomol. Exp. Appl.* **83:**317–322.
Ritchie, S. A., and Rochester, W. (2001). Wind-blown mosquitoes and introduction of Japanese encephalitis into Australia. *Emerging Infectious Diseases* **7**(5)**:** (September–October 2001):900–903 (http://www.cdc.gov/ncidod/eid/vol7no5/ritchie.htm).
Robert, Y. (1987). Aphid vector monitoring in Europe. *In* "Current Topics in Vector Research," (K. F. Harris, ed.), "Current Topics in Vector Research," Vol. 3, pp. 81–129. Springer-Verlag, New York.
Rose, D. J. W. (1972). Dispersal and quality in populations of *Cicadulina* species (Cicadellidae). *J. Anim. Ecol.* **41:**589–609.

Rose, D. J. W. (1978). Epidemiology of maize streak disease. *Ann. Rev. Entomol.* **23:**259–282.
Rosen, L. (1986). The natural history of Japanese encephalitis virus. *Ann. Rev. Entomol.* **40:**395–414.
Rosen, L., Shroyer, D. A., and Lien, J. C. (1980). Transovarial transmission of Japanese encephalitis virus by *Culex tritaeniorhynchus* mosquitoes. *Am. J. Trop. Med. Hyg.* **29:**711–712.
Rosenberg, L. J., and Magor, J. I. (1983). Flight duration of the brown planthopper, *Nilaparvata lugens* (Homoptera: Delphacidae). *Ecol. Entomol.* **8:**341–350.
Ruesink, W. G., and Irwin, M. E. (1986). Soybean mosaic virus epidemiology: A model and some implications. *In* "Plant Virus Epidemics: Monitoring, Modelling and Predicting Outbreaks" (G. D. McLean, R. G. Garrett, and W. G. Ruesink, eds.), pp. 295–313. Academic Press, Sydney, Australia.
Russell, G. E. (1963). Some factors affecting the relative incidence, distribution and importance of beet yellows and sugar beet mild yellowing virus in eastern England, 1955–1962. *Ann. Appl. Biol.* **52:**405–413.
Schofield, C. J., Lehane, M. J., McEwen, P., Catala, S. S., and Gorla, D. E. (1992). Dispersive flight by *Triatoma infestans* under natural climatic conditions in Argentina. *Med. Vet. Entomol.* **6:**51–56.
Scott, R. W., and Achtemeier, G. L. (1987). Estimating pathways of migrating insects carried in atmospheric winds. *Environ. Entomol.* **16:**1244–1254.
Sellers, R. F. (1980). Weather, host and vector: Their interplay in the spread of insect-borne animal virus diseases. *J. Hyg. (Cambridge)* **85:**65–102.
Sellers, R. F., and Maarouf, A. R. (1988). Impact of climate on western equine encephalitis in Manitoba, Minnesota and North Dakota, 1980–1983. *Epidemiol. Infect.* **101:**511–535.
Sellers, R. F., and Maarouf, A. R. (1993). Weather factors in the prediction of western equine encephalitis in Manitoba. *Epidemiol. Infect.* **111:**375–390.
Sellers, R. F., Pedgley, D. E., and Tucker, M. R. (1977). Possible spread of African horse sickness on the wind. *J. Hyg. (Cambridge)* **79:**279–298.
Sellers, R. F., Pedgley, D. E., and Tucker, M. R. (1982). Rift Valley fever, Egypt 1977: Disease spread by windborne insect vectors? *Vet. Record* **110:**73–77.
Service, M. W. (1997). Mosquito (Diptera: Culicidae) dispersal: The long and short of it. *J. Med. Entomol.* **34:**579–588.
Sherlock, P. L., Bowden, J., and Digby, P. G. N. (1986). Studies of elemental composition as a biological marker in insects: V. The elemental composition of *Rhopalosiphum padi* (L.) (Hemiptera: Aphididae) from *Prunus padus* at different localities. *Bull. Entomol. Res.* **76:**621–632.
Shields, E. J., and Testa, A. M. (1999). Fall migratory flight initiation of the potato leafhopper, *Empoasca fabae* (Homoptera: Cicadellidae): Observations in the lower atmosphere using remotely piloted vehicles. *Agric. Forest Meteorol.* **97:**317–330.
Sigvald, R. (1998). Forecasting aphid-borne virus diseases. *In* "Plant Virus Disease Control" (A. Hadidi, R. K. Khetarpal, and H. Koganezawa, eds.), pp. 172–187. APS Press, St. Paul, Minnesota.
Simon, J.-C., Baumann Sunnucks, P., Hebert, P. D. N., Pierre, J. S., Le Gallic, J. F., and Dedryver, C. A. (1999). Reproductive mode and population genetic structure of the cereal aphid *Sitobion avenae* studied using phenotypic and microsatellite markers. *Molecular Ecol.* **8:**531–546.
Simon, J.-C., Carre, S., Boutin, M., Prunier-Leterme, N., Sabater-Munoz, B., Latorre, A., and Bournoville, R. (2003). Host-based divergence in populations of the pea aphid:

Insights from nuclear markers and the prevalence of facultative symbionts. *Proc. Roy. Soc. Lond. B* **270:**1703–1712.

Solbreck, C. (1985). Insect migration strategies and population dynamics. *In* "Migration: Mechanisms and Adaptive Significance" (M. A. Rankin, ed.)., *Contrib. Marine Sci.* **27** (Suppl.), pp. 641–662.

Southwood, T. R. E., and Henderson, P. A. (2000). "Ecological Methods," 3rd Ed. Blackwell Science, Oxford.

Sparks, A. N., Jackson, R. D., Carpenter, J. E., and Muller, R. A. (1986). Insects captured in light traps in the Gulf of Mexico. *Ann. Entomol. Soc. Am.* **79:**132–139.

Tang, J. Y., Cheng, J. A., and Norton, G. A. (1994). HOPPER: An expert system for forecasting the risk of white-backed planthopper attack in the first crop season in China. *Crop Prot.* **13:**463–473.

Taylor, L. R. (1974). Insect migration, flight periodicity and the boundary layer. *J. Anim. Ecol.* **43:**225–238.

Taylor, L. R. (1986). The distribution of virus disease and the migrant vector aphid. *In* "Plant Virus Epidemics: Monitoring, Modelling and Predicting Outbreaks" (G. D. McLean, R. G. Garrett, and W. G. Ruesink, eds.), pp. 35–57. Academic Press, Sydney.

Taylor, L. R., and Palmer, J. M. P. (1972). Aerial sampling. *In* "Aphid Technology" (H. F. van Emden, ed.), pp. 189–234. Academic Press, London.

Taylor, R. A. J., and Reling, D. (1986). Preferred wind direction of long-distance leafhopper (*Empoasca fabae*) migrants and its relevance to the return migration of small insects. *J. Anim. Ecol.* **55:**1103–1114.

Thomas, A. A. G., Ludlow, A. R., and Kennedy, J. S. (1977). Sinking speeds of falling and flying *Aphis fabae* Scopoli. *Ecol. Entomol.* **2:**315–326.

Thresh, J. M. (1980). An ecological approach to the epidemiology of plant virus disease. *In* "Comparative Epidemiology" (J. Palti and J. Kranz, eds.), pp. 57–70. Pudoc, Wageningen, The Netherlands.

Thresh, J. M. (1983). The long-range dispersal of plant viruses by arthropod vectors. *Phil. Trans. Roy. Soc. Lond. B* **302:**497–528.

Thresh, J. M. (1985). Plant virus dispersal. *In* "The Movement and Dispersal of Agriculturally Important Biotic Agents" (D. R. MacKenzie, C. S. Barfield, G. C. Kennedy, R. D. Berger, and D. J. Taranto, eds.), pp. 51–106. Claitor's Publishing Division, Baton Rouge, Louisiana.

Torina, A., Caracappa, S., Mellor, P. S., Baylis, M., and Purse, B. V. (2004). Spatial distribution of bluetongue virus and its *Culicoides* vectors in Sicily. *Med. Vet. Entomol.* **18:**1–9.

Turner, R. H., and Bowden, J. (1983). X-ray microanalysis applied to the study of insect migration with special reference to the rice bug, *Nilaparvata lugens. Scanning Elcctron. Microsc.* **1983**(2)**:**873–878.

Turner, R., Song, Y.-H., and Uhm, K.-B. (1999). Numerical model simulations of brown planthopper *Nilaparvata lugens* and the white-backed planthopper *Sogatella furcifera* (Hemiptera: Delphacidae) migration. *Bull. Entomol. Res.* **89:**557–568.

Wada, T., Ito, K., and Takahashi, A. (1994). Biotype comparisons of the brown planthopper, *Nilaparvata lugens* (Homoptera: Delphacidae) collected in Japan and the Indochina Peninsula. *Appl. Entomol. Zool.* **29:**477–484.

Walker, T. J. (1980). Migrating Lepidoptera: Are butterflies better than moths? *Florida Entomol.* **63:**79–98.

Watanabe, T. (1995). Forecasting systems for migrant pests. II. The rice planthoppers *Nilaparvata lugens* and *Sogatella furcifera* in Japan. *In* "Insect Migration: Tracking

Resources through Space and Time" (V. A. Drake and A. G. Gatehouse, eds.), pp. 365–376. Cambridge University Press, Cambridge.
Wiktelius, S. (1977). The importance of southerly winds and other weather data on the incidence of sugar beet yellowing viruses in southern Sweden. *Swed. J. Agric. Res.* **7**:89–95.
Wiktelius, S. (1984). Long-range migration of aphids into Sweden. *Int. J. Biomet.* **28**:185–200.
Woiwod, I. P., and Harrington, R. (1994). Flying in the face of change: The Rothamsted Insect Survey. *In* "Long-Term Experiments in Agricultural and Ecological Sciences" (R. A. Leigh and A. E. Johnson, eds.), pp. 321–342. CAB International, Wallingford, UK.
Woiwod, I. P., Reynolds, D. R., and Thomas, C. D. (2001). Introduction and overview. *In* "Insect Movement: Mechanisms and Consequences" (I. P. Woiwod, D. R. Reynolds, and C. D. Thomas, eds.), pp. 1–18. CABI Publishing, Wallingford, UK.
Wolf, W. W., Westbrook, J. K., Raulston, J., Pair, S. D., and Hobbs, S. E. (1990). Recent airborne radar observations of migrant pests in the United States. *Phil. Trans. Roy. Soc. B* **328**:619–630.
Worner, S. P., Lankin, G. O., Samarasinghe, S., and Teulon, D. A. J. (2001). Improving prediction of aphid flights by temporal analysis of input data for an artificial neural networks. *N.Z. Plant Prot.* **55**:312–316.
Wynne, I. R., Howard, J. J., Loxdale, H. D., and Brookes, C. P. (1994). Population genetic structure during aestivation in the sycamore aphid *Drepanosiphum platanoidis* (Hemiptera: Drepanosiphidae). *Eur. J. Entomol.* **91**:375–383.
Zeyen, R. J., Stromberg, E.-L., and Kuehnast, E. L. (1987). Long-range aphid transport hypothesis for maize dwarf mosaic virus: History and distribution in Minnesota, USA. *Ann. Appl. Biol.* **111**:325–336.
Zhang, S.-Y., Dou, G.-L., and Wang, M.-X. (1985). "Symposium of Medical Entomology" [in Chinese]. Editorial Board of Chinese Journal of Epidemiology and Office of National Patriotic Health Movement Committee, Beijing.
Zhang, X.-S., Holt, J., and Colvin, J. (2000). A general model of plant-virus disease infection to incorporate vector aggregation. *Plant Pathol.* **49**:435–444.
Zhou, B.-H., Wang, H.-K., and Cheng, X.-N. (1995). Forecasting systems for migrant pests. I. The brown planthopper in China. *In* "Insect Migration: Tracking Resources through Space and Time" (V. A. Drake and A. G. Gatehouse, eds.), pp. 353–364. Cambridge University Press, Cambridge.

INDEX

A

D

E

F

G

H

I

Q

R

S

T

W

X

Y

Z